# Atmospheric Science: Chemistry and Physics

# Atmospheric Science: Chemistry and Physics

Editor: Bruce Mullan

R CALLISTO
REFERENCE

www.callistoreference.com

**Callisto Reference,**
118-35 Queens Blvd., Suite 400,
Forest Hills, NY 11375, USA

Visit us on the World Wide Web at:
www.callistoreference.com

ISBN: 978-1-64116-044-5 (Hardback)

**Trademark Notice:** Registered trademark of products or corporate names are used only for explanation and identification without intent to infringe.

**Cataloging-in-Publication Data**

Atmospheric science : chemistry and physics / edited by Bruce Mullan.
        p. cm.
Includes bibliographical references and index.
ISBN 978-1-64116-044-5
1. Atmosphere. 2. Atmospheric chemistry. 3. Atmospheric physics. 4. Meteorology. 5. Earth sciences.
I. Mullan, Bruce.
QC861.3 .A86 2019
551.5--dc23

# Table of Contents

Preface.........................................................................................................................................VII

Chapter 1 **Particulate sulfur in the upper troposphere and lowermost stratosphere – sources and climate forcing**................................................................................................................1
Bengt G. Martinsson, Johan Friberg, Oscar S. Sandvik, Markus Hermann, Peter F. J. van Velthoven and Andreas Zahn

Chapter 2 **Disentangling fast and slow responses of the East Asian summer monsoon to reflecting and absorbing aerosol forcings**...................................................................18
Zhili Wang, Lei Lin, Meilin Yang, Yangyang Xu, and Jiangnan Li

Chapter 3 **Fungi diversity in PM$_{2.5}$ and PM1 at the summit of Mt. Tai: abundance, size distribution, and seasonal variation**........................................................................32
Caihong Xu, Min Wei, Jianmin Chen, Chao Zhu, Jiarong Li, Ganglin Lv, Xianmang Xu, Lulu Zheng, Guodong Sui, Weijun Li, Bing Chen, Wenxing Wang, Qingzhu Zhang, Aijun Ding and Abdelwahid Mellouki

Chapter 4 **Pre-monsoon air quality over Lumbini, a world heritage site along the Himalayan foothills** ....................................................................................................................45
Dipesh Rupakheti, Bhupesh Adhikary, Puppala Siva Praveen, Maheswar Rupakheti, Shichang Kang, Khadak Singh Mahata, Manish Naja, Qianggong Zhang, Arnico Kumar Panday and Mark G. Lawrence

Chapter 5 **Re-evaluating black carbon in the Himalayas and the Tibetan Plateau: concentrations and deposition**...........................................................................................68
Chaoliu Li, Fangping Yan, Shichang Kang, Pengfei Chen, Xiaowen Han, Zhaofu Hu, Guoshuai Zhang, Ye Hong, Shaopeng Gao, Bin Qu, Zhejing Zhu, Jiwei Li, Bing Chen and Mika Sillanpää

Chapter 6 **Comparison of large-scale dynamical variability in the extratropical stratosphere among the JRA-55 family data sets: impacts of assimilation of observational data in JRA-55 reanalysis data**.................................................................82
Masakazu Taguchi

Chapter 7 **Cross-polar transport and scavenging of Siberian aerosols containing black carbon during the 2012 ACCESS summer campaign** ........................................................97
Jean-Christophe Raut, Louis Marelle, Jerome D. Fast, Jennie L. Thomas, Bernadett Weinzierl, Katharine S. Law, Larry K. Berg, Anke Roiger, Richard C. Easter, Katharina Heimerl, Tatsuo Onishi, Julien Delanoë, and Hans Schlager

Chapter 8 **Impacts of large-scale atmospheric circulation changes in winter on black carbon transport and deposition to the Arctic** .........................................................124
Luca Pozzoli, Srdan Dobricic, Simone Russo, and Elisabetta Vignati

Chapter 9    **Near-real-time processing of a ceilometer network assisted with sun-photometer data: monitoring a dust outbreak over the Iberian Peninsula**.....................140
Alberto Cazorla, Juan Andrés Casquero-Vera, Roberto Román,
Juan Luis Guerrero-Rascado, Carlos Toledano, Victoria E. Cachorro,
José Antonio G. Orza, María Luisa Cancillo, Antonio Serrano, Gloria Titos,
Marco Pandolfi, Andres Alastuey, Natalie Hanrieder and Lucas Alados-Arboledas

Chapter 10    **Impacts of stratospheric sulfate geoengineering on tropospheric ozone**.................156
Lili Xia, Peer J. Nowack, Simone Tilmes and Alan Robock

Chapter 11    **Projected global ground-level ozone impacts on vegetation under different emission and climate scenarios** ...........................................171
Pierre Sicard, Alessandro Anav, Alessandra De Marco and Elena Paoletti

Chapter 12    **Classification of Arctic, midlatitude and tropical clouds in then mixed-phase temperature regime** ...........................................190
Anja Costa, Jessica Meyer, Armin Afchine, Anna Luebke, Gebhard Günther,
James R. Dorsey, Martin W. Gallagher, Andre Ehrlich, Manfred Wendisch,
Darrel Baumgardner, Heike Wex and Martina Krämer

**Permissions**

**List of Contributors**

**Index**

# Preface

This book brings forth some of the most innovative concepts and elucidates the unexplored aspects of atmospheric science with respect to topics of atmospheric chemistry and physics. As a field of scientific study, atmospheric chemistry and physics is related to the study of Earth's atmosphere. It also examines the chemical and physical processes taking place in the atmosphere. Climatology and meteorology are the two main branches of atmospheric science. This book presents information about some upcoming concepts and theories related to this field. It strives to provide a fair idea about this discipline and develop a better understanding of the latest advances within this field. Scientists and students actively engaged in this field will find this book full of crucial and unexplored concepts.

Significant researches are present in this book. Intensive efforts have been employed by authors to make this book an outstanding discourse. This book contains the enlightening chapters which have been written on the basis of significant researches done by the experts.

Finally, I would also like to thank all the members involved in this book for being a team and meeting all the deadlines for the submission of their respective works. I would also like to thank my friends and family for being supportive in my efforts.

Editor

# Particulate sulfur in the upper troposphere and lowermost stratosphere – sources and climate forcing

**Bengt G. Martinsson**[1], **Johan Friberg**[1], **Oscar S. Sandvik**[1], **Markus Hermann**[2], **Peter F. J. van Velthoven**[3], and **Andreas Zahn**[4]

[1]Division of Nuclear Physics, Lund University, Lund, Sweden
[2]Leibniz Institute for Tropospheric Research, Leipzig, Germany
[3]Royal Netherlands Meteorological Institute (KNMI), De Bilt, the Netherlands
[4]Institute of Meteorology and Climate Research, Institute of Technology, Karlsruhe, Germany

*Correspondence to:* Bengt G. Martinsson (bengt.martinsson@nuclear.lu.se)

**Abstract.** This study is based on fine-mode aerosol samples collected in the upper troposphere (UT) and the lowermost stratosphere (LMS) of the Northern Hemisphere extratropics during monthly intercontinental flights at 8.8–12 km altitude of the IAGOS-CARIBIC platform in the time period 1999–2014. The samples were analyzed for a large number of chemical elements using the accelerator-based methods PIXE (particle-induced X-ray emission) and PESA (particle elastic scattering analysis). Here the particulate sulfur concentrations, obtained by PIXE analysis, are investigated. In addition, the satellite-borne lidar aboard CALIPSO is used to study the stratospheric aerosol load. A steep gradient in particulate sulfur concentration extends several kilometers into the LMS, as a result of increasing dilution towards the tropopause of stratospheric, particulate sulfur-rich air. The stratospheric air is diluted with tropospheric air, forming the extratropical transition layer (ExTL). Observed concentrations are related to the distance to the dynamical tropopause. A linear regression methodology handled seasonal variation and impact from volcanism. This was used to convert each data point into stand-alone estimates of a concentration profile and column concentration of particulate sulfur in a 3 km altitude band above the tropopause. We find distinct responses to volcanic eruptions, and that this layer in the LMS has a significant contribution to the stratospheric aerosol optical depth and thus to its radiative forcing. Further, the origin of UT particulate sulfur shows strong seasonal variation. We find that tropospheric sources dominate during the fall as a result of downward transport of the Asian tropopause aerosol layer (ATAL) formed in the Asian monsoon, whereas transport down from the Junge layer is the main source of UT particulate sulfur in the first half of the year. In this latter part of the year, the stratosphere is the clearly dominating source of particulate sulfur in the UT during times of volcanic influence and under background conditions.

## 1 Introduction

The global mean surface temperature has increased considerably in the last 2 years (NOAA, 2016), which followed on a 15-year period with slow temperature evolution. CMIP5 (Coupled Model Intercomparison Project) models predict stronger-than-observed temperature increases in this period (Fyfe et al., 2013, 2016). Reasons for these differences were sought, and the Interdecadal Pacific Oscillation connected with increased subduction and upwelling (England et al., 2014; Meehl and Teng, 2014), variations in volcanic aerosol (Solomon et al., 2011; Santer et al., 2014) and solar (Myhre et al., 2013) forcings were identified as main causes of the discrepancies. These phenomena are all elements of natural climate variability, thus highlighting the importance of these influences for assessing the human impact on the climate (Ramanathan and Feng, 2008).

Air from the tropical troposphere containing aerosol precursor gases is lifted into the tropical stratosphere in the Brewer–Dobson circulation. Carbonyl sulfide (OCS) is the most abundant sulfur-containing gas in the atmosphere.

When lofted into the stratosphere, OCS is converted into sulfur dioxide ($SO_2$) at approximately 25 km altitude aided by UV radiation (Crutzen, 1976) and next into sulfuric acid, giving rise to the background stratospheric aerosol layer. The upward tropical flow is connected with a downward flow in the extratropics, where stratospheric air flows back to the troposphere.

The stratospheric aerosol concentration varies strongly due to volcanic influence, from events like the strong 1991 Mt. Pinatubo eruption, inducing negative radiative forcing in excess of $1\,W\,m^{-2}$ (McCormick et al., 1995), to close to background conditions at around the turn of the millennium (Bauman et al., 2003; Martinsson et al., 2005). Overall, anthropogenic influence on stratospheric aerosol is small compared to that from volcanism (Neely III et al., 2013). Recent volcanic eruptions, such as in 2008 and 2009 of the Kasatochi and Sarychev in the extratropics and in 2011 of the Nabro in the tropics, significantly altered stratospheric aerosol (Vernier et al., 2011a; Bourassa et al., 2012). These eruptions also had a profound impact on the Northern Hemisphere lowermost stratosphere (LMS; Martinsson et al., 2009) and the global aerosol optical depth (AOD) of the stratosphere (Andersson et al., 2015). After reaching the LMS, stratospheric aerosol is transported to the upper troposphere (UT). A recent study indicates that volcanic aerosol also has an indirect climate effect by affecting the reflectance of cirrus clouds in the UT (Friberg et al., 2015).

Upper troposphere particulate sulfur, in addition, has tropospheric sources. In the tropics, convection lofts aerosol and aerosol precursor gases from low altitudes into the UT (Hess, 2005). During the monsoon season, primarily the Asian monsoon, appreciable amounts of aerosol and gaseous aerosol precursors are lifted to the tropopause region, extending from the UT to approximately 420 K potential temperature in the stratosphere, forming the Asian tropopause aerosol layer (ATAL; Vernier et al., 2011b). The ATAL is a seasonal and regional phenomenon which, according to atmospheric modeling, is strongly connected with Asian pollution sources (Neely III et al., 2014). In the extratropics, warm conveyor belts (WCBs; Stohl, 2001) and deep convection are the main lofting channels, the latter especially in the summer season. Large amounts of pollutants are transported long-range across the Pacific and Atlantic oceans, where sources are affecting the background concentrations of various species over remote continents. This transport is most efficient above the boundary layer in the pressure interval 700–900 hPa (Luan and Jaeglé, 2013). Warm conveyor belts reaching the UT have their maximum frequency in the winter (Eckhardt et al., 2004) and transport of boundary layer air into the LMS maximizes in the winter and spring (Škerlak et al., 2014), whereas deep convection dominates in the summer (Kiley and Fuelberg, 2006).

Aerosol particles in the UT and LMS contain a large fraction of sulfur compounds, mostly sulfates (Dibb et al., 2000; Martinsson et al., 2001; Xu et a., 2001; Kojima et al., 2004).

These particles also contain a considerable fraction of carbonaceous aerosol (Murphy et al., 1998; Nguyen et al., 2008; Martinsson et al., 2009), of which a minor fraction is black carbon (Schwarz et al., 2010; Friberg et al., 2014). In some periods, particularly during spring, crustal particles are also found in this part of the atmosphere (Papaspiropoulos et al., 2002).

In this study the concentration of particulate sulfur in the UT and LMS is investigated based on aerosol samples collected from the IAGOS-CARIBIC platform in the time period 1999–2014. In this period the UT and LMS aerosol was affected by volcanism from several eruptions as demonstrated in our previous studies based on this data set (Martinsson et al., 2009, 2014; Andersson et al., 2013, 2015; Friberg et al., 2014, 2015). The IAGOS-CARIBIC aerosol elemental concentration measurements from the LMS are taken in strong concentration gradients that are affected by mixing tropospheric air into the lowest part of the LMS. Each measurement flight results in a small number of samples, being insufficient to reconstruct the gradient. Therefore, we have frequently relied on concurrent IAGOS-CARIBIC measurements, mostly by relating the particulate sulfur measurements to ozone concentrations to express, for example, volcanic influence on the aerosol concentration (Martinsson et al., 2009). Satellite-based measurements do not usually provide specific chemical information about aerosol particles. This lack of chemical information in, for example, lidar measurements can cause a bias in LMS particle concentrations close to the extratropical tropopause from non-volcanic species such as crustal particles and enhanced signal caused by particle hygroscopic growth. Here we present stand-alone estimates of the stratospheric sulfur aerosol in terms of mass concentration profiles and column concentrations based on a new method, which is then used to study the AOD of the lowest part of Northern Hemisphere LMS and its radiative forcing. This study also comprises a discussion on the relative importance of stratospheric and the tropospheric sources to the UT of particulate sulfur, seasonal dependences and different modes of tropospheric transport involved.

## 2 Methods

### 2.1 Sampling, analysis and classification

This study is based on measurements of particulate sulfur taken from the IAGOS-CARIBIC platform (Brenninkmeijer et al., 2007; www.caribic-atmospheric.com/), where the atmosphere is studied using modified passenger aircraft (March 1999–April 2002: Boeing 767–300 ER from LTU International Airways; May 2005–present: Airbus 340-600 from Lufthansa) during monthly sets of usually four intercontinental flights. A large number of trace gases and aerosol parameters are measured from this platform during flights in the altitude range 8.8–12 km, including gaseous and condensed water, $O_3$, CO, NO/NO$_y$, volatile organic compounds (VOCs), greenhouse gases, halo-carbons, mercury, particle number concentrations, size distributions and elemental concentrations (Brenninkmeijer et al., 2007; Hermann et al., 2003; Schuck et al., 2009; Baker et al., 2010; Oram et al., 2012; Zahn et al., 2012; Martinsson et al., 2014; Dyroff et al., 2015; Slemr et al., 2016; Hermann et al., 2016).

Aerosol sampling from the IAGOS-CARIBIC platform in the time period 1999–2014 resulted in 1198 samples analyzed for aerosol elemental concentrations. The measurements were mainly taken in the Northern Hemisphere (NH) extratropics and the tropics, while only a small fraction of the samples were taken in the Southern Hemisphere. Here the focus is on the extratropical LMS and UT of the NH. Aerosol particles with aerodynamic diameter in the range 0.08–2 µm were collected with a multi-channel impactor with a collection efficiency of $97\% \pm 4\%$ (Nguyen et al., 2006). The typical time required to collect one sample is 100 min. Accelerator-based methods were used to analyze the collected samples with respect to elemental concentrations, using particle-induced X-ray emission (PIXE) to analyze the concentration of elements with atomic number larger than 15 (Martinsson et al., 2001). Concentrations of hydrogen, carbon, nitrogen and oxygen were investigated using particle elastic scattering analysis (PESA; Nguyen and Martinsson, 2007). Here the particulate sulfur concentrations are used. The accuracy of the analyses is estimated to be 10 % and the combined uncertainty in sampling and analysis is estimated to be 12 %. Further analytical details are found in Martinsson et al. (2014). Finally, the concentration of particulate sulfur is mostly given as a mixing ratio by normalization to STP (standard temperature, 273.15 K, and pressure, 101 300 Pa). When computing column concentration and AOD of the lowest 3 km of the LMS, the STP concentrations are converted into volume concentrations using the pressure and temperature of the measurement, and the altitude dependence in the LMS of the molar volume obtained from ECMWF (European Centre for Medium-Range Weather Forecasts).

The dynamical tropopause (Gettelman et al., 2011) at the potential vorticity (PV) of 1.5 PVU (potential vorticity units; $1\,\text{PVU} = 10^{-6}\,\text{K}\,\text{m}^2\,\text{kg}^{-1}\,\text{s}^{-1}$) was used to classify samples with respect to tropospheric and stratospheric air. The PV along the flight track was obtained from archived analyses from ECMWF with a resolution of $1 \times 1°$ in the horizontal and 91 vertical hybrid sigma-pressure model levels. The PV was interpolated linearly by latitude, longitude, log pressure and time to the position of the IAGOS-CARIBIC aircraft.

## 2.2 Altitude

The UT usually holds significantly lower particulate sulfur concentration than the LMS. Combined with bi-directional exchange of tropospheric and stratospheric air across the tropopause, this leads to a gradient of increasing concentra-

tion in the LMS from the tropopause. In addition, the concentration of particulate sulfur in the LMS varies due to the influence from volcanism (Martinsson et al., 2009), which has been shown to cause significant radiative forcing (Andersson et al., 2015). In order to study the particulate sulfur gradient in the LMS, fine aerosol elemental concentration measurements from the IAGOS-CARIBIC platform were used in relation to the distance between measurement position and the tropopause ($Z$):

$$Z = Z_a - Z_{tp}, \tag{1}$$

where $Z_a$ and $Z_{tp}$ are the altitudes of the aircraft and the tropopause. $Z_{tp}$ refer to the dynamical tropopause of 1.5 PVU, which is a low limit to ensure that very little LMS air will be considered as tropospheric. The $Z_{tp}$ was obtained from the ERA-Interim data of ECMWF, whereas the altitude of the aircraft was obtained from pressure measurement which was converted into altitude using the ECMWF data. The position of the tropopause was obtained using the aircraft as the starting position. If the position is in the UT, i.e., it has a potential vorticity lower than 1.5 PVU, then the tropopause is found by searching upwards in the potential vorticity field. Tropopause folds in a small number of cases induce multiple tropopauses in the vertical direction, complicating the analysis of the stratospheric samples. In order to handle that problem, searches were undertaken both upwards and downwards to find the tropopause closest to the aircraft. That distance is assigned a positive value irrespective of whether the tropopause is below or above the aircraft, because positive sign indicates stratospheric air.

The study of the aerosol concentration gradient deals primarily with the LMS. The data set, however, contains observations both in the LMS and the UT and one sample sometimes contains particles from both regions. These concentrations are connected by the exchange across the tropopause. Figure 1 shows the samples that were taken in the UT during the entire sampling time. It is clear that the dependence on the distance from the tropopause is non-existent ($R^2 = 0.02$). Further evaluation (not shown) by normalization to seasonal average concentrations ($R^2 = 0.01$) or to individual groups of concentration data that will be explained below ($R^2 = 0.002$) did not reveal a dependence of the UT particulate sulfur concentration on the distance from the tropopause either. It should be noted that the variability in distance to the tropopause is mainly caused by variability in the altitude of the tropopause because the altitude of the measurement aircraft is fairly constant. This implies that the distance from the tropopause in Fig. 1 does not reflect the sampling altitude, and hence not an altitude-typical degree of cloud processing. The distance to the tropopause for some summer measurements was very large due to the seasonality of the position of the tropics. Approximately one-third of all stratospheric air masses transported across the extratropical tropopause reach the 500 hPa level of the atmosphere, corresponding to approximately 5000 m transport, in 4–5 days (Škerlak et

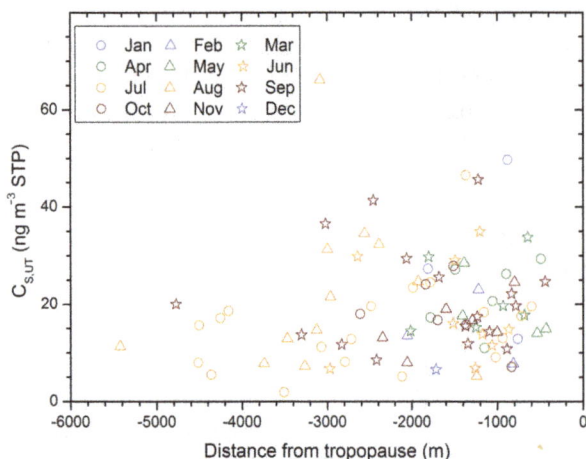

**Figure 1.** Upper tropospheric particulate sulfur concentration related to distance from the tropopause, where the symbols indicate sampling month.

al., 2014). This illustrates that the exchange from the stratosphere goes deep into the troposphere in a rather short time. Based on these observations and arguments, the UT particulate sulfur concentration is considered independent of the distance from the tropopause under presented conditions, and thus indicative of the particulate sulfur concentrations of the air mixed into the LMS.

The time required to collect one sample was subdivided into 10 time intervals of equal length, where $Z$ was computed along the flight route. For samples taken in the LMS, the sample was represented by the average distance to the tropopause of the 10 time intervals. Those time intervals when the sampling was undertaken in the UT, i.e., with $Z < 0$, $Z$ was set to zero, since no $Z$ dependence of the UT particulate sulfur concentration could be identified. This implies that samples collected entirely in the UT are found at $Z = 0$. Some of the samples were collected both in the LMS and in the UT. In these cases, the average $Z$ over the 10 time intervals of each sample was computed with all UT parts of a sample set to $Z = 0$. That way all samples could be utilized to study the particulate sulfur concentration in the LMS.

### 2.3 Methodology to evaluate the particulate sulfur gradient around the tropopause

The particulate sulfur concentration in the LMS is investigated using linear regression in two steps. To that end we need to consider that the LMS shows a seasonal dependence induced by variability in stratospheric circulation and exchange across the tropopause over the year. Within one season differences between years can be large, mainly due to varying influence from volcanism. Further, the onset of volcanic influence on the particulate sulfur concentration can cause large variability within the season in the same year,

and patchiness of young volcanic clouds can further affect the data analysis.

This study was limited to samples taken north of 30° N, where in total 765 samples are available. The highest latitude of sampling was 77° N, and 90 % of the samples (5 % removed each side) were taken in the range 32–64° N. Most of the LMS samples (95 %) were taken within 3000 m from the tropopause. In terms of altitude, the results of this study therefore can be considered representative of the range 0–3000 m above the tropopause.

A large number of measurements is needed to obtain high statistical significance. For that, data averaged over 3 months were used. However, to allow for exchange of data in the analyses in order to test the stability of the results, as well as catch smooth seasonal variations in the upper troposphere–lower stratosphere (UTLS), a seasonal overlapping technique is used. This technique uses 3-month seasons shifted by 1 month; e.g., the season MAM is followed by a season AMJ, where data from April and May are used in both "seasons". Data from one season were grouped with respect to concentration levels of the different years, resulting in 4 to 5 groups of data for each season, and in total 52 groups of data from the 12 seasons were analyzed. Some data were excluded from these analyses, which reduced the number of samples from 765 to 694. Of the excluded 71 samples, 60 pertained to periods when fresh volcanism induced strong patchiness in the particulate sulfur concentration. Eleven samples were considered outliers for other reasons, e.g., single samples affected by volcanism during a season or recent uptransport from strongly polluted regions. The remaining data of each year were tested for systematical differences. Those years where the data overlapped in particulate sulfur by altitude above the tropopause space were grouped together. This way, groups with varying degrees of volcanic influence were formed. Groups typical of "background" conditions were primarily based on data obtained during the 1999–2002 period characterized by low volcanic influence on the stratospheric aerosol (Bauman et al., 2003; Deshler, 2008) and data from mid-2013 to mid-2014, when the LMS was back to near-background conditions in the NH extratropics; see Table 1 for relevant volcanic eruptions. Another period that we will return to later is mid-2005 to mid-2008, when the stratosphere was affected by three tropical eruptions in 2005–2006: Manam, Soufriere Hills and Rabaul (Vernier et al., 2009), which also affected the NH LMS (Friberg et al., 2014). There was some variability during these years, implying that years were not always grouped together. They could be grouped with other years; e.g., 2011 and 2013 often had similar concentrations and gradients during the winter and spring seasons. These groups were handled individually in the regression procedures described next, but in a context of the discussion section these groups of "moderate influence from tropical volcanism" were averaged to describe the UT aerosol along with the "background" group described above.

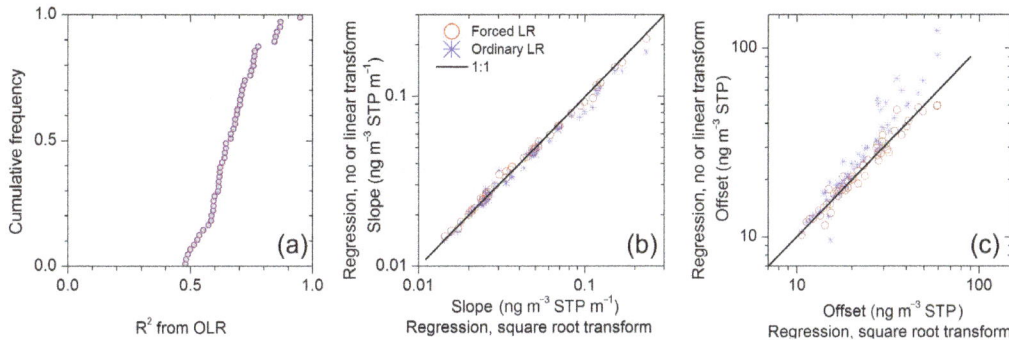

**Figure 2. (a)** Cumulative frequency of the coefficient of determination ($R^2$) of ordinary linear regression (OLR) between particulate sulfur concentration and distance from the tropopause of the 52 data groups used in this study. **(b, c)** Comparison of slope (Fig. 2b) and offset **(c)** between a square root transformed dependent variable and both OLR and forced linear regression.

**Table 1.** Most significant volcanic eruptions for the aerosol concentration in the Northern Hemisphere LMS in the time period studied.

| Volcano | Date | Lat, long | SO$_2$ (Tg) |
|---|---|---|---|
| Manam | 27 Jan 2005 | 4° S, 145° E | 0.1[a] |
| Soufriere Hills | 20 May 2006 | 17° N, 62° W | 0.2[b] |
| Rabaul | 7 Oct 2006 | 4° S, 152° E | 0.2[a] |
| Jebel at Tair | 30 Sep 2007 | 16° N, 42° E | 0.1[c] |
| Okmok | 12 Jul 2008 | 53° N, 168° W | 0.1[c] |
| Kasatochi | 7 Aug 2008 | 52° N, 176° W | 1.7[c] |
| Redoubt | 23 Mar 2009 | 60° N, 153° W | 0.1[d] |
| Sarychev | 12 Jun 2009 | 48° N, 153° E | 1.2[e] |
| Grimsvötn | 21 May 2011 | 64° N, 17° W | 0.4[f] |
| Nabro | 12 Jun 2011 | 13° N, 42° E | 1.5[f] |

[a] Prata and Bernardo (2007). [b] Carn and Prata (2010). [c] Thomas et al. (2011).
[d] Brühl et al. (2015). [e] Haywood et al. (2010). [f] Clarisse et al. (2012).

First linear vertical concentration gradients of particulate sulfur in the lower LMS was calculated. Ordinary linear regression (OLR) was undertaken for the 52 groups of data mentioned above of particulate sulfur concentration as a function of the altitude above the tropopause. Figure 2a shows the cumulative frequency of the coefficient of determination ($R^2$) of the 52 OLRs undertaken. $R^2$ spans 0.48 to 0.95, 72 % of the groups having $R^2$ exceeding 0.6 and 50 % exceeding 0.66. The deviations from the OLR models consists of scatter that does not show any trends.

When investigating the variance of the dependent variable (the particulate sulfur concentration, $C_S$) along the independent variable (altitude above the tropopause, $Z$) it is clear that the variance of $C_S$ increases with increasing $Z$. This heteroscedastic nature of the data, which is shared by most natural science data sets, could unfavorably affect in particular the offset of the OLR. In order to further investigate the effects of this problem, two variable transformations of the dependent variable were tested: logarithmic and square root transformations. The logarithmic transformation turned the problem around, i.e., the logarithm of $C_S$ has large

variance for small $Z$ and small variance for large $Z$. The square root transformation of $C_S$, on the other hand, shows rather constant variance along the $Z$ axis, thus making this transformation more suitable for regression. The transformation $C'_S = \sqrt{C_S}$ was applied to the data. As mentioned, $C_S$ and $Z$ have a linear relationship implying that the following expression should be minimized with respect to slope ($a$) and offset ($b$): $(C'_S - \sqrt{aZ+b})^2$. This results in rather tedious expressions that we solved numerically for all the 52 data groups. Figure 2b and c show comparisons between the slopes and the offsets obtained by the square root transformed regression results and the OLRs. It is clear that the heteroscedastic nature of the data causes large deviations, especially in the offset.

Yet another transformation, based on forcing the regression to comply with the data for small $Z$, was investigated. To that end, the average concentration ($C_{S,0}$) of the data points closest to the tropopause at the average distance $Z_0$ from the tropopause was formed. $C_{S,0}$ of the 52 data groups is on average based on 13 measurements, the minimum being 5, and the average $Z_0$ is 88 m and the largest is 273 m. The data were transformed linearly to place $C_{S,0}$ and $Z_0$ in the origin by forming $C'_S = C_S - C_{S,0}$ and $Z' = Z - Z_0$ followed by linear regression forced through the origin, i.e., $C'_S = aZ'$. Finally, the regression results are transformed back to the form $C_S = aZ + b$, where the slope $a$ is not changed by the translation, and the offset is obtained by $b = C_{S,0} - aZ_0$. These results of the forced linear regressions are compared with the square root transformed regressions in Fig. 2b and c. As can be seen, the forced linear regressions, in contrast to the OLRs, show only small deviations from the square root transformed data. Due to the more direct determination of the offset as well as the simplicity of forced linear regression compared with square root transformation, the forced linear regression method is chosen for the analyses. Thus, for each season ($s$) and year ($y$) in total 52 forced linear regressions were undertaken:

$$C_S(y,s,Z) = a(y,s)Z + b(y,s), \qquad (2)$$

where $a$ and $b$ varies with season and the strength of the volcanic influence.

The sulfur concentration at the tropopause and in the UT, expressed by $b$ in Eq. (2), is dependent on the stratospheric concentration which is affected by volcanism (Friberg et al., 2015). This means that the offset of the regressions is affected. This is expressed here by

$$b(y,s) = C_{S,UT}(y,s) = a(y,s)k(s) + C_{S,UTtrop}(s), \qquad (3)$$

where the first term expresses the contribution from stratospheric sources and the second expresses that of tropospheric sources. This results in the combined equation:

$$C_S(y,s,Z) = a(y,s)Z + a(y,s)k(s) + C_{S,UTtrop}(s), \qquad (4)$$

where $k(s)$ reflects stratospheric influence on the UT particulate sulfur concentration and $C_{S,UTtrop}(s)$ is the particulate sulfur concentration of tropospheric origin.

In order to obtain estimates of $k$ and $C_{S,UTtrop}$, a second regression for each season is undertaken where the offsets ($b$) are related to the slopes ($a$); see Fig. 3. The relative variance of the slopes was much smaller than that of the offsets (average variance ratio of 0.19). For simplicity, the variance of the slope was therefore neglected in these regressions. The groups of data of a season differ in $a$ and $b$ mainly due to volcanic influence. A zero slope ($a = 0$) would be obtained should the stratospheric concentration become as low as the UT concentration. The UT aerosol of tropospheric origin ($C_{S,UTtrop}$) thus can be estimated as the offset of the $b - a$ regression. The slope of that regression shows how the offset $b$ changes with increased slope $a$, hence expressing the sensitivity ($k$) of the UT concentration to changes in stratospheric concentration. With access to these two parameters, $C_{S,UTtrop}(s)$ and $k(s)$, the LMS concentration gradient and tropopause concentration can be estimated based on a single measurement of the particulate sulfur concentration.

The uncertainties of the forced regression results of a given season vary among the data groups. In order to account for this variability, weights are used in the regression between the offsets and slopes of a season. The weights are based on 70 % double-sided Student $t$ distribution ($t70\%$) estimates because some of the estimated $C_{S,0}$ and $Z_0$ rely on few observations. The $t70\%$ estimate of $b$ ($b = C_{S,0} - aZ_0$; see above) is obtained by combining the upper $t70\%$ limit of $C_{S,0}$ with the weakest $t70\%$ slope ($a$) and the strongest $t70\%$ slope at the lower limit multiplied with $Z_0$. The inverse of the squared $t70\%$ estimates of $b$ obtained in this way are then used as weights in the second step regressions.

## 2.4 Lidar data from the CALIPSO satellite

The evaluation of the particulate sulfur concentrations from the IAGOS-CARIBIC aircraft was aided by the use of lidar data from the CALIOP sensor aboard the CALIPSO (Cloud-aerosol lidar and infrared pathfinder satellite observation)

satellite from NASA (Winker et al., 2010), performing 15 orbits per day covering the globe from 82° S to 82° N with a repeat cycle of 16 days during night and day. The data evaluation was based on the methodology developed by Vernier et al. (2009). Here only the nighttime data of the 532 nm wavelength lidar signals were used. The level 1 data of version 4–10 (averaged in a grid of 180 m vertically and $1 \times 1$ horizontally) were used to obtain scattering ratios, i.e., the ratio of the measured, combined air and aerosol scattering to the modeled air scattering based on molecule and ozone number concentrations from the GMAO (Global Modeling and Assimilation Office). Cloud pixels were removed by rejecting pixels of a depolarization ratio greater than 5 %. This cloud mask was extended 360 m upwards to remove faint cloud residues, and pixels from underneath clouds were removed to avoid bias from cloud absorption; see Andersson et al. (2015) for further details. Removal of cloud pixels as well as periods of instrument failure sometimes resulted in few observations in these pixels. In order to avoid statistical noise, at least 20 % of the maximum possible observation data was required for any given pixel (latitude, altitude). The pixels were generated by averaging in the longitude interval 60 to 120° E, the main longitude region of the ATAL (Vernier et al., 2015).

## 3   Results

The linear regression methodology described in the previous section was applied to particulate sulfur concentrations ($C_S$) as a function of the distance from the tropopause ($Z$) for seasons comprising 3 months. Data from different years of one season were grouped according to their concentrations, resulting in four or five groups differing with respect to the $C_S - Z$ relationship. The resulting groups of data from each season were modeled by forced linear regression. These model results are used in a second regression step to model seasonal influences from transport, which in turn are used to obtain the response of the LMS and UT particulate sulfur concentrations to changes induced mainly by volcanic eruptions.

This methodology was applied to all seasons, having a duration of 3 months. The final product of the regression methodology, i.e., the second regression step, of all 12 overlapping seasons is shown in Fig. 3. It is clear that slopes and offsets of the season groups obtained in the first step of forced linear regression, differing in particulate sulfur concentration related to distance from the tropopause, can readily be described by linear regressions for all twelve 3-month seasons. The relationship between slopes and offsets has a very strong seasonal dependence. In some seasons small changes in the LMS sulfur concentration slope is connected with a strong change in tropopause sulfur concentration. This is most pronounced for the seasons centered in February, March and April. At the other end we find the seasons centered in September, October and November, where a change in the

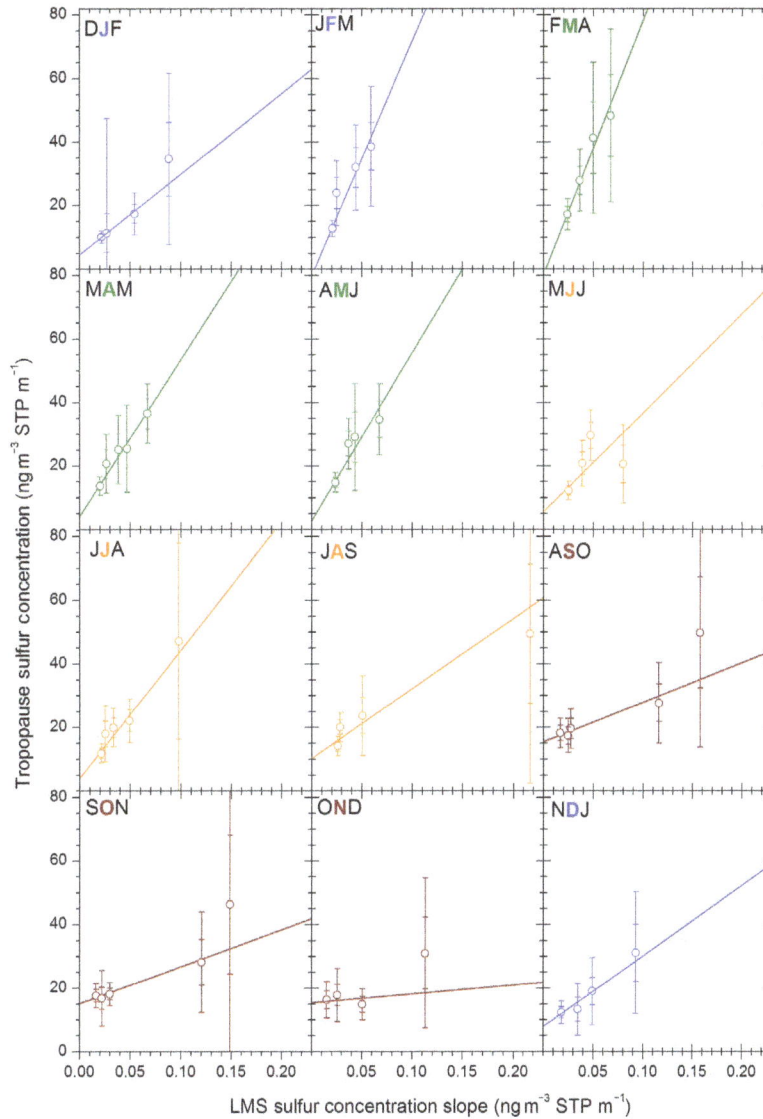

**Figure 3.** Weighted regression between slopes and offsets of all groups of each season. The offset indicates the tropopause (and UT) concentration of tropospheric origin ($C_{S,UTtrop}$) and the slope of the fit ($k$) expresses the sensitivity of the UT concentration to changes in the stratospheric concentration slope ($a$). The error bars show Student $t$ 70 and 95 % confidence interval, respectively.

slope of the first regression step due to varying influence from volcanism has little effect on the tropopause concentration. To further illustrate the connection to the measurements, Fig. 4 shows the four data groups of the seasons most (Fig. 3: FMA) and least (Fig. 3: OND) susceptible to change in the tropopause concentration due to changed stratospheric concentrations. In Fig. 4a the data group of least volcanic influence includes years 1999–2002 and 2014, and the most influenced years are 2009 and 2012 which are the springs after the Kasatochi (7 August 2008) and Nabro (12 June 2011) eruptions (we have no late winter/spring data after the most powerful eruption of the period studied, Sarychev, 12 June 2009, due to maintenance of the measurement aircraft). The season centered in November has the weakest volcanic influ-

ence years 1999–2001, 2012 and 2013 (Fig. 4b), whereas the group with strongest slope includes the three strongest eruptions of the period studied (Kasatochi, Sarychev and Nabro), a few months after respective eruption. Despite the very strong volcanic influence of the latter group, the tropopause concentration ($Z = 0$) remains close to that of the closely distributed tropopause concentrations of the other groups of that season.

The sensitivity of the tropopause and UT concentration to changes in LMS concentration slope is denoted $k$ in Eqs. (3) and (4), which thus is obtained for each regression depicted in Fig. 3. These results are collected in Fig. 5a. The salient features of the seasonal dependence can be described by two Gaussian distributions (Fig. 5a). The maximum sensitivity

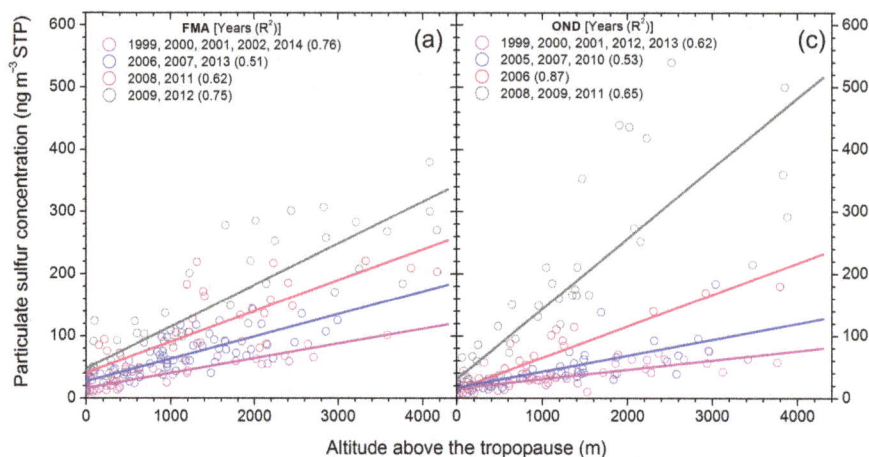

**Figure 4.** Examples of first step forced linear regression between the particulate sulfur concentration and the altitude above the tropopause for the data groups of two seasons: **(a)** FMA (February, March and April) and **(b)** OND (October, November and December). The legend shows which years are included in the respective data groups followed by the coefficient of determination in parentheses.

appears in late winter and early spring when the seasonal variation in tropopause altitude has its maximum rate of upward motion (Appenzeller et al., 1996). The maximum in rate of downward motion of the tropopause appears in the fall. This delays transport from the stratosphere to the troposphere, which is reflected by a low-sensitivity $k$. This sensitivity can, in addition, be affected by the residence time of particulate sulfur in the UT. Interestingly, the downward transport in association with the Brewer–Dobson circulation from deeper, aerosol-rich stratospheric layers through the LMS takes place in the same season as the maximum in $k$, thus further enhancing the stratospheric influence on the UT by the term $a(y, s) \times k(s)$ in Eq. (4).

The offsets of the regression lines shown in Fig. 3 (when $a = 0$) expresses the case when the stratospheric and tropospheric concentrations are equal, implying that this offset expresses the particulate sulfur concentration of tropospheric origin ($C_{S,\mathrm{UTtrop}}$) that is mixed into the LMS; see Fig. 5b. Two Gaussian distributions were used as fits to the seasonal variation of $C_{S,\mathrm{UTtrop}}$. As for $k$, the seasonal variation of $C_{S,\mathrm{UTtrop}}$ is strong. The seasonal variation of $C_{S,\mathrm{UTtrop}}$ will be elaborated in the discussion section.

After obtaining $C_{S,\mathrm{UTtrop}}$ and $k$, the data that are needed for conversion of every measurement of the concentration to an estimate of the slope and offset of the LMS concentration are available. Thus, for each individual measurement (i) in the LMS, consisting of the particulate sulfur concentration ($C_{S,i}$) and the altitude $Z_i$ above the tropopause, the slope and offset of Eqs. (3) and (4) are obtained from

$$a_i = \frac{C_{S,i} - C_{S,\mathrm{UTtrop}}(s)}{Z_i + k(s)}, \tag{5}$$

$$b_i = a_i k(s) + C_{S,\mathrm{UTtrop}}(s). \tag{6}$$

The estimated slopes and offsets are shown in Fig. 6a and b, where the dots are individual measurements and the his-

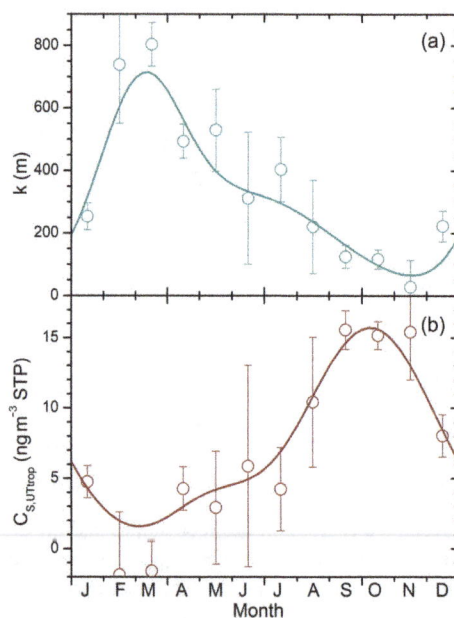

**Figure 5.** Seasonal variation of **(a)** the sensitivity of the UT particulate sulfur concentration to the concentration slope in the LMS ($k$) and **(b)** the UT particulate sulfur concentration of tropospheric origin ($C_{S,\mathrm{UTtrop}}$).

togram monthly averages. Here measurements taken at an altitude of less than 50 m above the tropopause are not shown, because they were judged to have too small a stratospheric character for an estimate of the concentration slope in the LMS. Both the slope and the tropopause concentration are affected by volcanism (Table 1), but the relative response of the slope is much stronger than that of the tropopause concentration; see, for example, the falls of 2008 and 2009 affected by the Kasatochi and Sarychev eruptions.

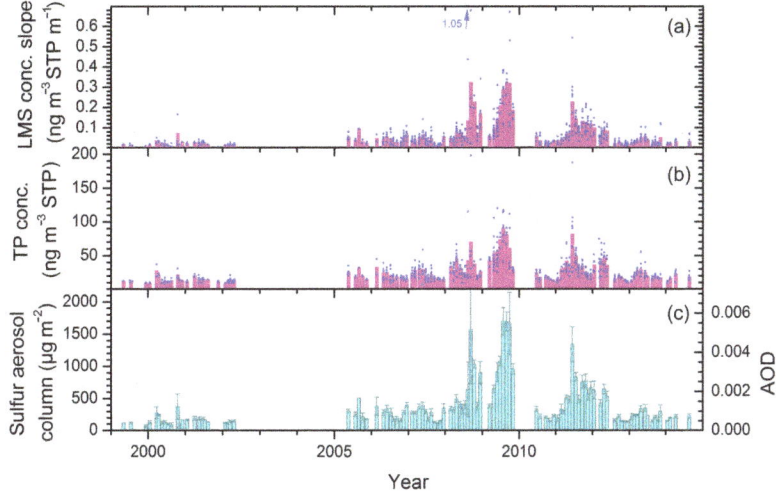

**Figure 6.** Estimated (**a**) slopes and (**b**) offsets based on individual measurements (dots) and monthly averages (magenta bars). (**c**) Sulfur aerosol column of the lowest 3000 m of the LMS with standard errors, assuming particles of 75 % sulfuric acid and 25 % water, and AOD (right $y$ axis) assuming stratospheric background aerosol particle size distribution.

Observations at various altitudes above the tropopause are difficult to compare, due to the concentration gradient in the LMS. With the estimates of the tropopause concentration and the slope in the particulate sulfur LMS concentration, each measurement becomes an estimate of the total amount of particulate sulfur in the altitude interval investigated through the integration of Eq. (4). However, first the STP concentrations ($C_{S,STP} = C_S = aZ + b$) need to be converted into volume concentrations ($C_{S,V}$), which are related by $C_{S,V} = C_{S,STP} Q_{STP}/Q_V(Z)$, where $Q$ is molar volumes. For that purpose, the altitude dependence of the molar volume from the tropopause up to 5 km above the tropopause was extracted from temperatures and pressures obtained from ECMWF for each sample. The molar volume can be expressed as $Q_V(Z) = Q_V(0)e^{wZ}$, where the tropopause is at $Z = 0$ and $w = 0.0001535\,\mathrm{m}^{-1}$ obtained as the average of all samples. For a measurement the molar volume is $Q_m$ obtained at distance $Z_m$ from the tropopause, and the tropopause molar volume is computed by $Q_V(0) = Q_m e^{-wZ_m}$. Finally, the column concentration of particulate sulfur for the first 3 km above the dynamical tropopause of 1.5 PVU is obtained by integration of the volume concentration:

$$C_{S,col} = \int_0^{3000} \frac{Q_{STP}}{Q_V(Z)} C_{S,STP}(Z)\,dZ$$

$$= \int_0^{3000} \frac{Q_{STP}}{Q_V(0)} e^{-wZ} (aZ+b)\,dZ. \quad (7)$$

The particulate sulfur column is calculated for altitudes above the tropopause ($Z$) in the range of 0 to 3000 m, the upper limit set where too few measurements (5 %) were taken

above that level. After integrating the column, it is also interesting to estimate the total amount of sulfur-connected aerosol. To that end, it was assumed that the aerosol consists of 75 % sulfuric acid and 25 % water, which is a commonly used stratospheric composition (Rosen, 1971; Arnold et al., 1998). This means that the total sulfur column was multiplied by a stoichiometric factor of $h = 4.084$ to obtain the column of the sulfuric acid–water aerosol:

$$C_{A,col} = hC_{S,col}. \quad (8)$$

The measurements were taken in the Northern Hemisphere latitudes higher than 30° with the highest latitude of 77°, with 90 % of the data in the latitude range 32–64° N and 70 % of the data between 37 and 57° N. The northern midlatitudes sulfur aerosol columns are shown in Fig. 6c as monthly averages with standard errors (the few months with only one measurement available are shown without error bars). The sulfur aerosol column of the LMS shows large variability primarily caused by volcanism. The lowest columns are found in the period 1999–2002, when the volcanic influence on the stratospheric aerosol was small (see Table 1 for relevant volcanic eruptions). The time period mid-2005 to mid-2008 was affected by tropical volcanism (Vernier et al., 2011a), which also caused elevated concentrations in the NH LMS (Friberg et al., 2014). The eruptions of the extratropical volcano Kasatochi in August 2008 placed two volcanic clouds in the stratosphere (Andersson et al., 2015), one in the LMS causing strongly elevated aerosol column of the LMS that ceased by November the same year, and the other above the LMS. The latter cloud was transported downward, causing a rise of the lower LMS aerosol column in December 2008. After some influence from several eruptions of the extratropical volcano Redoubt in the spring of 2009, the eruption of

Sarychev strongly affected the Northern Hemisphere stratosphere from June 2009. The eruption of the Icelandic volcano Grimsvötn in May 2011 had a strong and short impact
on the northern LMS, which is reflected by a peak in June
to July 2011 (Fig. 6c), before the tropical volcano Nabro
reached the northern LMS in the early fall of the same year.
After that eruption a gradual decrease of the aerosol load can
be seen. The concentrations after mid-2013 approach those
of the period 1999–2002, which was close to stratospheric
background conditions.

The LMS aerosol contains a significant fraction of carbonaceous material (Martinsson et al., 2009), mainly organic
in nature (Friberg et al., 2014), which adds to the aerosol
columns of Fig. 6c and affect the refractive index of the
particles. However, this work is dealing with the sulfurous
fraction, the main fraction of the stratospheric aerosol. To
put the results presented in Fig. 6c into perspective, the
AOD is estimated using a simplified aerosol (which thus
likely is an underestimation of the AOD). Thus, the "standard" stratospheric 75 % sulfuric acid, 25 % water composition, $1.669\,\mathrm{g\,cm^{-3}}$ particle density and 1.44 refractive index
will be used. Furthermore, particle size distribution measurements from IAGOS-CARIBIC have been taken since 2010
(Hermann et al., 2016). The changes of the size distribution induced by the moderate 2011 eruptions of Grimsvötn
and Nabro were small (Martinsson et al., 2014), and agree
well with previous measurements (Andersson et al., 2015)
of the stratospheric background particle size distribution by
Jäger and Deshler (2002). Thus, for the estimation of the
AOD the latter particle size distribution was used for the
entire time period studied. For fixed composition and particle size distribution the AOD is obtained as a fixed relationship to aerosol column: $\mathrm{AOD} = f C_{\mathrm{A,col}}$, where $f$ contains
the relationships between mass and area/extinction, in this
case $f = 3.29 \times 10^{-6}\,\mathrm{m^2\,\mu g^{-1}}$. Finally, converting the AOD
into radiative forcing (RF) using the global average relation
(Hansen et al., 2005; Solomon et al., 2011) of

$$\mathrm{RF} = -25 \times \mathrm{AOD}; \quad \text{in } \mathrm{W\,m^{-2}}, \tag{9}$$

to obtain an estimate of the climate influence of the sulfate
aerosol of the lower LMS. The peak AOD of the Kasatochi
and Sarychev eruptions is approximately 0.006, correspond-
ing to $-0.15\,\mathrm{W\,m^{-2}}$ in regional radiative forcing of the lowest 3000 m of the Northern Hemisphere LMS. Although no
detailed comparisons will be made here, we find that the
AOD and radiative forcing obtained from the particulate sulfur measurements show similar tendencies to satellite-based
measurements (Andersson et al., 2015). The findings presented here also corroborate the findings of Andersson et
al. (2015) on the importance of the LMS for the total stratospheric AOD and radiative impact.

## 4   Discussion

The results presented here are based on measurements in
the extratropical UT and the LMS of the NH, where the
latter includes the extratropical transition layer (ExTL). Bi-
directional exchange across the tropopause affects strong
gradients in the ExTL for species having clearly different
stratospheric and tropospheric concentrations (Hoor et al.,
2002), such as particulate sulfur (Martinsson et al., 2005).
In the previous section, the UT concentration, the gradient
in the LMS and column amount of particulate sulfur in the
ExTL were investigated, with seasonal dependence (Fig. 5)
and influence from volcanism (Fig. 6). In the processing of
the data to obtain these results, one feature stands out in particular: the seasonal dependence of the particulate sulfur concentration from tropospheric sources that is mixed into the
ExTL (Fig. 5b). It shows a broad maximum from August to
December, peaking in September to November and a deep
minimum in the late winter and early spring (February and
March).

Let us now compare these concentrations in the UT of tropospheric origin ($C_{\mathrm{S,UTtrop}}$) with average concentrations of
particulate sulfur in the UT for two cases: "background conditions" dominated by data from mid-1999 to mid-2002 and
"moderate influence from tropical volcanism" dominated by
data from mid-2005 to mid-2008; see Sect. 2.3 for details.
The seasonal dependence of these two categories are shown
in Fig. 7a together with $C_{\mathrm{S,UTtrop}}$. It is clear that, in line with
the findings of Friberg et al. (2015), the UT particulate sulfur
concentration is affected by volcanism. In Fig. 7a we see that
the main differences in UT concentrations of the two cases
appears from January to July, coinciding with the season of
transport down from the Junge layer into to the LMS and
the shrinkage of the LMS due to tropopause upward motion
(Appenzeller et al., 1996; Gettelman et al., 2011). In the period September to November the influence from volcanism
on the UT particulate sulfur concentrations is small (compare the background and moderate volcanism cases; Fig. 7a).
When comparing these cases to the concentration of particulate sulfur found to be of tropospheric origin, it is clear
that there is strong agreement between all three categories
in the fall months, whereas differences are large during the
remainder of the year. The UT particulate sulfur concentration of stratospheric origin can be estimated by subtracting
$C_{\mathrm{S,UTtrop}}$ from the two cases of UT concentrations. The results are shown in Fig. 7b, where peak stratospheric influences of 19 and $36\,\mathrm{ng\,m^{-3}}$ STP are found in the spring, and
minimum contributions of approximately $1\,\mathrm{ng\,m^{-3}}$ STP are
found in the fall for the "background" and "moderate volcanism" cases. Summing the observations up (Fig. 7c), a clear
seasonal dependence in the fraction of the UT particulate sulfur concentration originating in the stratosphere was found,
from close to 100 % in late winter/spring to approximately
10 % in the fall. On a yearly average the fraction of the UT
particulate sulfur that originates in the stratosphere is approx-

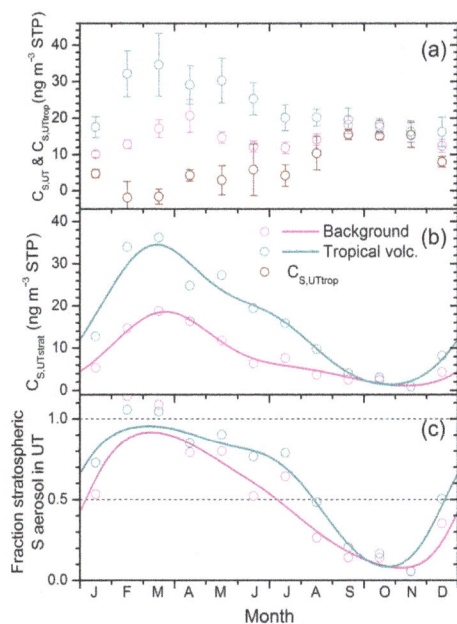

**Figure 7.** Seasonal variation of **(a)** the UT particulate sulfur concentration during LMS background conditions and moderate volcanic influence, together with the estimated UT particulate sulfur concentration of tropospheric origin ($C_{S,UTtrop}$), **(b)** UT particulate sulfur concentration of stratospheric origin ($C_{S,UTstrat}$) obtained by subtracting the $C_{S,UTtrop}$ from the two UT concentration cases in **(a)** and the ratio $C_{S,UTstrat}$ / $C_{S,UT}$ for the two cases **(c)**.

imately 50 % during background conditions and 70 % during moderate influence from tropical volcanism.

The tropospheric source of UT particulate sulfur could be transported from the planetary boundary layer either in the form of particulate sulfur or precursor gases, in the latter case primarily sulfur dioxide ($SO_2$). In fall, winter and spring, with maximum in the winter, vertical transport by WCBs from the boundary layer to the UT is strong (Eckhardt et al., 2004), whereas in the summer deep convection is the most important mode (Hess, 2005; Kiley and Fuelberg, 2006) in the extratropics. For sulfate, the most common chemical form of particulate sulfur, the concentration usually shows a rapid decline with altitude in the troposphere (Heald et al., 2011) associated with formation of precipitation. Cloud processing also tends to strongly reduce $SO_2$ concentrations with altitude, where the relative availability of $SO_2$ and hydrogen peroxide ($H_2O_2$) is important for the $SO_2$ lifetime in the cloud.

We have found some very distinctive characteristics of the particulate sulfur concentration of tropospheric origin in the UT, with low concentrations in February and March, increasing concentrations during the summer and maximum in the sources of tropospheric origin in the fall. Carbon monoxide (CO) is often used as a tracer of air pollution. Zbinden et al. (2013) found winter/spring maximum in the CO concentration in the UT of the NH in contrast to the tropospheric

component of the UT particulate sulfur. This contrast can, at least in part, be explained by the oxidizing capacity in the UT. The summer abundance of the hydroxyl (OH) radical (Bahm and Khalil, 2004) induces a decline in CO (Bergamaschi et al., 2000; Osman et al., 2016). High abundance of this radical, on the other hand, and availability of $SO_2$ can lead to production of particulate sulfur. $SO_2$ measurements by the satellite-based instrument MIPAS have recently become available (Höpfner et al., 2015). These data have large uncertainties (Höpfner et al., 2015), and likely overestimate the $SO_2$ concentration (Rollins et al., 2017). However, here we only qualitatively use the seasonal variation. The MIPAS results indicate a UT seasonal variation in the NH midlatitudes with low concentrations in December to March and the highest concentrations in June to September (Höpfner et al., 2015). Deep convection provides a rapid route upwards in the atmosphere, favoring transport of short-lived species like $SO_2$ (TF-HTAP-2010, 2010; Dickerson et al., 2007). Low abundance of both $SO_2$ and OH in the NH UT thus could explain the weak tropospheric contribution to the UT particulate sulfur during late winter and early spring. The increase in $C_{S,UTtrop}$ during the spring and summer months coincides with increases in convective activity as well as in the concentration of both $SO_2$ and OH, thus offering a plausible explanation. However, the quantitative understanding of the $SO_2$ transport paths and UT seasonality requires further study. Still, we need to consider the late maximum in the fall of the tropospheric source of UT particulate sulfur.

The Asian monsoon is an important feature of global circulation in June to September, which has been found to reach deep into the lower stratosphere. This intrusion of tropospheric air into the tropical transition layer (Sunilkumar et al., 2017) has been manifested by gas-phase components including water and ozone (Gettelman et al., 2004; Randel and Park, 2006) and hydrogen cyanide (Randel et al., 2010). Later an aerosol layer extending from the UT to the potential temperature of 420 K in the lower stratosphere was found at 14 to 18 km altitude (Vernier et al., 2011b; Thomason and Vernier, 2013; Vernier et al., 2015). Figure 8a–e show monthly means of the scattering ratio obtained from the CALIPSO sensor CALIOP for the months June to October 2013, when the volcanic influence was low. Formation of the ATAL can be identified in July (Fig. 8b) with maximum intensity in August (Fig. 8c). Even in September can a weakened ATAL be identified (Fig. 8d), whereas in June and October the scattering in the ATAL region at 14–18 km altitude is very weak (Fig. 8a and e).

The altitude range of the ATAL is above the measurement altitudes of IAGOS-CARIBIC (9–12 km). However, the poleward circulation along isentropes bending downwards, further amplified by an extratropical cross-isentrope, downward component, brings the ATAL down to lower altitudes. The ATAL is too weak to be traced by CALIOP in the downward transport (Fig. 8a–e). We therefore illustrate the subsidence using the eruption of the tropical volcano Nabro in

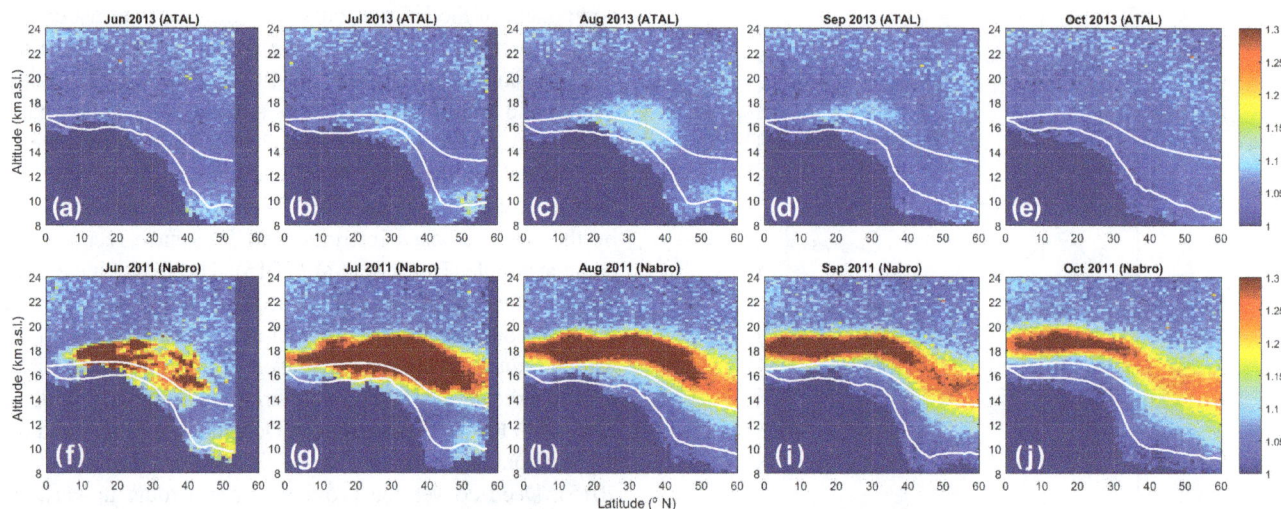

**Figure 8.** Latitude and altitude distribution of scattering from aerosol in the stratosphere and the UT. Two series, from 2013 (upper row, **a–e**) and 2011 (lower row, **f–j**), show the monthly mean 532 nm wavelength scattering ratio from June to October without and with, respectively, fresh volcanic influence. The former series allows identification of the comparably weak ATAL in July to September. The lower series includes two volcanic eruptions: the high-altitude cloud from the tropical volcano Nabro (eruption in 12 June 2011) and the low-altitude and midlatitude faint volcanic cloud from Grimsvötn (21 May 2011). The two white lines in every graph show the monthly averaged positions of 380 K potential temperature (upper line) and the 1.5 PVU dynamical tropopause.

summer 2011 (Table 1), with effluents that occupied approximately the same region as the ATAL (see Fig. 3 in Bourassa et al., 2012). The effluents of this eruption were rapidly transported to the north in the tropical tropopause layer. Figure 8f–j actually includes two volcanic eruptions; besides the tropical volcano Nabro, the Icelandic volcano Grimsvötn injected a volcanic cloud into the tropopause region at midlatitudes which is visible in June and July 2011. The time series of the Nabro volcanic cloud in Fig. 8f–j unambiguously demonstrates the transport down to the IAGOS-CARIBIC flight altitudes in the season of the ATAL. In Fig. 6c we see a clear increase in the particulate sulfur column in September 2011, following the decline of the June–July 2011 short peak from the of Grimsvötn eruption in May 2011, thus demonstrating agreement between CALIOP and IAGOS-CARIBIC measurements of particulate sulfur. The volcanic cloud from Nabro initially resided at somewhat higher altitude than the ATAL, implying that the ATAL can be expected to reach IAGOS-CARIBIC flight altitudes somewhat earlier in the year. The transport in this season contains little particulate sulfur during periods without fresh volcanic aerosol; compare Fig. 7 "tropical volcanism" (the "tropical volcanism" category of Fig. 7 does not include the Nabro eruption – see Sect. 2.3) with "background" and the particulate sulfur of tropospheric origin.

The downward transport allows delayed detection of the ATAL in the fall from altitudes above the measurement altitude range of IAGOS-CARIBIC in the same way as the spring detection of aerosol transported from the Junge layer, with or without volcanic influence, as shown in Fig. 7. A

difference between these two seasons is that in the spring the strong source of particulate sulfur is stratospheric, whereas in the fall the source is tropospheric extending from the UT into the stratosphere. We therefore conclude that the UT particulate sulfur concentration of tropospheric origin that starts to increase rapidly in August, peaking in September to November, and is back at low concentration in January (Fig. 5b) is most likely caused by the ATAL formation from the Asian monsoon.

Model studies find similar UT/stratosphere distribution of the ATAL as the experimental studies, that Asian pollution sources strongly contribute to the ATAL and that sulfate is an important component of that aerosol layer (Neely III et al., 2014). Besides this component, primary and secondary organic aerosol constituents are predicted to be important components of the ATAL (Yu et al., 2015). The present study is, as far as we know, the first observation of a chemical component of the ATAL. In agreement with modeling results we find that particulate sulfur is a component of the Asian tropopause aerosol layer. This component affects the tropopause region from July to September at 14 to 18 km altitude (Fig. 8) and the extratropical tropopause region from August to December (Fig. 5b).

## 5   Conclusions

Particulate sulfur (usually sulfate) in the upper troposphere (UT) and the lowermost stratosphere (LMS) obtained from the IAGOS-CARIBIC platform was investigated at northern midlatitudes in the time period 1999–2014, which covers several tropical and extratropical volcanic eruptions. The study is based on the use of linear regression models, where individual measurements in the strong gradient of the extratropical transition layer (ExTL) can be converted into an estimate of the column of particulate sulfur in a layer 3000 m above the dynamical tropopause (here defined at 1.5 PVU). The obtained time series in particulate sulfur column concentration shows distinct response to extratropical volcanism and delayed elevation of the column concentration following tropical eruptions. Assuming the stratospheric background particle size distribution and composition (75 % sulfuric acid and 25 % water) the AOD and radiative forcing were estimated; e.g., the peak values following the 2009 Sarychev eruptions were estimated to be 0.006 and $-0.15 \, \mathrm{W \, m^{-2}}$, respectively. These estimates refer mainly to the ExTL, i.e., lowest part of the LMS, thus highlighting the importance of the lowest part of the stratosphere for the overall climate impact of volcanism.

As part of this investigation the sources of UT particulate sulfur were explored. A distinct pattern emerges where tropospheric sulfur sources dominate the supply of particulate sulfur to the UT in the fall, whereas stratospheric sources dominate in January to July, the main season of transport from the Junge layer into the LMS. The smallest contributions from the troposphere are found in February and March in conjunction with low sulfur dioxide ($SO_2$) and oxidant concentrations. As the concentrations of these species increase, the UT particulate sulfur concentration of tropospheric origin increases somewhat in April to July. The particulate sulfur concentration shows a threefold increase during the fall, with maximum concentration in September to November. Making use of lidar data from the CALIPSO satellite together with the in situ measurements we find that the Asian tropopause aerosol layer (ATAL) resulting from the Asian monsoon is the cause of the increase. The ATAL is formed at 14–18 km altitude and extends from the UT to approximately 420 K potential temperature in the stratosphere, with main extension in July to September. The ATAL is transported downwards and affects the extratropical tropopause region in August to December. As far as we know, this is the first measurement of a chemical species in particles connected with the ATAL. The stratospheric and tropospheric contributions to the UT particulate sulfur concentrations thus have strong and opposite seasonal dependences. On annual average the stratospheric contribution to the UT particulate sulfur is estimated to be 50 % during stratospheric background conditions. During influence from moderate tropical volcanism the stratospheric fraction rises to 70 %. The particulate sulfur concentration in the UT is thus to a large degree governed by the stratosphere and volcanism.

*Competing interests.* The authors declare that they have no conflict of interest.

*Acknowledgements.* We acknowledge all members of the IAGOS-CARIBIC project, Lufthansa and Lufthansa Technik for enabling the IAGOS-CARIBIC observatory. Financial support from the Swedish National Space Board (contract 130/15) and the Swedish Research Council for Environment, Agricultural Sciences and Spatial Planning (contract 942-2015-995) is gratefully acknowledged. Moreover, the German Federal Ministry of Education and Research (BMBF) is acknowledged for financing the instruments' operation as part of the joint project IAGOS-D. Aerosol measurements from CALIPSO were produced by NASA Langley Research Center.

Edited by: Kostas Tsigaridis

## References

Andersson, S. M., Martinsson, B. G., Friberg, J., Brenninkmeijer, C. A. M., Rauthe-Schöch, A., Hermann, M., van Velthoven, P. F. J., and Zahn, A.: Composition and evolution of volcanic aerosol from eruptions of Kasatochi, Sarychev and Eyjafjallajökull in 2008–2010 based on CARIBIC observations, Atmos. Chem. Phys., 13, 1781–1796, https://doi.org/10.5194/acp-13-1781-2013, 2013.

Andersson, S. M., Martinsson, B. G., Vernier, J. P., Friberg, J., Brenninkmeijer, C. A. M., Hermann, M., van Velthoven, P. F. J., and Zahn, A.: Significant radiative impact of volcanic aerosol in the lowermost stratosphere, Nat. Commun., 6, 7692, https://doi.org/10.1038/ncomms8692, 2015.

Appenzeller, C., Holton, J. R., and Rosenlof, K. H.: Seasonal variation of mass transport across the tropopause, J. Geophys. Res., 101, 15071–15078, 1996.

Arnold, F., Curtis, J., Spreng, S., and Deshler, T.: Stratospheric aerosol sulfuric acid: First direct in situ measurement using a novel balloon-based mass spectrometer apparatus, J. Atmos. Chem., 30, 3–10, 1998.

Bahm, K. and Khalil, M. A. K.: A new model of tropospheric hydroxyl radical concentrations, Chemosphere, 54, 143–166, 2004.

Baker, A. K., Slemr, F., and Brenninkmeijer, C. A. M.: Analysis of non-methane hydrocarbons in air samples collected aboard the CARIBIC passenger aircraft, Atmos. Meas. Tech., 3, 311–321, https://doi.org/10.5194/amt-3-311-2010, 2010.

Bauman, J. J., Russel, P. B., Geller, M. A., and Hamill, P.: A stratospheric aerosol climatology from SAGE II and CLAES measurements: 2. Results and comparisons, 1984–1999, J. Geophys. Res., 108, 4383, https://doi.org/10.1029/2002JD002993, 2003.

Bergamaschi, P., Heim, R., Heimann, M., and Crutzen, P. J.: Inverse modeling of the global CO cycle 1. Inversion of CO mixing ratios, J. Geophys. Res., 105, 1909–1927, 2000.

Bourassa, A. E., Robock, A., Randel, W. J., Deshler, T., Rieger, L. A., Lloyd, N. D., Llewellyn, E. J., and Degenstein, D. A.: Large volcanic aerosol load in the stratosphere linked to Asian monsoon transport, Science, 337, 78–81, 2012.

Brenninkmeijer, C. A. M., Crutzen, P., Boumard, F., Dauer, T., Dix, B., Ebinghaus, R., Filippi, D., Fischer, H., Franke, H., Frieß, U.,

Heintzenberg, J., Helleis, F., Hermann, M., Kock, H. H., Koeppel, C., Lelieveld, J., Leuenberger, M., Martinsson, B. G., Miemczyk, S., Moret, H. P., Nguyen, H. N., Nyfeler, P., Oram, D., O'Sullivan, D., Penkett, S., Platt, U., Pupek, M., Ramonet, M., Randa, B., Reichelt, M., Rhee, T. S., Rohwer, J., Rosenfeld, K., Scharffe, D., Schlager, H., Schumann, U., Slemr, F., Sprung, D., Stock, P., Thaler, R., Valentino, F., van Velthoven, P., Waibel, A., Wandel, A., Waschitschek, K., Wiedensohler, A., Xueref-Remy, I., Zahn, A., Zech, U., and Ziereis, H.: Civil Aircraft for the regular investigation of the atmosphere based on an instrumented container: The new CARIBIC system, Atmos. Chem. Phys., 7, 4953–4976, https://doi.org/10.5194/acp-7-4953-2007, 2007.

Brühl, C., Lilievelt, J., Tost, H., Höpfner, M., and Glatthor, N.: Stratospheric sulfur and its implications for radiative forcing simulated by the chemistry climate model EMAC, J. Geophys. Res., 120, 2103–2118, https://doi.org/10.1002/2014JD022430, 2015.

Carn, S. A. and Prata, F. J.: Satellite-based constraints on explosive $SO_2$ release from Soufrière Hills Volcano, Montserrat, Geophys. Res. Lett., 37, L00E22, https://doi.org/10.1029/2010GL044971, 2010.

Clarisse, L., Hurtmans, D., Clerbaux, C., Hadji-Lazaro, J., Ngadi, Y., and Coheur, P.-F.: Retrieval of sulphur dioxide from the infrared atmospheric sounding interferometer (IASI), Atmos. Meas. Tech., 5, 581–594, https://doi.org/10.5194/amt-5-581-2012, 2012.

Crutzen, P. J.: The possible importance of CSO for the sulfate layer of the stratosphere, Geophys. Res. Lett., 3, 73–76, 1976.

Deshler, T.: A review of global stratospheric aerosol: Measurements, importance, life cycle, and local stratospheric aerosol, Atmos. Res., 90, 223–232, 2008.

Dibb, J. E., Talbot, R. W., and Scheuer, E. M.: Composition and distribution of aerosols over the North Atlantic during the subsonic assessment ozone and nitrogen oxide experiment (SONEX), J. Geophys. Res., 105, 3709–3717, https://doi.org/10.1029/1999JD900424, 2000.

Dickerson R. R., Li, C., Li, Z., Marufu, L. T., Stehr, J. W., McClure, B., Krotkov, N., Chen, H., Wang, P., Xia, X., Ban, X., Gong, F., Yuan, J., and Yang, J.: Aircraft observations of dust and pollutants over northeast China: Insight into the meteorological mechanisms of transport, J. Geophys. Res., 112, D24S90, https://doi.org/10.1029/2007JD008999, 2007.

Dyroff, C., Zahn, A., Christner, E., Forbes, R., Tompkins, A. M., and van Velthoven, P. F. J.: Comparison of ECMWF analysis and forecast humidity data with CARIBIC upper troposphere and lower stratosphere observations, Q. J. Roy. Meteor. Soc., 141, 833–844, https://doi.org/10.1002/qj.2400, 2015.

Eckhardt, S., Stohl, A., Wernli, H., James, P., Forster, C., and Spichtinger, N.: A 15-year climatology of Warm Conveyor Belts, J. Climate, 17, 218–237, 2004.

England, M. H., McGregor, S., Spence, P., Meehl, G. A., Timmermann, A., Cai, W., Gupta, A. S., McPhaden, M. J., Purich, A., and Santoso, A.: Recent intensification of wind-driven circulation in the Pacific and the ongoing warming hiatus, Nat. Clim. Change, 4, 222–227, 2014.

Friberg, J., Martinsson, B. G., Andersson, S. M., Brenninkmeijer, C. A. M., Hermann, M., van Velthoven, P. F. J., and Zahn, A.: Sources of increase in lowermost stratospheric sulphurous and carbonaceous aerosol background concentrations during

1999–2008 derived from CARIBIC flights, Tellus B, 66, 23428, https://doi.org/10.3402/tellusb.v66.23428, 2014.

Friberg, J., Martinsson, B. G., Sporre, M. K., Andersson, S. M., Brenninkmeijer, C. A. M., Hermann, M., van Velthoven, P. F. J., and Zahn, A.: Influence of volcanic eruptions on midlatitude upper tropospheric aerosol and consequences for cirrus clouds, Earth Space Sci., 2, 285–300, https://doi.org/10.1002/2015EA000110, 2015.

Fyfe, J. C., Gillett, N. P., and Zwiers, F. W.: Overestimated global warming over the past 20 years, Nat. Clim. Change, 3, 767–769, 2013.

Fyfe, J. C., Meehl, G. A., England, M. H., Mann, M. E., Santer, B. D., Flato, G. M., Hawkins, E., Gillett, N. P., Xie, S.-P., Kosaka, Y., and Swart, N. C.: Making sense of the early-2000s warming slowdown, Nat. Clim. Change, 6, 224–228, 2016.

Gettelman, A., Kinnison, D. E., Dunkerton, T. J., and Brasseur, G. P.: Impact of monsoon circulations on the upper troposphere and lower stratosphere, J. Geophys. Res., 109, D22101, https://doi.org/10.1029/2004JD004878, 2004.

Gettelman, A., Hoor, P., Pan, L. L., Randel, W. J., Hegglin, M. I., and Birner, T.: The extratropical upper troposphere and lower stratosphere, Rev. Geophys., 49, RG3003, https://doi.org/10.1029/2011RG000355, 2011.

Hansen, J., Sato, M., Ruedy, R., Nazarenko, L., Lacis, A., Schmidt, G. A., Russell, G., Aleinov, I., Bauer, M., Bauer, S., Bell, N., Cairns, B., Canuto, V., Chandler, M., Cheng, Y., Del Genio, A., Faluvegi, G., Fleming, E., Friend, A., Hall, T., Jackman, C., Kelley, M., Kiang, N., Koch, D., Lean, J., Lerner, J., Lo, K., Menon, S., Miller, R., Minnis, P., Novakov, T., Oinas, V., Perlwitz, Ja., Perlwitz, Ju., Rind, D., Romanou, A., Shindell, D., Stone, P., Sun, S., Tausnev, N., Thresher, D., Wielicki, B., Wong, T., Yao, M., and Zhang, S.: Efficacy of climate forcings, J. Geophys. Res., 110, D18104, https://doi.org/10.1029/2005JD005776, 2005.

Haywood, J. M., Jones, A., Clarisse, L., Bourassa, A., Barnes, J., Telford, P., Bellouin, N., Boucher, O., Agnew, P., and Clerbaux, C.: Observations of the eruption of the Sarychev volcano and simulations using the HadGEM2 climate model, J. Geophys. Res., 115, D21212, https://doi.org/10.1029/2010JD014447, 2010.

Heald, C. L., Coe, H., Jimenez, J. L., Weber, R. J., Bahreini, R., Middlebrook, A. M., Russell, L. M., Jolleys, M., Fu, T.-M., Allan, J. D., Bower, K. N., Capes, G., Crosier, J., Morgan, W. T., Robinson, N. H., Williams, P. I., Cubison, M. J., DeCarlo, P. F., and Dunlea, E. J.: Exploring the vertical profile of atmospheric organic aerosol: comparing 17 aircraft field campaigns with a global model, Atmos. Chem. Phys., 11, 12673–12696, https://doi.org/10.5194/acp-11-12673-2011, 2011.

Hermann, M., Heintzenberg, J., Wiedensohler, A., Zahn, A., Heinrich, G., and Brenninkmeijer, C. A. M.: Meridional distributions of aerosol particle number concentrations in the upper troposphere and lower stratosphere obtained by Civil Aircraft for Regular Investigation of the Atmosphere Based on an Instrument Container (CARIBIC) flights, J. Geophys. Res., 108, 4114, https://doi.org/10.1029/2001JD001077, 2003.

Hermann, M., Weigelt, A., Assmann, D., Pfeifer, S., Müller, T., Conrath, T., Voigtländer, J., Heintzenberg, J., Wiedensohler, A., Martinsson, B. G., Deshler, T., Brenninkmeijer, C. A. M., and Zahn, A.: An optical particle size spectrometer for aircraft-borne measurements in IAGOS-CARIBIC, Atmos. Meas. Tech., 9, 2179–2194, https://doi.org/10.5194/amt-9-2179-2016, 2016.

Hess, P. G.: A comparison of two paradigms: The relative global roles of moist convective versus non-convective transport, J. Geophys. Res., 110, D20302, https://doi.org/10.1029/2004JD005456, 2005.

Hoor, P., Fischer, H., Lange, L., and Lelieveld, J.: Seasonal variations of a mixing layer in the lowermost stratosphere as identified by the CO-$O_3$ correlation from in situ measurements, J. Geophys. Res., 107, ACL 1-1–ACL 1-11, https://doi.org/10.1029/2000JD000289, 2002.

Höpfner, M., Boone, C. D., Funke, B., Glatthor, N., Grabowski, U., Günther, A., Kellmann, S., Kiefer, M., Linden, A., Lossow, S., Pumphrey, H. C., Read, W. G., Roiger, A., Stiller, G., Schlager, H., von Clarmann, T., and Wissmüller, K.: Sulfur dioxide ($SO_2$) from MIPAS in the upper troposphere and lower stratosphere 2002–2012, Atmos. Chem. Phys., 15, 7017–7037, https://doi.org/10.5194/acp-15-7017-2015, 2015.

Jäger, H. and Deshler, T.: Lidar backscatter to extinction, mass and area conversions for stratospheric aerosols based on midlatitude ballonborne size distribution measurements, Geophys. Res., Lett., 29, 1929, https://doi.org/10.1029/2002GL015609, 2002.

Kiley, C. M. and Fuelberg, H. E.: An examination of summertime cyclone transport processes during Intercontinental Chemical Transport Experiment (INTEX-A), J. Geophy. Res., 111, D24S06, https://doi.org/10.1029/2006JD007115, 2006.

Kojima, T., Buseck, P. R., Wilson, J. C., Reeves, J. M., and Mahoney, M. J.: Aerosol particles from tropical convective systems: Cloud tops and cirrus anvils, J. Geophys. Res., 109, D12201, https://doi.org/10.1029/2003JD004504, 2004.

Luan, Y. and Jaeglé, L.: Composite study of aerosol export events from East Asia and North America, Atmos. Chem. Phys., 13, 1221–1242, https://doi.org/10.5194/acp-13-1221-2013, 2013.

Martinsson, B. G., Papaspiropoulos, G., Heintzenberg, J., and Hermann, M.: Fine mode particulate sulphur in the tropopause region from intercontinental commercial flights, Geophys. Res. Lett., 28, 1175–1178, 2001.

Martinsson, B. G., Nguyen, H. N., Brenninkmeijer, C. A. M., Zahn, A., Heintzenberg, J., Hermann, M., and van Velthoven, P. F. J.: Characteristics and origin of lowermost stratospheric aerosol at northern midlatitudes under volcanically quiescent conditions based on CARIBIC observations, J. Geophys. Res., 110, D12201, https://doi.org/10.1029/2004JD005644, 2005.

Martinsson, B. G., Brenninkmeijer, C. A. M., Carn, S. A., Hermann, M., Heue, K.-P., van Velthoven, P. F. J., and Zahn, A.: Influence of the 2008 Kasatochi volcanic eruption on sulfurous and carbonaceous aerosol constituents in the lower stratosphere, Geophys. Res. Lett., 36, L12813, https://doi.org/10.1029/2009GL038735, 2009.

Martinsson, B. G., Friberg, J., Andersson, S. M., Weigelt, A., Hermann, M., Assmann, D., Voigtländer, J., Brenninkmeijer, C. A. M., van Velthoven, P. J. F., and Zahn, A.: Comparison between CARIBIC Aerosol Samples Analysed by Accelerator-Based Methods and Optical Particle Counter Measurements, Atmos. Meas. Tech., 7, 2581–2596, https://doi.org/10.5194/amt-7-2581-2014, 2014.

McCormick, M. P., Thomason, L. W., and Trepte, C. R.: Atmospheric effects of the Mt Pinatubo eruption, Nature, 373, 399–404, 1995.

Meehl, G. A. and Teng, H.: CMIP5 multi-model hindcasts for the mid-1970s shift and early 2000s hiatus and predictions for 2016–2035, Geophys. Res. Lett., 41, 1711–1716, 2014.

Murphy, D. M., Thomson, D. S., and Mahoney, M. J.: In situ measurements of organics, meteoritic material, mercury, and other elements in aerosols at 5 to 19 kilometers, Science, 282, 1664–1669, https://doi.org/10.1126/science.282.5394.1664, 1998.

Myhre, G., Shindell, D., Bréon, F.-M., Collins, W., Fuglestvedt, J., Huang, J., Koch, D., Lamarque, J.-F., Lee, D., Mendoza, B., Nakajima, T., Robock, A., Stephens, G., Takemura, T., and Zhang, H.: Anthropogenic and Natural Radiative Forcing. In: Climate Change 2013: The Physical Science Basis. Contribution of Working Group I to the Fifth Assessment Report of the Intergovernmental Panel on Climate Change, edited by: Stocker, T. F., Qin, D., Plattner, G.-K., Tignor, M., Allen, S. K., Boschung, J., Nauels, A., Xia, Y., Bex, V., and Midgley, P. M., Cambridge University Press, Cambridge, UK and New York, NY, USA, 2013.

Neely III, R. R., Toon, O. B., Solomon, S., Vernier, J. P., Alvarez, C., English, J. M., Rosenlof, K. H., Mills, M. J., Bardeen, C. G., Daniel, J. S., and Thayer, J. P.: Recent anthropogenic increases in $SO_2$ from Asia have minimal impact on stratospheric aerosol, Geophys. Res. Lett., 40, 999–1004, https://doi.org/10.1002/grl.50263, 2013.

Neely III, R. R, Yu, P., Rosenlof, K. H., Toon, O. B., Daniel, J. S., Solomon, S., and Miller, H. L.: The contribution of anthropogenic $SO_2$ emissions to the Asian tropopause aerosol layer, J. Geophys. Res.-Atmos., 119, 1571–1579, https://doi.org/10.1002/2013JD020578, 2014.

Nguyen, H. N. and Martinsson, B. G.: Analysis of C, N and O in aerosol collected on an organic backing using internal blank measurements and variable beam size, Nucl. Instr. Meth. B, 264, 96–102, 2007.

Nguyen, H. N., Gudmundsson, A., and Martinsson, B. G.: Design and calibration of a multi-channel aerosol sampler for studies of the tropopause region from the CARIBIC platform, Aerosol Sci. Tech., 40, 649–655, 2006.

Nguyen, H. N., Martinsson, B. G., Wagner, J. B., Carlemalm, E., Ebert, M., Weinbruch, S., Brenninkmeijer, C. A. M., Heintzenberg, J., Hermann, M., Schuck, T., van Velthoven, P. F. J., and Zahn, A.: Chemical composition and morphology of individual aerosol particles from a CARIBIC fight at 10 km altitude between 50° N and 30° S. J. Geophys. Res., 113, D23209, https://doi.org/10.1029/2008JD009956, 2008.

NOAA National Centers for Environmental Information, State of the Climate: Global Analysis for October 2016, available at: http://www.ncdc.noaa.gov/sotc/global/201610 (last access: 5 December 2016), 2016.

Oram, D. E., Mani, F. S., Laube, J. C., Newland, M. J., Reeves, C. E., Sturges, W. T., Penkett, S. A., Brenninkmeijer, C. A. M., Röckmann, T., and Fraser, P. J.: Long-term tropospheric trend of octafluorocyclobutane (c-$C_4F_8$ or PFC-318), Atmos. Chem. Phys., 12, 261–269, https://doi.org/10.5194/acp-12-261-2012, 2012.

Osman, M. K., Tarasick, D. W., Liu, J., Moeini, O., Thouret, V., Fioletov, V. E., Parrington, M., and Nédélec, P.: Carbon monoxide climatology derived from the trajectory mapping of global MOZAIC-IAGOS data, Atmos. Chem. Phys., 16, 10263–10282, https://doi.org/10.5194/acp-16-10263-2016, 2016.

Papaspiropoulos, G., Martinsson, B. G., Zahn, A., Brenninkmeijer, C. A. M., Hermann M., Heintzenberg J., Fischer, H., and van Velthoven, P. F. J.: Aerosol elemental concentrations in the tropopause region from intercontinental flights with the CARIBIC platform, J. Geophys. Res., 107, 4671, https://doi.org/10.1029/2002JD002344, 2002.

Prata, A. J. and Bernardo, C.: Retrieval of volcanic $SO_2$ column abundance from atmospheric infrared sounder data, J. Geophys. Res., 112, D20204, https://doi.org/10.1029/2006JD007955, 2007.

Ramanathan, V. and Feng, Y.: On avoiding dangerous anthropogenic interference with the climate system: formidable chalenges ahead, P. Natl. Acad. Sci. USA, 105, 14245–14250, 2008.

Randel, W. J. and Park, M.: Deep convective influence on the Asian summer monsoon anticyclone and associated tracer variability observed with the Atmospheric Infrared Sounder (AIRS), J. Geophys. Res., 111, D12314, https://doi.org/10.1029/2005JD006490, 2006.

Randel, W. J., Park, M., Emmons, L., Kinnison, D., Bernath, P., Walker, K. A., Boone, C., and Pumphrey, H.: Asian monsoon transport of pollution to the stratosphere, Science, 238, 611–613, 2010.

Rollins, A. W., Thornberry, T. D., Watts, L. A., Yu, P., Rosenlof, K. H., Mills, M., Baumann, E., Giorgetta, F. R., Bui, T. V., Höpfner, M., Walker, K. A., Boone, C., Bernath, P. F., Colarco, P. R., Newman, P. A., Fahey, D. W., and Gao, R. S.: The role of sulfur dioxide in stratospheric aerosol formation evaluated by using in situ measurements in the tropical lower stratosphere, Geophys. Res. Lett., 44, 4280–4286, https://doi.org/10.1002/2017GL072754, 2017.

Rosen, J. M.: The boiling point of stratospheric aerosols, J. Appl. Meteorol., 10, 1044–1045, 1971.

Santer, B. D., Bonfils, C., Painter, J. F., Zelinka, M. D., Mears, C., Solomon, S., Schmidt, G. A., Fyfe, J. C., Cole, J. N. S., Nazarenko, L., Taylor, K. E., and Wentz, F. J.: Volcanic contribution to decadal changes in tropospheric temperature, Nat. Geosci., 7, 185–189, 2014.

Schuck, T. J., Brenninkmeijer, C. A. M., Slemr, F., Xueref-Remy, I., and Zahn, A.: Greenhouse gas analysis of air samples collected onboard the CARIBIC passenger aircraft, Atmos. Meas. Tech., 2, 449–464, https://doi.org/10.5194/amt-2-449-2009, 2009.

Schwarz, J. P., Spackman, J. R., Gao, R. S., Watts, L. A., Stier, P., Schulz, M., Davis, S. M., Wofsy, S. C., and Fahey, D. W.: Global-scale black carbon profiles observed in the remote atmosphere and compared to models, Geophys. Res. Lett., 37, L18812, https://doi.org/10.1029/2010GL044372, 2010.

Škerlak, B., Sprenger, M., and Wernli, H.: A global climatology of stratosphere–troposphere exchange using the ERA-Interim data set from 1979 to 2011, Atmos. Chem. Phys., 14, 913–937, https://doi.org/10.5194/acp-14-913-2014, 2014.

Slemr, F., Weigelt, A., Ebinghaus, R., Kock, H. H., Bödewadt, J., Brenninkmeijer, C. A. M., Rauthe-Schöch, A., Weber, S., Hermann, M., Becker, J., Zahn, A., and Martinsson, B.: Atmospheric mercury measurements onboard the CARIBIC passenger aircraft, Atmos. Meas. Tech., 9, 2291–2302, https://doi.org/10.5194/amt-9-2291-2016, 2016.

Solomon, S., Daniel, J. S., Neely III, R. R., Vernier, J.-P., Dutton, E. G., and Thomason, L. W.: The persistently variable "background" stratospheric aerosol layer and global climate change, Science, 333, 866–870, 2011.

Stohl, A.: A 1-year Lagrangian "climatology" of airstreams in the northern hemisphere troposphere and lowermost stratosphere, J. Geophys. Res., 106, 7263–7279, 2001.

Sunilkumar, S. V., Muhsin, M., Venkam Ratnam, M., Parameswaran, K., Krishna Murthy, B. V., and Emmanuel, M.: Boundaries of tropopause layer (TTL): A new perspective based on thermal and stability profiles, J. Geophys. Res.-Atmos., 122, 741–754, https://doi.org/10.1002/2016JD025217, 2017.

TF-HTAP-2010: Task Force on Hemispheric Transport of Air Pollution: Hemispheric Transport of Air Pollution 2010, Air Pollut. Stud. 17, edited by: Dentener, F., Keating, T., and Akimoto, H., UNECE, Geneva, Switzerland, available at: http://www.htap.org/ (last access: November 2016), 2010.

Thomas, H. E., Watson, I. M., Carn, S. A., Prata, A. J., and Realmuto, V. J.: A comparison of AIRS, MODIS and OMI sulphur dioxide retrievals in volcanic clouds, Geomatics, Natural Hazards and Risk, 2, 217–232, https://doi.org/10.1080/19475705.2011.564212, 2011.

Thomason, L. W. and Vernier, J.-P.: Improved SAGE II cloud/aerosol categorization and observations of the Asian tropopause aerosol layer: 1989–2005, Atmos. Chem. Phys., 13, 4605–4616, https://doi.org/10.5194/acp-13-4605-2013, 2013.

Vernier, J. P., Pommereau, J. P., Garnier, A., Pelon, J., Larsen, N., Niesen, J., Christensen, T., Cairo, F., Thomason, L. W., Leblanc, T., and McDermid, I. S.: Tropical stratospheric aerosol layer from CALIPSO lidar observations. J. Geophys. Res., 114, D00H10, https://doi.org/10.1029/2009JD011946, 2009.

Vernier, J.-P., Thomason, L. W., Pommereau, J.-P., Bourassa, A. E., Pelon, J., Garnier, A., Hauchecorne, A., Blanot, L., Trepte, C., Degenstein, D., and Vargas, F.: Major influence of tropical volcanic eruptions on the stratospheric aerosol layer during the last decade, Geophys. Res. Lett., 38, L12807, https://doi.org/10.1029/2011GL047563, 2011a.

Vernier, J.-P., Thomason, L. W., and Kar, J.: CALIPSO detection of an Asian tropopause aerosol layer, Geophys. Res. Lett., 38, L07804, https://doi.org/10.1029/2010GL046614, 2011b.

Vernier, J.-P., Farlie, T. D., Natarajan, M., Wienhold, F. G., Bian, J., Martinsson, B. G., Crumeyrolle, S., Thomason, L. W., and Bedka, K. M.: Increase in upper tropospheric and lower stratospheric aerosol levels and its potential connection with Asian pollution, J. Geophys. Res.-Atmos., 120, 1608–1619, https://doi.org/10.1002/2014JD022372, 2015.

Winker, D. M., Pelon, J., Coakley Jr., J. A., Ackerman, S. A., Charlson, R. J., Colarco, P. R., Flamant, P., Fu, Q., Hoff, R. M., Kittaka, C., Kubar, T. L., Le Treut, H., Mccormick, M. P., Mégie, G., Poole, L., Powell, K., Trepte, C., Vaughan, M. A., and Wielicki, B. A.: The CALIPSO mission – A 3D view of aerosols and clouds, B. Am. Meteorol. Soc., 91, 1211–1229, 2010.

Xu, L., Okada, K., Iwasaka, Y., Hara, K., Okuhara, Y., Tsutsumi, Y., and Shi, G.: The composition of individual aerosol particle in the tropaphereand stratosphere over Xianghe (39.45° N, 117.0° E), China, Atmos. Environ., 35, 3145–3153, https://doi.org/10.1016/S1352-2310(00)00532-X, 2001.

Yu, P., Toon, O. B., Neely, R. R., Martinsson, B. G., and Brenninkmeijer, C. A. M.: Composition and physical properties of the Asian tropopause aerosol layer and the North American tropospheric aerosol layer, Geophys. Res. Lett., 42, 2540–2546, https://doi.org/10.1002/2015GL063181, 2015.

Zahn, A., Weppner, J., Widmann, H., Schlote-Holubek, K., Burger, B., Kühner, T., and Franke, H.: A fast and precise chemiluminescence ozone detector for eddy flux and airborne application, Atmos. Meas. Tech., 5, 363–375, https://doi.org/10.5194/amt-5-363-2012, 2012.

Zbinden, R. M., Thouret, V., Ricaud, P., Carminati, F., Cammas, J.-P., and Nédélec, P.: Climatology of pure tropospheric profiles and column contents of ozone and carbon monoxide using MOZAIC in the mid-northern latitudes (24° N to 50° N) from 1994 to 2009, Atmos. Chem. Phys., 13, 12363–12388, https://doi.org/10.5194/acp-13-12363-2013, 2013.

# Disentangling fast and slow responses of the East Asian summer monsoon to reflecting and absorbing aerosol forcings

Zhili Wang[1], Lei Lin[2], Meilin Yang[3], Yangyang Xu[4], and Jiangnan Li[5]

[1]State Key Laboratory of Severe Weather and Key Laboratory of Atmospheric Chemistry of CMA,
Chinese Academy of Meteorological Sciences, Beijing, 100081, China
[2]School of Atmospheric Sciences and Guangdong Province Key Laboratory for Climate Change and Natural Disaster Studies,
Sun Yat-sen University, Zhuhai, 519000, China
[3]Institute of Urban Meteorology, China Meteorological Administration, Beijing, 100089, China
[4]Department of Atmospheric Sciences, Texas A&M University, College Station, Texas 77843, USA
[5]Canadian Centre for Climate Modelling and Analysis, Science and Technology Branch, Environment Canada, Victoria,
V8P5C2, Canada

*Correspondence to:* Lei Lin (linlei3@mail.sysu.edu.cn)

**Abstract.** We examine the roles of fast and slow responses in shaping the total equilibrium response of the East Asian summer monsoon (EASM) to reflecting (sulfate, $SO_4$) and absorbing (black carbon, BC) aerosol forcings over the industrial era using the Community Earth System Model version 1 (CESM1). Our results show that there is a clear distinction between fast and slow responses of the EASM to aerosol forcings and the slow climate response due to aerosol-induced change in sea surface temperature (SST) plays an important role in the impacts of aerosols on the EASM. The EASM is weakened by a decrease in land–sea surface thermal contrast in the fast response (FR) component to $SO_4$ forcing, whereas the weakening is more intensive due to the changes in tropospheric thermodynamic and dynamic structures in the slow response (SR) component to $SO_4$. The total climate adjustment caused by $SO_4$ is a significant weakening of the EASM and a decrease in precipitation. The BC-induced fast adjustment strengthens the EASM both by increasing the local land–sea surface thermal contrast and shifting the East Asian subtropical jet (EASJ) northwards. The BC-induced slow climate adjustment, however, weakens the EASM through altering the atmospheric temperature and circulation. Consequently, the EASM is slightly enhanced, especially north of 30° N, in the total response (TR) to BC. The spatial patterns of precipitation change over East Asia due to BC are similar in the total response and slow response.

This study highlights the importance of ocean response to aerosol forcings in driving the changes of the EASM.

## 1 Introduction

The East Asian summer monsoon (EASM) is one of the most complex and influential monsoon systems over the globe (Ding and Chan, 2005). The activities of about 20 % of world's population would be affected by rainfall change due to the variation of the EASM (Lei et al., 2011). Further understanding of the features of the EASM change has important implications for social economics, agriculture, ecosystem, and water resource management (Hong and Kim, 2011; Auffhammer et al., 2012).

The long-term variation of the EASM is possibly attributed to the influence of various factors, including natural factors (e.g., internal climate variability, volcanic eruptions, and solar variability) and anthropogenic factors (e.g., anthropogenic aerosols and greenhouse gases, GHGs) (Wang et al., 2001, 2015; Li et al., 2010, 2016; Salzmann et al., 2014). Among them, aerosol forcing has been recognized as an important contributor to the long-term change. The analyses based on the Coupled Model Intercomparison Project Phase 5 multi-model simulations indicated that aerosol forcing dominantly contributed to the weakening of the Asian summer monsoon during the second half of the 20th century (Salzmann et al., 2014; Song et al., 2014). Other previous studies based on individual climate models also showed that the increases in anthropogenic aerosols could decrease the land–sea surface thermal contrast, thereby leading to a weakening of the EASM (e.g., Liu et al., 2011, 2017; Zhang et al., 2012; Jiang et al., 2013; Wang et al., 2017).

Despite the modeling and observational evidence, there is still debate over whether the total aerosols enhance or weaken the EASM (Guo et al., 2013; Yan et al., 2015), which could be related to the complicated nature of aerosol chemical compositions, an issue we aim to address in this study. Aerosols in the atmosphere consist of optically reflecting and absorbing components. Reflecting aerosols (e.g., sulfate, $SO_4$; and organic carbon) can cool the surface by decreasing the amount of sunlight arriving at the top of the atmosphere (TOA) and surface and cause weak cooling inside the atmosphere due to a weakened solar absorption (Myhre et al., 2013). However, absorbing aerosols (e.g., black carbon, BC; dust; and some components of organic carbon) are able to not only change the radiation budget at the TOA and surface but also directly heat the atmospheric column (Koch and Del Genio, 2010; Huang et al., 2014). Consequently, BC affects the atmospheric stability, cloud cover, and convection. Therefore, the impact of aerosols on climate derived from modeling studies is likely to be substantially different when various aerosol species are accounted for (Ocko et al., 2014). Using a Goddard Institute for Space Studies model, Menon et al. (2002) suggested that the "wetter-south–dryer-north" phenomenon that has appeared frequently in summer over eastern China during the past decades may be related to the increase in BC emission. However, Zhang et al. (2009) showed responses that are opposite to those in Menon et al. (2002) when considering the integrated effects of carbonaceous aerosols.

Several studies attempted to contrast the $SO_4$ and BC responses and indicated that scattering and absorbing aerosols would have markedly different effects on regional temperature, atmospheric circulation, and precipitation over East Asia (e.g., Guo et al., 2013; Jiang et al., 2013; Persad et al., 2014). However, these studies all only considered the fast adjustments of atmosphere and land surface to aerosol forcings, without considering the response of oceans. Climate response to a forcing agent can be regarded as a synthesis of fast and slow responses (Andrews et al., 2010; Ganguly et al., 2012). The response to direct effects of aerosols on radiation, cloud, atmospheric heating rate, and land surface is treated as the fast response, while the response to change in global surface temperature, especially sea surface temperature (SST), caused by the aerosol forcing is identified as the slow response (SR). The latter can have a more important effect on the climate system (Allen and Sherwood, 2010; Ganguly et al., 2012; Xu and Xie, 2015; Voigt et al., 2017). A general circulation model study by Hsieh et al. (2013) showed that aerosols could lead to different spatial responses of climate over the global scale when using an interactive ocean model as opposed to fixed SST as the ocean boundary conditions. Ganguly et al. (2012) also indicated that the slow component played a more critical role in shaping the total equilibrium response of the South Asian summer monsoon to aerosol forcing.

The East Asian monsoon is considered as a more complex monsoon system. What role does the feedback of oceans to aerosol forcings play in driving the changes of the EASM? This study explores the roles of fast and slow responses in forming the total equilibrium response of the EASM to both reflecting and absorbing aerosol forcings over the industrial era using a state-of-the-art Earth system model. We take $SO_4$ and BC as the representatives of reflecting and absorbing aerosols separately. To our knowledge, no previous study has partitioned the fast and slow responses of the East Asian monsoon to various aerosol species using a fully coupled climate model.

The paper is organized as follows. The model and simulations performed are described in Sect. 2. The total, fast, and slow responses of the EASM to various aerosol forcings are presented in Sect. 3. Our discussion and conclusions are summarized in Sect. 4. We primarily focus on the variation of the EASM over the region 20–40° N, 100–140° E. The summer includes the months of June, July, and August (JJA).

## 2 Method

### 2.1 Global climate model

We used the Community Earth System Model version 1 (CESM1), a fully coupled ocean–atmosphere–land–sea-ice model, created by the National Center for Atmospheric Research of the US (Hurrell et al., 2013). The model is a version with a finite-volume approximation 1° horizontal resolution (latitude 0.9° × longitude 1.25° for the atmosphere and land, and 1° × 1° for the ocean) and 30-level vertical resolution, with a rigid lid at 4 hPa. CESM1 includes the primary anthropogenic forcing agents, such as GHGs, tropospheric and stratospheric ozone, sulfate, and black and primary organic carbon. The three-mode modal aerosol model that contains the Aitken, accumulation, and coarse modes has been implemented in the model (Liu et al., 2012). It can provide the number and mass concentrations of internally mixed aerosols for the three modes. The model also includes the physical representations of aerosol direct, semi-direct, and indirect effects for both liquid- and ice-phase clouds (Morrison and Gettelman, 2008; Gettelman et al., 2010; Ghan et al., 2012).

Anthropogenic and biomass burning emissions of aerosols and their precursors are based on Lamarque et al. (2010). However, the BC emission at the present day is adjusted due to the potential underestimation of BC heating in the atmosphere in CESM1 (Xu et al., 2013; Xu and Xie, 2015). BC emissions over East Asia and South Asia are increased by a

**Table 1.** Simulation setups.

| Simulation | Aerosol emissions | Ocean | Ensembles |
|---|---|---|---|
| PI | Year 1850 $SO_2$ and BC | Dynamic ocean model | 1 |
| $PDSO_4$ | Year 2000 $SO_2$ and 1850 BC | Dynamic ocean model | 1 |
| PDBC | Year 1850 $SO_2$ and 2000 BC | Dynamic ocean model | 5 |
| PI_FSST | Year 1850 $SO_2$ and BC | Fixed SST from PI | 1 |
| $PDSO_4$_FSST | Year 2000 $SO_2$ and 1850 BC | Fixed SST from PI | 1 |
| PDBC_FSST | Year 1850 $SO_2$ and 2000 BC | Fixed SST from PI | 1 |

factor of 2 and 4, respectively. The emissions are changed in all economic sectors (industrial, energy, etc.) and all seasons by the same ratio. Such an adjustment significantly improved the simulated radiative forcing compared to the direct observations.

## 2.2 Simulations

This study used a series of simulations (Table 1):

- *Fully Coupled CESM1 simulations.* The control case was a 394-year preindustrial simulation (referred to as PI). Two perturbed simulations, sulfur dioxide ($SO_2$) (a precursor of $SO_4$) and BC emissions, were increased instantaneously from preindustrial to present-day (PD) levels, but the GHG concentrations were unchanged (referred to as $PDSO_4$ and PDBC). Starting from the end of the 319th year, the perturbed simulations were run for 75 years, with the last 60 years being analyzed. To increase the signal-to-noise ratio caused by BC forcing (a smaller forcing), we performed an ensemble of five perturbed simulations by altering the atmospheric initial conditions by an air temperature difference at round-off level (order of $10^{-14}\,°C$). The long averaging time (394 years for the PI case, 60 years in the $SO_4$-perturbed simulations, and $60 \times 5$ years in the BC-perturbed simulations) can restrain the impact of decadal natural climate variability and obtain a clear effect due to aerosol forcings.

- *Atmosphere-only model simulations with fixed SST.* The model settings were same as those in the coupled simulations, but the SST was always fixed at the preindustrial level, with only seasonal variability. The SST data are from the outputs of the PI-coupled simulation. Three simulations were performed – using the preindustrial aerosol emissions (referred to as PI_FSST), present-day $SO_2$ emission (referred to as $PDSO_4$_FSST), and present-day BC emission (referred to as PDBC_FSST), respectively. Each simulation was run for 75 years, with the last 60 years being analyzed. These three atmosphere-only simulations were also used to calculate the effective radiative forcings (ERFs) of $SO_4$ and BC at the present day following Myhre et al. (2013).

Those sets of simulations mentioned above have been adopted to examine the responses of the tropospheric atmosphere (Xu and Xie, 2015), mountain snow cover (Xu et al., 2016), and terrestrial aridity (Lin et al., 2016) to various forcing agents. The total response (TR) of the EASM to $SO_4$ or BC forcing was defined as the difference between $PDSO_4$ or PDBC and PI:

$$TRSO_4 = PDSO_4 - PI, \tag{1}$$

$$TRBC = PDBC - PI. \tag{2}$$

The fast response (FR) of the EASM to $SO_4$ or BC forcing was expressed as the difference between $PDSO_4$_FSST or PDBC_FSST and PI_FSST:

$$FRSO_4 = PDSO_4\_FSST - PI\_FSST, \tag{3}$$

$$FRBC = PDBC\_FSST - PI\_FSST. \tag{4}$$

Note that the slow response of the EASM to aerosol forcing, defined as the climate response to aerosol-induced SST change, was calculated by subtracting the FR from the TR (Andrews et al., 2010; Ganguly et al., 2012; Samset et al., 2016) rather than by performing the simulations with the perturbed SST pattern caused by aerosol forcing:

$$SR = TR - FR. \tag{5}$$

Hsieh et al. (2013) and Xu and Xie (2015) indicated that this approximate method was a legitimate metric to obtain the slow response of climate to aerosol forcing.

## 3 Results

### 3.1 Aerosol ERFs and their induced SST responses

Figure 1 shows the changes in aerosol optical depths (AODs) at 550 nm from PI to PD induced by $SO_4$ and BC. The AOD increases significantly over most of the globe except for some oceans due to the increase in anthropogenic aerosol loading. The change in AOD induced by $SO_4$ is larger than that induced by BC. The prominent increase in AOD caused by $SO_4$ appears over eastern China, the USA, India, and western Europe, while the AOD decreases over the tropical and subtropical oceans of the Southern Hemisphere (SH). The

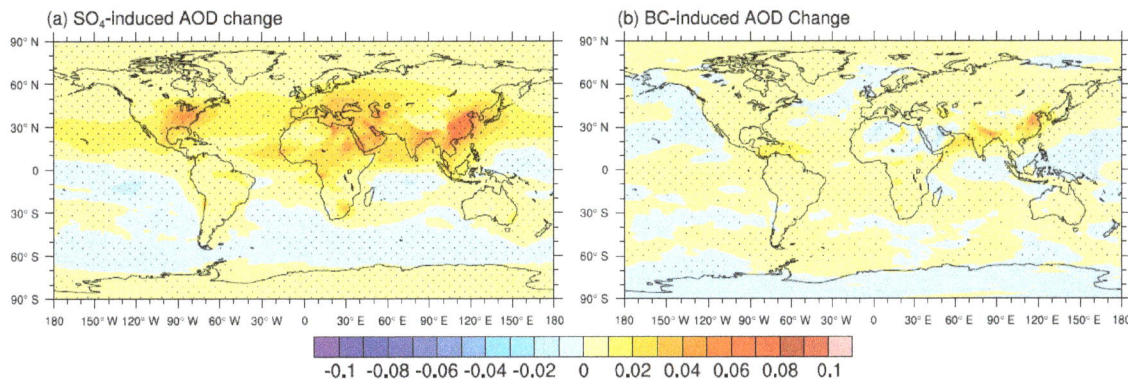

**Figure 1.** Annual mean distributions of changes in aerosol optical depths at 550 nm from PI to PD induced by (**a**) $SO_4$ and (**b**) BC. The dots represent significance at $\geq 95\,\%$ confidence level from the $t$ test.

change in BC leads to a large increase in AOD over eastern China and South Asia but a slight reduction over the tropical Pacific and Indian Ocean of the Northern Hemisphere (NH), northern Atlantic, and high latitudes of the SH.

The fifth assessment report of the Intergovernmental Panel on Climate Change provided a new definition of radiative forcing named ERF, which is a better indicator of the climate responses (Myhre et al., 2013). The global distributions of simulated $SO_4$ and BC ERFs at the TOA are shown in Fig. 2. The ERFs are calculated using the atmosphere-only model simulations with fixed SST by subtracting the net radiative flux at the TOA. There are fundamental differences between both aerosol ERFs. Reflecting $SO_4$ gives rise to large negative ERFs, especially in East and Southeast Asia, Central Africa, western Europe, and the subtropical oceans. However, absorbing BC leads to marked positive ERFs over East and South Asia and Central Africa, where the BC emission is large. The simulated global annual mean $SO_4$ and BC ERFs are $-0.98$ and $+0.36\,\mathrm{W\,m^{-2}}$, respectively. The simulated $SO_4$ forcing is close to those estimated by Zelinka et al. (2014) and Forster et al. (2016), while our results show a larger BC forcing. This is attributed to the correction of BC emission in our simulations (Xu et al., 2013). The difference between reflecting and absorbing aerosol forcings implies the substantially different climate responses.

Aerosol-induced SST change is an important part of the climatic effect of aerosols (Xu and Xie, 2015). Figure 3 shows the changes in SST caused by various aerosol species from the fully coupled simulations. Despite the essential difference between both types of forcings, the spatial pattern of SST change caused by $SO_4$ is found to be similar to that caused by BC (opposite in sign). It is characterized by a large SST change over the mid- and high-latitude oceans of the NH but only a slight SST change in the SH. The $SO_4$ forcing leads to a significant decrease in SST over the northern Pacific, northwestern Atlantic, and NH high-latitude oceans, with the largest cooling exceeding 1.5 K. However, the opposite occurs over those regions in response to BC, with the

**Table 2.** Simulated changes in annual and JJA mean sea surface temperatures caused by $SO_4$ and BC averaged over globe, Northern Hemisphere (NH), and Southern Hemisphere (SH; unit: K).

|  | Globe | NH | SH |
|---|---|---|---|
| $SO_4$ | $-0.44/-0.42$ | $-0.7/-0.64$ | $-0.24/-0.26$ |
| BC | $0.12/0.11$ | $0.17/0.16$ | $0.07/0.08$ |

largest warming reaching 1 K. A unique characteristic of the SST response to BC is the obvious warming over the Indian Ocean–western Pacific warm pool. Similar patterns in SST changes were found by Chung and Seinfeld (2005), Friedman et al. (2013), and Ocko et al. (2014). However, Ocko et al. (2014) showed a weaker SST change in the NH high latitudes induced by $SO_4$ or BC and a warming in the SH high latitudes caused by $SO_4$, which was not seen in other studies. The simulated global annual mean SST changes caused by $SO_4$ and BC are $-0.44$ K (NH: $-0.7$ K, SH: $-0.24$ K) and $+0.12$ K (NH: $+0.17$ K, SH: $+0.07$ K), respectively (Table 2). Such an interhemispheric asymmetric adjustment in SST has been used as a crucial index of climate change (Ocko et al., 2014).

### 3.2 Response of the EASM to $SO_4$ forcing

The sign of the change in surface temperature is consistent with that of the forcing. Negative $SO_4$ forcing leads to a marked surface cooling in summer over the East Asian monsoon region (EAMR), which increases with latitude (Fig. 4a). In particular, the cooling exceeds 1 K over most of the NH subtropical oceans. The anomalous northerly winds prevail over eastern China and the surrounding oceans between 20 and 40° N due to $SO_4$ forcing (Fig. 4d), which signifies the weakening of the EASM circulation. As seen in Fig. 4, the slow responses of surface air temperature and winds at 850 hPa to $SO_4$-induced SST change closely resemble their total responses to $SO_4$.

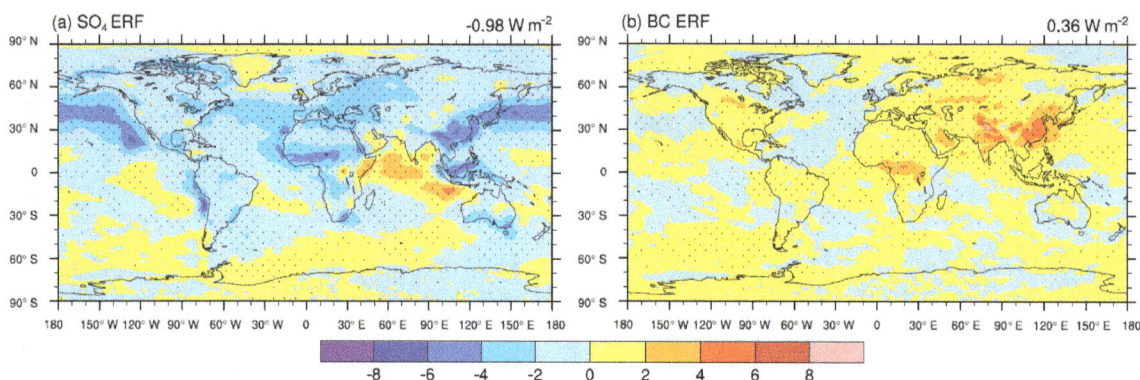

**Figure 2.** Annual mean distributions of **(a)** $SO_4$ and **(b)** BC ERF from PI to PD (unit: $W\,m^{-2}$). ERF is defined as the perturbation of net radiative flux at the TOA caused by aerosols. The dots represent significance at $\geq 95\,\%$ confidence level from the $t$ test.

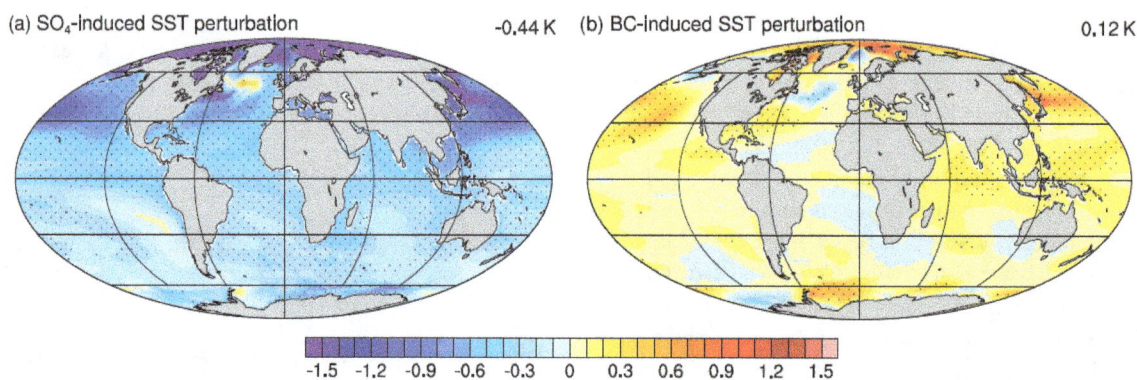

**Figure 3.** Annual mean distributions of SST responses to **(a)** $SO_4$ and **(b)** BC forcings (unit: K). The dots represent significance at $\geq 95\,\%$ confidence level from the $t$ test.

The fast response of surface air temperature to $SO_4$ forcing primarily features a cooling over continental East Asia, with the values being less than $-0.3\,K$ over most of continental East Asia (Fig. 4b), because the SST is fixed in these simulations and changes in $SO_2$ emissions are concentrated over land. Such a change in surface temperature decreases the land–sea surface thermal contrast over the EAMR, thus weakening the EASM circulation (Fig. 4e). This is consistent with previous studies using other general circulation models with fixed SST (e.g., Jiang et al., 2013; Dong et al., 2016). However, note that the weakening of the EASM in the fast response to $SO_4$ is too weak to explain the total response of the EASM to $SO_4$, especially over eastern China (Fig. 4d and e). Therefore, we next elaborate the physical mechanism behind the slow response of the EASM to $SO_4$.

Figure 5 shows the JJA mean responses of zonally averaged atmospheric temperature between 100 and $140^\circ\,E$ to $SO_4$ forcing over the EAMR. The $SO_4$-induced slow climate response leads to a significant cooling in the whole troposphere (Fig. 5c), though $SO_4$ does not largely affect the radiation in the atmosphere. It is responsible for a large fraction of the atmospheric cooling in the total response to $SO_4$

(Fig. 5a and c). This is because the prominent decrease in the JJA mean SST caused by $SO_4$ also occurs in the NH midlatitude oceans, with the values being less than $-1\,K$ over most of the northern Pacific and northwestern Atlantic (Fig. S1a in the Supplement). The interhemispheric asymmetric change in SST may distinctly affect the free troposphere by alerting the tropical circulations and midlatitude eddies (Ming et al., 2011; Hsieh et al., 2013; Ocko et al., 2014; Xu and Xie, 2015). The most remarkable feature of change in atmospheric temperature in the slow response to $SO_4$ is a deep tropospheric cooling between 30 and $45^\circ\,N$ (Fig. 5c). A similar temperature response to aerosol forcings was found by Rotstayn et al. (2014) based on the multi-model ensemble simulations, which indicates that this is a robust feature of climate response to aerosols. There is an anomalous cooling center in the upper troposphere ($200$–$500\,hPa$), with the cooling exceeding $1\,K$ (Fig. 5c), which leads to a prominent decrease in geopotential height at those altitudes (Fig. 5f). The geopotential height at about $200\,hPa$ is reduced by more than $35\,m$.

The East Asian subtropical jet (EASJ) that is located around $40^\circ\,N$ at $200\,hPa$ is an important component of the

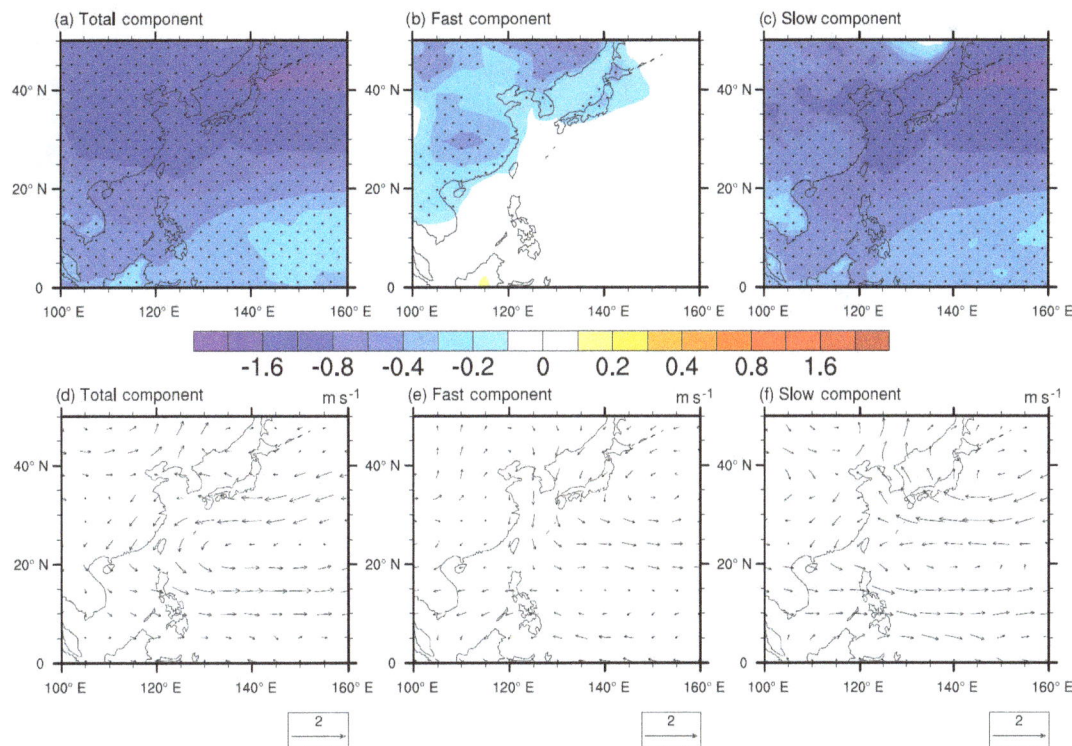

**Figure 4.** JJA mean total, fast, and slow responses of **(a, b, c)** surface air temperature (unit: K) and **(d, e, f)** wind vectors at 850 hPa (unit: m s$^{-1}$) to SO$_4$ forcing. The dots represent significance at $\geq 95\%$ confidence level from the $t$ test.

East Asian monsoon. The change in geopotential height in the slow response to SO$_4$ increases the poleward (equatorward) pressure gradient force to the south (north) of the cooling center region. Such a change in pressure gradient force leads to an increase (decrease) in westerlies to the south (north) of the EASJ center through the geostrophic balance between the Coriolis force and pressure gradient force (Yu et al., 2004). It is shown in Fig. 6a that the largest increase and decrease of more than $\pm 1$ m s$^{-1}$ in westerlies occur at about 25 and 45° N, respectively. Consequently, the EASJ shifts southwards in response to SO$_4$. The slow response dominates over the total response of the EASJ to SO$_4$ (Fig. 6).

The north and south flanks of the jet axis correspond generally to the divergence and convergence areas in the lower atmosphere, respectively. The southward displacement of the EASJ center implies the southward spread of divergence areas, thereby resulting in an anomalous surface anticyclone over continental East Asia between 30 and 40° N. The anomalous subsidence motion in the lower atmosphere around 40° N (Fig. S2c) due to the large surface cooling also intensifies the anomalous surface anticyclone. To the east of the anticyclonic center, anomalous northerlies increase prominently (Fig. 4f). In addition, the interhemispheric SST gradient caused by SO$_4$ (Fig. S1a) strengthens the ascending branch of the local Hadley cell between 20 and 35° N in the summer (Figs. S2c and S3c), thereby resulting in an

anomalous cyclonic vortex over southeastern China and the western Pacific (Fig. 4f). To the west of the cyclonic center, anomalous northerly winds are further increased. Finally, the SO$_4$-induced slow climate response leads to a more intense weakening of the EASM circulation than its fast response. Dai et al. (2013) also suggested that the thermal contrast in the mid-upper troposphere played a more important role than that in the mid-lower troposphere in impacting the strength and variations of the Asian summer monsoon circulations. Drop in tropopause height over the EAMR can suppress the convection and weaken the EASM. As seen in Fig. 7a, the tropopause north of 40° N in the summer declines significantly in the total response to SO$_4$, which primarily contributed by its slow response. The sharp drop also coincides with the southward displacement of the NH subtropical jet, as the jet approximately divides the tropics (with higher tropopause) and extratropics (Ming et al., 2011). The above analyses indicate the importance of ocean response to SO$_4$ forcing in driving the changes of the EASM circulation.

Note that the changes in atmospheric temperature and geopotential height due to the adjustments in clouds and atmospheric states in the fast response to SO$_4$ (Fig. 5b and e) lead to the increase of more than 0.8 m s$^{-1}$ in westerlies at the north of the jet center (Fig. 6b). The positive change of westerlies in the fast response is comparable to the negative change of westerlies in the slow response to SO$_4$ due to the

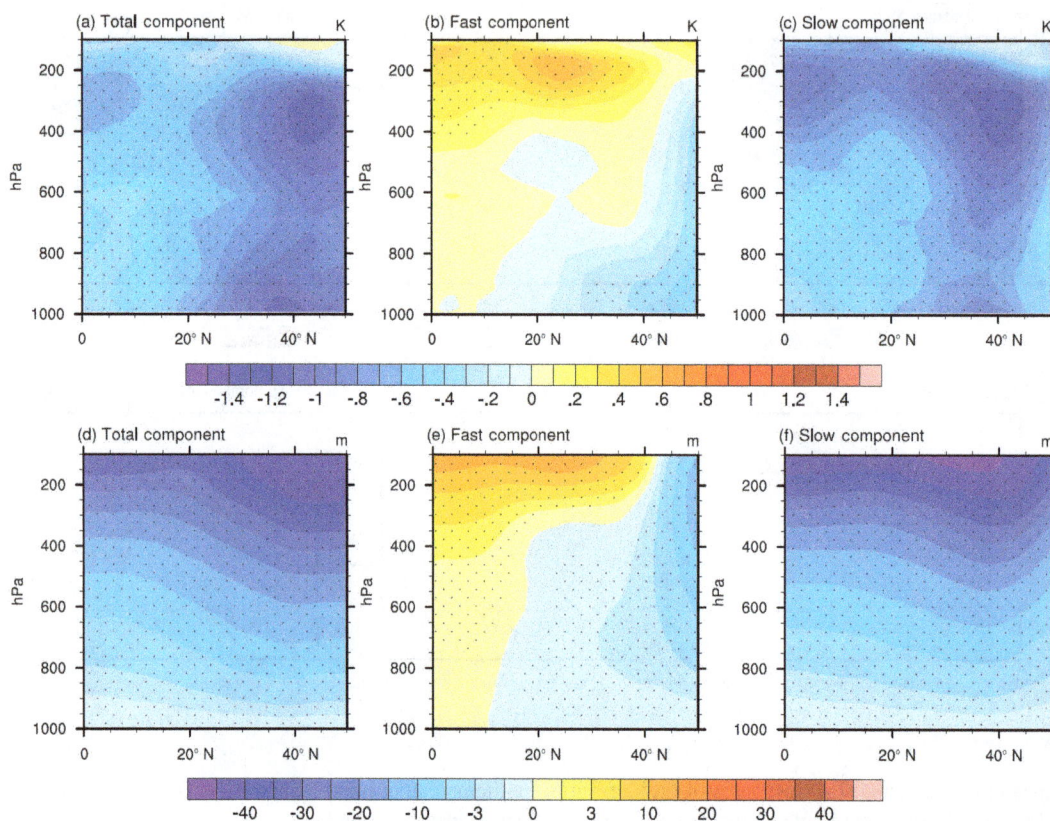

**Figure 5.** JJA mean total, fast, and slow responses of zonally averaged (**a, b, c**) atmospheric temperature (unit: K) and (**d, e, f**) geopotential height (unit: m) between 100 and 140° E to SO$_4$ forcing. The dots represent significance at $\geq 95\%$ confidence level from the $t$ test.

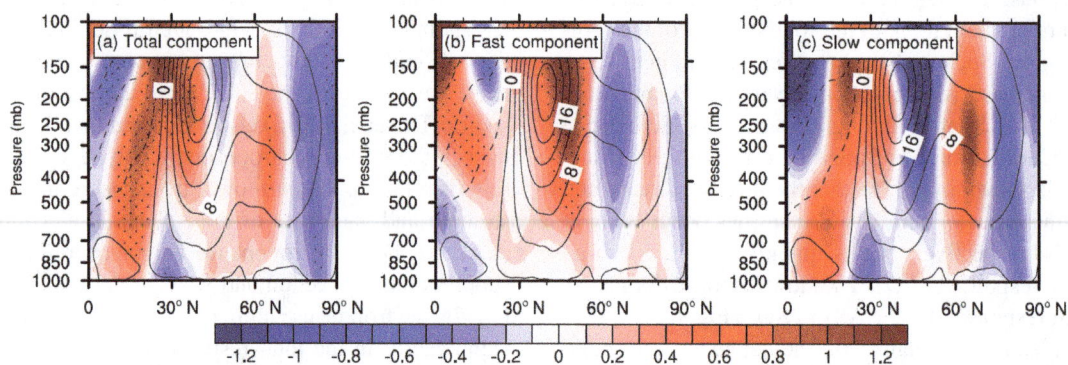

**Figure 6.** JJA mean total, fast, and slow responses of zonally averaged zonal wind between 100 and 140° E to SO$_4$ forcing (unit: m s$^{-1}$). The dashed and solid lines represent the climatological JJA mean easterly and westerly winds in PI, respectively. The dots represent significance at $\geq 95\%$ confidence level from the $t$ test.

comparable changes in temperature and geopotential height. The change of the jet in the fast response to SO$_4$ (Fig. 6b) is conducive to the enhancement of the EASM, which partially offsets the weakening of the EASM due to the decrease of the land–sea surface thermal contrast.

The weakening of the EASM circulation caused by SO$_4$ forcing suppresses the transport of surface warm and moist air northwards and upwards, which results in a significant de-

crease in precipitation over eastern and southern China and the ambient oceans (Fig. 8a). In particular, the precipitation is decreased by more than 0.6 mm day$^{-1}$ over most of southern China. The cooling in the lower troposphere and warming in the upper troposphere due to the fast response north of 20° N (Fig. 5b) can suppress the vertical ascending motion (Fig. S2b) and moisture transfer, thereby also contributing to the decrease in precipitation. However, the precipitation

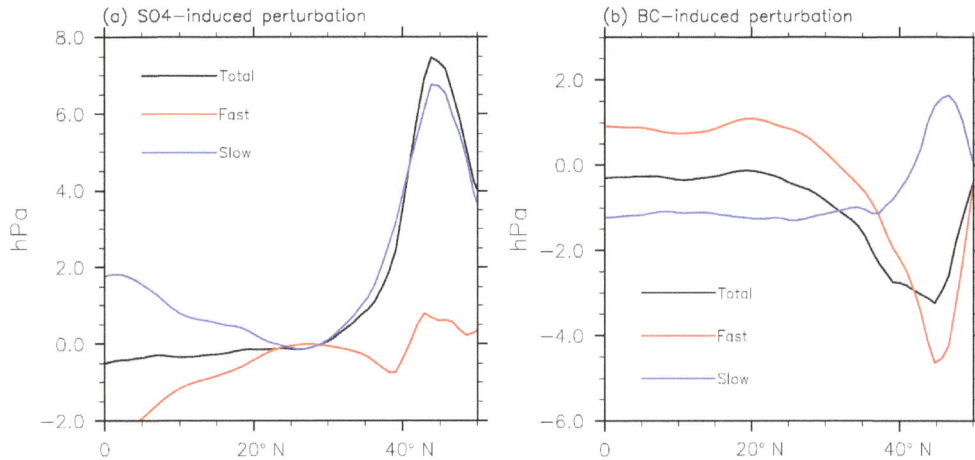

**Figure 7.** JJA mean total, fast, and slow responses of zonally averaged tropopause heights between 100 and 120° E to (a) SO$_4$ and (b) BC forcing (unit: hPa). The positive values indicate a drop in the tropopause.

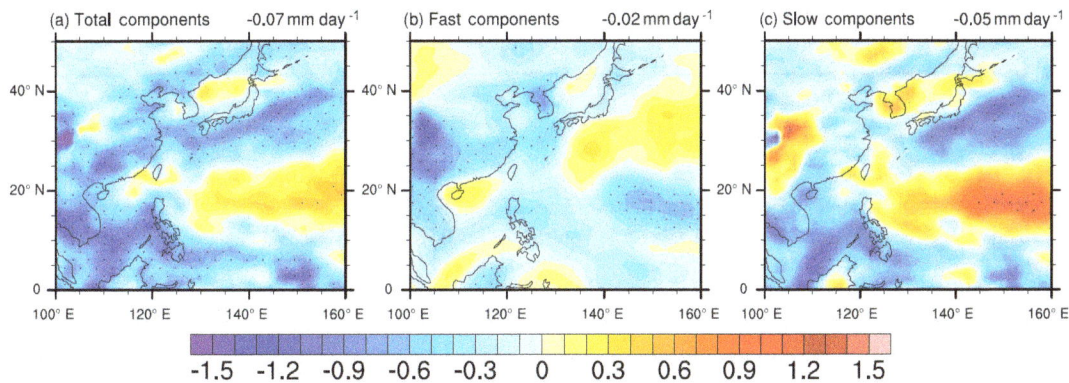

**Figure 8.** JJA mean total, fast, and slow responses of precipitation rate to SO$_4$ forcing (unit: mm day$^{-1}$). The values in the top right corner of the figures represent the responses averaged over the region 0–50° N, 100–140° E. The dots represent significance at $\geq 95\,\%$ confidence level from the $t$ test.

increases (yet not significantly) over some of the western Pacific due to the enhanced convection in the slow response to SO$_4$ (Fig. S2c), with the maximum exceeding 1 mm day$^{-1}$. Note that the SO$_4$-induced slow climate response leads to a large increase in precipitation over western China, which is the opposite compared to its fast response (Fig. 8b and c). This is because the enhanced easterly anomalies in the lower troposphere between 25 and 35° N in the slow response (Fig. 4f and 6c) bring about more moisture into the inland regions of China, which is beneficial to the formation of clouds and precipitation. In a word, the decrease in precipitation over land in East Asia in the total response to SO$_4$ is dominated by the fast response, while the change in precipitation over the adjacent ocean is dominated by the slow response.

### 3.3  Response of the EASM to BC forcing

Figure 9 shows JJA mean responses of surface air temperature and wind vectors at 850 hPa to BC forcing. Absorbing

BC increases the surface air temperature over the EAMR, with the largest warming appearing at the NH midlatitudes, especially over the northwestern Pacific, with the maximum exceeding 0.8 K (Fig. 9a). This is mainly from the contribution of the slow climate response to BC (Fig. 9c). The small anomalous southerly winds at 850 hPa prevail and the EASM circulation is slightly enhanced, especially north of 30° N, in the total climate response to BC (Fig. 9d), which is mainly attributed to their fast responses to BC. The fast and slow responses of surface winds to BC over the EAMR are inverses of each other. The anomalous northerlies in the slow response that tend to weaken the EASM greatly offset the anomalous southerlies in the fast response to BC (Fig. 9e and f).

Now we explain why the enhancement of the monsoon in the fast response to BC forcing is strong. Firstly, the large surface warming in the fast response to BC occurs over continental East Asia, especially north of 30° N, with the warming exceeding 0.2 K (Fig. 9b). This increases the land–sea surface thermal contrast over the EAMR, thereby enhancing

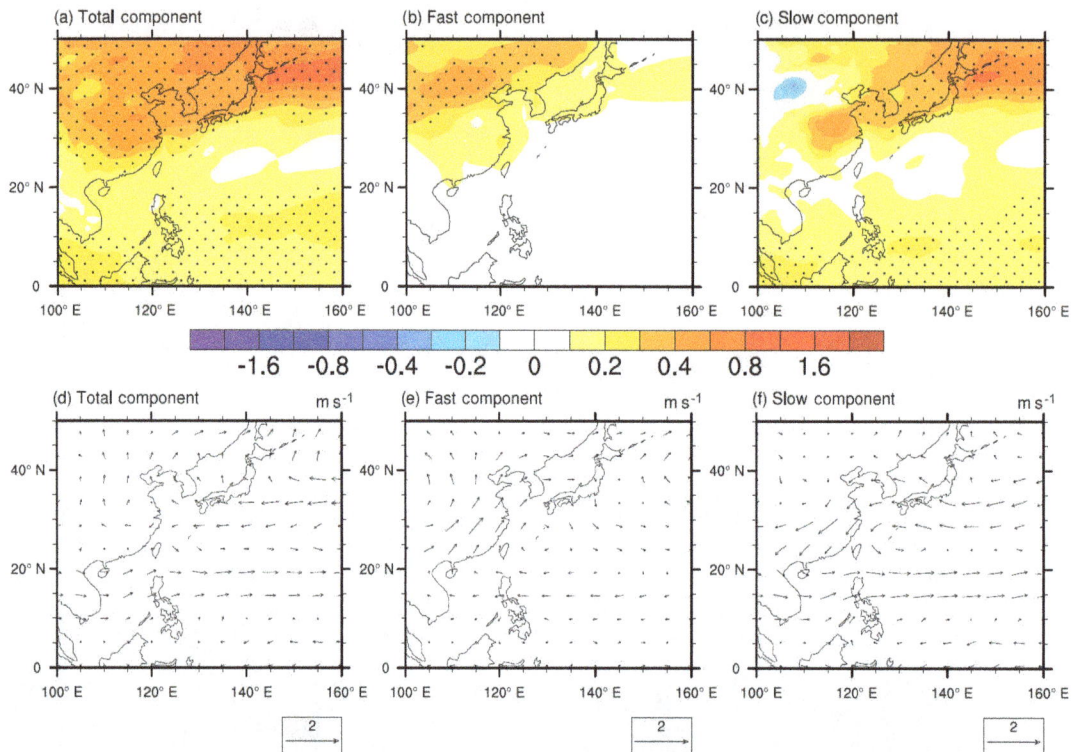

**Figure 9.** JJA mean total, fast, and slow responses of (**a, b, c**) surface air temperature (unit: K) and (**d, e, f**) wind vectors at 850 hPa (unit: m s$^{-1}$) to BC forcing. The dots represent significance at $\geq 95\%$ confidence level from the $t$ test.

the EASM circulation (Fig. 9e). This mechanism is also at work in the fast response to SO$_4$. Secondly and unique to the BC case, the direct absorption of solar radiation by BC leads to a deep tropospheric warming of 0.2 to 1 K at the NH midlatitudes (Fig. 10), which dominates over the tropospheric warming in the total response to BC. An anomalous warming center of more than 0.6 K appears and the geopotential height is increased by more than 10 m in the upper troposphere around 40° N (Fig. 10c). Consequently, the pressure increases in the uppermost troposphere, which strengthens the poleward (equatorward) pressure gradient force in the north (south) flank of the warming region. This results in an increase of 0.2 to 1 m s$^{-1}$ (a decrease of 0.8 to 1.2 m s$^{-1}$) in westerly winds in the north (south) flank of the EASJ center and the northward movement of the EASJ (Fig. 11b). The total response of the jet is consistent with the fast response of it to BC. With the change of the EASJ, an anomalous cyclonic vortex is formed over land in East Asia and anomalous southerly winds increase over eastern China. This second mechanism involving the EASJ change further magnifies the enhancement of the EASM caused by the increase in land–sea surface thermal contrast in the fast response to BC. The elevation of the tropopause between 40 and 50° N in the summer due to the fast response also implies the strengthening of the EASM (Fig. 7b). The fast response dominates over the total response of tropopause height to BC.

The BC-induced slow response is in the opposite direction of the fast response. Like the annual mean SST response, the significant increase in the JJA mean SST caused by BC occurs not only in the NH midlatitude oceans but also in the Indian Ocean–western Pacific warm pool, with the warming exceeding 0.2 K over most areas (Fig. S1b). This results in the deep tropospheric warming north of 40° N and a larger warming in the upper troposphere between 20 and 30° N, respectively (Fig. 10c). Keshavamurty (1982) found that the warming over tropical western Pacific could significantly enhance the convection motion in the western Pacific and that it was more efficient in producing atmospheric circulation anomalies. Therefore, the BC-induced slow climate response also strengthens the ascending branch of the local Hadley cell between 15 and 30° N in the summer (Figs. S4c and S5c). This leads to an anomalous cyclone in the lower atmosphere over these regions, thus increasing the anomalous northerly winds over eastern China (Fig. 9f). While the tropospheric temperature increases in the slow response to BC, the warming in the upper troposphere of around 40° N is less than that on both of its flanks (Fig. 10c). Such an adjustment in tropospheric temperature is conducive to a southward shifting of the EASJ (Fig. 11c). These EASJ changes cause the BC-induced slow response to weaken the EASM circulation, which even overcomes the strengthening of the EASM due to the increase in land–sea surface thermal contrast in the slow

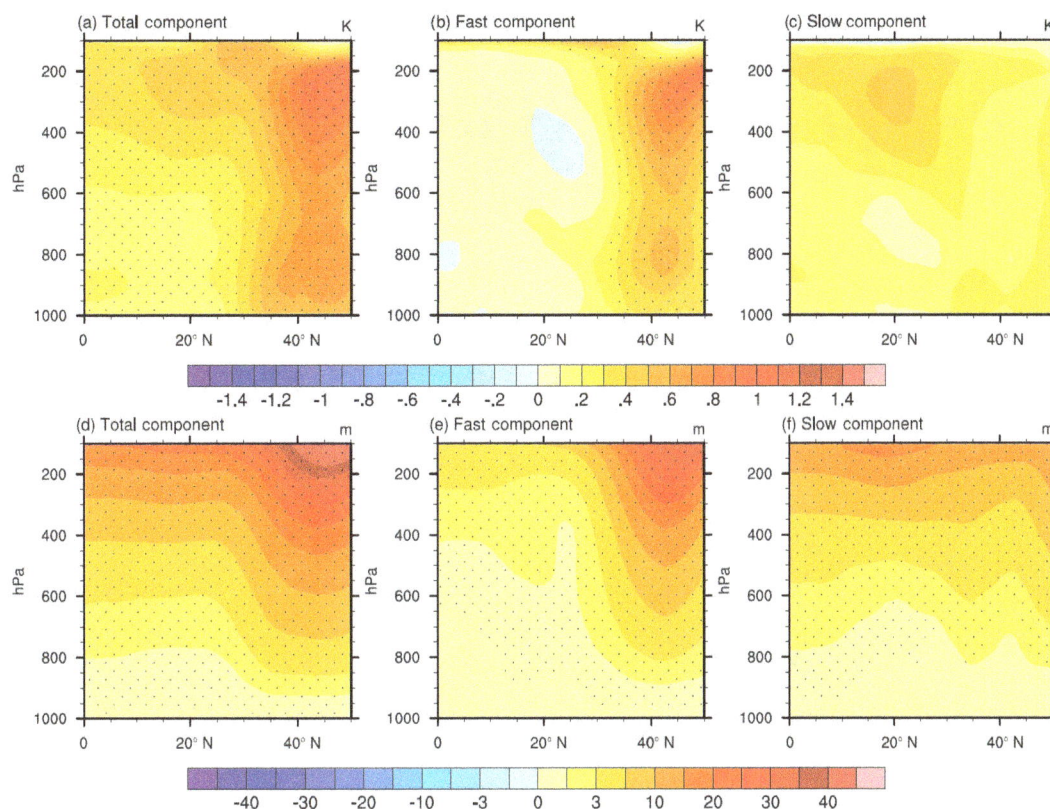

**Figure 10.** JJA mean total, fast, and slow responses of zonally averaged **(a, b, c)** atmospheric temperature (unit: K) and **(d, e, f)** geopotential height (unit: m) between 100 and 140° E to BC forcing. The dots represent significance at ≥ 95 % confidence level from the $t$ test.

response (Fig. 9c). The opposite fast and slow responses of tropopause height to BC also indicate their adverse impact on the EASM circulation (Fig. 7b).

Lastly, the JJA mean response of precipitation to BC forcing over the EAMR is weaker than that found in the response to $SO_4$ with less area in which the change is significant (Fig. 12), mainly because of a smaller radiative forcing. The total response of precipitation to BC manifests a spatial pattern of wetting–drying–wetting from north to south over the EAMR, with an increase of 0.1 to 0.6 mm day$^{-1}$ over most of southeastern China and north of 40° N and a decrease of 0.1 to 0.5 mm day$^{-1}$ over the Yangtze–Huai River valley (Fig. 12a). This is not consistent with that reported by Menon et al. (2002), which indicated that BC forcing primarily contributed to the wetter-south–dryer-north phenomenon in eastern China during the past decades. The change in precipitation caused by BC forcing is mainly in line with the change in monsoon circulation. The fast and slow responses of precipitation to BC are almost opposite over the EAMR (Fig. 12b and c) due to the opposite circulation changes. The deep tropospheric warming north of 40° N due to the fast response (Fig. 10b) can intensify the vertical ascending motion (Fig. S4b) and moisture transfer, which dominates over the increase in precipitation here. However, the warming in the troposphere and anomalous surface easterly winds between

20 and 30° N due to the slow response (Figs. 10c and 11c) are conducive to the development of ascending motion (Fig. S4c) and moisture transport from the oceans, which contributes to the increase in precipitation over these regions. In addition, the southward shifting of the EASJ in the slow response leads to an increase in surface divergence (Fig. 9f) and a decrease in precipitation over the Yangtze–Huai River valley. In general, the spatial distribution of the total precipitation response agrees well with that of the slow precipitation response to BC except north of 40° N, which also shows the significance of the SST change induced by BC forcing in impacting the EASM.

## 4 Discussion and conclusions

This study investigates the roles of fast and slow components in shaping the total equilibrium response of the EASM to reflecting $SO_4$ and absorbing BC forcings using an Earth system model with a fully coupled dynamic ocean, in contrast to most of the previous studies that adopted a slab ocean model (e.g., Allen and Sherwood, 2010; Ganguly et al., 2012). Such a decomposition of the total response will be helpful in better understanding the mechanisms by which aerosols impact the EASM. Our results show that reflecting $SO_4$ produces

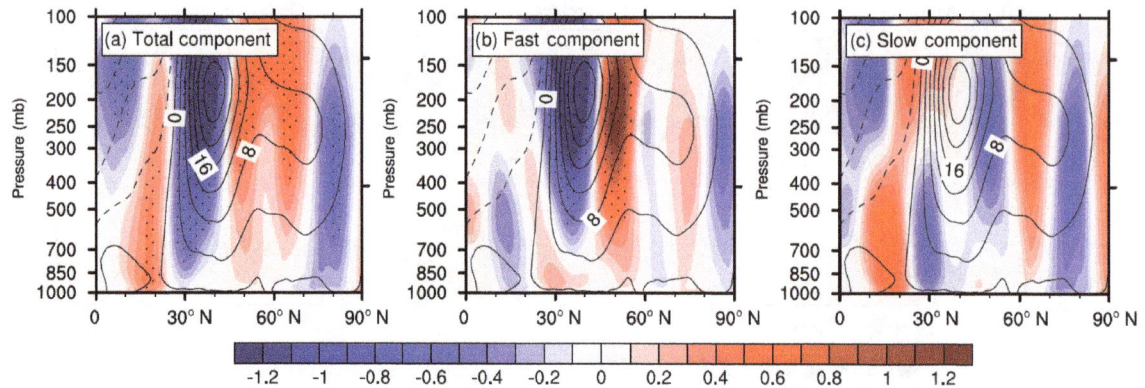

**Figure 11.** JJA mean total, fast, and slow responses of zonally averaged zonal wind between 100 and 140° E to BC forcing (unit: m s$^{-1}$). The dashed and solid lines represent the climatological JJA mean easterly and westerly winds in PI, respectively. The dots represent significance at $\geq 95\%$ confidence level from the $t$ test.

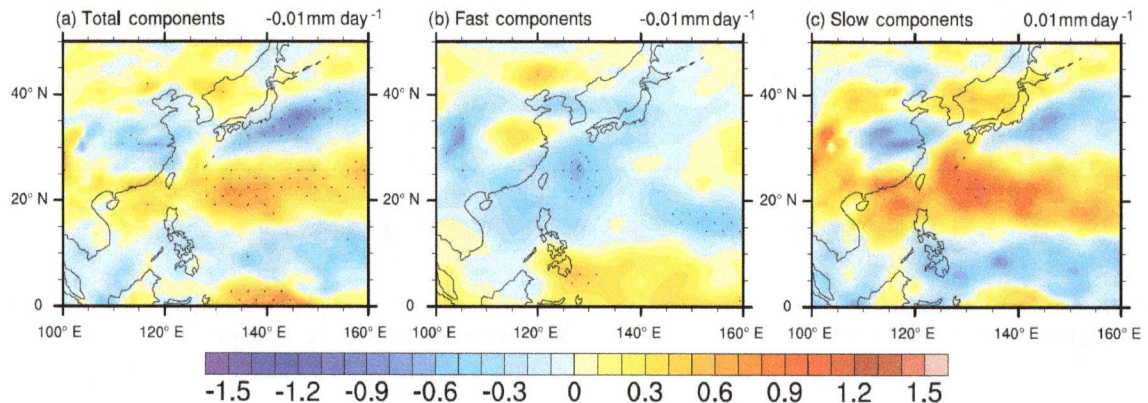

**Figure 12.** JJA mean total, fast, and slow responses of precipitation rate to BC forcing (unit: mm day$^{-1}$). The values in the top right corner of the figures represent the responses averaged over the region 0–50° N, 100–140° E. The dots represent significance at $\geq 95\%$ confidence level from the $t$ test.

a global mean ERF of $-0.98\,\mathrm{W\,m^{-2}}$, while absorbing BC leads to an ERF of $+0.36\,\mathrm{W\,m^{-2}}$. Despite the essential difference in forcings, the spatial distribution of the SST response at the global scale is prominently similar between SO$_4$ and BC forcings. However, a unique characteristic of the SST response to BC is the obvious warming over the Indian Ocean–western Pacific warm pool.

There are significantly different mechanisms between fast and slow responses of the EASM to different aerosol forcings. Table 3 provides a summary of the responses of the EASM in various cases. The SO$_4$-induced fast climate response weakens the EASM through decreasing the land–sea surface thermal contrast over the EAMR. This has been shown in many earlier studies (e.g., Jiang et al., 2013; Wang et al., 2015). However, we show here that the SO$_4$-induced SST change (i.e., slow climate response) further weakens the EASM by changing the tropospheric thermodynamic and circulation structures, especially through a southward shifting of the EASJ. Eventually, the EASM circulation is significantly weakened, and the precipitation is reduced over the

EAMR in the total response to SO$_4$. However, the decrease in precipitation over land in East Asia in the total response to SO$_4$ is dominated by the fast response, while the change in precipitation over the adjacent ocean is dominated by the slow response.

The BC-induced changes are weaker and more complicated. The fast climate response significantly strengthens the EASM both by increasing the land–sea surface thermal contrast over the EAMR and moving the EASJ northwards. However, the BC-induced slow climate response weakens the EASM by strongly affecting the atmospheric temperature and circulation. The role of the EASJ has not been clearly shown in previous studies, which often only considered the fast adjustment of climate to BC forcing. As a result of the competing factors of the land–sea contrast and ESAJ shift, the EASM in the total response to BC is weaker and less significant, with a slight enhancement north of 30° N. As for the precipitation responses, the total response to BC shows a spatial pattern of wetting–drying–wetting from north to south over the EAMR. This differs from the results in Menon et al. (2002), which suggested that the increased BC emission contributed to the wetter-south–dryer-north phenomenon in summer over eastern China in the past decades. The spatial pattern of the total precipitation response is similar to that of the slow precipitation response to BC.

This study elaborates on the mechanisms of the impacts of various aerosol species on the EASM system, highlighting the importance of ocean response to aerosol forcings (i.e., slow response component) in driving the changes of the EASM. Given a larger negative ERF due to $SO_4$, it can be speculated that the integrated effect of total anthropogenic aerosols likely tends to weaken the EASM over the industrial era, as suggested by earlier works (e.g., Song et al., 2014; Salzmann et al., 2014).

Our results clearly suggest that one pathway aerosol forcings have to affect the EASM is by changing the land–sea surface thermal contrast, as shown in previous studies (e.g., Liu et al., 2011; Zhang et al., 2012; Salzmann et al., 2014; Wang et al., 2015, 2016). However, we also emphasize the role of the EASJ, which could amplify or offset the effects of the surface thermal contrast. The response of the EASJ to aerosols needs further studies, preferably using multi-model ensembles, because (1) it is quite sensitive to the atmospheric forcing component (Fig. 11b) that is altitude dependent and (2) as a component of the larger NH westerly jet stream, it is more subject to the influence of non-local (outside Asia) aerosols that could undergo a different emission pathway than local aerosol emissions in a shorter time.

*Author contributions.* ZW and LL conceived the study and performed the analysis. ZW, LL, MY, and YX wrote the paper. All authors provided comments and contributed to the text.

*Competing interests.* The authors declare that they have no conflict of interest.

*Acknowledgements.* The authors thank the two anonymous reviewers for valuable comments and suggestions. This study was supported by the National Key R&D Program of China (2017YFA0603503 and 2016YFC0203306), (key) National Natural Science Foundation of China (41575139 and 91644211), and Jiangsu Collaborative Innovation Center for Climate Change.

Edited by: Rolf Müller

## References

Allen, R. J. and Sherwood, S. C.: The impact of natural versus anthropogenic aerosols on atmospheric circulation in the Community Atmosphere Model, Clim. Dynam., 36, 1959–1978, 2010.

**Table 3.** Summary of the fast and slow responses of the EASM to $SO_4$ and BC forcings.

| | Fast response | | | | Slow response | | | |
| --- | --- | --- | --- | --- | --- | --- | --- | --- |
| | Thermal contrast | Southerly winds | Subtropical jet | Precipitation | Thermal contrast | Southerly winds | Subtropical jet | Precipitation |
| $SO_4$ | Decrease | Decrease (weakly) | Northward shift (weakly) | Decrease over land | Decrease (weakly) | Decrease | Southward shift | Decrease in eastern and southern China but increase in western China |
| BC | Increase | Increase | Northward shift | Pattern of drying-wetting-drying | Increase (weakly) | Decrease | Southward shift (weakly) | Pattern of wetting-drying-wetting |

Andrews, T., Forster, P. M., Boucher, O., Bellouin, N., and Jones, A.: Precipitation, radiative forcing and global temperature change, Geophys. Res. Lett., 37, L14701, https://doi.org/10.1029/2010GL043991, 2010.

Auffhammer, M., Ramanathan, V., and Vincent, J. R.: Climate change, the monsoon, and rice yield in India, Climatic Change, 111, 411–424, 2012.

Chung, S. H. and Seinfeld, J. H.: Climate response of direct radiative forcing of anthropogenic black carbon, J. Geophys. Res., 110, D11102, https://doi.org/10.1029/2004JD005441, 2005.

Dai, A., Li, H., Sun, Y., Hong, L.-C., Ho, L., Chou, C., and Zhou, T.: The relative roles of upper and lower tropospheric thermal contrasts and tropical influences in driving Asian summer monsoons, J. Geophys. Res.-Atmos., 118, 7024–7045, 2013.

Ding, Y. H. and Chan, J. C. L.: The East Asian summer monsoon: an overview, Meteorol. Atmos. Phys., 89, 117–142, 2005.

Dong, B., Sutton, R. T., Highwood, E. J., and Wilcox, L. J.: Preferred response of the East Asian summer monsoon to local and non-local anthropogenic sulphur dioxide emissions, Clim. Dynam., 46, 1733–1751, 2016.

Forster, P. M., Richardson, T., Maycock, A. C., Smith, C. J., Samset, B. H., Myhre, G., Andrews, T., Pincus, R., and Schulz, M.: Recommendations for diagnosing effective radiative forcing from climate models for CMIP6, J. Geophys. Res.-Atmos., 121, 12460–12475, 2016.

Friedman, A. R., Hwang, Y.-T., Chiang, J. C., and Frierson, D. M.: Interhemispheric temperature asymmetry over the twentieth century and in future projections, J. Climate, 26, 5419–5433, 2013.

Ganguly, D., Rasch, P. J., Wang, H., and Yoon, J.: Fast and slow responses of the South Asian monsoon system to anthropogenic aerosols, Geophys. Res. Lett., 39, L18804, https://doi.org/10.1029/2012GL053043, 2012.

Gettelman, A., Liu, X., Ghan, S. J., Morrison, H., Park, S., Conley, A. J., Klein, S. A., Boyle, J., Mitchell, D. L., and Li, J.-L. F.: Global simulations of ice nucleation and ice supersaturation with an improved cloud scheme in the community atmosphere model, J. Geophys. Res., 115, D18216, https://doi.org/10.1029/2009JD013797, 2010.

Ghan, S. J., Liu, X., Easter, R. C., Zaveri, R., Rasch, P. J., Yoon, J.-H., and Eaton, B.: Toward a minimal representation of aerosols in climate models: comparative decomposition of aerosol direct, semi-direct and indirect radiative forcing, J. Climate, 25, 6461–6476, 2012.

Guo, L., Highwood, E. J., Shaffrey, L. C., and Turner, A. G.: The effect of regional changes in anthropogenic aerosols on rainfall of the East Asian Summer Monsoon, Atmos. Chem. Phys., 13, 1521–1534, https://doi.org/10.5194/acp-13-1521-2013, 2013.

Hong, J. and Kim, J.: Impact of the Asian monsoon climate on ecosystem carbon and water exchanges: a wavelet analysis and its ecosystem modeling implications, Glob. Change Biol., 17, 1900–1916, 2011.

Hsieh, W.-C., Collins, W. D., Liu, Y., Chiang, J. C. H., Shie, C.-L., Caldeira, K., and Cao, L.: Climate response due to carbonaceous aerosols and aerosol-induced SST effects in NCAR community atmospheric model CAM3.5, Atmos. Chem. Phys., 13, 7489–7510, https://doi.org/10.5194/acp-13-7489-2013, 2013.

Huang, J., Wang, T., Wang, W., Li, Z., and Yan, H.: Climate effects of dust aerosols over East Asian arid and semiarid regions, J. Geophys. Res.-Atmos., 119, 11398–11416, 2014.

Hurrell, J., Holland, M., Ghan, S., Lamarque, J., Lawrence, D., Lipscomb, W., Mahowald, N., Marsh, D. R., Neale, R. B., Rasch, P., Vavrus, S., Vertenstein M., Bader D., Collins, W. D., Hack, J. J., Kiehl, J., and Marshall, S.: The Community Earth System Model: a framework for collaborative research, B. Am. Meteorol. Soc., 94, 1339–1360, 2013.

Jiang, Y., Liu, X., Yang, X.-Q., and Wang, M.: A numerical study of the effect of different aerosol types on East Asian summer clouds and precipitation, Atmos. Environ., 70, 51–63, 2013.

Keshavamurty, R. N.: Response of the atmosphere to sea surface temperature anomalies over the equatorial Pacific and the teleconnections of the Southern Oscillation, J. Atmos. Sci., 39, 1241–1259, 1982.

Koch, D. and Del Genio, A. D.: Black carbon semi-direct effects on cloud cover: review and synthesis, Atmos. Chem. Phys., 10, 7685–7696, https://doi.org/10.5194/acp-10-7685-2010, 2010.

Lamarque, J.-F., Bond, T. C., Eyring, V., Granier, C., Heil, A., Klimont, Z., Lee, D., Liousse, C., Mieville, A., Owen, B., Schultz, M. G., Shindell, D., Smith, S. J., Stehfest, E., Van Aardenne, J., Cooper, O. R., Kainuma, M., Mahowald, N., McConnell, J. R., Naik, V., Riahi, K., and van Vuuren, D. P.: Historical (1850–2000) gridded anthropogenic and biomass burning emissions of reactive gases and aerosols: methodology and application, Atmos. Chem. Phys., 10, 7017–7039, https://doi.org/10.5194/acp-10-7017-2010, 2010.

Lei, Y., Hoskins, B., and Slingo, J.: Exploring the interplay between natural decadal variability and anthropogenic climate change in summer rainfall over china. Part I: observational evidence, J. Climate, 24, 4584–4599, 2011.

Li, J., Wu, Z., Jiang, Z., and He, J.: Can global warming strengthen the East Asian summer monsoon?, J. Climate, 23, 6696–6705, 2010.

Li, Z., Lau, W. K.-M., Ramanathan, V., Wu, G., Ding, Y., Manoj, M. G., Liu, J., Qian, Y., Li, J., Zhou, T., Fan, J., Rosenfeld, D., Ming, Y., Wang, Y., Huang, J., Wang, B., Xu, X., Lee, S.-S., Cribb, M., Zhang, F., Yang, X., Zhao, C., Takemura, T., Wang, K., Xia, X., Yin, Y., Zhang, H., Guo, J., Zhai, P. M., Sugimoto, N., Babu, S. S., and Brasseur, G. P.: Aerosol and monsoon climate interactions over Asia, Rev. Geophys., 54, 119–161, 2016.

Lin, L., Gettelman, A., Xu, Y., and Fu, Q.: Simulated responses of terrestrial aridity to black carbon and sulfate aerosols, J. Geophys. Res.-Atmos., 121, 785–794, 2016.

Liu, J., Rühland, K. M., Chen, J., Xu, Y., Chen, S., Chen, Q., Huang, W., Xu, Q., Chen, F., and Smol, J. P.: Aerosol-weakened summer monsoons decrease lake fertilization on the Chinese Loess Plateau, Nat. Clim. Chang., 7, 190–194, 2017.

Liu, X., Xie, X., Yin, Z., Liu, C., and Gettelman, A.: A modeling study of the effects of aerosols on clouds and precipitation over East Asia, Theor. Appl. Climatol., 106, 343–354, 2011.

Liu, X., Easter, R. C., Ghan, S. J., Zaveri, R., Rasch, P., Shi, X., Lamarque, J.-F., Gettelman, A., Morrison, H., Vitt, F., Conley, A., Park, S., Neale, R., Hannay, C., Ekman, A. M. L., Hess, P., Mahowald, N., Collins, W., Iacono, M. J., Bretherton, C. S., Flanner, M. G., and Mitchell, D.: Toward a minimal representation of aerosols in climate models: description and evaluation in the Community Atmosphere Model CAM5, Geosci. Model Dev., 5, 709–739, https://doi.org/10.5194/gmd-5-709-2012, 2012.

Menon, S., Hansen, J., Nazarenko, L., and Luo, Y. F.: Climate effects of black carbon aerosols in China and India, Science, 297, 2250–2253, 2002.

Ming, Y., Ramaswamy, V., Chen, G.: A model investigation of aerosol-induced changes in boreal winter extratropical circulation, J. Climate, 24, 5125–5133, 2011.

Morrison, H. and Gettelman, A.: A new two-moment bulk stratiform cloud microphysics scheme in the community atmosphere model, version 3 (CAM3). Part I: description and numerical tests, J. Climate, 21, 3642–3659, 2008.

Myhre, G., Shindell, D., Bréon, F.-M., Collins, W., Fuglestvedt, J., Huang, J., Koch, D., Lamarque, J.-F., Lee, D., Mendoza, B., Nakajima, T., Robock, A., Stephens, G., Takemura, T., and Zhang, H.: Anthropogenic and natural radiative forcing, in: Climate Change 2013: The Physical Science Basis. Contribution of Working Group I to the Fifth Assessment Report of the Intergovernmental Panel on Climate Change, edited by: Stocker, T. F., Qin, D., Plattner, G.-K., Tignor, M., Allen, S. K., Boschung, J., Nauels, A., Xia, Y., Bex, V., and Midgley, P. M., Cambridge Univ. Press, Cambridge, UK, New York, NY, USA, 659–740, 2013.

Ocko, I. B., Ramaswamy, V., and Ming, Y.: Contrasting climate responses to the scattering and absorbing features of anthropogenic aerosol forcings, J. Climate, 27, 5329–5345, 2014.

Persad, G. G., Ming, Y., and Ramaswamy, V.: The role of aerosol absorption in driving clear-sky solar dimming over East Asia, J. Geophys. Res.-Atmos., 119, 10410–10424, 2014.

Rotstayn, L. D., Plymin, E. L., Collier, M. A., Boucher, O., Dufresne, J.-L., Luo, J.-J., von Salzen, K., Jeffrey, S. J., Foujols, M.-A., Ming, Y., and Horowitz, L. W.: Declining aerosols in CMIP5 projections: effects on atmospheric temperature structure and midlatitude jets, J. Climate, 27, 6960–6977, 2014.

Salzmann, M., Weser, H., and Cherian, R.: Robust response of Asian summer monsoon to anthropogenic aerosols in CMIP5 models, J. Geophys. Res.-Atmos., 119, 11321–11337, 2014.

Samset, B. H., Myhre, G., Forster, P. M., Hodnebrog, Ø., Andrews, T., Faluvegi, G., Fläschner, D., Kasoar, M., Kharin, V., Kirkevåg, A., Lamarque, J.-F., Olivié, D., Richardson, T., Shindell, D., Shine, K. P., Takemura, T., and Voulgarakis, A.: Fast and slow precipitation responses to individual climate forcers: a PDRMIP multimodel study, Geophys. Res. Lett., 43, 2782–2791, 2016.

Song, F., Zhou, T., and Qian, Y.: Responses of East Asian summer monsoon to natural and anthropogenic forcings in the 17 latest CMIP5 models, Geophys. Res. Lett., 41, 596–603, 2014.

Voigt, A., Pincus, R., Stevens, B., Bony, S., Boucher, O., Bellouin, N., Lewinschal, A., Medeiros, B., Wang, Z., and Zhang, H.: Fast and slow shifts of the zonal-mean intertropical convergence zone in response to an idealized anthropogenic aerosol, J. Adv. Model. Earth Syst., 9, https://doi.org/10.1002/2016MS000902, 2017.

Wang, B., Wu, R., and Lau, K. M.: Interannual variability of Asian summer monsoon: contrast between the Indian and western North Pacific-East Asian monsoons, J. Climate, 14, 4073–4090, 2001.

Wang, Q. Y., Wang, Z. L., and Zhang, H.: Impact of anthropogenic aerosols from global, East Asian, and non-East Asian sources on East Asian summer monsoon system, Atmos. Res., 183, 224–236, 2017.

Wang, T. J., Zhuang, B. L., Li, S., Liu, J., Xie, M., Yin, C. Q., Zhang, Y., Yuan, C., Zhu, J. L., Ji, L. Q., and Han, Y.: The interactions between anthropogenic aerosols and the East Asian summer monsoon using RegCCMS, J. Geophys. Res.-Atmos., 120, 5602–5621, 2015.

Wang, Z. L., Zhang, H., and Zhang, X. Y.: Projected response of East Asian summer monsoon system to future reductions in emissions of anthropogenic aerosols and their precursors, Clim. Dynam., 47, 1455–1468, 2016.

Xu, Y., Bahadur, R., Zhao, C., Leung, L., and Ramanathan, V.: Estimating the radiative forcing of carbonaceous aerosols over California based on satellite and ground observations, J. Geophys. Res.-Atmos., 118, 11148–11160, 2013.

Xu, Y. and Xie, S.-P.: Ocean mediation of tropospheric response to reflecting and absorbing aerosols, Atmos. Chem. Phys., 15, 5827–5833, https://doi.org/10.5194/acp-15-5827-2015, 2015.

Xu, Y., Ramanathan, V., and Washington, W. M.: Observed high-altitude warming and snow cover retreat over Tibet and the Himalayas enhanced by black carbon aerosols, Atmos. Chem. Phys., 16, 1303–1315, https://doi.org/10.5194/acp-16-1303-2016, 2016.

Yan, H., Qian, Y., Zhao, C., Wang, H., Wang, M., Yang, B., Liu, X., and Fu, Q.: A new approach to modeling aerosol effects on East Asian climate: parametric uncertainties associated with emissions, cloud microphysics, and their interactions, J. Geophys. Res.-Atmos., 120, 8905–8924, 2015.

Yu, R., Wang, B., and Zhou, T.: Tropospheric cooling and summer monsoon weakening trend over East Asia, Geophys. Res. Lett., 31, L22212, https://doi.org/10.1029/2004GL021270, 2004.

Zelinka, M. D., Andrews, T., Forster, P. M., and Taylor, K. E.: Quantifying components of aerosol-cloud-radiation interactions in climate models, J. Geophys. Res.-Atmos., 119, 7599–7615, 2014.

Zhang, H., Wang, Z. L., Guo, P. W., and Wang, Z. Z.: A modeling study of the effects of direct radiative forcing due to carbonaceous aerosol on the climate in East Asia, Adv. Atmos. Sci., 26, 57–66, 2009.

Zhang, H., Wang, Z. L., Wang, Z. Z., Liu, Q. X., Gong, S., Zhang, X.-Y., Shen, Z. P., Lu, P., Wei, X., Che, H., and Li, L.: Simulation of direct radiative forcing of typical aerosols and their effects on global climate using an online AGCM-aerosol coupled model system, Clim. Dynam., 38, 1675–1693, 2012.

# Fungi diversity in PM$_{2.5}$ and PM$_1$ at the summit of Mt. Tai: abundance, size distribution, and seasonal variation

Caihong Xu[1], Min Wei[1,a], Jianmin Chen[1,2,3], Chao Zhu[1], Jiarong Li[1], Ganglin Lv[1], Xianmang Xu[1], Lulu Zheng[2], Guodong Sui[2], Weijun Li[1], Bing Chen[1], Wenxing Wang[1], Qingzhu Zhang[1], Aijun Ding[3], and Abdelwahid Mellouki[1,4]

[1]Environment Research Institute, School of Environmental Science and Engineering, Shandong University, Jinan 250100, China
[2]Shanghai Key Laboratory of Atmospheric Particle Pollution and Prevention (LAP3), Fudan Tyndall Centre, Department of Environmental Science & Engineering, Fudan University, Shanghai 200433, China
[3]Institute for Climate and Global Change Research, School of Atmospheric Sciences, Nanjing University, Nanjing 210023, Jiangsu, China
[4]Institut de Combustion, Aérothermique, Réactivité et Environnement, CNRS, 45071 Orléans CEDEX 02, France
[a]now at: College of Geography and Environment, Shandong Normal University, Jinan 250100, China

*Correspondence to:* Jianmin Chen (jmchen@fudan.edu.cn, jmchen@sdu.edu.cn)

**Abstract.** Fungi are ubiquitous throughout the near-surface atmosphere, where they represent an important component of primary biological aerosol particles. This study combined internal transcribed spacer region sequencing and quantitative real-time polymerase chain reaction (qPCR) to investigate the ambient fungi in fine (PM$_{2.5}$, 50 % cutoff aerodynamic diameter $D_{a50} = 2.5\,\mu m$, geometric standard deviation of collection efficiency $\sigma_g = 1.2$) and submicron (PM$_1$, $D_{a50} = 1\,\mu m$, $\sigma_g = 1.2$) particles at the summit of Mt. Tai located in the North China Plain, China. Fungal abundance values were $9.4 \times 10^4$ and $1.3 \times 10^5$ copies m$^{-3}$ in PM$_{2.5}$ and PM$_1$, respectively. Most of the fungal sequences were from Ascomycota and Basidiomycota, which are known to actively discharge spores into the atmosphere. The fungal community showed a significant seasonal shift across different size fractions according to Metastats analysis and the Kruskal–Wallis rank sum test. The abundance of *Glomerella* and *Zasmidium* increased in larger particles in autumn, whereas *Penicillium*, *Bullera*, and *Phaeosphaeria* increased in smaller particles in winter. Environmental factors, namely Ca$^{2+}$, humidity, and temperature, were found to be crucial for the seasonal variation in the fungal community. This study might serve as an important reference for fungal contribution to primary biological aerosol particles.

## 1 Introduction

Inhaled particulate matter (PM), categorized as PM$_{2.5}$ and PM$_1$ (aerodynamic equivalent diameters of $\leq 2.5$ and $\leq 1\,\mu m$, respectively), has proven to be associated with the increasing morbidity and mortality from cardiovascular and respiratory diseases (Brauer et al., 2013; Wang et al., 2014). Primary biological aerosol particles (PBAPs; about $10^4$–$10^8$ cells cm$^{-2}$) constitute an important component of PM. They can actively metabolize in the atmosphere with their mass concentrations ranging from 5.49 to 102 ng m$^{-3}$ (Zhong et al., 2016). Furthermore, they play an important role in agriculture, the biosphere, cloud formation, global climate, and atmospheric dynamics (Brodie et al., 2007; Despres et al., 2012; Christner et al., 2008; Zhou et al., 2014; Jaenicke et al., 2005). Fungi, the primary group of PBAPs, include 1.5 million unique species, distributed across rural and urban environments (Hawksworth et al., 2001). They actively eject their spores with aqueous jets or droplets into the atmosphere. The global emissions of fungal spores are estimated as the largest source of bioaerosols (Elbert et al., 2007). Pioneering studies have reported global fungal emissions to reach 28 Tg per year and contribute to about 4–13 % of the mass concentration of PM$_{2.5}$ (Heald et al., 2009; Womiloju et al., 2003). More recently, some specific fungal species have been verified to be linked with the occurrence of public health problems (Morris et al., 2002; Yadav et al., 2004; Bowers et al., 2012, 2013; Cao et al., 2014; Ryan et al., 2009). Despite their importance, the abundance, diversity, and community structure of fungi associated with PM have received limited attention in terms of research.

Earlier studies on airborne fungal communities, primarily based on culturing methods, found the dominant phyla to be Ascomycota (AMC) and Basidiomycota (BMC). Some of the species are considered major pathogens and allergens of plants, animals, and humans, e.g., *Hemileia vastatrix*, *Aspergillus*, *Cryptococcus*, and *Pneumocystis* spp. (Despres et al., 2012; Smets et al., 2016). While most of the fungal species remain unknown because cultivable species (typically less than 100) occupy only a tiny minority of all existing species, advances in nucleic acid sequencing allow the accurate determination of both cultured and uncultured microbial communities in environmental samples. For bacterial community composition, Xu et al. (2017a) investigated the abundance and community of bacteria in submicron particles during severe haze episodes in Jinan, China. Later, they discussed the diurnal variation of diverse bacterial communities in cloud water at Mt. Tai, China (Xu et al., 2017b). For diverse fungi in Mainz, Germany, Frohlich-Nowoisky et al. (2009) described the fungal community in coarse (> 3 μm) and fine (≤ 3 μm) PM using internal transcribed spacer (ITS) region sequencing. Yamamoto et al. (2012) reported the crucial influence of aerodynamic diameter and season on the fungal taxonomic composition in the northeastern United States by 454 pyrosequencing. The fungal allergens clustered in the largest size ranges (> 9 μm) in the fall season, whereas the pathogens were most abundant in the spring season and were typically observed in particles with aerodynamic diameters of < 4.7 μm. Subsequently, DeLeon-Rodriguez et al. (2013) discussed the effect of tropical storm or hurricane periods on the shift of airborne fungal species over the upper troposphere. Gou et al. (2016) described the fungal abundance and taxonomic composition of fungi in $PM_1$ and $PM_{10}$ in winter in China by 18S rRNA gene sequencing. However, that study focused on the ambient fungi in total suspended particles (TSP), $PM_{10}$, and $PM_{2.5}$, and was primarily conducted over the ground's surface; therefore, fungal populations in $PM_1$ at high-elevation sites were not well accounted for. Diverse microbes at high altitudes (such as in cloud water and precipitation) can act as nucleating agents for cloud and ice condensation, influence precipitation patterns (Xu et al., 2017b; Pratt et al., 2009; Creamean et al., 2013; Bower et al., 2013), and drive the biogeochemical cycling of elements in ecosystem processes. Hence, it is essential to advance the knowledge of microbes in PM, especially across the East Asian regions which are frequently ravished by dust, haze or other weather phenomenon. During 2013, 2014, and 2015, serious air pollution events associated with the inadequate use of clean energy in the transport, domestic, and industrial sectors affected northern China, which includes several areas with severe air pollution, namely Beijing, Tian-

jin, Shijiazhuang, Jinan, and Qingdao. Most researchers focus their attention on the case study of bacterial abundance and diversity (Gao et al., 2014, 2017a; Xu, et al., 2017a; Wei et al., 2017). The various physical, chemical, and biological factors caused by the severe haze or dust episodes may cause shifts in the bacterial community structure. Moreover, the airborne microbial abundance and diversity are also effected by seasonal and meteorological factors; however, the investigations into the seasonal variation of fungal characteristics in aerosol particles have been very limited.

Mt. Tai (36°15′ N, 117°06′ E; 1534 m a.s.l.), the highest site in the North China Plain, is a tilted fault block mountain, its height increasing from the north to the south, facing the Japanese islands, Korean Peninsula, East China Sea, and the Yellow Sea. The vegetation cover is 80 %, with nearly 1000 kinds of plants growing in the area. The number of tourists, from both China and abroad, visiting this mountain increased from 5.5 million in 2014 to 5.9 million in 2015. Past investigations in this region mainly concentrated on the physicochemical characteristics of aerosol particles and cloud water and their influence on air quality and human health. Thus far, there have been no studies addressing the diverse fungal community in aerosol particles at Mt. Tai, necessitating the development of a reliable knowledge base on the atmospheric aerosols in such scenic destinations.

The objectives of the present study were: (i) to fill the knowledge gaps regarding the ambient fungi of $PM_{2.5}$ and $PM_1$ at a high-elevation site of East Asia, (ii) to elucidate the size-based differences between the data of ambient fungal concentration and viable fungal community structure across different seasons, and (iii) to estimate whether environmental factors play a role in the variation of fungal characteristics at Mt. Tai.

## 2 Materials and methods

### 2.1 Sample collection

At Mt Tai, spring occurs from March to May; summer, June to August; fall, September to November; and winter, December to February, according to the environmental temperature. Two middle-volume (100 L min$^{-1}$) samplers (TH-150A; Wuhan Tianhong Instruments Co. Ltd., Wuhan, China) were deployed with particles larger than 2.5 and 1 μm trapped by the impactors and particles smaller than 2.5 and 1 μm collected on the quartz filters. The 50 % cutoff aerodynamic diameters are 2.5 and 1 μm. The smaller the aerosol particles, the higher the collection efficiency. Sixty quartz membrane filters (Pall, NY, USA; 88 mm) were obtained for 23 h (09:00 to 08:00 the next day) over 8–13 days during each season from 2014 to 2015 at the summit of Mt. Tai (Table 1). The blank filters were obtained by placing sterilized quartz microfiber filters inside the sampler without any operation. Before sampling, all the filters were baked in a muffle fur-

**Table 1.** Sample descriptions and the associated meteorological characteristics of the atmosphere, including the temperature ($T$), relative humidity (RH), $PM_{2.5}$ mass concentration (MC), $PM_1$ mass concentration, and fungal cell concentrations on the basis of qPCR analysis of rRNA copy numbers in $PM_{2.5}$ and $PM_1$.

| Season | Date of collection | $T$ | RH | $PM_{2.5}$ | | | | $PM_1$ | | | |
|---|---|---|---|---|---|---|---|---|---|---|---|
| | | | | MC | Fungal SSU rRNA gene copy number | Fungal spore OC MC | Fungal spore MC | MC | Fungal SSU rRNA gene copy number | Fungal spore OC MC | Fungal spore MC |
| Unit | | °C | % | µg m⁻³ | $10^4$ copy m⁻³ | ng C m⁻³ | µg m⁻³ | µg m⁻³ | $10^4$ m⁻³ | ng C m⁻³ | µg m⁻³ |
| Summer | 06/25/15 | 12.6 | 98 | 5.5 | 11.00 | 7.15 | 0.02 | BDL | 18.00 | 11.69 | 0.03 |
| | 06/26/15 | 14 | 97 | 2.8 | 3.58 | 2.33 | 0.01 | BDL | 6.51 | 4.23 | 0.01 |
| | 06/27/15 | 15.1 | 94.6 | 52.7 | 1.01 | 0.66 | 0.00 | 18.1 | 0.40 | 0.26 | 0.00 |
| | 06/28/15 | 16.9 | 84.4 | 91.1 | 6.85 | 4.45 | 0.01 | 40.0 | 3.10 | 2.02 | 0.01 |
| | 06/29/15 | 17.3 | 62.6 | 16.8 | 4.71 | 3.06 | 0.01 | 13.3 | 2.79 | 1.81 | 0.00 |
| | 07/03/15 | 17.7 | 31.0 | 15.3 | 12.20 | 7.92 | 0.02 | 12.7 | 11.40 | 7.42 | 0.02 |
| | 07/07/15 | 16.9 | 84.4 | 94.0 | 47.70 | 31.00 | 0.08 | 39.9 | 4.20 | 2.73 | 0.01 |
| | 07/08/15 | 17.3 | 62.6 | 110.9 | 22.60 | 14.71 | 0.04 | 42.0 | 5.31 | 3.45 | 0.01 |
| | 08/07/15 | 17.4 | 97.6 | 13.5 | 5.64 | 3.67 | 0.01 | 11.4 | 6.94 | 4.51 | 0.01 |
| Autumn | 10/22/14 | 6.7 | 60.7 | 40.1 | 9.34 | 6.07 | 0.02 | 28.1 | 0.37 | 0.24 | 0.00 |
| | 10/25/14 | 10.3 | 80.7 | 48.1 | 7.28 | 4.73 | 0.01 | 34.6 | 45.70 | 29.73 | 0.08 |
| | 10/26/14 | 11.4 | 73.6 | 50.8 | 6.78 | 4.41 | 0.01 | 31.9 | 8.26 | 5.37 | 0.01 |
| | 11/03/14 | 0.3 | 21.6 | 4.9 | 1.57 | 1.02 | 0.00 | BDL | 22.20 | 14.44 | 0.04 |
| | 11/04/14 | 2.6 | 33.7 | 31.6 | 7.95 | 5.17 | 0.01 | 24.5 | 6.18 | 4.01 | 0.01 |
| | 11/05/14 | 4.1 | 30.9 | 33.1 | 3.70 | 2.41 | 0.01 | 25.6 | 103.00 | 66.75 | 0.17 |
| | 11/06/14 | 5.1 | 19.3 | 22.7 | 12.70 | 8.24 | 0.02 | 18.0 | 8.86 | 5.76 | 0.01 |
| | 11/07/14 | 2.6 | 34.0 | 19.8 | 7.89 | 5.13 | 0.01 | 16.0 | 6.39 | 4.15 | 0.01 |
| | 11/08/14 | 2.4 | 45.7 | 22.4 | 14.70 | 9.54 | 0.02 | 17.8 | 2.92 | 1.90 | 0.00 |
| | 11/09/14 | 1.1 | 73.1 | 77.1 | 5.56 | 3.61 | 0.01 | 33.5 | 3.61 | 2.34 | 0.01 |
| | 11/10/14 | 3.0 | 49.0 | 49.2 | 9.38 | 6.10 | 0.02 | 37.2 | 16.70 | 10.87 | 0.03 |
| | 11/11/14 | 2.7 | 65.4 | 32.7 | 27.50 | 17.85 | 0.05 | 25.3 | 26.30 | 17.07 | 0.04 |
| | 11/12/14 | 1.0 | 50.1 | 51.7 | 7.50 | 4.87 | 0.01 | 25.7 | 18.30 | 11.87 | 0.03 |
| Winter | 12/03/14 | −8.9 | 24.4 | 13.7 | 5.03 | 3.27 | 0.01 | 9.7 | 6.84 | 4.45 | 0.01 |
| | 12/04/14 | −11 | 39.1 | 35.0 | 8.68 | 5.64 | 0.01 | 30.6 | 2.78 | 1.81 | 0.00 |
| | 12/05/14 | −10.6 | 23.4 | 14.5 | 1.09 | 0.71 | 0.00 | 13.3 | 16.20 | 10.52 | 0.03 |
| | 12/06/14 | −5.7 | 11.0 | 9.1 | 6.32 | 4.11 | 0.01 | 8.3 | 4.15 | 2.70 | 0.01 |
| | 12/07/14 | −5.4 | 45.7 | 38.8 | 7.90 | 5.14 | 0.01 | 30.9 | 9.36 | 6.08 | 0.02 |
| | 12/08/14 | −7.9 | 35.7 | 36.5 | 1.33 | 0.86 | 0.00 | 29.0 | 7.17 | 4.66 | 0.01 |
| | 12/09/14 | −5.3 | 16.1 | 16.5 | 10.10 | 6.55 | 0.02 | 13.5 | 5.73 | 3.72 | 0.01 |
| | 12/10/14 | −5.6 | 58.3 | 9.3 | 3.24 | 2.10 | 0.01 | 8.1 | 7.10 | 4.62 | 0.01 |

C – carbon, MC – mass concentration, $T$ – temperature, RH – relative humidity, BDL – below the detection line.

nace at 500 °C for 5 h, placed into sterilized aluminum foil, and then deposited into a sealed bag. To avoid contamination, the sampling filter holder and materials used for changing filters were treated with 75 % ethanol every day. After sampling, the samples were stored at −80 °C until the next analysis. $PM_{2.5}$ and $PM_1$ mass concentrations were monitored by a synchronized hybrid ambient real-time particulate monitor (Model 5030; Thermo Fisher Scientific, Wilmington, DE, USA). Half of the $PM_{2.5}$ and $PM_1$ filters were used to analyze water-soluble inorganic ions ($NO_3^-$, $SO_4^{2-}$, $NH_4^+$, $K^+$, $Ca^{2+}$, $Na^+$, and $Mg^{2+}$) by an ambient ion monitor (URG-9000; URG Corporation, Chapel Hill, NC, USA). The remaining filters were analyzed in the same batch of laboratory experiments, including DNA extraction, PCR amplification, quantitative real-time PCR (qPCR), and Illumina sequencing, except for sample A29 on 9 December 2014 (ac-cidentally omitted in the first batch of Illumina sequencing). Considering that a part of the sequences in the two batches of experiments differed, we removed this sample before quality control. Meteorological data, including relative humidity, wind speed, wind direction, and temperature, were obtained from http://www.underground.com at a resolution of 3 h during the sampling period. The visibility was monitored online by a visibility sensor (Model PWD22; Vaisala, Finland) with a maximum limit of 20 km.

## 2.2   DNA extraction and PCR amplification

The sample pretreatment and DNA extraction experiments were performed following a protocol optimized by Jiang et al. (2015). This protocol can extract sufficient DNA from low-biomass environmental samples (e.g., aerosol particles) and boosted the DNA extraction efficiency by more than twice as compared to the non-optimized extraction method. Besides, it has been applied for studying airborne microbial diversity in different environments (Cao et al., 2014; Deng et al., 2016; Tong et al., 2017; Gao et al., 2017b). Half of the filters (about $121.64\,cm^2$ in area) were cut into small pieces, inserted into 50 mL Falcon tubes that were filled with sterilized $1\times$ PBS buffer, and centrifuged at $200 \times g$ for 3 h at 4 °C. The resuspension was collected into a $0.2\,\mu m$ Supor 200 PES membrane disc filter. We cut the PES membrane disc filter into small pieces, heated the pieces to 65 °C in PowerBead tubes for 15 min and then vortexed them for 15 min. DNA was extracted according to the standard PowerSoil DNA isolation protocol (Judd et al., 2016) and purified by AMPure XP bead purification. A parallel extraction procedure was performed with the blank filter to check for sample contamination. DNA concentrations were quantified by a NanoDrop 2000 spectrophotometer (Thermo Fisher Scientific). The fragments of ITS1 regions were amplified from genomic DNA by PCR using the forward primer ITS1F (5′-CTTGGTCATTTAGAGGAAGTAA-3′) and the reverse primer ITS4 (5′-TCCTCCGCTTATTGATATGC-3′), which target the fungal ITS region of the rRNA gene (Manter et al., 2007). The experiment was conducted using the Gene Amp® PCR System 9700 (Applied Biosystems, CA, USA) in a total volume of $50\,\mu L$ PCR mix containing PCR buffer $(1\times)$, $1.5\,\mu M$ $MgSO_4$, $0.4\,\mu M$ of each deoxynucleotide triphosphate, $0.3\,\mu M$ each of the forward and reverse primers, 0.5 U Ex Taq (TaKaRa, Dalian, China), 100 ng template DNA, and double distilled $H_2O$. The thermal cycling profile was 94 °C for 1 min; 35 cycles of denaturation at 98 °C for 20 s, annealing at 68 °C for 30 s, and elongation at 72 °C for 45 s; and final extension at 72 °C for 5 min. Three replicates of PCR for each sample were combined together. The final products were separated by 1.5 % agarose gel electrophoresis and purified using the Qiaquick PCR purification kit (Qiagen, Valencia, CA, USA). Purified amplicons were quantified by a Qubit 2.0 fluorometer (Thermo Scientific) and pooled with equal molar amounts. Sequencing libraries were generated using the Truseq DNA PCR-Free Sample Prep Kit following manufacturer's instructions. Sequencing was performed on an Illumina MiSeq instrument (Illumina, San Diego, CA, USA) with the MiSeq reagent kit V3 (Illumina) according to the standard protocols.

## 2.3 Sequence analyses

After high-throughput sequencing, we removed the chimeric and low-quality sequences using the FASTX-ToolKit (http://hannonlab.cshl.edu/fastx_toolkit) and UCHIME algorithm (Edge et al., 2011) before statistical analysis. The remaining high-quality sequences were normalized to 7973 reads to compare the different samples effectively. They were then clustered into operational taxonomic units (OTUs) at a 97 % similarity cutoff using USEARCH software (Version 7.1, http://drive5.com/uparse/). We used the OTUs as the basis

for estimating the alpha diversity and beta diversity. Alpha diversity estimators, including Chao1, Simpson's index, and Shannon's index, were performed by the Quantitative Insights into Microbial Ecology software (Version 1.8.0, http://qiime.org/scripts/assign_taxonomy.html; Kuczynski et al., 2011). The taxonomy of ITS sequences was analyzed by RDP Classifier against the UNITE database (release 7.0, http://unite.ut.ee/index.php; Koljalg et al., 2013) using a confidence threshold of 70 %. RDP Classifier was used to determine the taxonomic composition at the phylum, class, order, family, genus, and species levels (Koiv et al., 2015; Miettinen et al., 2015). The raw reads were deposited into the NCBI Sequence Read Archive database under accession number SRR5146156.

## 2.4 qPCR for ITS regions

To determine the fungal biomass, we performed qPCR (Gao et al., 2017a; Yamaguchi et al., 2016; Lee et al., 2010) using a CFX96 real-time PCR detection system (Bio-Rad, Hercules, CA, USA) in $25\,\mu L$ reaction mixtures containing $12.5\,\mu L$ TransStart Green qPCR SuperMix, $1\,\mu L$ ITS3-KYO2 (5′-GATGAAGAACGYAGYRAA-3′), $1\,\mu L$ ITS4 (5′-TCCTCCGCTTATTGATATGC-3′), $5\,\mu L$ sample DNA, and $5.5\,\mu L$ double-distilled $H_2O$. The amplification followed a three-step PCR for fungal ITS regions: 40 cycles of denaturation at 95 °C for 30 s, primer annealing at 52 °C for 30 s, and extension at 72 °C for 30 s. A standard curve was created using tenfold dilution series of fungal ITS region plasmids. Assuming that the average fungal genome has about 30–200 rRNA copies, the fungal concentrations were calculated using the methods described by Lee et al. (2010) and van Doorn et al. (2007).

## 2.5 Fungal contribution to atmospheric organic carbon

The contributions of fungal spores to organic carbon (OC) were calculated using mannitol as a biotracer. We assumed 1.7 pg mannitol and 13 pg OC per spore. To assess the contribution of fungal spores to the OC and to the mass balance of atmospheric aerosol particles quantitatively, we used the weighted-average carbon (C) conversion factor of 13 pg C per spore and of 33 pg fresh weight per spore, which had been obtained earlier as the average carbon content of spores from airborne fungal species (Bauer et al., 2008; Zhu et al., 2016; Liang et al., 2017).

## 2.6 Statistical analyses

To determine the differences in the fungal community variations among different size fractions, meta-analyses based on the permutation $t$-test were conducted using Mothur software (version 1.35.1). The program Metastats can produce a tab-delimited table to display the mean relative abundance of the mean, variance, and standard error, together with the $p$ values and $q$ values. Values were considered significant

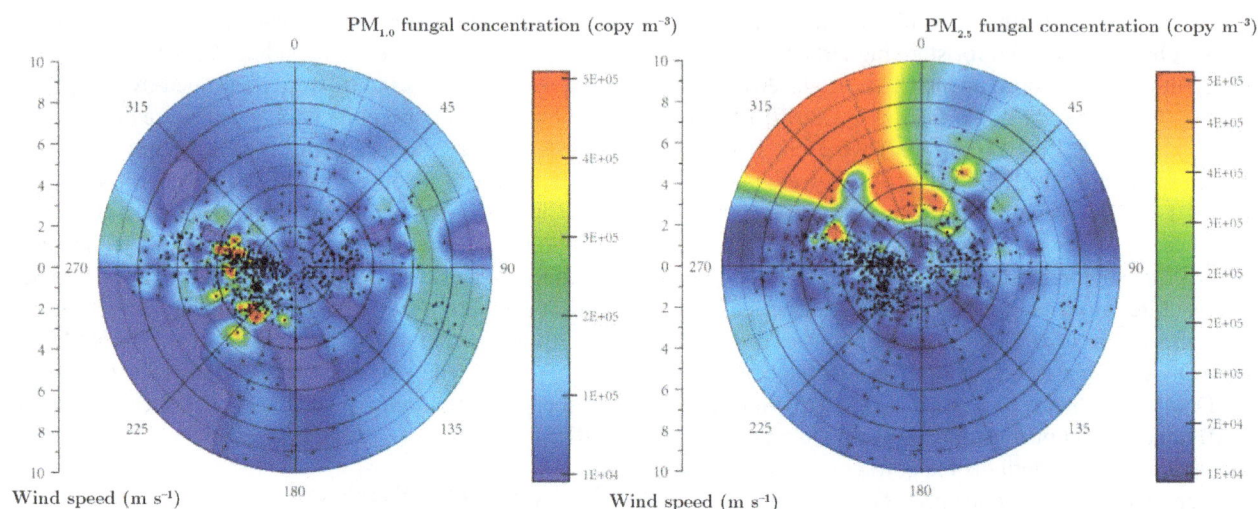

**Figure 1.** Relationships between fungal number concentrations of $PM_{2.5}$ and $PM_1$ with wind speed and wind direction.

if $p \leq 0.05$ and $q \leq 0.05$. The Kruskal–Wallis rank sum test was used to evaluate the seasonal variation of the microbial community. Boxplots and $q$ values have been provided for illustration. The relationship between the ambient microbial community and environmental factors, including PM concentrations and chemical compositions, was assessed with nonparametric Spearman's rank correlation coefficients by SPSS 16.0.

## 3  Results and discussion

### 3.1  Concentration of fungal spores in $PM_{2.5}$ and $PM_1$

$PM_{2.5}$ and $PM_1$ samples were collected during summer, autumn, and winter at the summit of Mt. Tai. Temporal variations of the mass concentration and corresponding fungal spore numbers of $PM_{2.5}$ and $PM_1$ are summarized in Table 1. $PM_1$ mass concentration was stable over different seasons, while $PM_{2.5}$ demonstrated a high seasonal variation, with higher average concentrations in summer ($44.7\,\mu g\,m^{-3}$) than in autumn ($37.2\,\mu g\,m^{-3}$) and winter ($21.7\,\mu g\,m^{-3}$). The values were much lower than that in the summer of 2006 ($123.1\,\mu g\,m^{-3}$; Deng et al., 2011) and comparable with that in the summer of 2007 ($59.3\,\mu g\,m^{-3}$; Zhou et al., 2009). The average $PM_1 / PM_{2.5}$ ratios were 0.45 in summer, 0.65 in autumn, and 0.84 in winter, implying that fine particles dominated in summer, while submicron particles dominated in autumn and winter.

qPCR revealed an average fungal gene copy number of $9.4 \times 10^4$ copies $m^{-3}$ (ranging from $1.0 \times 10^4$ to $4.8 \times 10^5$ copies $m^{-3}$) and $1.3 \times 10^5$ copies $m^{-3}$ (ranging from $3.7 \times 10^3$ to $1.0 \times 10^6$ copies $m^{-3}$) in $PM_{2.5}$ and $PM_1$, respectively. There is no significant differences between $PM_{2.5}$ and $PM_1$ based on the uncertainty estimate (95 % confidence intervals). Assuming an average rRNA gene copy

number of 200 per fungal genome (van Doorn et al., 2007; Lee et al., 2010), we obtained an average fungal concentration of 467 and 644 spores $m^{-3}$ in $PM_{2.5}$ and $PM_1$, respectively. The concentrations at Mt. Tai were lower than those at surface ground sites, including those in South Korea (ranging from $9.56 \times 10^1$ to $4.2 \times 10^4$ cells $m^{-3}$; Lee et al., 2010), Austria ($1.8 \times 10^4$ cells $m^{-3}$ in urban sites and $2.3 \times 10^4$ cells $m^{-3}$ in suburban sites; Bauer et al., 2008), Portugal (ranging from 891 to 964 spores $m^{-3}$; Oliveira et al., 2009), and the United States (6450 spores $m^{-3}$; Tsai et al., 2007). Our lower values might be ascribed to an underestimation of the fungal numbers. We used a higher gene copy number of 200 for each microbe studied, whereas DeLeon-Rodriguez et al. (2013) employed a lower number of rRNA copies of fungal genomes (30–100 copies per genome). The discrepancy between our results and those of Lee et al. (2010) might be because of the differences in sample type, sampling time, and altitude. Lee et al. (2010) focused on the fungal concentration in TSP by a high-volume TSP sampler ($0.225\,m^3\,min^{-1}$) 15 m above the ground in autumn and winter, whereas we obtained the $PM_{2.5}$ and $PM_1$ by middle-volume samplers ($0.1\,m^3\,min^{-1}$) 1534 m above the ground in summer, autumn, and winter. It is difficult to explain the disparity between different studies without uniform guidelines for the sampling and quantitative assessment of bioaerosols.

Fungal abundance varied seasonally with different size particles in the near-surface atmosphere. Saari et al. (2015) found that coarse fluorescent bioaerosol particles (1.5–5 µm) increased in summer, whereas in winter, these particles primarily existed in smaller particles (0.5–1.5 µm). The snow cover and decreased biological activity in winter resulted in the disappearance of microbes from the coarse fluorescent bioaerosol particles. In this study, the highest fungal concentration in $PM_{2.5}$ was observed in summer (641 spores $m^{-3}$), whereas the highest value in $PM_1$ was found in autumn

(1033 spores m$^{-3}$), indicating different origins of fungal spores. Huffman et al. (2010) found that long-range transport of aerosols and anthropogenic sources such as combustion influence the fluorescent biological aerosol particles having diameters less than 1 µm. During the autumn sampling, no obvious straw combustion phenomena occurred, and we detected some long-range transportation events in November 2014. Long-range transported airborne PM were mainly derived from the outer Mongolia regions, well-known to be one of the dustiest places in East Asia (6 November), Siberia (3 and 12 November), and the Taklimakan and Gobi desert regions (5 November). Influenced by the air movements from the desert region, the corresponding fungal abundance increased from $6.18 \times 10^4$ to $10.3 \times 10^5$ copies m$^{-3}$ (about 16.7-fold). Similarly, the corresponding fungal abundance influenced by air parcels from the Siberian, and Taklimakan and Gobi desert regions increased to $22.2 \times 10^4$ copies m$^{-3}$ and $18.3 \times 10^4$ copies m$^{-3}$, respectively. Hence, we hypothesized that the long-range transport of air parcels from north China might have contributed to the fungal enrichment of PM$_1$. In addition, the increased fungal abundance might be explained by meteorological diversity (Abdel Hameed et al., 2012). Low wind speed hinders fungal dispersal owing to the accumulation effect. According to Almaguer et al. (2014), in Cuba, the calm winds coming from the southwest direction induce the accumulation of fungal spores over the northern coast of the island. Lin et al. (2000) observed a strongly negative correlation between wind speeds of $< 4$ m s$^{-1}$ and fungal concentration; the fungal concentration increased as the wind speed became higher than 5 m s$^{-1}$ in the Taipei area. In our present study, the fungal abundance in PM$_1$ showed no obvious increase under breezy conditions (wind speed $< 2$ m s$^{-1}$) mainly from the southern direction (Fig. 1). When the wind speed was higher than 2 m s$^{-1}$, the fungal abundance increased markedly under the influence of westerly winds. As the westerly wind velocity increased, the fungal concentration increased slowly. Meanwhile, in PM$_{2.5}$, the fungal abundance increased with wind velocities higher than 2 m s$^{-1}$, mainly from the northwest direction of the continental areas, where diverse vegetation grows. The phenomenon implies that westerly and northwesterly winds might highly induce fungal growth and abundance in PM at Mt. Tai.

### 3.2 Contribution of spores to OC concentrations and PM mass

OC, accounting for 7–80 % of PM mass, constitutes a significant fraction of atmospheric aerosols (Yu et al., 2004; Ram et al., 2012; Ho et al., 2012). Ambient fungi are considered a possible source of OC in PMs. Cheng et al. (2009) estimated the mean fungal OC concentrations in Hong Kong to be 3.7, 6.0, and 9.7 ng m$^{-3}$, corresponding to 0.1, 1.2, and 0.2 % of the total OC in PM$_{2.5}$, PM$_{2.5-10}$, and PM$_{10}$, respectively. In the present study, the range and average concentrations of fungal contribution to atmospheric OC and mass

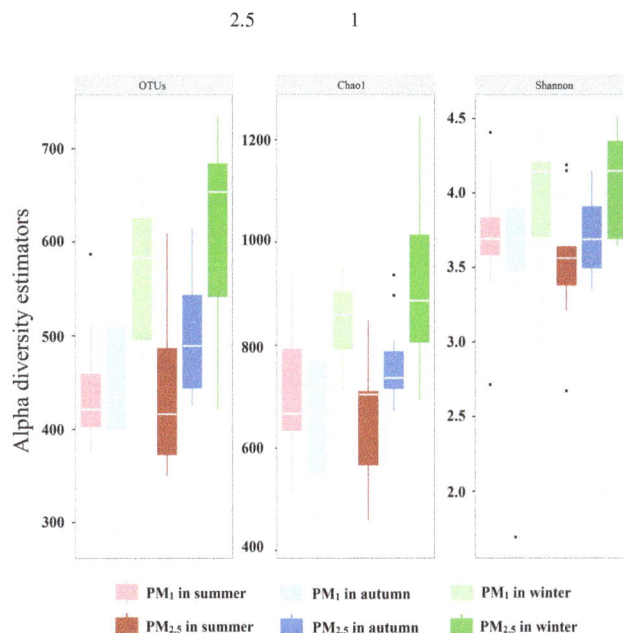

**Figure 2.** Statistical comparisons of OTUs, and Chao1 and Shannon indices among summer, autumn, and winter in PM$_{2.5}$ and PM$_1$.

concentration for PM$_{2.5}$ and PM$_1$ are listed in Table 1. The daily averaged concentrations of fungal OC in PM$_{2.5}$ and PM$_1$ were 6.1 and 8.3 ng C m$^{-3}$, respectively, with the respective contributions to PM being 0.067 and 0.096 %, indicating that airborne fungal spores as a minor source of carbonaceous aerosols cannot be ignored at Mt. Tai. The fungal contribution to OC obtained at Mt. Tai was comparable with that observed at an urban site in Hong Kong (3.7 ng C m$^{-3}$; Cheng et al., 2009) but lower than that obtained at an urban site in Austria (117.9 ng C m$^{-3}$; Bauer et al., 2008) and a forest site on Hainan Island (147–923 ng C m$^{-3}$; Zhang et al., 2015). The discrepancy between the abovementioned studies can be justified by the difference in particle type studied (TSP, PM$_{10}$, PM$_{2.5}$, and PM$_1$), fungal concentration, spore carbon content, and assessment method (e.g., sugar alcohol, cultivation, mannitol, and light microscopy). On the basis of the same conversion factor of 13 pg C spore$^{-1}$ by mannitol, the results were much lower than that obtained at an urban site in Beijing ($0.3 \pm 0.2$ µ C m$^{-3}$; Liang et al., 2017), implying a lower fungal concentration at Mt. Tai than that in Beijing. More studies are needed to better understand the spatial, temporal, and size distributions of fungal OC contributions to atmospheric particles in urban areas in the North China Plain.

### 3.3 Taxonomic diversity and composition of ambient fungi

On average, 509 and 475 OTUs were obtained in PM$_{2.5}$ and PM$_1$, respectively, which were higher than those obtained in earlier airborne fungal studies at the ground level in Beijing, China (34–285; Yan et al., 2016) and Rehovot, Israel (121–178; Dannemiller et al., 2014). The OTUs associated with

**Table 2.** The relative abundance of the top five orders and two genera in TSP, $PM_{10}$, $PM_{2.5}$, and $PM_1$.

| Taxonomic level | Common fungi | RAS[a] | RAF[b] | References | Samplers | Sample type | Concentration or abundance |
|---|---|---|---|---|---|---|---|
| Genera | Alternaria | 11.7 | 6.2 | Adhikari et al. (2004) | Andersen sampler (Thermo Andersen, Smyrna, 300082-5211, USA) | TSP | 2.6 % |
| | | | | Dannemiller et al. (2014) | High volume PM10 samplers (Ecotech, Knoxfield, VIC, Australia) | $PM_{10}$ | >1 % |
| | | | | Alghamdi et al. (2014) | PM2.5 samplers (Staplex Air Sampler Division, USA) | $PM_{2.5}$ | 2.6 % |
| | | | | Gou et al. (2016) | Low volume air sampler (BGI, USA) | $PM_1$ | >1 % |
| | Aspergillus | 2.3 | 1.9 | Cao et al. (2014) | Air samplers (Thermo Electron Corp., MA, USA) | $PM_{10}$ and $PM_{2.5}$ | abundant |
| | | | | Gou et al. (2016) | Low volume air sampler (BGI, USA) | $PM_{10}$ and $PM_1$ | abundant |
| Order | Pleosporales | 18.4 | 45.4 | Rittenour et al. (2014) | Buck Bioaire sampler (A.P. Buck, Inc, Orlando, FL, USA) | TSP | 46 % |
| | | | | Yan et al. (2016) | Air samplers (Air Metrics, USA, 5 L min$^{-1}$) | $PM_{10}$ and $PM_{2.5}$ | 29.4 % |
| | | | | Gou et al. (2016) | Low volume air sampler (BGI,USA) | $PM_{10}$ and $PM_1$ | 10–15 % |
| | Xylariales | 5.0 | 14.4 | Womack et al. (2015) | SKC Biosamplers (BioSampler SKC Inc.) | TSP | abundant |
| | | | | Gou et al. (2016) | Low volume air sampler (BGI, USA) | $PM_{10}$ and $PM_1$ | 0–5 % |
| | Eurotiales | 4.8 | 13.3 | Yan et al. (2016) | Air samplers (Air Metrics, USA, 5 L min$^{-1}$) | $PM_{10}$ and $PM_{2.5}$ | 10.6 % |
| | | | | Gou et al. (2016) | Low volume air sampler (BGI, USA) | $PM_{10}$ and $PM_1$ | 10–15 % |
| | Capnodiales | 4.4 | 12.5 | Yan et al. (2016) | Air samplers (Air Metrics, USA, 5 L min$^{-1}$) | $PM_{10}$ and $PM_{2.5}$ | 27.96 % |
| | | | | Gou et al. (2016) | Low volume air sampler (BGI, USA) | $PM_{10}$ and $PM_1$ | ~25 % |
| | Polyporales | 2.5 | 6.4 | Womack et al. (2015) | SKC Biosamplers (BioSampler SKC Inc.) | TSP | abundant |
| | | | | Yan et al. (2016) | Air samplers (Air Metrics, USA, 5 L min$^{-1}$) | $PM_{10}$ and $PM_{2.5}$ | 3.6 % |
| | | | | Yamamoto et al. (2012) | Eight-stage Andersen sampler (New Star Environmental, Roswell, GA, USA) | PM with aerodynamic diameter is 2.1–3.3, 3.3–4.7, 4.7–5.8, 5.8–9.0 and >9.0 μm | abundant |

RAS[a] indicates relative abundance in submicron particles and RAF[b] indicates relative abundance in fine particles.

$PM_{2.5}$ in summer, autumn, and winter were higher than those associated with $PM_1$, implying more diverse fungal spores in $PM_{2.5}$. However, the Shannon and Chao1 indices showed different trends in $PM_{2.5}$ and $PM_1$ (Fig. 2). The ambient fungi showed the highest richness and diversity in winter, followed by autumn and summer. Although $PM_1$ mass concentration dominated in autumn and winter, the corresponding fungal diversity was lower than that in $PM_{2.5}$. Similarly, the dominant $PM_{2.5}$ mass concentration in summer presented lower diversity than that in $PM_1$.

In the fungal community, AMC (89.7 %) and BMC (7.0 %) were the predominant phyla, and they are known to actively discharge spores into the atmosphere (Fig. 3a). The remaining phyla were Zygomycota (ZMC) and Glomeromycota. AMC and BMC present a global pattern across continental (Austria, Arizona, Brazil, and Germany), coastal (Taiwan, Puerto Rico, and UK), and marine sites (Pacific, Indian, Atlantic, and Southern Ocean) (Frohlich-Nowoisky et al., 2012). In continental samples, BMC (64 %) seems to be more abundant than AMC (34 %), whereas in marine sites, AMC (72 %) is about 2.6 times more abundant than BMC. Herein, the abundance of AMC was approximately 12.8 times higher than that of BMC. Members of AMC have single-celled or filamentous vegetative growth forms that are easily aerosolized, unlike BMC (Womack et al., 2015). Furthermore, 10 classes belonging to AMC, 10 to BMC, and 1 to ZMC were observed (Fig. 3b). The preponderant classes belonging to AMC were Dothideomycetes (37.3 %),

Sordariomycetes (15.0 %), and Eurotiomycetes (6.1 %). The dominant orders in Dothideomycetes included Pleosporales (14.9 %), Capnodiales (5.3 %), and Botryosphaeriales (1.6 %) (Fig. 3c). Pleosporales has been reported to include fungi allergenic to local residents (Rittenour et al., 2014). The values were lower than those reported in Beijing's PM (Pleosporales: 29.39 % and Capnodiales: 27.96 %) (Yan et al., 2016). Likewise, the dominant classes in BMC were Agaricomycetes (4.4 %) and Tremellomycetes (1.5 %), including the orders Polyporales (2.5 %), Agaricales (1.6 %), and Tremellales (1.2 %). About 291 taxa from the genus level were determined, including *Alternaria, Glomerella, Zasmidium, Pestalotiopsis, Aspergillus,* and *Phyllosticta.* The distribution was discrepant with that at the ground level, wherein *Cladosporium* occupied more than 50 % of total fungi, followed by *Alternaria, Didymella,* and *Khuskia* (Oh et al., 2014). The top five orders (Pleosporales, Xylariales, Eurotiales, Capnodiales, Polyporales) and genera (*Alternaria* and *Aspergillus*) were commonly observed in suspended aerosol particles (including TSP, $PM_{10}$, $PM_{2.5}$, and $PM_1$) but showed variable relative abundances, as shown in Table 2. We attribute this disparity to the different sampling approaches, instruments, and analysis methods. This aspect needs to be probed and studied in depth in the future.

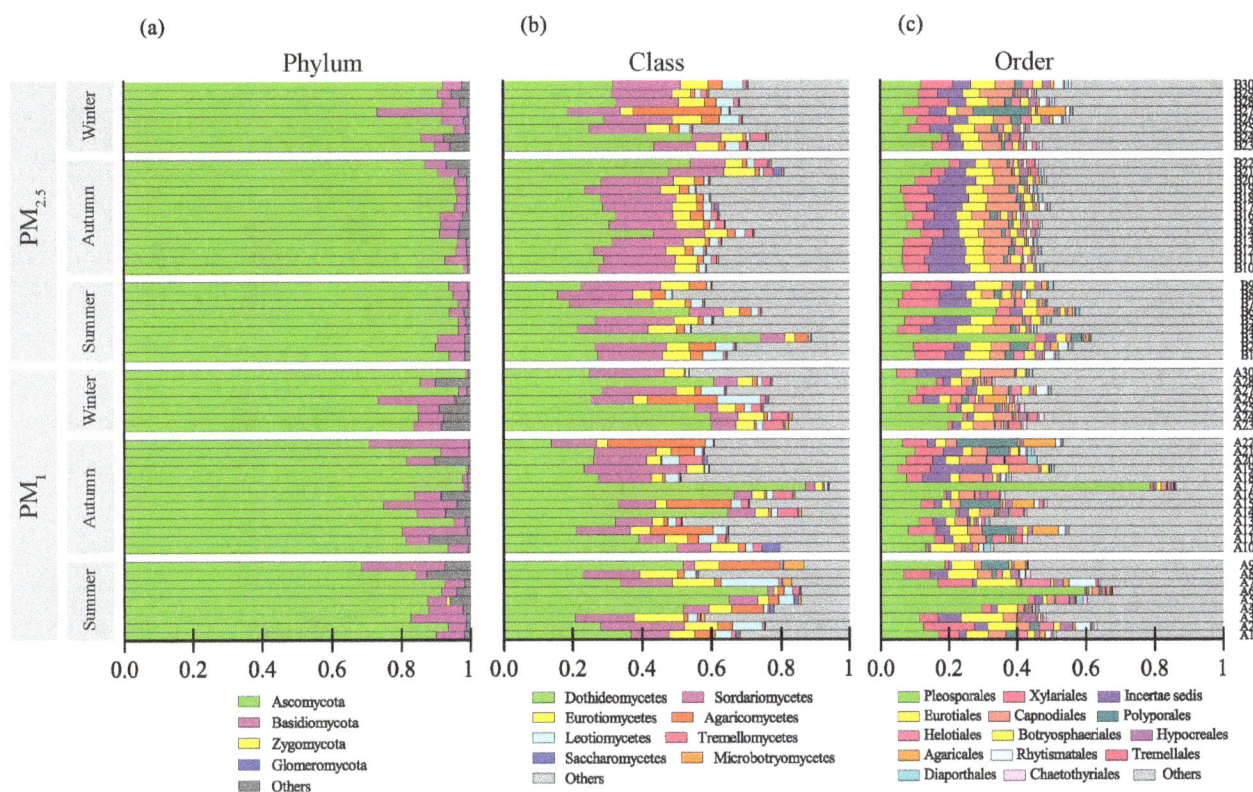

**Figure 3.** Relative abundances of fungal communities of $PM_{2.5}$ and $PM_1$ at phylum **(a)**, class **(b)**, and order level **(c)**.

### 3.4 Implication of the allergenic and pathogenic fungi

To date, about 123 fungal genera (mainly belonging to the phylum AMC) have been identified to be human allergens (Simon-Nobbe et al., 2008). Of the 11 potentially allergy-inducing AMC species and 1 potentially allergy-inducing BMC species found at Mt. Tai, the 3 most common species were *Aspergillus flavus*, *Blumeria graminis*, and *Saccharomyces cerevisiae*. *Aspergillus flavus* is a common human pathogen found in air, and it is also a human allergen and mycotoxin producer (Adhikari et al., 2004). It is associated with invasive aspergillosis and superficial infections (Hedayati et al., 2007). *Blumeria graminis*, found on the surface of plant leaves, causes powdery mildew on cereal plants (Belanger et al., 2003). Such pathogens and allergens are expected to be widely spread around the atmospheric environment in temperate and tropical zones (Vermani et al., 2010). Our results also revealed that the abundance of potential allergenic and pathogenic fungal spores in summer were the highest compared to those in autumn and winter. Clinicians should consider the fungal spores described herein as a possible cause of human and plant disease under long exposure to airborne particles throughout the year, especially in the summer season. Furthermore, the abundance of the above-mentioned allergenic and pathogenic fungal spores in $PM_1$ was about 3.8 times higher than that in $PM_{2.5}$ in summer, implying relatively higher health risks for smaller particles.

Residents and even visitors at Mt. Tai should be warned about this phenomenon.

### 3.5 Size distribution and seasonal variation of fungal communities

Both fungal abundance and fungal community show a seasonal trend across different size fractions (Awad et al., 2013). Yamamoto et al. (2012) observed that the pathogenic fungi were mainly detected at $PM_{4.7}$ (PM with aerodynamic diameter < 4.7 µm), while the allergenic fungi existed primarily at PM sizes with aerodynamic diameter > 9 µm. In the present study, a discrepant size distribution of the fungal community was observed according to the Metastat analysis by permutation $t$-tests (Table 3). *Glomerella*, *Zasmidium*, and *Phyllosticta* were abundantly enriched in $PM_{2.5}$, while the abundance of *Preussia*, *Truncatella*, *Umbelopsis*, *Sebacina*, and *Cordyceps* increased in $PM_1$. The Kruskal–Wallis rank sum test showed that 6 fungal genera had apparent seasonal variation (Fig. 4). *Glomerella* and *Zasmidium* increased in autumn and decreased as the particle size increased. *Glomerella* was widely found on the surface of leaves, suggesting that leaf senescence is an important source of fungi in $PM_{2.5}$ in autumn (Wang et al., 2015). Some crucial environmental factors having a potential influence on fungal release and growth, such as temperature; $NO_2$; $PM_{10}$; $SO_2$; CO; relative humidity (Yan et al., 2016); radiation; vegetation

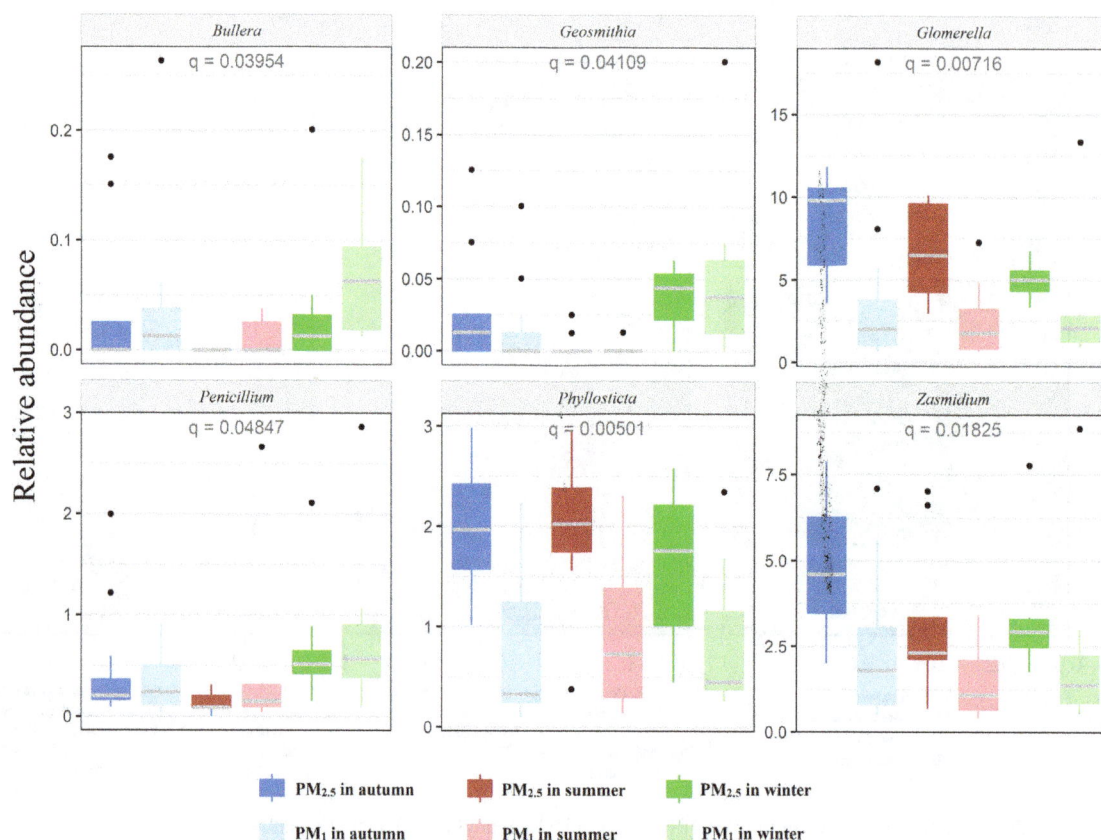

**Figure 4.** Variance analysis of fungal genera based on the Kruskal–Wallis rank sum test.

**Table 3.** Metastats analysis showing the fungal genera that are significantly different among $PM_{2.5}$ and $PM_1$.

| Taxa | $PM_1$ | | | $PM_{2.5}$ | | | $p$ value | $q$ value |
|---|---|---|---|---|---|---|---|---|
| | mean | SE | variance | mean | SE | variance | | |
| Glomerella | 10.51984 | 0.021813 | 0.013798 | 22.49025 | 0.01807 | 0.009796 | 0.000999 | 0.025543 |
| Zasmidium | 6.523201 | 0.011769 | 0.004017 | 12.71881 | 0.012239 | 0.004494 | 0.000999 | 0.025543 |
| Phyllosticta | 2.507228 | 0.004038 | 0.000473 | 5.948659 | 0.004366 | 0.000572 | 0.000999 | 0.025543 |
| Preussia | 0.039161 | 0.000195 | 1.10E-06 | 0.009109 | 6.81E-05 | 1.39E-07 | 0.002322 | 0.042885 |
| Truncatella | 0.030579 | 0.00024 | 1.67E-06 | 0.005152 | 2.44E-05 | 1.79E-08 | 0.002784 | 0.046274 |
| Umbelopsis | 0.027549 | 0.000252 | 1.84E-06 | 0.005369 | 2.55E-05 | 1.95E-08 | 0.001669 | 0.034675 |
| Sebacina | 0.021306 | 0.000196 | 1.11E-06 | 0.001261 | 1.26E-05 | 4.77E-09 | 0.000550 | 0.022857 |
| Cordyceps | 0.020939 | 0.000137 | 5.45E-07 | 0.002518 | 1.75E-05 | 9.18E-09 | 0.001392 | 0.030848 |

(Moreau et al., 2016); urbanization; and accidental events, e.g., dust storms (Prospero et al., 2005), rainfall (Zhang et al., 2015), hurricanes (DeLeon-Rodriguez et al., 2013), and haze (Yan et al., 2016), have been identified. Herein Spearman's rank coefficient analysis indicated that $Ca^{2+}$, a typical water-soluble inorganic ion from dust, was negatively related to the prevalence of *Glomerella* and *Zasmidium* in autumn (Fig. 5). The increase of *Penicillium*, *Bullera*, and *Geosmithia* in winter is ascribed to their sensitivity to low temperature (Sousa et al., 2008; Abdel Hameed et al., 2012). The results based on Spearman's rank correlation test analy-

sis support this notion (Fig. 5, $p < 0.01$). Humidity, another important factor for fungal release into the atmosphere either by active or passive modes, is a crucial factor for the variation in fungal spores such as *Lophium* ($p < 0.01$), *Cenococcum* ($p < 0.05$), *Tricholoma* ($p < 0.05$), and *Candida* ($p < 0.05$). In summer, no distinct difference was observed based on the top 40 fungal genera (Fig. 4). However, some trace fungal genera presented an inverse correlation with temperature (*Coccomyces*, $p < 0.01$; and *Dictyosporium*, $p < 0.01$), humidity (*Botryosphaeria*, $p < 0.001$; *Coccomyces*, $p < 0.01$; and *Dictyosporium*, $p < 0.01$), $PM_{2.5}$ (*Acremonium*, $p < 0.01$;

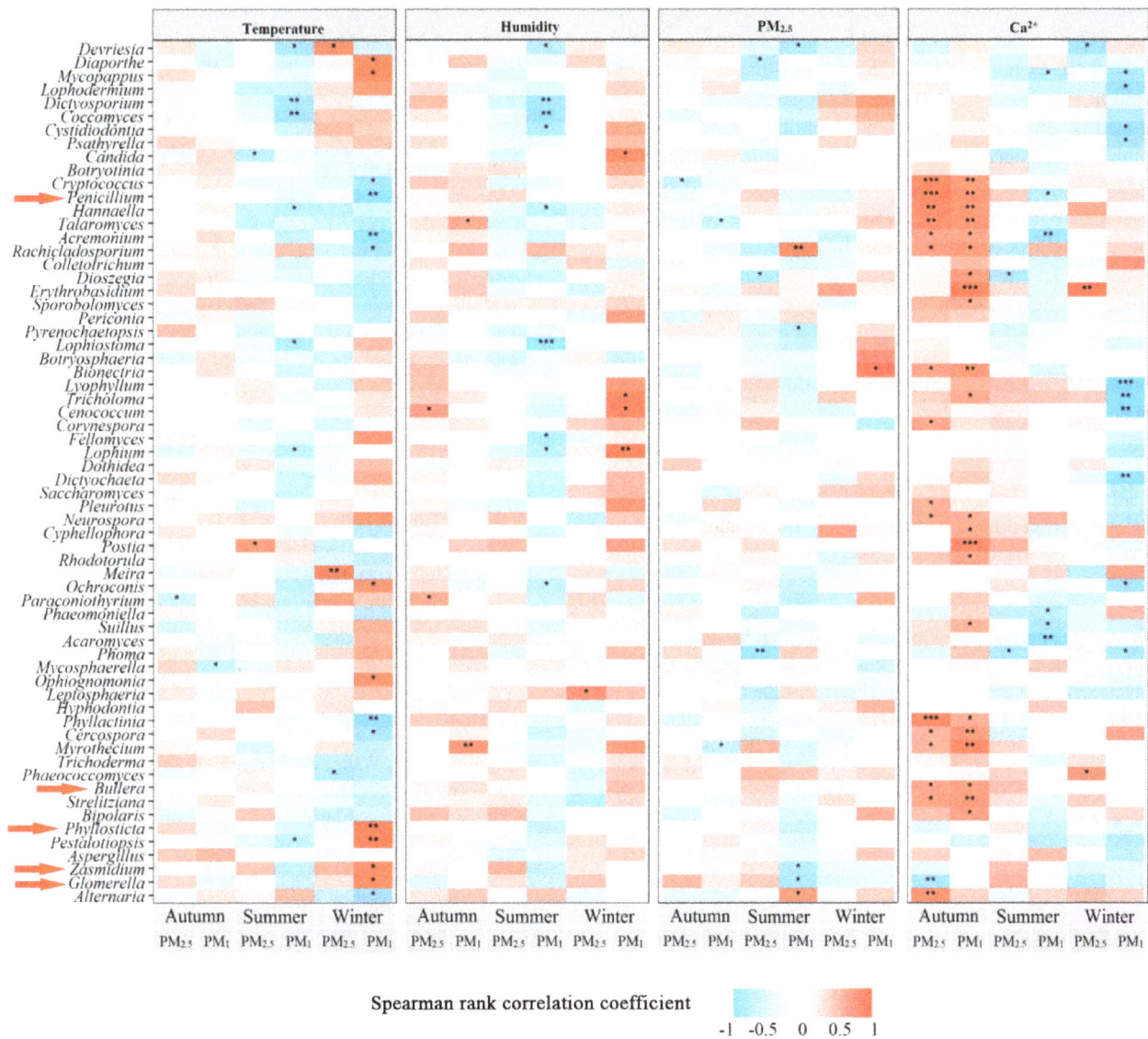

**Figure 5.** Heatmap analysis of the top 64 fungal genera based on Spearman's rank correlations (*** $p < 0.001$; ** $p < 0.01$; * $p < 0.05$). Red arrows indicate that the specific fungi varied significantly in different seasons.

*Phoma*, $p < 0.01$), and $Ca^{2+}$ (*Talaromyces*, $p < 0.01$; *Acaromyces*, $p < 0.01$). The crucial environmental factors we identified contributed to the variation in the fungal community. Due to the limited culture studies on the mechanism for the effects of environmental factors on specific fungal spores, the relationship between bioaerosols and environmental factors still needs to be surveyed over a longer duration.

## 4 Conclusions

Diverse airborne fungal spores are relevant for studies on the atmosphere, biogeoscience, climate and ecology, environmental hygiene, agriculture, and bioengineering. As the details of fungal spores present at high-elevation sites re-

main unknown, the detection and characterization of ambient fungi can help elucidate the regional and global distribution of diverse fungi. To that end, we provide a comprehensive framework of the fungal abundance and communities associated with different seasons across $PM_{2.5}$ and $PM_1$ at Mt. Tai. The results revealed that the concentration and fungal community structure at Mt. Tai differs considerably from those reported for surface ground sites. Over the sampling period, average fungal concentrations of 467 and 644 spores $m^{-3}$ in $PM_{2.5}$ and $PM_1$, respectively, were calculated. In addition to long-distance air-mass movement events, westerly and northwesterly winds also favor the increase in fungal abundance. Diverse fungal communities presented significant seasonal variation across different size particles. The prevalence of *Glomerella* and *Zasmidium* increased in autumn and decreased as the particle size increased. In winter, the prevalence of *Penicillium*, *Bullera*, and *Geosmithia* increased with the decrease in particle size. No distinct disparity was observed in summer. The variation in fungal profile can be influenced by environmental factors including humidity, temperature, wind speed, $PM_{2.5}$, and some chemical components in PMs (including $Ca^{2+}$). Nevertheless, the detailed specific effects of environmental factors on the ambient fungal community remain poorly explained. In further studies, a combination of traditional culture-based methods and metagenomics may help answer the various unresolved questions.

*Competing interests.* The authors declare that they have no conflict of interest.

*Special issue statement.* This article is part of the special issue "Regional transport and transformation of air pollution in eastern China". It is not associated with a conference.

*Acknowledgements.* This work was supported by the National Natural Science Foundation of China (no. 41375126, 21527814), Taishan Scholar Grant (grant number ts20120552), Cyrus Tang Foundation (no. CTF-FD2014001), the Ministry of Science and Technology of China (no. 2016YFC0202701, 2014BAC22B01), and the European Union's Horizon 2020 research and innovation programme under grant agreement no. 690958 (MARSU project).

Edited by: Tong Zhu

# References

Abdel Hameed, A. A., Khoder, M. I., Ibrahim, Y. H., Saeed, Y., Osman, M. E., and Ghanem, S.: Study on some factors affecting survivability of airborne fungi, Sci. Total Environ., 414, 696–700, 2012.

Adhikari, A., Sen, M. M., Gupta-Bhattacharya, S., and Chanda, S.: Airborne viable, non-viable, and allergenic fungi in a rural agricultural area of India: a 2-year study at five outdoor sampling stations, Sci. Total Environ., 326, 123–141, 2004.

Almaguer, M., Aira, M. J., Rodriguez-Rajo, F. J., and Rojas, T. I.: Temporal dynamics of airborne fungi in Havana (Cuba) during dry and rainy seasons: influence of meteorological parameters, Int. J. Biometeorol., 58, 1459–1470, 2014.

Awad, A. H., Gibbs, S. G., Tarwater, P. M., and Green, C. F.: Seasonal evaluation of fine and coarse culturable bacterial aerosols from residences within a rural and an urban city in Egypt, Int. J. Environ. Heal. R., 23, 269–280, 2013.

Bauer, H., Schueller, E., Weinke, G., Berger, A., Hitzenberger, R., Marr, I. L., and Puxbaum, H.: Significant contributions of fungal spores to the organic carbon and to the aerosol mass balance of the urban atmospheric aerosol, Atmos. Environ., 42, 5542–5549, 2008.

Belanger, R. R., Benhamou, N., and Menzies, J. G.: Cytological evidence of an active role of silicon in wheat resistance to powdery mildew (*Blumeria graminis f. sp. tritici*), Phytopathology, 93, 402–412, 2003.

Bowers, R. M., Lauber, C. L., Wiedinmyer, C., Hamady, M., Hallar, A. G., Fall, R., Knight, R., and Fierer, N.: Characterization of airborne microbial communities at a high-elevation site and their potential to act as atmospheric ice nuclei, Appl. Environ. Microbiol., 75, 5121–5130, 2009.

Bowers, R. M., McCubbin, I. B., Hallar, A. G., and Fierer, N.: Seasonal variability in airborne bacterial communities at a high-elevation site, Atmos. Environ., 50, 41–49, 2012.

Bowers, R. M., Clements, N., Emerson, J. B., Wiedinmyer, C., Hannigan, M. P., and Fierer, N.: Seasonal variability in bacterial and fungal diversity of the near-surface atmosphere, Environ. Sci. Technol., 47, 12097–12106, 2013.

Brauer, M., Freedman, G., Frostad, J., van Donkelaar, A., Martin, R. V., Dentener, F., van Dingenen, R., Estep, K., Amini, H., Apte, J. S., Balakrishnan, K., Barregard, L., Broday, D., Feigin, V., Ghosh, S., Hopke, P. K., Knibbs, L. D., Kokubo, Y., Liu, Y., Ma, S., Morawska, L., Sangrador, J. L., Shaddick, G., Anderson, H. R., Vos, T., Forouzanfar, M. H., Burnett, R. T., and Cohen, A.: Ambient air pollution exposure estimation for the global burden of disease 2013, Environ. Sci. Technol., 50, 79–88, 2013.

Brodie, E. L., DeSantis, T. Z., Parker, J. P., Zubietta, I. X., Piceno, Y. M., and Andersen, G. L.: Urban aerosols harbor diverse and dynamic bacterial populations, P. Natl. Acad. Sci. USA, 104, 299–304, 2007.

Cao, C., Jiang, W., Wang, B., Fang, J., Lang, J., Tian, G., Jiang, J., and Zhu, T. F.: Inhalable microorganisms in Beijing's $PM_{2.5}$ and $PM_{10}$ pollutants during a severe smog event, Environ. Sci. Technol., 48, 1499–1507, 2014.

Cheng, J. Y. W., Chan, C. K., Lee, C. T., and Lau, A. P. S.: Carbon content of common airborne fungal species and fungal contribution to aerosol organic carbon in a subtropical city, Atmos. Environ., 43, 2781–2787, 2009.

Christner, B. C., Morris, C. E., Foreman, C. M., Cai, R., and Sands, D. C.: Ubiquity of biological ice nucleators in snowfall, Science, 319, 1214, https://doi.org/10.1126/science.1149757, 2008.

Creamean, J. M., Suski, K. J., Rosenfeld, D., Cazorla, A., DeMott, P.J., Sullivan, R. C., White, A. B., Ralph, F. M., Minnis, P., Comstock, J. M., Tomlinson, J. M., and Prather, K. A.: Dust and biological aerosols from the Sahara and Asia influence precipitation in the western US, Science, 339, 1572–1578, 2013.

Dannemiller, K. C., Lang-Yona, N., Yamamoto, N., Rudich, Y., and Peccia, J.: Combining real-time PCR and next-generation DNA sequencing to provide quantitative comparisons of fungal aerosol populations, Atmos. Environ., 84, 113–121, 2014.

DeLeon-Rodriguez, N., Lathem, T. L., Rodriguez, R. L., Barazesh, J. M., Anderson, B. E., Beyersdorf, A. J., Ziemba, L. D., Bergin, M., Nenes, A., and Konstantinidis, K. T.: Microbiome of the upper troposphere: species composition and prevalence, effects of tropical storms, and atmospheric implications, P. Natl. Acad. Sci. USA, 110, 2575–2580, 2013.

Deng, W. J., Chai, Y. M., Lin, H. Y., So, W. W. M., Ho, K. W. K., Tsui, A. K. Y., and Wong, P. K. S.: Distribution of bacteria in inhalable particles and its implications for health risks in kindergarten children in Hong Kong, Atmos. Environ., 128, 268–275, 2016, .

Després, V. R., Nowoisky, J. F., Klose, M., Conrad, R., Andreae, M. O., and Pöschl, U.: Characterization of primary biogenic aerosol particles in urban, rural, and high-alpine air by DNA sequence and restriction fragment analysis of ribosomal RNA genes, Biogeosciences, 4, 1127–1141, https://doi.org/10.5194/bg-4-1127-2007, 2007.

Despres, V. R., Huffman, J. A., Burrows, S. M., Hoose, C., Safatov, A. S., Buryak, G., Fröhlich-Nowoisky, J., Elbert, W., Andreae, M. O., and Pöschl, U.: Primary biological aerosols in the atmosphere: A review of observations and relevance, Tellus B, 64, 1–58, 2012.

Edgar, R. C., Haas, B. J., Clemente, J. C., Quince, C., and Knight, R.: UCHIME improves sensitivity and speed of chimera detection, Bioinformatics, 27, 2194–2200, 2011.

Elbert, W., Taylor, P. E., Andreae, M. O., and Pöschl, U.: Contribution of fungi to primary biogenic aerosols in the atmosphere: wet and dry discharged spores, carbohydrates, and inorganic ions, Atmos. Chem. Phys., 7, 4569–4588, https://doi.org/10.5194/acp-7-4569-2007, 2007.

Fröhlich-Nowoisky, J., Pickersgill, D. A., Despres, V. R., and Pöschl, U.: High diversity of fungi in air particulate matter, P. Natl. Acad. Sci. USA, 106, 12814–12819, 2009.

Fröhlich-Nowoisky, J., Burrows, S. M., Xie, Z., Engling, G., Solomon, P. A., Fraser, M. P., Mayol-Bracero, O. L., Artaxo, P., Begerow, D., Conrad, R., Andreae, M. O., Després, V. R., and Pöschl, U.: Biogeography in the air: fungal diversity over land and oceans, Biogeosciences, 9, 1125–1136, https://doi.org/10.5194/bg-9-1125-2012, 2012.

Gandolfi, I., Bertolini, V., Bestetti, G., Ambrosini, R., Innocente, E., Rampazzo, G., Papacchini, M., and Franzetti, A.: Spatiotemporal variability of airborne bacterial communities and their correlation with particulate matter chemical composition across two urban areas, Appl. Microbiol. Biotechnol., 99, 4867–4877, 2015.

Gao, J. F., Fan, X. Y., Li, H. Y., and Pan, K. L.: Airborne bacterial communities of $PM_{2.5}$ in Beijing-Tianjin-Hebei megalopolis, China as revealed by Illumina MiSeq sequencing: a case study, Aerosol Air Qual. Res, 17, 778–798, 2017a.

Gao, M., Jia, R. Z., Qiu, T. L., Han, M. L., and Wang, X. M.: Size-related bacterial diversity and tetracycline resistance gene abundance in the air of concentrated poultry feeding operations, Environ. Pollut., 220, 1342–1348, 2017b.

Gou, H., Lu, J. J., Li, S., Tong, Y. B., Xie, C. B., and Zheng, X.: Assessment of microbial communities in $PM_1$ and $PM_{10}$ of Urumqi during winter, Environ. Pollut., 214, 202–210, 2016.

Hawksworth, D. L.: The magnitude of fungal diversity: the 1.5 million species estimate revisited, Mycol. Res., 105, 1422–1432, 2001.

Heald, C. L. and Spracklen, D. V.: Atmospheric budget of primary biological aerosol particles from fungal spores, Geophys. Res. Lett., 36, 269–277, 2009.

Hedayati, M. T., Pasqualotto, A. C., Warn, P. A., Bowyer, P., and Denning, D. W.: Aspergillus flavus: human pathogen, allergen and mycotoxin producer, Microbiology, 153, 1677–1692, 2007.

Ho, K. F., Lee, S. C., Yu, J. C., Zou, S. C., and Fung, K.: Carbonaceous characteristics of atmospheric particulate matter in Hong Kong, Sci. Total Environ., 300, 59–67, 2002.

Huffman, J. A., Treutlein, B., and Pöschl, U.: Fluorescent biological aerosol particle concentrations and size distributions measured with an Ultraviolet Aerodynamic Particle Sizer (UV-APS) in Central Europe, Atmos. Chem. Phys., 10, 3215–3233, https://doi.org/10.5194/acp-10-3215-2010, 2010.

Jiang, W. J., Liang, P, Wang, B. Y., Fang, J. H., Lang, J. D., Tian, G., Jiang, J. K., and Zhu, T. F.: Optimized DNA extraction and metagenomic sequencing of airborne microbial communities, Nat. Protoc., 10, 768–779, 2015.

Jaenicke, R.: Abundance of cellular material and proteins in the atmosphere, Science, 308, p. 73, 2005.

Judd, C. R., Koyama, A., Simmons, M. P., Brewer, P. E., and von Fischer, J. C.: Co-variation in methanotroph community composition and activity in three temperate grassland soils, Soil Biol. Biochem., 95, 78–86, 2016.

Koljalg, U., Nilsson, R. H., and Abarenkov, K.: Towards a unified paradigm for sequence-based identification of fungi, Mol. Ecol., 22, 5271–5277, 2013.

Koiv, V., Roosaare, M., Vedler, E., Kivistik, P. A., Toppi, K., Schryer, D. W., Remm, M., Tenson, T., and Mae, A.: Microbial population dynamics in response to Pectobacterium atrosepticum infection in potato tubers, Sci. Rep-UK, 5, 11606, https://doi.org/10.1038/srep11606, 2015.

Kuczynski, J., Stombaugh, J., Walters, W. A., GonzAlez, A., Gregory Caporaso, J., and Knight, R.: Using QIIME to analyze 16S rRNA gene sequences from microbial communities, Curr. Protoc. Bioinformatics., Chapter 10, Unit 10.7, https://doi.org/10.1002/0471250953.bi1007s36, 2011.

Lee, S. H., Lee, H. J., Kim, S. J., Lee, H. M., Kang, H., and Kim, Y. P.: Identification of airborne bacterial and fungal community structures in an urban area by T-RFLP analysis and quantitative real-time PCR, Sci. Total Environ., 408, 1349–1357, 2010.

Liang, L., Engling, G., Du, Z., Duan, F., Cheng, Y., Liu, X., and He, K.: Contribution of fungal spores to organic carbon in ambient aerosols in Beijing, China, Atmos. Pollut. Res., 8, 351–358, 2017.

Lin, W. H. and Li, C. S.: Associations of fungal aerosols, air Pollutants, and meteorological factors, Aerosol Sci. Technol., 32, 359–368, 2000.

Manter, D. K. and Vivanco, J. M.: Use of the ITS primers, ITS1F and ITS4, to characterize fungal abundance and diversity in mixed-template samples by qPCR and length heterogeneity analysis, J. Microbiol. Meth., 71, 7–14, 2007.

Miettinen, H., Kietavainen, R., Sohlberg, E., Numminen, M., Ahonen, L., and Itavaara, M.: Microbiome composition and geochemical characteristics of deep subsurface high-pressure environment, Pyhasalmi mine Finland, Front. Microbiol., 6, 1203, https://doi.org/10.3389/fmicb.2015.01203, 2015.

Moreau, C. S., Weikl, F., Tischer, C., Probst, A. J., Heinrich, J., Markevych, I., Jochner, S., and Pritsch, K.: Fungal and bacterial communities in indoor dust follow different environmental determinants, Plos One, 11, e0154131, https://doi.org/10.1371/journal.pone.0154131, 2016.

Morris, C. E., Kinkel, L.L., Lindow, S. E., Hechtpoinar, E. I., and Elliott, V. J.: Fifty years of phyllosphere microbiology: significant contributions to research in related fields, Phyllosphere Microbiology, St. Paul, MN, USA, APS Press, 365–375, 2002.

Oh, S. Y., Fong, J. J., Park, M. S., Chang, L., and Lim, Y. W.: Identifying airborne fungi in Seoul, Korea using metagenomics, J. Microbiol., 52, 465–472, 2014.

Oliveira, M., Ribeiro, H., Delgado, J. L., and Abreu, I.: The effects of meteorological factors on airborne fungal spore concentration in two areas differing in urbanisation level, Int. J. Biometeorol., 53, 61–73, 2009.

Pratt, K. A., DeMott, P. J., French, J. R., Wang, Z., Westphal, D. L., Heymsfield, A. J., Twohy, C. H., Prenni, A. J., and Prather, K. A.: In situ detection of biological particles in cloud ice-crystals, Nat. Geosci., 2, 398–401, 2009.

Prospero, J. M., Blades, E., Mathison, G., and Naidu, R.: Interhemispheric transport of viable fungi and bacteria from Africa to the Caribbean with soil dust, Aerobiologia, 21, 1–19, 2005.

Ram, K., Sarin, M. M., and Tripathi, S. N.: Temporal trends in atmospheric $PM_{2.5}$, $PM_{10}$, elemental carbon, organic carbon, water-soluble organic carbon, and optical properties: impact of biomass burning emissions in the Indo Gangetic Plain, Environ. Sci. Technol., 46, 686–695, 2012.

Rittenour, W. R., Ciaccio, C. E., Barnes, C. S., Kashon, M. L., Lemons, A. R., Beezhold, D. H., and Green, B. J.: Internal transcribed spacer rRNA gene sequencing analysis of fungal diversity in Kansas City indoor environments, Environ. Sci. Proc. Impacts, 16, 33–43, 2014.

Ryan, P. H., Bernstein, D. I., Lockey, J., Reponen, T., Levin, L., Grinshpun, S., Villareal, M., Hershey, G. K. K., Burkle, J., and Lemasters, G.: Exposure to traffic-related particles and endotoxin during infancy is associated with wheezing at age 3 years, Am. J. Resp. Crit. Care, 180, 1068–1075, 2009.

Saari, S.: Seasonal and diurnal variations of fluorescent bioaerosol concentration and size distribution in the urban environment, Aerosol Air Qual. Res., 15, 572–581, 2015.

Simon-Nobbe, B., Denk, U., Poll, V., Rid, R., and Breitenbach, M.: The spectrum of fungal allergy, Int. Arch. Allergy Imm., 145, 58–86, 2008.

Smets, W., Moretti, S., Denys, S., and Lebeer, S.: Airborne bacteria in the atmosphere: presence, purpose, and potential, Atmos. Environ., 139, 214–221, 2016.

Sousa, S. I. V., Martins, F. G., Pereira, M .C., Alvim-Ferraz, M. C. M., Ribeiro, H., Oliveira, M., and Abreu, I.: Influence of atmospheric ozone, $PM_{10}$ and meteorological factors on the concentration of airborne pollen and fungal spores, Atmos. Environ., 42, 7452–7464, 2008.

Tong, X. L., Xu, H. T., Zou, L. H., Cai, M., Xu, X. F., Zhao, Z. T., Xiao, F., and Li, Y. M.: High diversity of airborne fungi in the hospital environment as revealed by meta-sequencing-based microbiome analysis, Sci. Rep.-UK, 7, 39606, https://doi.org/10.1038/srep39606, 2017.

Tsai, F. C., Macherb, J. M., and Hung, Y. Y.: Biodiversity and concentrations of airborne fungi in large US office buildings from the BASE study, Atmos. Environ., 41, 5181–5191, 2007.

van Doorn, R., Szemes, M., Bonants, P., Kowalchuk, G. A., Salles, J. F., Ortenberg, E., and Schoen, C. D.: Quantitative multiplex detection of plant pathogens using a novel ligation probe-based system coupled with universal, high-throughput real-time PCR on OpenArrays, BMC genomics, 8, 1–4, 2007.

Vermani, M., Vijayan, V. K., Kausar, M. A., and Agarwal, M. K.: Quantification of airborne Aspergillus allergens: redefining the approach, J. Aasthma., 47, 754–761, 2010.

Wang, L., Song Z. W., Xu, A. L., Wu, D. D., and Yan, X.: Phylogenetic diversity of airborne microbes in Qingdao downtown in autumn, J. Appl. Ecol., 26, 1121–1129, 2015 (in Chinese).

Wang, X., Chen, J., Sun, J., Li, W., Yang, L., Wen, L., Wang, W., Wang, X., Collett Jr, J. L., Shi, Y., Zhang, Q., Hu, J., Yao, L., Zhu, Y., Sui, X., Sun, X., and Mellouki, A.: Severe haze episodes and seriously polluted fog water in Ji'nan, China, Sci. Total Environ., 493, 133–137, 2014.

Wei, M., Xu, C., Chen, J., Zhu, C., Li, J., and Lv, G.: Characteristics of bacterial community in cloud water at Mt Tai: similarity and disparity under polluted and non-polluted cloud episodes, Atmos. Chem. Phys., 17, 5253–5270, doi:10.5194/acp-17-5253-2017, 2017.

Womack, A. M., Artaxo, P. E., Ishida, F. Y., Mueller, R. C., Saleska, S. R., Wiedemann, K. T., Bohannan, B. J. M., and Green, J. L.: Characterization of active and total fungal communities in the atmosphere over the Amazon rainforest, Biogeosciences, 12, 6337–6349, https://doi.org/10.5194/bg-12-6337-2015, 2015.

Womiloju, T. O., Miller, J. D., Mayer, P. M., and Brook, J. R.: Methods to determine the biological composition of particulate matter collected from outdoor air, Atmos. Environ., 37, 4335–4344, 2003.

Xu, C. H., Wei, M., Chen, J. M., Wang, X. F., Sui, X., Zhu, C., Li, J. R., Zheng, L. L., Sui, G. D., Li, W. J., Wang, W. X., Zhang, Q. Z., and Mellouki, A.: Bacterial characterization in ambient submicron particles during severe haze episodes at Ji'nan, China, Sci. Total Environ., 580, 188–1196, 2017a.

Xu, C. H., Wei, M., Chen, J. M., Sui, X., Zhu, C., Li, J. R., Zheng, L. L., Sui, G. D., Li, W. J., Wang, W. X., Zhang, Q. Z., and Mellouki, A.: Investigation of diverse bacteria in cloud water at Mt. Tai, China, Sci. Total Environ., 580, 258–265, 2017b.

Yadav, R. K. P., Halley, J. M., Karamanoli, K., Constantinidou, H. I., and Vokou, D.: Bacterial populations on the leaves of Mediterranean plants: quantitative features and testing of distribution models, Environ. Exp. Bot., 52, 63–77, 2004.

Yamaguchi, N., Baba, T., Ichijo, T., Himezawa, Y., Enoki, K., Saraya, M., Li, P. F., and Nasu, M.: Abundance and community structure of bacteria on Asian dust particles collected in Beijing, China, during the Asian dust season, Biological and Pharmaceutical Bulletin, 39, 68–77, 2016.

Yamamoto, N., Bibby, K., Qian, J., Hospodsky, D., Rismani-Yazdi, H., Nazaroff, W. W., and Peccia, J.: Particle-size distributions and seasonal diversity of allergenic and pathogenic fungi in outdoor air, ISME J., 6, 1801–1811, 2012.

Yan, D., Zhang, T., Su, J., Zhao, L. L., Wang, H., Fang, X. M., Zhang, Y. Q., Liu, H. Y., and Yu, L. Y.: Diversity and composition of airborne fungal community associated with particulate matters in Beijing during haze and non-haze days, Front. Microbiol., 7, 1–12, 2016.

Yu, J. Z., Tung, J. W. T., Wu, A. W. M., Lau, A. K. H., Louie, P. K. K., and Fung, J. C. F.: Abundance and seasonal characteristics of elemental and organic carbon in Hong Kong $PM_{10}$, Atmos. Environ., 38, 1511–1521, 2004.

Zhang, Z., Engling, G., Zhang, L., Kawamura, K., Yang, Y., Tao, J., Zhang, R., Chan, C. Y., and Li, Y.: Significant influence of fungi on coarse carbonaceous and potassium aerosols in a tropical rainforest, Environ. Res. Lett., 10, 034015, https://doi.org/10.1088/1748-9326/10/3/034015, 2015.

Zhong, X., Qi, J., Li, H., Dong, L., and Gao, D.: Seasonal distribution of microbial activity in bioaerosols in the outdoor environment of the Qingdao coastal region, Atmos. Environ., 140, 506–513, 2016.

# Pre-monsoon air quality over Lumbini, a world heritage site along the Himalayan foothills

**Dipesh Rupakheti**[1,2]**, Bhupesh Adhikary**[3]**, Puppala Siva Praveen**[3]**, Maheswar Rupakheti**[4,5]**, Shichang Kang**[2,6,7]**, Khadak Singh Mahata**[4]**, Manish Naja**[8]**, Qianggong Zhang**[1,7]**, Arnico Kumar Panday**[3]**, and Mark G. Lawrence**[4]

[1]Key Laboratory of Tibetan Environment Changes and Land Surface Processes, Institute of Tibetan Plateau Research, Chinese Academy of Sciences, Beijing 100101, China
[2]University of Chinese Academy of Sciences, Beijing 100049, China
[3]International Centre for Integrated Mountain Development (ICIMOD), Kathmandu, Nepal
[4]Institute for Advanced Sustainability Studies (IASS), Potsdam 14467, Germany
[5]Himalayan Sustainability Institute (HIMSI), Kathmandu, Nepal
[6]State Key Laboratory of Cryospheric Science, Cold and Arid Regions Environmental and Engineering Research Institute (CAREERI), Lanzhou 730000, China
[7]Center for Excellence in Tibetan Plateau Earth Sciences, Chinese Academy of Sciences, Beijing 100085, China
[8]Aryabhatta Research Institute of Observational Sciences (ARIES), Nainital, India

*Correspondence to:* Dipesh Rupakheti (dipesh.rupakheti@itpcas.ac.cn) and Shichang Kang (shichang.kang@lzb.ac.cn)

**Abstract.** Lumbini, in southern Nepal, is a UNESCO world heritage site of universal value as the birthplace of Buddha. Poor air quality in Lumbini and surrounding regions is a great concern for public health as well as for preservation, protection and promotion of Buddhist heritage and culture. We present here results from measurements of ambient concentrations of key air pollutants (PM, BC, CO, $O_3$) in Lumbini, first of its kind for Lumbini, conducted during an intensive measurement period of 3 months (April–June 2013) in the pre-monsoon season. The measurements were carried out as a part of the international air pollution measurement campaign; SusKat-ABC (Sustainable Atmosphere for the Kathmandu Valley – Atmospheric Brown Clouds). The main objective of this work is to understand and document the level of air pollution, diurnal characteristics and influence of open burning on air quality in Lumbini. The hourly average concentrations during the entire measurement campaign ranged as follows: BC was 0.3–30.0 µg m$^{-3}$, PM$_1$ was 3.6–197.6 µg m$^{-3}$, PM$_{2.5}$ was 6.1–272.2 µg m$^{-3}$, PM$_{10}$ was 10.5–604.0 µg m$^{-3}$, $O_3$ was 1.0–118.1 ppbv and CO was 125.0–1430.0 ppbv. These levels are comparable to other very heavily polluted sites in South Asia. Higher fraction of coarse-mode PM was found as compared to other nearby sites in the Indo-Gangetic Plain region. The $\Delta BC / \Delta CO$ ratio obtained in Lumbini indicated considerable contributions of emissions from both residential and transportation sectors. The 24 h average PM$_{2.5}$ and PM$_{10}$ concentrations exceeded the WHO guideline very frequently (94 and 85 % of the sampled period, respectively), which implies significant health risks for the residents and visitors in the region. These air pollutants exhibited clear diurnal cycles with high values in the morning and evening. During the study period, the worst air pollution episodes were mainly due to agro-residue burning and regional forest fires combined with meteorological conditions conducive of pollution transport to Lumbini. Fossil fuel combustion also contributed significantly, accounting for more than half of the ambient BC concentration according to aerosol spectral light absorption coefficients obtained in Lumbini. WRF-STEM, a regional chemical transport model, was used to simulate the meteorology and the concentrations of pollutants to understand the pollutant transport pathways. The model estimated values were $\sim$ 1.5 to 5 times lower than the observed concentrations for CO and PM$_{10}$, respectively. Model-simulated regionally tagged CO tracers showed that the majority of CO came from the upwind region of Ganges Valley. Model performance needs significant improvement in simulating aerosols in the region. Given the high air pollution level, there is a clear and urgent need for setting up a network of long-term air quality monitoring stations in the greater Lumbini region.

---

## 1 Introduction

The Indo-Gangetic Plain (IGP) stretches over 2000 km encompassing a vast area of land in northern South Asia: the eastern parts of Pakistan, most of northern and eastern India, southern part of Nepal and almost all of Bangladesh. The Himalayan mountains and their foothills stretch along the northern edge of IGP. The IGP region is among the most fertile and most intensely farmed region of the world. It is a heavily populated region with about 900 million residents or 12 % of the world's population. Four megacities – Lahore, Delhi, Kolkata and Dhaka – are located in the IGP region, with dozens more cities with populations exceeding 1 million. The region has witnessed impressive economic growth in recent decades but unfortunately it has also become one of the most polluted, and an air pollution "hotspot" of local, regional and global concern (Ramanathan et al., 2007). Main factors contributing to air pollution in the IGP and surrounding regions include emissions from vehicles, thermal power plants, industries, biomass and fossil fuel used in cooking and heating activities, agricultural activities, crop residue burning and forest fires. Air pollution gets transported long distances away from emission sources and across national borders. As a result, the IGP and adjacent regions get shrouded with a dramatic annual buildup of regional-scale plumes of air pollutants, known as atmospheric brown clouds (ABC), during the long and dry winter and pre-monsoon seasons each year (Ramanathan and Carmichael, 2008). Figure 1 shows monthly synoptic wind and mean aerosol optical depth during April–June 2013 over South Asia. Very high aerosol optical depth along the entire stretch of IGP reflects the severity of air pollution over large areas in the region.

Poor air quality continues to pose a significant threat to human health in the region. In a new study of global burden of disease released recently, Forouzanfar et al. (2015) estimated that in 2013 around 1.7 million people died prematurely in Pakistan, India, Nepal and Bangladesh as a result of air pollution exposure, nearly 30 % of global total premature deaths due to air pollution. Air pollution also affects precipitation (e.g., South Asian monsoon), agricultural productivity, ecosystems, tourism, climate and broadly socioeconomic and national development goals of the countries in the region (Burney and Ramanathan, 2014; Shindell et al., 2012; Ramanathan and Carmichael, 2008). It has also been linked to intensification of cold wave and winter fog in the IGP region over recent decades (Lawrence and Lelieveld, 2010, and references therein; Safai et al., 2009; Ganguly et al., 2006). Besides high levels of aerosol loading as shown in

Fig. 1, IGP also has very high levels of ground-level ozone or tropospheric ozone ($O_3$) (e.g., Ramanathan and Carmichael, 2008), which is toxic to plant and human health and a major greenhouse gas (IPCC, 2013; Shindell et al., 2012; Mohnen et al., 1993). South Asia, in particular IGP, has been projected to be the most ozone-polluted region in the world by 2030 (Stevenson et al., 2006). The majority of crop loss in different parts of the world results from effects of ozone on crop health and productivity (Shindell et al., 2012). Burney and Ramanathan (2014) also reported a significant loss in wheat and rice yields in India from 1980 to 2010 due to direct effects of black carbon (BC) and $O_3$. BC and $O_3$ are two key short-lived climate pollutants. Similarly, species like fine particles and carbon monoxide (CO) are potent to health damages by posing impacts upon the respiratory and cardiovascular system and even also to the climate system (Singh et al., 2017, and references therein). Because of the IGP's close proximity to the Himalaya–Tibetan plateau region, this once relatively clean region is now subjected to increasing air pollution transported from regions such as the IGP, which can exert additional risks to sensitive ecosystems in the mountain region (e.g., Lüthi et al., 2015; Marinoni et al., 2013; Duchi et al., 2011). However, air pollution transport pathways to Himalayas are still not yet fully understood.

Monuments and buildings made with stones are vulnerable to air pollution damage (Brimblecombe, 2003; Gauri and Holdren, 1981). The damage to the monuments and buildings could be in various forms like corrosion, soiling, abrasion and discoloration. For example, a recent study has reported that deposition of light absorbing aerosol particles (black carbon, brown carbon) and dust is responsible for the discoloration of Taj Mahal, a world-famous monument in India (Bergin et al., 2015). Lumbini, located near the northern edge of the central IGP, is famous as the birthplace of the Lord Buddha and thus a UNESCO world heritage site of outstanding universal value to humanity. Since the study area is renowned for its historical and archaeological significance, Lumbini is also getting worldwide attention for poor air quality in the region. There was no regular air quality monitoring in Lumbini at the time of our measurement campaign.

Through this study, we want to understand the level of air pollution, its diurnal characteristic and the influence of open burning on air quality in Lumbini. We carried out continuous measurements of ambient concentrations of key air pollutants (PM, BC, CO, $O_3$) and meteorological parameters during an intensive measurement period of 3 months (April–June) in the year 2013. These are the first reported pollutant measurements for Lumbini. A regional chemical transport model called Sulfur Transport and dEposition Model (STEM) was used to simulate the variations of meteorological parameters and air pollutants during the observation period to examine the extent to which a state-of-the-art, widely used air quality model is able to simulate the observations, as an indication for where there are still gaps in our knowledge and what further measurements and emissions dataset de-

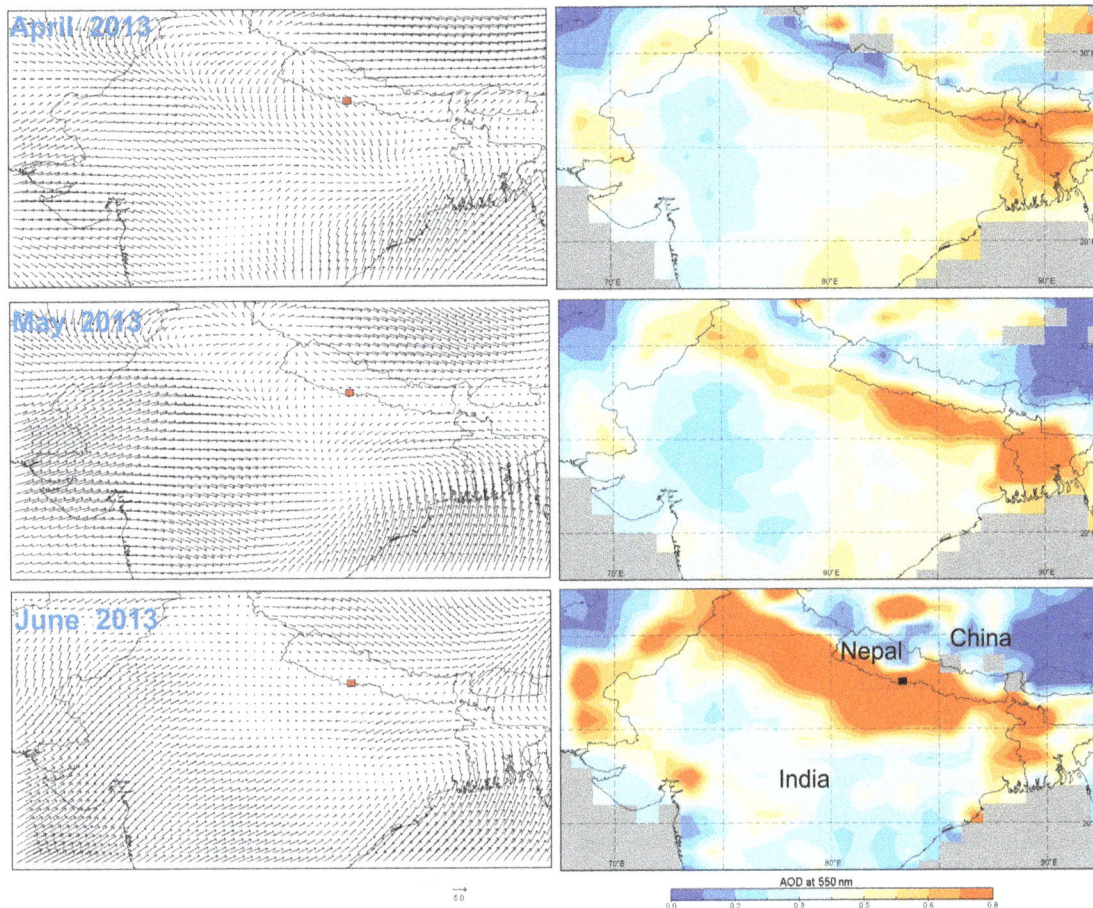

**Figure 1.** Monthly synoptic wind (at 1000 hPa) for April, May and June 2013, based on NCEP/NCAR reanalysis data where the orientations of arrows refer to wind direction and the length of arrows represents the magnitude of wind ($m s^{-1}$). Red square box in the figure (left) represents the location of Lumbini. Figures on the right side represent monthly aerosol optical depth acquired with the MODIS instrument aboard TERRA satellite. High aerosol loading can be seen over the entire Indo-Gangetic Plain (IGP). Light gray color used in the figure represents the absence of data.

velopments are needed. Model-simulated regionally tagged CO tracers were used to identify emission source regions impacting pollutant concentration observed at Lumbini. Satellite data have also been used to understand the high-pollution events during the monitoring period. These measurements were carried out as a part of the SusKat-ABC international air pollution measurement campaign (Rupakheti et al., 2017) jointly led by the International Centre for Integrated Mountain Development (ICIMOD), Kathmandu, Nepal, and Institute for Advanced Sustainability Studies (IASS), Potsdam, Germany.

## 2 Experimental setup

### 2.1 Sampling site

The Lumbini measurement site ($27°29.387'$ N, $83°16.745'$ E; elevation: $\sim 100$ m a.s.l.) is located at the premise of the Lumbini International Research Institute (LIRI), a Buddhist library in Lumbini in the Rupandehi district. According to the national census conducted in 2011, the total population of Rupandehi district is about 900 000 with a population density of about 650 people $km^{-2}$, which is the fourth most densely populated district in the country. Over 130 000 tourists visited Lumbini in 2014 (http://www.tourism.gov.np/ne/). A local road (asphalt) lies about 200 m north of the sampling site and experiences intermittent passing of vehicles. About 25 km north of Lumbini the foothills begin, while the main peaks of the Himalayas are 140 km to the north. The remaining three sides are surrounded by flat plain land of Nepal and India. The site is only about 8 km from the Nepal–India border in the south. A three-storied 10 m tall water tower was used as the platform for the automatic weather station (AWS), while remaining instruments were placed inside a room near the base of the tower. An uninterrupted power back up was set up in order to assure the regular power supply even during hours with scheduled power cuts during the monitoring period. Figure S1 in the Supplement shows the location of Lumbini, the Kenzo Tange Master Plan area of the Lumbini development project, the sampling tower and a brief discussion on the surroundings of the site. Outside of the Master Plan area lies a vast area of agricultural fields, village pockets, several brick kilns and cement industries.

## 2.2 Monitoring instruments

The summary of instruments deployed in Lumbini is presented in Table 1. All data were collected in Nepal standard time (NST), which is GMT + 05:45 h. $PM_1$, $PM_{2.5}$ and $PM_{10}$ mass concentrations were monitored continuously with the EDM 164 (Grimm Aerosol Technik, Germany), which uses the light scattering at 655 nm to derive mass concentrations. Similarly, aerosol light absorptions at seven wavelengths (370, 470, 520, 590, 660, 880, 950 nm) were measured continuously with an Aethalometer (model AE-42, Magee Scientific, USA), averaging and reporting data every 5 min. It was operated at a flow rate of $5 \, L \, min^{-1}$. No cutoff was applied for inlet; hence the reported concentration of BC is total suspended BC particles. As described by the manufacturer, ambient BC concentration is derived from light absorption at 880 nm using a specific mass absorption cross section. To obtain BC concentration in Lumbini, we used a specific mass absorption cross-section value of $8 \, m^2 \, g^{-1}$ for the 880 nm channel. A similar value has been previously used for BC measurement in the IGP (Praveen et al., 2012). To remove the filter loading effect, we used the correction method suggested by Schmid et al. (2006), which was also used by Praveen et al. (2012) for BC measurements at a rural site in the IGP. Surface $O_3$ concentration was measured continuously with an ozone analyzer (model 49i, Thermo Scientific, USA) which utilizes UV (254 nm wavelength) photometric technology to measure ozone concentration in ambient air. A CO analyzer (model 48i, Thermo Scientific, USA) was used to monitor ambient CO concentration. The ambient air was drawn through a 6 μm pore-size Savillex 47 mm filter at the inlet that removed the particles before sending the air into the CO and $O_3$ analyzers using a Teflon tube. The filters were replaced every 7–10 days depending on particle loading, based on manual inspection. The CO instrument was set to auto-zero at a regular interval of 6 h. Local meteorological parameters (temperature, relative humidity (RH), wind speed (WS), wind direction (WD), precipitation and global solar radiation) were monitored with an AWS (Campbell Scientific, Loughborough, UK), recording data every minute.

## 2.3 Regional chemical transport model

Aerosol and trace gas distributions were simulated using a regional chemical transport model. STEM, a 3-D Eulerian model that has been used extensively in the past to characterize air pollutants in South Asian region (Adhikary

et al., 2010, 2007), was used to understand observations at Lumbini. The Weather Research and Forecasting (WRF) model (Skamarock et al., 2008) version 3.5.1 was used to generate the required meteorological variables necessary for simulating pollutant transport in STEM. The model domain was centered at 24.94° N latitude and 82.55° E longitude, covering a region from 3.390 to 43.308° N latitude and 34.880 to 130.223° E longitude. The model has $425 \times 200$ horizontal grid cells with grid resolution of $25 \, km \times 25 \, km$ and 41 vertical layers with the top of the model set at 50 mbar. The WRF model was run from 1 November 2012 to 30 June 2013. However, for this study, modeled data from April to June 2013 only have been used. The WRF model was initialized with FNL data available from NCAR/UCAR site (http://rda.ucar.edu/datasets/ds083.2/).

The tracer version of STEM provides mass concentration of sulfate, BC (hydrophilic and hydrophobic), organic carbon (OC), sea salt (fine and coarse mode), dust (fine $PM_{2.5}$ and $PM_{10}$), CO (open burning and anthropogenic) and regionally tagged CO tracers. STEM domain size, resolution and projection are those of the WRF model. Details about the tracer version of STEM are outlined elsewhere (Kulkarni et al., 2015; Adhikary et al., 2007). Anthropogenic emission of various pollutants ($CH_4$, CO, $SO_2$, $NO_x$, NMVOC, $NH_3$, $PM_{10}$, $PM_{2.5}$, BC and OC) used in this analysis were taken from EDGAR HTAP v2 (http://edgar.jrc.ec.europa.eu/htap_v2/index.php?SECURE=123) for 2010. Annual emissions given in $kg \, m^{-2} \, sec^{-1}$ at $0.1° \times 0.1°$ resolution were converted to molecules $cm^{-2} \, sec^{-1}$ and re-gridded to $25 \, km \times 25 \, km$ resolution using four-point interpolation techniques available in the STEM emission preprocessor. The emissions were given a diurnal profile using previously used parameterization available in the preprocessor. Open biomass burning emissions on a daily basis during the simulated period were taken from data obtained from the FINN model (Wiedinmyer et al., 2011). As with the WRF model, STEM was run from 2 November 2012 to 30 June 2013 but data presented here are only during the intensive field campaign period.

## 3 Results and discussions

### 3.1 Meteorology

Hourly average time series of various meteorological parameters like precipitation in $mm \, h^{-1}$ (Prec), temperature in °C ($T$), relative humidity in %, WS in $m \, s^{-1}$ and WD in degree during the monitoring period are shown in Fig. 2. Meteorological parameters were obtained with the sensors at the height of $\sim 12 \, m$ from the ground. Meteorology results from WRF model simulations have been used to indicate any significantly different air-mass type present during the measurement campaign after the meteorological observations malfunctioned. Precipitation data were derived from

**Table 1.** Summary of instruments deployed during monitoring in Lumbini.

| Instrument (model) | Manufacturer | Parameters | Inlet/sensor height (above ground) | Sampling interval | Sampled period |
|---|---|---|---|---|---|
| Environmental dust monitor (EDM 164) | Grimm Aerosol Technik, Germany | $PM_1$, $PM_{2.5}$, $PM_{10}$ | 5 m | 5 min | 2 Apr–10 May, 2–13 Jun |
| Aethalometer (AE42) | Magee Scientific, USA | Aerosol light absorption at seven wavelengths and BC concentration | 3 m | 5 min | 1 Apr–5 Jun |
| CO analyzer (48$i$) | Thermo Scientific, USA | CO concentration | 3 m | 1 min | 1 Apr–15 Jun |
| $O_3$ analyzer (49$i$) | Thermo Scientific, USA | $O_3$ concentration | 3 m | 1 min | 1 Apr–15 Jun |
| Automatic weather station (AWS) | Campbell Scientific, UK | $T$, RH, WS, WD, global radiation, precipitation | 12 m | 1 min | 1 Apr–15 Jun |

TRMM satellite (TRMM_3B42_007 at a horizontal resolution of 0.25°) from the Giovanni platform (http://giovanni.gsfc.nasa.gov/giovanni/) as the rain gauge malfunctioned during the sampling period. Precipitation data from TRMM (Fig. 2) show that Lumbini was relatively dry in the early portion of the measurement campaign, while the site did experience some rainfall events as the pre-monsoon edged closer to the monsoon onset. This lowered aerosol loading in the later half of the measurement campaign due to washout and less biomass open burning. Comparison of WRF model outputs with TRMM data shows that the model underpredicts rainfall through out the campaign.

Average observed temperature for the sampling period until the sensor stopped working (on 8 May 2013, i.e., for 38 days of measurement) was 28.1 °C (minimum: 16.5 °C, maximum: 40 °C). Average temperature from the model, during same period, was 31 °C, with values ranging between 19 and 40 °C. As shown in Fig. 2, the model captures the variability of temperature and is mostly within the range of daily values. However, the model has a high bias and does not capture well daily minimum temperature values. The model data were interpolated to match the observation site's latitude, longitude and altitude for all variables discussed in this paper. In addition, the model does not show any large variation in temperature for the campaign period after the sensors stopped working. This insight will be useful to interpret pollution data later on. For the same period (until the sensor stopped working), the average (observed) RH was ∼ 50 % (ranging from 10.5 to 97.5 %) whereas the model showed the average RH to be ∼ 23 % with values ranging between 6 and 78 %. RH values are highly underestimated by the model but, as previously mentioned, the model does not show significant changes in RH during the measurement campaign after the observations stopped working.

Average observed wind speed during the study period was 2.4 m s$^{-1}$, with hourly values ranging between 0.03 and 7.4 m s$^{-1}$, whereas from the WRF model average wind speed was found to be 3.2 m s$^{-1}$ (range: 0.06–11.1 m s$^{-1}$). Diurnal variation of observed hourly average wind speed suggested that wind speeds were lower during nights and mornings while higher wind speed prevailed during daytime, with average winds > 3 m s$^{-1}$ up to ∼ 3.3 m s$^{-1}$ between 09:00 and 13:00 LT (Supplement, Fig. S2, lower panel). High-speed strong winds (> 4 m s$^{-1}$) were from the NW direction during the month of April and later switched to the almost opposite direction, i.e., SE direction, from the month of May onwards. The monthly wind rose plot using the data from both observation and modeling, where the difference in the pattern could be potentially due to the data resolution, is shown in Fig. S3. Comparing modeled wind direction prediction skills at the surface with one point measurement is not sufficient. However, in the absence of other measurements, we also show the comparison of wind direction as an indication of model performance over this region and not as model validation where a more high-resolution modeling and sensitivity analysis of model physics and chemistry may be required. Discrepancy on model results might have occurred due to various factors inherently uncertain in a weather prediction using a model. Additionally, air pollution transport occurs via elevated layers and is not limited to surface winds. We show NCEP/NCAR reanalysis plots at 850 hPa in Fig. S3 to illustrate the distinctly differing wind direction compared to the surface winds seen from observations as well as NCEP/NCAR reanalysis plot at 1000 hPa shown in Fig. 1. There are no upper wind measurement data nearby Lumbini to show model performance. Regardless, we believe that air quality model data are vital for understand-

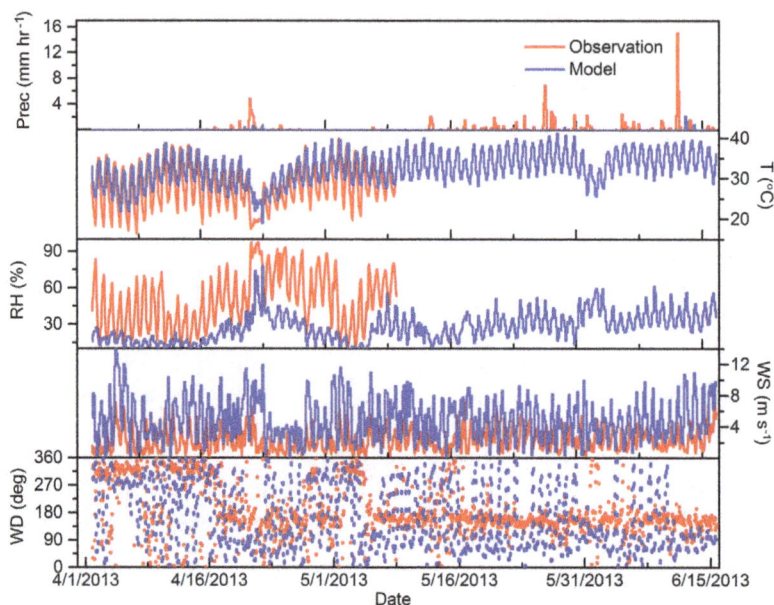

**Figure 2.** Time series of hourly average observed (red) and model estimated (blue) meteorological parameters at Lumbini, Nepal, for the entire sampling period from 1 April to 15 June 2013.

ing pollutant transport in an area where observation data are non-existent or incomplete.

## 3.2 Air quality

### 3.2.1 General overview, PM ratios and influence of meteorology on pollution concentrations

Figure 3 shows hourly averaged time series of observed BC, $PM_1$, $PM_{2.5}$, $PM_{10}$, $O_3$ and CO observed at Lumbini during the study period. Similar temporal behavior was shown by BC, particulate matter fractions ($PM_1$, $PM_{2.5}$ and $PM_{10}$) and CO. The gap in the figure (for PM time series) is due to the power interruption to the instrument. BC concentrations during the measurement period ranged between 0.3 and 29.9 $\mu g\,m^{-3}$ with a mean ($\pm$SD) value of 4.9 ($\pm$3.8) $\mu g\,m^{-3}$. BC concentrations in Lumbini during pre-monsoon months are lower compared to BC concentrations observed in the Kathmandu Valley because of high number of vehicles plying on the street, brick kilns and other industries in Kathmandu Valley (Sharma et al., 2012; Putero et al., 2015). The lowest concentration was observed during a rainy day (21–22 April) whereas the highest concentration was observed during a period of forest fire (detailed in Sect. 3.3). For the entire measurement period, we found average (of hourly average values) $PM_1$ of $35.8 \pm 25.6$ $\mu g\,m^{-3}$ (minimum–maximum range: 3.6–197.6 $\mu g\,m^{-3}$), $PM_{2.5}$ of $53.1 \pm 35.1$ $\mu g\,m^{-3}$ (6.1–272.2 $\mu g\,m^{-3}$), $PM_{10}$ of $128.9 \pm 91.9$ $\mu g\,m^{-3}$ (10.5–603.9 $\mu g\,m^{-3}$) and coarse-mode fraction of $75.7 \pm 61.7$ $\mu g\,m^{-3}$ (1.9–331.8 $\mu g\,m^{-3}$). The coarse-mode ($PM_{10-2.5}$) fraction was $\sim$ 60 % of the $PM_{10}$. The share of coarse-mode aerosol to $PM_{10}$ in Lumbini was higher than

that observed in other sites in the IGP, such as Guwahiti, India (42 %) (Tiwari et al., 2017), and Dibrugarh, India (9–16 %) (Pathak et al., 2013), both in eastern IGP and Delhi (38 %) (Tiwari et al., 2015) in western IGP, indicating the higher contribution of coarse aerosols in Lumbini, likely lifted from soils from nearby agricultural fields and construction materials by stronger winds during pre-monsoon season. Values of coarse-mode fraction, similar to Lumbini, have been reported by Misra et al. (2014) at Kanpur for dust dominated and mixed aerosols events.

The share of BC in PM fractions was found to be $\sim$ 13 % in $PM_1$, 9 % in $PM_{2.5}$ and $\sim$ 4 % in $PM_{10}$ but the correlation coefficients of BC with three PM fractions were found to be 0.89 ($PM_1$), 0.88 ($PM_{2.5}$) and 0.69 ($PM_{10}$), indicating the commonality in the sources of these pollutants. The contribution of BC in $PM_1$ was found to be $\sim$ 12 % in Kanpur during February–March (Kumar et al., 2016a), similar to Lumbini. Regarding the share of BC in $PM_{10}$, the share observed in Lumbini ($\sim$ 4 %) was similar to that observed over Varanasi ($\sim$ 340 km south of our site) in central IGP (5 %) (Tiwari et al., 2016) and Dibrugarh in eastern IGP ($\sim$ 5 %) (Pathak et al., 2013). Thus our results indicate that despite our station being located at the northern edge of the IGP along the foothills of the Himalayan range, the share of BC in PM is similar to that found in heavily polluted sites in the central and eastern IGP.

In Lumbini, the average (hourly) share of $PM_1$ in $PM_{2.5}$, $PM_1$ in $PM_{10}$ and $PM_{2.5}$ in $PM_{10}$ was found to be $\sim$ 70, 34 and 47 %, respectively. Regarding other sites in IGP region, $PM_{2.5}$ / $PM_{10}$ ratios were reported to be 56 % in Kanpur (Snider et al., 2016), 60 % in Varanasi (Kumar et al., 2015),

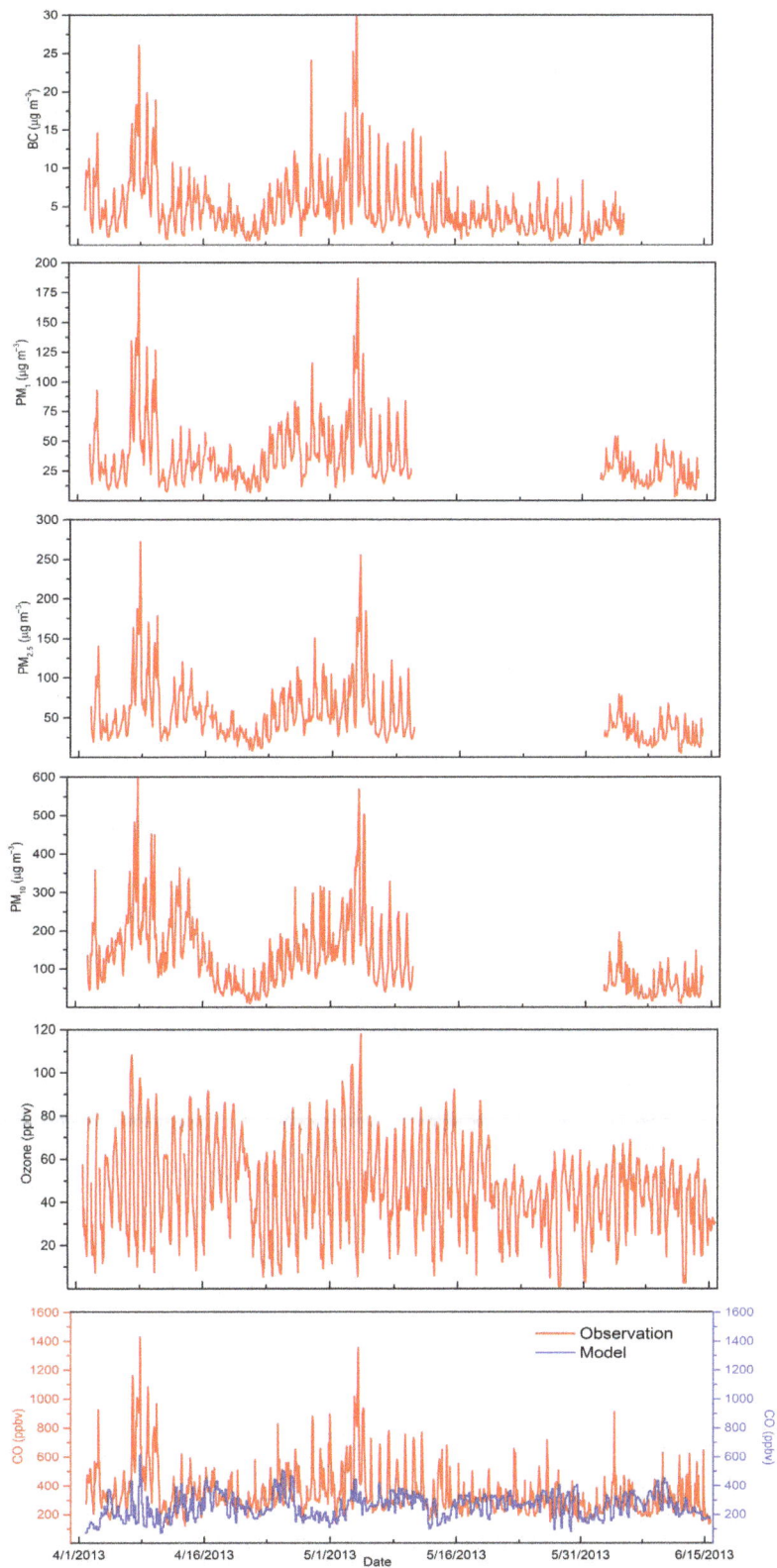

**Figure 3.** Time series of the observed (red line) and model estimated (blue line) hourly average concentrations of BC, $PM_1$, $PM_{2.5}$, $PM_{10}$, $O_3$ and CO at Lumbini, Nepal, for the entire sampling period from 1 April to 15 June 2013.

57 % in Guwahiti (Tiwari et al., 2017), 90 % in Dribugarh (Pathak et al., 2013) and 62 % in Delhi (Tiwari et al., 2015), indicating local differences within IGP as well as suggesting that the influence of combustion sources at Lumbini is still lower compared to other locations in the Indian section of the IGP. A recent study (Putero et al., 2015) reported that the $PM_1 / PM_{10}$ during pre-monsoon of 2013 was found to be 0.39 in the Kathmandu Valley of Nepal. Lumbini has significantly lower vehicle emissions and population than the Kathmandu Valley yet the ratios are similar, indicating the importance of regional combustion sources in Lumbini for finer aerosols ($PM_1$) and soil-based emissions such as road dust in the Kathmandu Valley. Future studies will need to explore the emission sources around Lumbini in much greater detail. Lower $PM_{2.5} / PM_{10}$ in Lumbini as compared to other regions mentioned earlier could be due to emissions from cement industries located within 15 km distance from the measurement site.

The observed 24 h average particulate matter concentrations ($PM_{2.5}$ and $PM_{10}$) were found frequently higher than the WHO prescribed guidelines for $PM_{2.5}$ ($25\,\mu g\,m^{-3}$) and $PM_{10}$ ($50\,\mu g\,m^{-3}$) with $PM_{2.5}$ exceeding 94 % and $PM_{10}$ of 85 % of the measurement period of 53 days in Lumbini.

Observed CO concentrations ranged between 124.9 and 1429.7 ppbv with an average value of $344.1 \pm 160.3$ ppbv. CO concentration observed in Lumbini is lower than that of Mohali, western India, where the average concentration was 566.7 ppbv during pre-monsoon season due to intense biomass and agro-residue burning over the region (Sinha et al., 2014). Temporal variation of CO concentrations is similar to that of BC, exhibiting very strong correlation ($r = 0.9$). Past studies have shown that the ratio of BC to CO depends upon multiple factors like site location, combustion characteristics (fuel and technology) at the sources and type of air mass (Girach et al., 2014; Pan et al., 2011; Zhou et al., 2009). Formation of the soot depends on the carbon-to-oxygen ratio of fuel whereas CO can also be produced naturally due to the oxidation of volatile organic compounds (Girach et al., 2014). Figure 4 shows the comparison of the average $\Delta BC/\Delta CO$ ratio (0.021) at Lumbini with that obtained from other sites. Please refer to Fig. S4 in the Supplement for the time series of $\Delta BC/\Delta CO$ ratio observed in Lumbini. We used the method described by Pan et al. (2011) to calculate the $\Delta BC/\Delta CO$ values. The ratio was calculated using the equation $(BC - BC_0)/(CO - CO_0)$, assuming the background values ($BC_0$ or $CO_0$) as 1.25 percentile of the data. The $\Delta BC/\Delta CO$ ratio in Lumbini is similar to that obtained at a suburban site, Pantnagar, in India (0.017) (Joshi et al., 2016) and in Maldives (0.017) (Dickerson et al., 2002), indicating the possibility of similar types of emission sources. However, the lower $\Delta BC/\Delta CO$ ratio obtained over megacities such as Beijing and Shanghai is due to the higher number of gasoline and diesel vehicles (Zhou et al., 2009). The ratios obtained at Lumbini are within the range of emission ratios from diesel used in transport sector

(0.0013–0.055), coal (0.0019–0.0572) and biofuels (0.0087–0.0266) for domestic activities (Verma et al., 2010, and references therein), implying that BC and CO observed are from mixed sources.

The hourly averaged observed ozone concentration ranged between 1.0 and 118.1 ppbv with a mean value of $46.6 \pm 20.3$ ppbv during the sampling period. The 8 h maximum $O_3$ concentration exceeded WHO guidelines (of $100\,\mu g\,m^{-3}$; (WHO, 2006) during $\sim 90$ % of the measurement period. Our results clearly indicate that the current pollution levels in Lumbini are of great concern to the health of the people living in the region, including over a million visitors who visit Lumbini, and agro-ecosystems.

The relationship of wind speed to aerosol and gaseous pollutants in Lumbini is shown in Fig. S5 (Supplement). We were interested in studying the relationship between wind speed and the pollutants since the wind governs the horizontal dilution of the pollutants (Huang et al., 2012) and also the likelihood of lifting soil dust. Except ozone, all other pollutants exhibited negative correlation with wind speed. BC shows negative correlation ($r = -0.42$, $P > 0.05$) with wind speed, which is similar with other pollutants as well (as can be seen from the figure). Past studies have also reported a similar negative correlation of BC with wind speed over urban and sub-urban areas (Huang et al., 2012; Cao et al., 2009; Ramachandran and Rajesh, 2007; Sharma et al., 2002; Tiwari et al., 2013), indicating that the locally generated BC can accumulate in the atmosphere during lower wind speed conditions (Cao et al., 2009). Tiwari et al. (2013) also reported similar negative correlation ($r = -0.45$) during the pre-monsoon season over Delhi. In contrast, secondary pollutants like ozone exhibited a positive relation to the WS ($r = 0.38$, $P > 0.05$), indicating the location of precursor emission sources at some distance away from the measurement site. Solar radiation is one of the most important factors for production of ozone in the atmosphere (Naja et al., 2003). The correlation of hourly ozone concentration with solar radiation (not shown here) was found to be 0.41 ($P > 0.05$), whereas wind speed during the daytime only (06:00–18:00) showed very weak correlation of 0.02 (nonsignificant) with ozone, possibly indicating transport of precursors during nighttime.

Interestingly, the highest concentrations of all measured pollutants were obtained when the wind speed was less than $1\,m\,s^{-1}$. In a separate analysis (not shown here), we considered only the $WS > 1\,m\,s^{-1}$ and calculated the correlation coefficients to investigate the influence of regional emissions. We found the similar correlation values as previous when all WS values were considered (BC vs. WS $= -0.41$, CO vs. WS $= -0.42$, $O_3$ vs. WS $= 0.29$, $PM_1$ vs. WS $= -0.40$, $PM_{2.5}$ vs. WS $= -0.38$, $PM_{10}$ vs. WS $= -0.33$; all at $P > 0.05$). The correlation of WS ($> 1\,m\,s^{-1}$) with concentration of air pollutants indicates that air pollution over Lumbini is not only of local origin but also from other nearby regions.

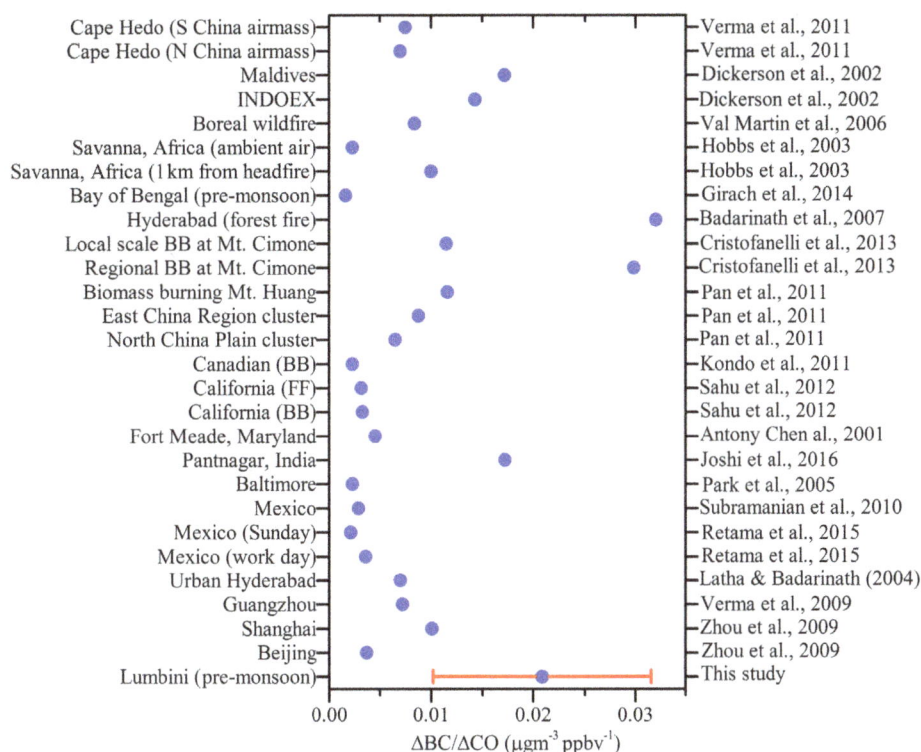

**Figure 4.** Comparison of black carbon concentrations to CO concentrations ($\Delta BC/\Delta CO$) ratios obtained for Lumbini with other sites. The red horizontal bar represents SD.

Past studies near this site have been focused on cities like Kathmandu (Sharma et al., 2012; Panday and Prinn, 2009; Putero et al., 2015) and Kanpur (Ram et al., 2010) and regions of IGP dominated by agro-residue burning (Rastogi et al., 2016; Sinha et al., 2014; Sarkar et al., 2013), all of which reported very high levels of pollution. Our study adds to the growing list of scientific observations in the IGP by providing data from the foothills of the central Himalayas. Very high aerosol loading is observed in South Asia during pre-monsoon, mostly over the IGP region (Supplement, Fig. S6). As this is the first study over an IGP site located in Nepal, pollution concentrations observed at Lumbini were compared with other sites in the region (Table 2). Different sites located at urban, semi-urban and remote locations were used for comparison to get a clear comparative picture of the situation at Lumbini amongst other locations in the region. Pre-monsoon seasonal average PM$_{2.5}$ concentration in Lumbini has been found to be lower than in the megacity Delhi (Bisht et al., 2015) and northwestern IGP (Sinha et al., 2014), possibly due to higher levels of emissions (from traffic and biomass burning, respectively) over those regions. In addition, average BC and CO concentrations in Lumbini fell between concentrations observed at rural sites (up to 6 times higher) and cities in the region (see Table 2), indicating that Lumbini, in a way, can still be considered a semi-urban location. The hourly average O$_3$ concentrations in Lumbini

were found to be higher than in cities like Kathmandu (Putero et al., 2015) and Kanpur during the pre-monsoon season (Gaur et al., 2014). However from a mesoscale perspective, the hourly average O$_3$ concentrations were lower at Lumbini as compared to the base camp of Mount Everest region due to the uplift of polluted air masses (Marinoni et al., 2013), stratospheric intrusion (Cristofanelli et al., 2010) and even the regional or long-range transport of the air pollutants (Bonasoni et al., 2010) to the high-altitude site.

Regarding the monthly average concentration, the concentrations of all measured pollutants decreased as the pre-monsoon months advanced. The monthly average concentrations of the monitored species are shown in Fig. S7 along with the monthly fire hotspots over the region. Reduction in concentration (except PM) during the month of May (as compared to April) could be attributed to the fewer fire events during May as well as previously discussed washout by rainfall. Two peak pollution episodes observed during the first half of April and May are discussed in more detail in the next section.

### 3.2.2 Observation–model intercomparison

Chemical transport models provide insight to observed phenomena; however, interpretation has to take into account model performance before arriving at any conclusion. This section describes pollution concentrations simulated by the

**Table 2.** Comparison of $PM_{2.5}$, BC, CO and $O_3$ concentrations at Lumbini with those at other sites in South Asia.

| Sites | Characteristics | Measurement period | $PM_{2.5}$ ($\mu g\,m^{-3}$) | BC ($\mu g\,m^{-3}$) | CO (ppbv) | $O_3$ (ppbv) | References |
|---|---|---|---|---|---|---|---|
| Lumbini, Nepal | Semi-urban | Pre-monsoon, 2013 | $53.1 \pm 35.1$ | $4.9 \pm 3.8$ | $344.1 \pm 160.3$ | $46.6 \pm 20.3$ | This study |
| Kathmandu, Nepal | Urban | Pre-monsoon, 2013 | – | $14.5 \pm 10$ | – | $38.0 \pm 25.6$ | Putero et al. (2015) |
| Mount Everest, Nepal | Remote | Pre-monsoon | – | $0.4 \pm 0.4$ | – | $61.3 \pm 7.7$ | Marinoni et al. (2013) |
| Delhi, India | Urban | Pre-monsoon (nighttime) | $82.3 \pm 50.5$ | $7.70 \pm 7.25$ | $1800 \pm 890$ | – | Bisht et al. (2015) |
| Kanpur, India | Urban | June 2009–May 2013, April–Jun | – | $2.1 \pm 0.9$ | $721 \pm 403$ | $27.9 \pm 17.8$ | Gaur et al. (2014) Ram et al. (2010) |
| Mohali, India | Semi-urban | May 2012 | $104 \pm 80.3$ | – | $566.7 \pm 239.2$ | $57.8 \pm 25.4$ | Sinha et al. (2014) |
| Mount Abu, India | Remote | January 1993–December 2000, pre-monsoon | – | $0.7 \pm 0.14$ | $131 \pm 36$ | $39.9 \pm 10.8$ | Naja et al. (2003) Das and Jayaraman (2011) |

WRF-STEM model. A comparison of model calculated pollutant concentration along with the minimum and maximum concentrations of various pollutants (with observation) is shown in Table 3. The model-based concentrations used here are values outputted for every third hour of the day (actual computation is carried out every 15 min). BC concentrations ranged between 0.4 and $3.7\,\mu g\,m^{-3}$ with a mean value of $1.8 \pm 0.7\,\mu g\,m^{-3}$ for the period of 1 April–15 June 2013. The average model BC concentration was $\sim 2.7$ times lower than the observed BC. Regarding $PM_1$, $PM_{2.5}$ and $PM_{10}$, the model-simulated average concentration was $12.3 \pm 5.5$ (0.9–41.7), $17.3 \pm 6.7$ (1.9–48.3) and $25.4 \pm 12.9$ (2.1–68.8) $\mu g\,m^{-3}$, respectively. The model estimated values were lower by a factor of 3 and 5, respectively, than the observed concentrations. The data show that the model needs much improvement in its ability to adequately predict observed aerosol characteristics at Lumbini given the input provided, e.g., emissions data. Since pollutant concentration is a function of emissions, transport, transformation and deposition, improvements in any of these areas would improve the model performance for this site. However, given observation insights by PM ratios, it seems that improvements are much needed in the emissions of primary aerosols. Current emissions (2010) do not account for trash burning, roadside dust and increasingly newer industries, especially emissions from cement factories that have popped up in recent years. We show sensitivity with emissions in a later section (Sect. 3.3.2) in the vicinity of Lumbini, but emission improvements are needed beyond Lumbini, which is outside the scope of this paper.

Average observed CO concentration was $255.7 \pm 83.5$ ppbv, ranging between 72.2 and 613.1 ppbv, with average model CO $\sim 1.35$ times lower than observed. Time series comparison of modeled CO vs. observation is shown in Fig. 3. Apart from two peak episodes the model does a better job in predicting CO concentration over Lumbini. A previous study using STEM over Kathmandu Valley showed that the model was able to capture the annual BC mean value but completely missed the concentrations during pre-monsoon and post-monsoon period (Adhikary et al.,

2007). Similar behavior is seen this time for CO where the model misses the peak values but reasonably captures CO concentration after mid-May when no biomass burning events are observed (model to observation ratio improves to 1.16). STEM CO performance can be significantly improved via better constraining emissions of open biomass burning as discussed in Sect. 3.3. This activity is beyond the scope of this current paper although improvements are underway for all these sectors.

### 3.2.3 Diurnal variations of air pollutants and boundary layer height

In the emission source region, diurnal variations of primary pollutants provide information about the time-dependent emission activities (Kumar et al., 2016b). Figure 5 shows the diurnal variation of hourly averaged concentrations of measured pollutants during the sampling period. Primary pollutants like BC, PM and CO showed typical characteristics of an urban environment, i.e., diurnal variation with a morning and an evening peak. However, Lumbini data show higher concentrations in the evenings compared to morning hours. Elevated concentrations can be linked to morning and evening cooking hours for BC and CO. Emission inventories for the region show that the residential sector has significant contribution to BC and CO. However, an explanation for the elevated evening concentration compared to the morning is needed. Increase in the boundary layer height, reduction in the traffic density on the roads, absence of cooking activities during mid-day and increase in wind speed often contribute to the dispersion of pollutants, resulting in lower concentrations during the afternoon. Diurnal variation of wind direction (Supplement, Fig. S2, upper panel) shows the dominance of wind coming from the south (mainly during the month of May until mid-June). Morning and evening periods experienced the winds coming from the southeastern direction while the winds were predominantly from the southwestern direction during late afternoon. The increase in CO concentrations in the evening hours might be due to transport of CO from source regions upwind of Lumbini, which, along

**Table 3.** Intercomparison of observed and model-simulated hourly average concentrations of air pollutants during the measurement campaign period. Units: BC and PM in $\mu g\,m^{-3}$ and CO in ppbv.

| Pollutants | Observed (mean and range) | Modeled (mean and range) | Ratio of mean (observed/modeled) |
|---|---|---|---|
| BC | 4.9 (0.3–29.9) | 1.8 (0.4–3.7) | 2.7 |
| $PM_1$ | 36.6 (3.6–197.6) | 12.3 (0.9–41.7) | 3 |
| $PM_{2.5}$ | 53.1 (6.1–272.2) | 17.3 (1.9–48.3) | 3 |
| $PM_{10}$ | 128.8 (10.5–604.0) | 25.4 (2.1–68.8) | 5 |
| CO | 344.1 (124.9–1429.7) | 255.7 (72.2–613.1) | 1.35 |

with the local emissions, get trapped under reduced planetary boundary layer (PBL) heights. Ozone concentration was lowest in the morning before the sunrise and highest in late afternoon around 15:00 after which concentrations started declining, exhibiting a typical characteristic of a polluted urban site. Photo-dissociation of accumulated $NO_x$ reservoirs (like HONO) provides sufficient NO concentration, leading to the titration of $O_3$ and resulting in minimum $O_3$ just before sunrise (Kumar et al., 2016b). The PBL height (in meters) was obtained from the WRF model as observations were not available. The study period's average PBL height over Lumbini was $\sim 910$ m (ranging between 24 and 3807 m, observed at 06:00 and 15:00, respectively). The daily average PBL height obtained from the model is compared with published values (Wan et al., 2017), as shown in Fig. 6, which indicate that the value is captured by our model during initial measurement period and overestimated in the months of mid-May onwards. As the pre-monsoon month advances, PBL height also increased. The monthly average PBL height was 799, 956 and 1014 m, respectively, during the month of April, May and (1–15) June. As presented in the figure, the monthly average diurnal variation also showed that the boundary layer height was at its maximum at 15:00 LT during each month, which coincides with the period of lowest concentration of the pollutants.

### 3.3 Influence of forest fires on Lumbini air quality

#### 3.3.1 Identification of large-scale forest fire influence using in situ observations, satellite and model data

Forest fires and agricultural biomass burning (mostly agro-residue burning on a large scale) are common over South Asia and the IGP region during pre-monsoon season. The northern Indo-Gangetic region is characterized by fires even during the monsoon and post-monsoon season (Kumar et al., 2016b; Putero et al., 2014). These activities not only influence air quality over nearby regions but also get transported towards high-elevation pristine environments like Mount Everest (Putero et al., 2014) and Tibet (Cong et al., 2015a,b). Thus, one of the main objectives of this study was to identify the influence of open burning on Lumbini air quality. Average wind speed during the whole measurement period was $2.4\,m\,s^{-1}$. Based on these data, open fire counts within the grid size of $200 \times 200$ km centering over Lumbini were used for this analysis, assuming that the emissions would take a maximum period of 1 day to reach our monitoring site. Forest fire counts were obtained from the MODIS satellite data product Fire Information for Resource Management System (FIRMS). Figure 7 shows the daily average $\Delta BC/\Delta CO$ ratio, aerosol absorption Ångström exponent (AAE), which is derived from Aethalometer data (by calculating the negative slope of absorption at 370 and 950 nm vs. wavelength in log-log plot), and daily open fire counts within the specified grid. The green box in the figure is used to show two peak events (presented earlier in Fig. 3) with the elevated BC and CO concentrations observed during the monitoring period. The first peak was observed during 7–9 April and second peak during 3–4 May 2013. Two pollutants having biomass burning as the potential primary source, BC and CO, were taken in consideration. High AAE values during these two events are also an indication of the presence of BC of biomass burning origin (Praveen et al., 2012; Bergstrom et al., 2007; Kirchstetter et al., 2004), with the value being $\sim 1.6$ for Lumbini. The chemical composition of TSP filter samples collected at Lumbini also showed higher concentration of Levoglucosan, a biomass burning tracer in Lumbini during the pre-monsoon season, compared to other seasons of the year (Wan et al., 2017). Wan et al. (2017) also reported that the higher correlation of $K^+$ with $Ca^{2+}$ and $Mg^{2+}$ indicates that dust is the main source of potassium in Lumbini.

Contrary to our expectation, we could not observe any significant influence of forest fire within the specified grid of $200 \times 200$ km (the influence of local forest fire on the air quality over Lumbini was not observed). Therefore, a wider area, covering South and Southeast Asia, was selected for the forest fire count. Figure 8a and b show the active fire hotspots from MODIS over the region during the peak events, indicating the first peak could have occurred due to the forest fire over the eastern India region whereas the second peak was influenced by the forest fire over the western IGP. Moreover, in order to strengthen our hypothesis, we have utilized satellite data products for various gaseous pollutants like CO and $NO_2$ (Atmospheric Infrared Sounder (AIRS) for CO and

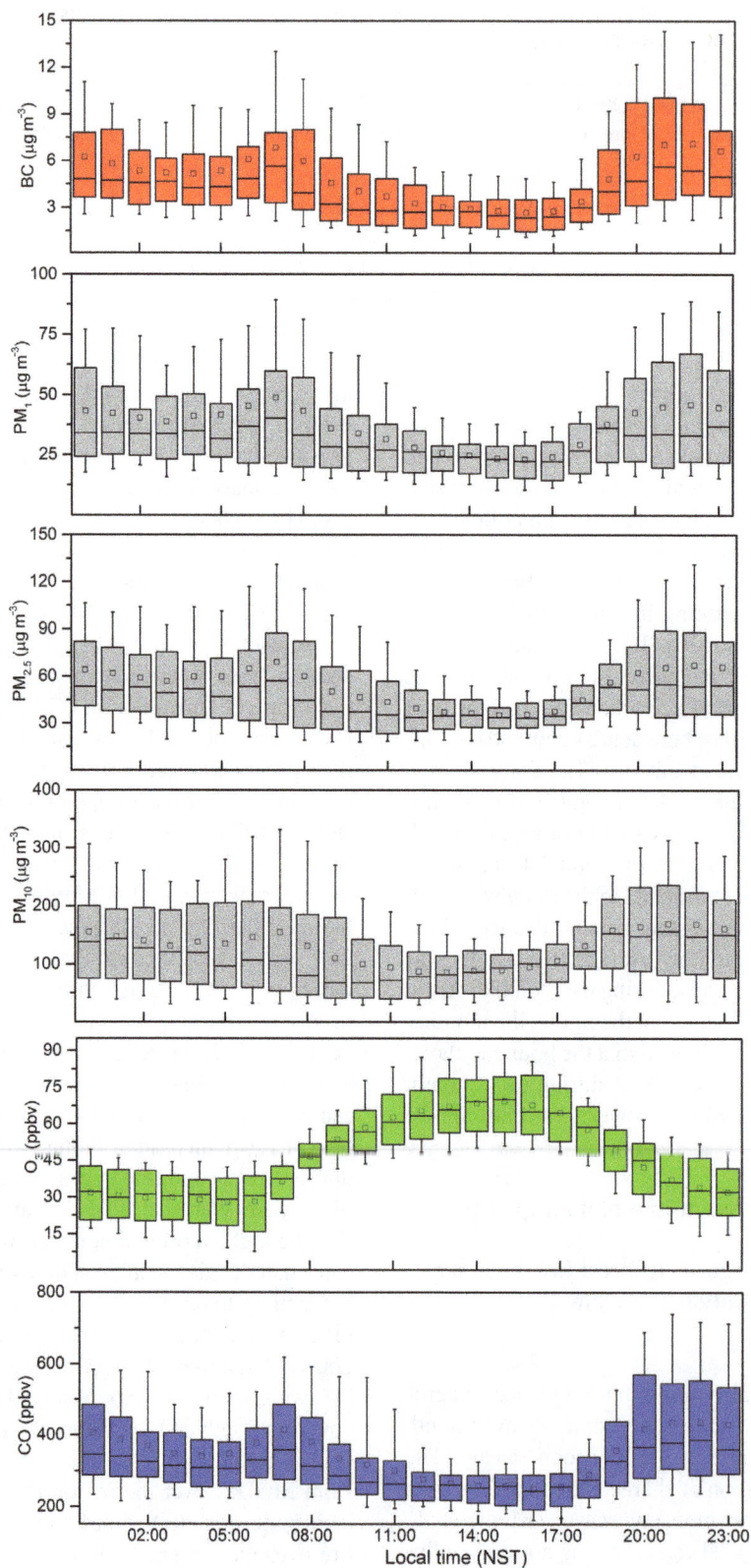

**Figure 5.** Diurnal variations of hourly average ambient concentrations of BC, $PM_1$, $PM_{2.5}$, $PM_{10}$, $O_3$ and CO at Lumbini during the monitoring period (1 April–15 June 2013). In each box, the lower and upper boundary of the box represent the 25th and 75th percentile, respectively; the top and bottom of the whisker represent the 90th and 10th percentile, respectively; the mid-line represents the median; and the square mark represents the mean for each hour.

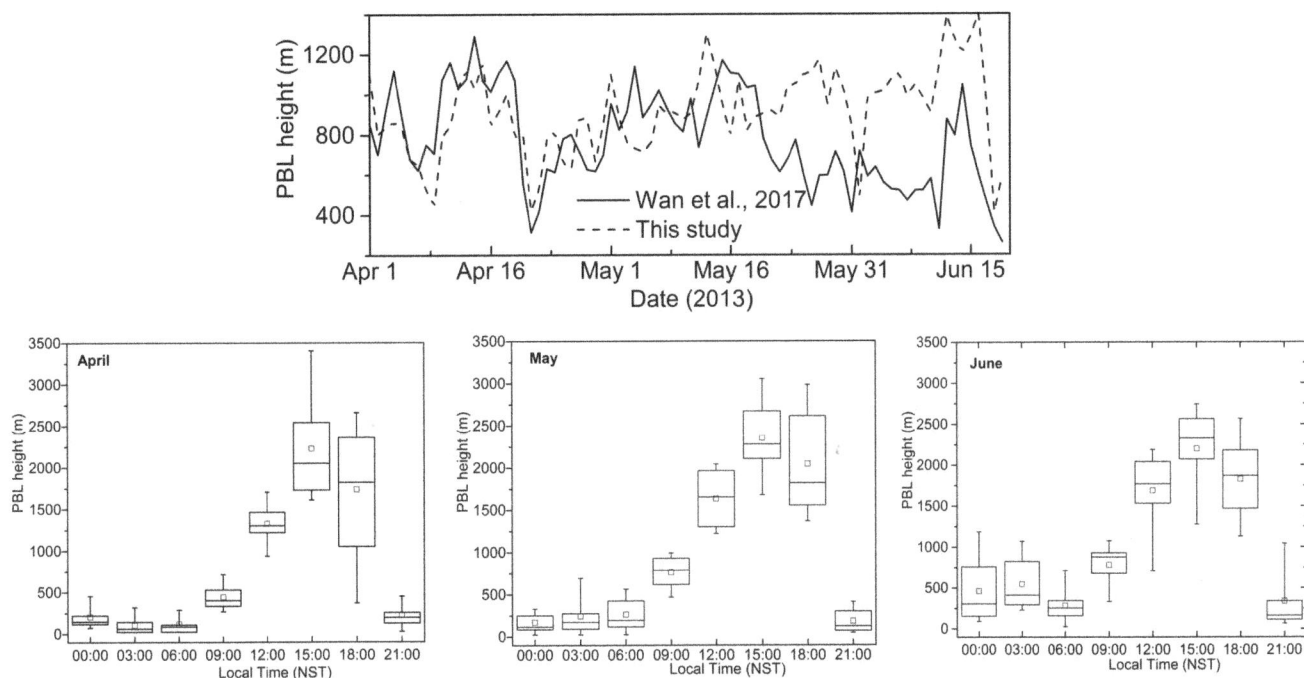

**Figure 6.** Daily time series of PBL height obtained from the model and reported values over Lumbini (obtained from Wan et al., 2017). The lower panel shows the monthly average diurnal variation of the PBL height. The square mark in each box represents the mean PBL height, the bottom and top of the box represent the 25th and 75th percentile and the top and bottom of the whisker represent the 90th and 10th percentile, respectively.

Ozone Monitoring Instrument (OMI) for $NO_2$, both obtained from Giovanni platform). Figure 8c–h show the daytime total column CO before, during and after occurrence of two events (peaks), as stated earlier. The AIRS satellite with daily temporal resolution and $1° \times 1°$ spatial resolution has been utilized to understand the CO concentration over the area. CO concentration over Lumbini during both of the peaks confirmed the role of open fires over the IGP region for elevated concentration of CO in Lumbini. To further strengthen our finding, HYSPLIT back-trajectory plots were used. Figure 8i and j represent the 6-hourly back trajectories for these two events only. However, the back trajectories (during both events) indicated that the air mass passed over the fire events in the northwestern IGP. We note that using back trajectories to identify source regions is also uncertain, as noted by Jaffe et al. (1997). Figure 8k shows model biomass CO peak coincident with observed CO. Although the magnitudes are significantly different, the timing of the peaks is captured well by the model. This, we believe, is due to the fact that satellite-based open fire detection has limitations because it cannot capture numerous small fires that are prevalent over South Asia but usually burn out before the next satellite overpass. More research is needed to assess the influence of these small fires on regional air quality.

In a separate analysis (not shown here), elevated $O_3$ concentrations during these two events were also observed. The average $O_3$ concentration before, during and after the events was found to be $46.2 \pm 20.3$, $53.5 \pm 31.1$ and $50.3 \pm 20.9$ ppbv, respectively (Event I), and $54.8 \pm 23.8$, $56.7 \pm 35$ and $55.6 \pm 13.4$ ppbv, respectively (Event II). Average ozone concentration outside these events was found to be $46 \pm 19$ ppbv. Increased ozone concentrations during the high peak events have been analyzed using the satellite $NO_2$ concentration over the region and considering the role of $NO_2$ as a precursor for ozone formation. Daily total column $NO_2$ was obtained from the OMI satellite (data available at the Giovanni platform; http://giovanni.gsfc.nasa.gov/giovanni/) at the spatial resolution of $0.25° \times 0.25°$. Figure 9 shows the $NO_2$ column value before, during and after both events. Even for the $NO_2$, maximum concentrations were observed during these two special events. It is likely that both local and regional pollution (transported from NW IGP region as indicated by synoptic wind in Fig. S8, Supplement) contributed to the elevated ozone levels. This remains a question to be investigated in future.

### 3.3.2 Identifying regional and local contribution

The WRF-STEM model has been used to identify the anthropogenic emission source region influencing the air quality over Lumbini. As previously explained, the model is able to capture the observed CO concentration when intense open burning events were not present. A recent study (Kulkarni et al., 2015) has explored the source region contribution of

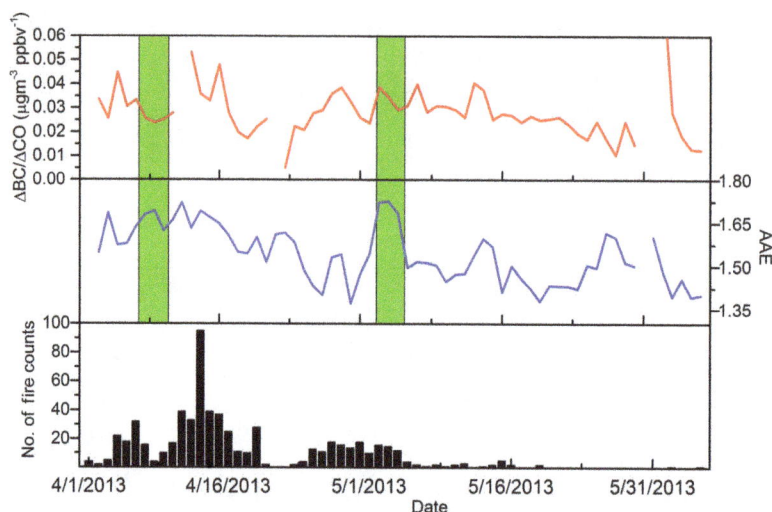

**Figure 7.** Time series of daily average $\Delta BC/\Delta CO$ ratio, absorption Ångström exponent (AAE) and fire counts acquired with the MODIS instrument on board the TERRA satellite for a $200 \times 200$ km grid centered at Lumbini. Two rectangular green boxes represent time of two episodes with high peaks in CO and BC concentrations as shown in Fig. 3.

various pollutants over central Asia using similar technique. Figure 10a shows the average contribution from different regions to CO concentration over Lumbini during the whole measurement period. The major share of CO was from the Ganges Valley (46 %), followed by the Nepal region (25 %) and the rest of India (~ 17.5 %). Contributions from other South Asian countries like Bangladesh and Pakistan were ~ 11 % whereas China contributed ~ 1 % of the CO concentration in Lumbini. Regarding the monthly average contribution, Ganges Valley's and Nepal's contributions were almost equal during the month of April (~ 34 and ~ 37 % respectively) but increased for the Ganges Valley during the month of May (~ 44 %) and were reduced for Nepal (~ 25 %) (Fig. S9, Supplement).

Figure 10b is the time series of percentage contribution to total CO concentration during the whole measurement period, showing different air masses arriving at 3-hourly intervals. During the whole measurement period, the majority of the CO reaching Lumbini was from the Ganges Valley (mainly the states of Punjab, Haryana, Uttar Pradesh, Bihar and West Bengal) region with the contribution sometimes reaching up to ~ 80 %. Other Indian (central, south, east and north) regions also contributed significantly. Bangladesh's contribution of CO loading was seen only after mid-April, lasting for only about a week and after the first week of May. The contribution from Bangladesh was sporadic compared to other regions. The highest contribution from Bangladesh was observed after the first week of June with the arrival of monsoonal air mass. Pakistan also contributed to the CO loading significantly. Others region as mentioned in the figure covered the regions like Afghanistan, the Middle East, western Asia, East Asia, Africa and Bhutan. Contributions from these regions were less than 5 %. Contribution from China was not

evident until the first week of June where a specific air-mass arrival shows contribution reaching up to 25 % of total CO loading.

A sensitivity analysis was performed for emission uncertainty in the model grid containing Lumbini. Lumbini and surrounding regions in the recent years have seen significant rise in urban activities and industrial activity and related emissions which may not be accurately reflected in the HTAP v2 emissions inventory. A month-long simulation was carried out with emissions from Lumbini and by switching off the four grids surrounding Lumbini, and another simulation with Lumbini and the surrounding four grids' emissions increased by 5 times the amount of HTAP v2 emissions inventory. The results are shown in Fig. 10c as percentage increase or decrease compared to model results using the current HTAP v2 emissions inventory. The black line shows the concentration as 100 % for the current HTAP v2 emissions inventory. Despite making Lumbini and the surrounding grids emissions zero, model calculation shows pollutant concentration on average is still about 78 % of the original value, indicating dominance of the background and regional sources compared to the local source in the model. Increasing emissions 5 times for Lumbini and surrounding four grids only increases the concentration on average by 151 %. Thus uncertainty in emissions is not a local uncertainty for Lumbini but rather for the whole region, which needs to be better understood for improving model performance against observations at Lumbini.

### 3.4 Does fossil fuel or biomass influence Lumbini air?

The aerosol spectral absorption is used to gain insight into nature and potential source of black carbon. This method

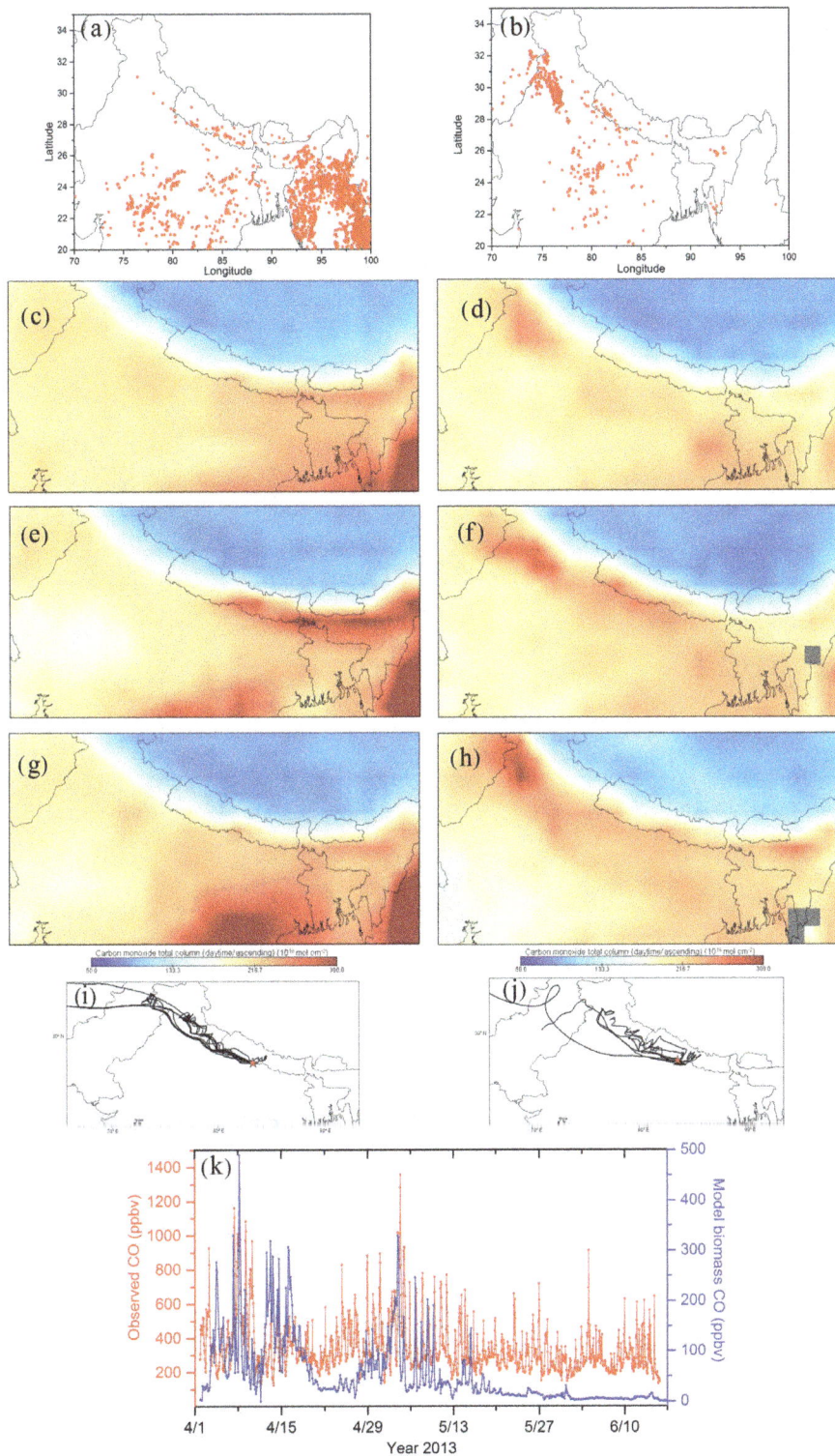

**Figure 8.** Active fire hotspots in the region acquired with the MODIS instrument on Aqua satellite during **(a)** Event I (7–9 April) and **(b)** Event II (3–4 May). CO emissions, acquired with AIRS satellite, in the region 2 days before (3–5 April), during (7–9 April) and 2 days after (10–12 April) the Event I are shown in panels **(c, e, g)**, respectively, while panels **(d, f, h)** show CO emissions 2 days before (1–2 May), during (3–4 May) and 2 days after (5–6 May) the Event II. Panels **(i, j)** represent the 6 h interval HYSPLIT back trajectories during Event I and II, respectively. Location of the Lumbini site is indicated by the red star in the panels **(i, j)**. Observed CO vs. model open burning CO illustrating the contribution of forest fires during peak CO loading is shown in panel **(k)**.

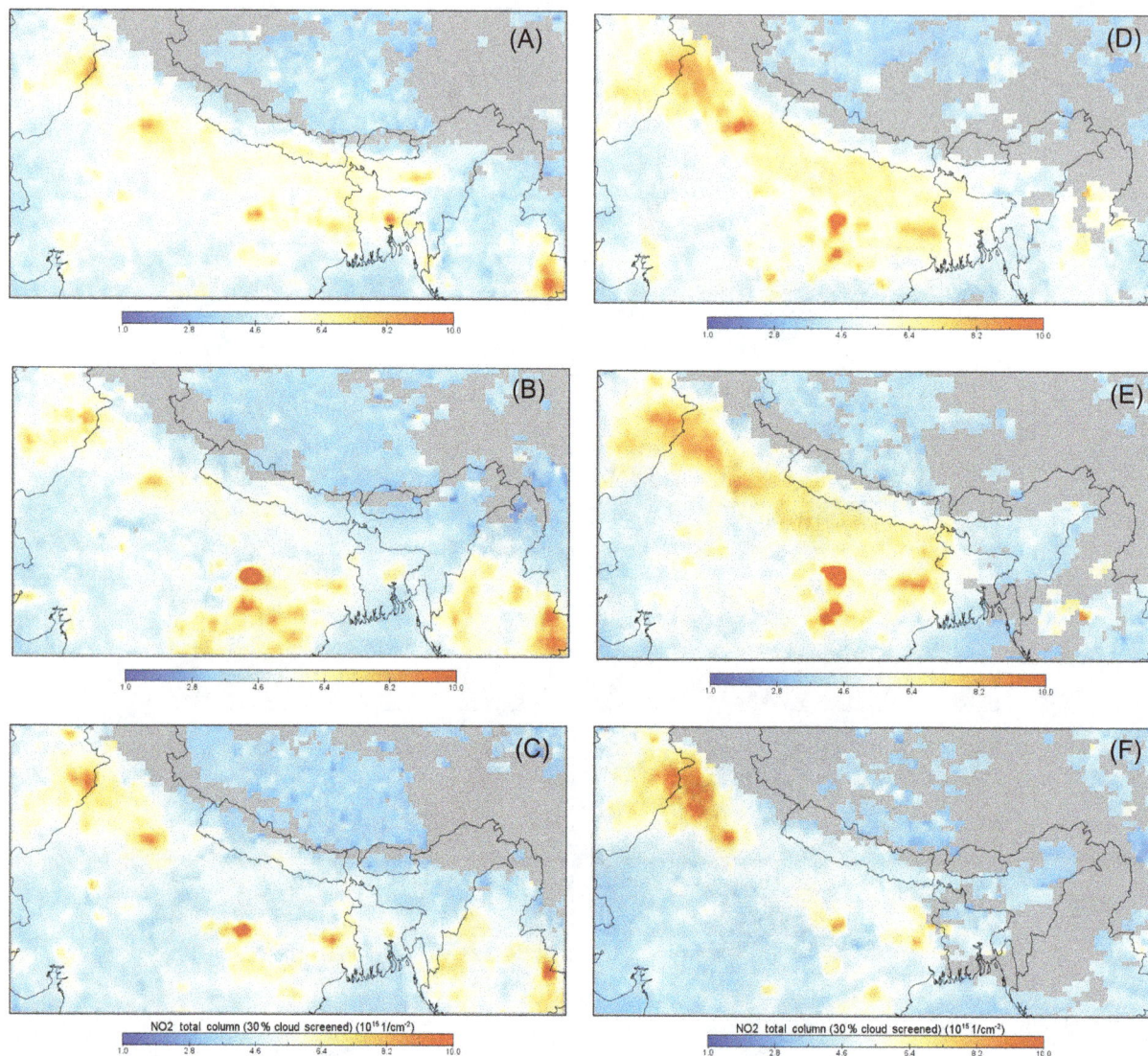

**Figure 9.** NO$_2$ total column obtained with OMI satellite over the region **(a)** before, **(b)** during and **(c)** after Event I. The panels **(d–f)** show NO$_2$ total column before, during and after Event II.

enables to analyze the contributions of fossil fuel combustion and biomass burning contributions to the observed BC concentration (Kirchstetter et al., 2004). Besides BC, other light absorbing (in the UV region) aerosols are also produced in course of combustion, collectively termed organic aerosols (often also called brown carbon) (Andreae and Gelencsér, 2006). Figure 11 shows the comparison of normalized light absorption as a function of the wavelength for BC observed at Lumbini during cooking and non-cooking hours and also for both events. Our results are compared with the published data of Kirchstetter et al. (2004) and that observed over a village site of Project Surya in the IGP (Praveen et al., 2012) (figure not shown). We discuss light absorption data from two distinct times of the day. The main reason behind using data from 07:00–08:00 and 16:00–17:00 is

these periods represent highest and lowest ambient concentration (Fig. 5). Also these periods represent cooking (07:00–08:00) and non-cooking (16:00–17:00) or high and low vehicular movement hours (Praveen et al., 2012). To understand the influence of biomass and fossil fuel we plotted normalized aerosol absorption at 700 nm wavelength for complete Aethalometer-measured wavelengths in Fig. 11. Kirchstetter et al. (2004) reported OC absorption efficiency at 700 nm to be zero. Thus we normalized measured absorption spectrum by 700 nm wavelength absorption. Since the Aethalometer does not provide 700 nm wavelength absorption values, we calculated the value using the absorption at nearby wavelengths and Ångström exponent following the methodology used by Praveen et al. (2012). Our results show that the normalized absorption for biomass burning aerosol is ∼ 3

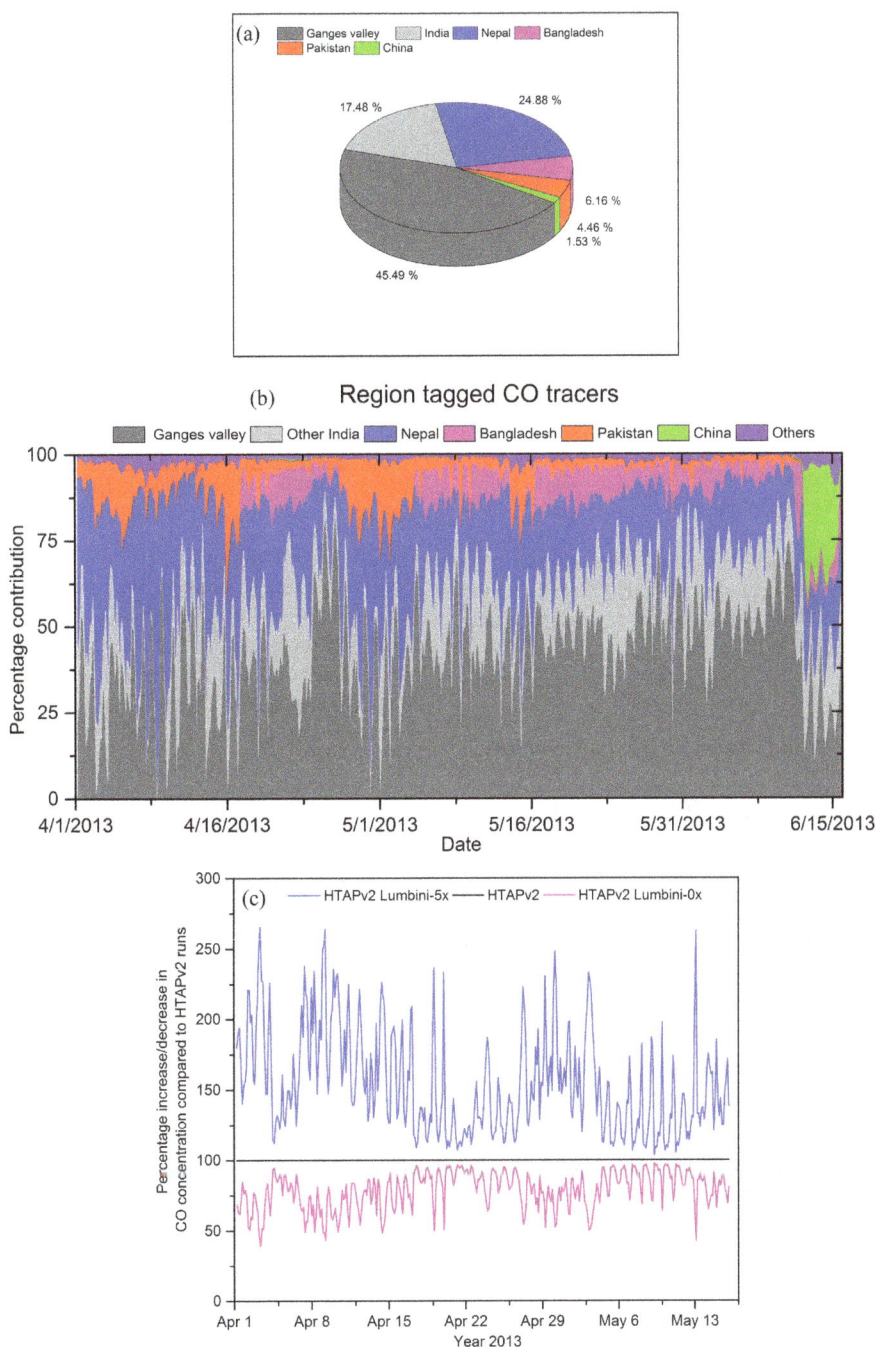

**Figure 10. (a)** WRF-STEM model estimated contributions of various source regions to average CO concentration in Lumbini for the sampling period; **(b)** time series of regionally tagged CO tracer during the whole measurement period using HTAP emission inventory; **(c)** percentage increase and decrease in CO concentration with different emissions scenario.

times higher at 370 nm than that at 700 nm whereas fossil fuel absorption is about 2.6 times higher at the same wavelength. In addition, the curve obtained for the both events is inclined towards the published biomass burning curve. The normalized curve obtained during both cooking and non-cooking period lies in between the standard curve of Kirchstetter et al. (2004). As shown in Fig. 11, the curve obtained

for the prime cooking time is closer to the published curve on biomass burning whereas that obtained during the non-cooking time is closer to the published fossil fuel curve. Similar results were also observed over the Project Surya village in the IGP region (Praveen et al., 2012; Rehman et al., 2011). This clearly indicates there is contribution from both sources,

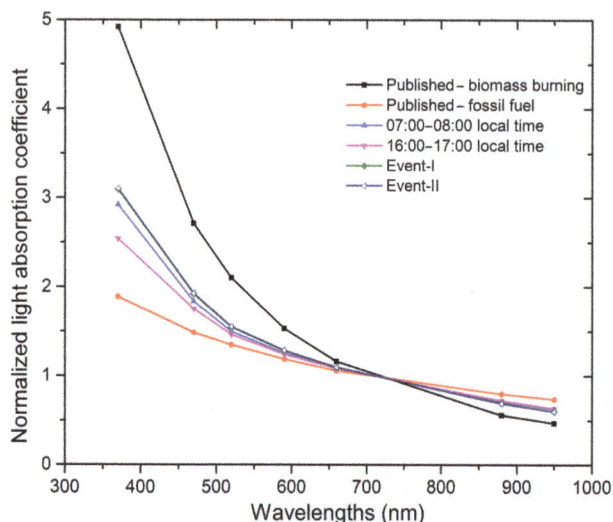

**Figure 11.** Comparison of normalized spectral light absorption co-efficients obtained during the prime cooking (07:00–08:00 LT) and non-cooking time (16:00–17:00 LT) at Lumbini with published data from Kirchstetter et al. (2004).

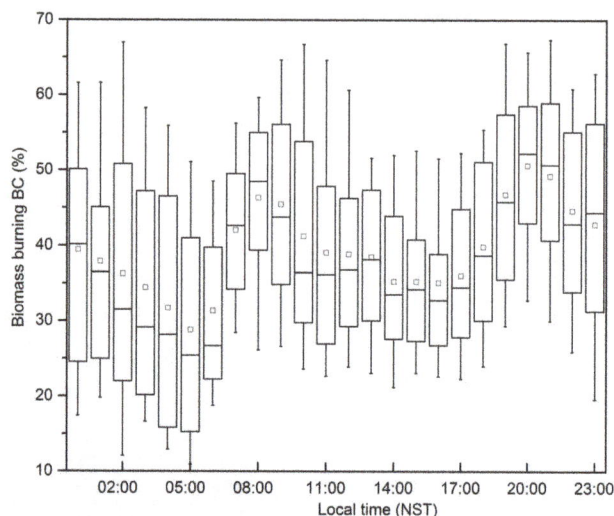

**Figure 12.** Diurnal variation of the fractional contribution of biomass burning to ambient BC concentration at Lumbini for the measurement period. In each box, the lower and upper boundary of the box represent the 25th and 75th percentile, respectively, and the top and bottom of the whisker represent the 90th and 10th percentile, respectively. The mid-line in each box represents the median while the square mark represents the mean for each hour.

biomass and fossil fuel, to the observed BC concentration over Lumbini.

In order to identify fractional contribution of biomass burning and fossil fuel combustion to observed BC aerosol, we adopted the method described by Sandradewi et al. (2008). Wavelength dependence of aerosol absorption coefficient ($b_{abs}$) is proportional to $\lambda^{-\alpha}$, where $\lambda$ is the wavelength and $\alpha$ is the absorption AAE. The $\alpha$ values ranges from 0.9 to 2.2 for fresh wood smoke aerosol (Day et al., 2006) and between 0.8 and 1.1 for traffic or diesel soot (references in Sandradewi et al., 2008). We have taken an $\alpha$ value of 1.86 for biomass burning and 1.1 for fossil fuel burning as suggested by previous literature (Sandradewi et al., 2008). Figure 12 shows diurnal variation of the biomass burning BC. Minimum contribution of biomass burning to total BC concentration was observed during 04:00–06:00 LT (only about 30 % of the total BC). As the cooking activities start in morning, the contribution of biomass BC starts to increase and reaches about 50 %. Similar pattern was repeated during evening cooking hours. Only during these two cooking periods, fossil fuel fraction BC was lower. Otherwise it remained significantly higher than biomass burning BC throughout the day. On average, $\sim 40$ % of BC was from biomass burning whereas the remaining 60 % was contributed by fossil fuel combustion during our measurement period. Interestingly, this is the opposite of the contributions that were concluded by Lawrence and Lelieveld (2010). Lawrence and Lelieveld (2010) concluded that $\sim 60$ % BC came from biomass vs. $\sim 40$ % from fossil fuel, based on a review of numerous previous studies to be likely for studies on the outflows from South Asia during the winter monsoon. When we compared observed AAE with Praveen et al. (2012), we noticed that

Lumbini values were lower than the Project Surya village site. This implies that Surya village center had a higher biomass fraction; additionally, it was observed absorption AAE exceeded 1.86 during cooking hours, which indicates 100 % biomass contribution. The difference is attributed to the fact that Lumbini sampling site is not a residential site like Surya, which can capture cooking influence efficiently. Further Lumbini sampling site is surrounded by commercial activities such as a local bus park, hotels, office buildings and industries and brick kilns slightly further away. Although the reason for this difference is not clear, it is an indication of the important role of diesel and coal emissions in the Lumbini and upwind regions.

## 4   Conclusions

Our measurements, a first for the Lumbini area, have shown very high air pollution at Lumbini. BC, CO, $O_3$ and particulate matter ($PM_{10}$, $PM_{2.5}$ and $PM_1$) were measured during the pre-monsoon of 2013 at a regional site of the SusKat-ABC campaign. Average pollutant concentrations during the monitoring period were found to be $4.9 \pm 3.8 \, \mu g \, m^{-3}$ for BC, $344.1 \pm 160.3$ ppbv for CO, $46.6 \pm 20.3$ ppbv for $O_3$, $128.8 \pm 91.9 \, \mu g \, m^{-3}$ for $PM_{10}$, $53.14 \pm 35.1 \, \mu g \, m^{-3}$ for $PM_{2.5}$ and $36.6 \pm 25.7 \, \mu g \, m^{-3}$ for $PM_1$, which is comparable with other urban sites like Kanpur and Delhi in the IGP region. However, our study finds a higher fraction of coarse-mode PM in Lumbini compared to other sites in the IGP region. In addition, the $\Delta BC/\Delta CO$ ratio obtained in Lumbini was within the range of emissions from both residential and transportation sectors, indicating them as potential key sources of BC and CO and likely most of $PM_1$ in Lumbini. The diurnal variation of the pollutants is similar to that of any urban location, with peaks during morning and evening. However, our results show higher evening concentration compared to morning concentration values, further research is needed to explain this behavior. During our measurement period, air quality in Lumbini was influenced by regional forest fires as shown by chemical transport model and satellite data analysis. A regional chemical transport model, WRF-STEM, was used to understand observations. Intercomparison of WRF-STEM model outputs with observations showed that the model underestimated the observed pollutant concentrations by a factor of $\sim 1.5$ to 5 but was able to capture the temporal variability. Model uncertainties are attributed mostly to uncertainties in meteorology and regional emissions as shown from sensitivity analysis with local emissions. Regionally tagged CO as air-mass tracers is employed in WRF-STEM model to understand the anthropogenic emission source region influencing Lumbini. Our analysis shows that the adjacent regions, mostly the Ganges Valley, other parts of India and Nepal, accounted for the highest contribution to pollutant concentration in the Lumbini. The normalized light absorption curve clearly indicated the contribution to BC in Lumbini from both sources: biomass as well as fossil fuel. On average, $\sim 40\%$ BC was found to be from the biomass burning and $\sim 60\%$ from fossil fuel burning.

Various improvements and extensions are possible in future studies. More reliable functioning of the AWS (temperature and RH sensor, rain gauge) would have allowed more in-depth analysis of the relationship between meteorological parameters and pollutants concentration. Continuous measurements of air pollutants throughout the year would allow for annual and seasonal variation study. Improvements in the model performance are much needed in its ability to simulate observed meteorology. Significant uncertainty lies with the regional emission inventory developed at national and continental scale vs. local bottom-up inventory and pollutant emissions from small-scale open burning not captured by satellites. There is a clear need to set up a continuous air quality monitoring station at Lumbini and the surrounding regions for long-term air quality monitoring.

*Author contributions.* MR and ML conceived the Lumbini portion of the SusKat experiment. MR and AKP coordinated the Lumbini field campaign. DR and KSM conducted the field observations at Lumbini. BA designed and ran the WRF-STEM model. PSP, BA and DR finalized the manuscript composition. DR, PSP, BA, MR and SK conducted the data analysis. DR and BA prepared the manuscript with inputs from all coauthors.

*Competing interests.* The authors declare that they have no conflict of interest.

*Special issue statement.* This article is part of the special issue "Atmospheric pollution in the Himalayan foothills: The SusKat-ABC international air pollution measurement campaign". It is not associated with a conference.

*Acknowledgements.* This study was partly supported by the Institute for Advanced Sustainability Studies (IASS), Germany, the International Centre for Integrated Mountain Development (ICIMOD), the National Natural Science Foundation of China (41630754, 41421061) and the State Key Laboratory of Cryospheric Science (SKLCS-ZZ-2016). Dipesh Rupakheti is supported by CAS-TWAS President's Fellowship for International PhD Students. The IASS is grateful for its funding from the German Federal Ministry for Education and Research (BMBF) and the Brandenburg Ministry for Science, Research and Culture (MWFK). ICIMOD authors would like to acknowledge that this study was partially supported by core funds of ICIMOD contributed by the governments of Afghanistan, Australia, Austria, Bangladesh, Bhutan, China, India, Myanmar, Nepal, Norway, Pakistan, Switzerland and the UK. The views and interpretations in this publication are those of the authors and are not necessarily attributable to the institutions they are associated with. We thank Bhogendra Kathayat, Bhoj Raj Bhatta, and Venerable Vivekananda and his colleagues (Panditarama Lumbini International Vipassana Meditation Center) for providing logistical support which was vital in setting up and running the site. We also thank Christoph Cüppers and Michael Pahlke of the Lumbini International Research Institute (LIRI) for proving the space and power to run the instruments at the LIRI premises. Satellite data providers (MODIS, AIRS, OMI) and HYSPLIT team are also equally acknowledged.

Edited by: Kim Oanh Nguyen Thi

## References

Adhikary, B., Carmichael, G. R., Tang, Y., Leung, L. R., Qian, Y., Schauer, J. J., Stone, E. A., Ramanathan, V., and Ramana, M. V.: Characterization of the seasonal cycle of south Asian aerosols: a regional-scale modeling analysis, J. Geophys. Res., 112, 1–22, https://doi.org/10.1029/2006jd008143, 2007.

Adhikary, B., Carmichael, G. R., Kulkarni, S., Wei, C., Tang, Y., D'Allura, A., Mena-Carrasco, M., Streets, D. G., Zhang, Q., Pierce, R. B., Al-Saadi, J. A., Emmons, L. K., Pfister, G. G., Avery, M. A., Barrick, J. D., Blake, D. R., Brune, W. H., Cohen, R. C., Dibb, J. E., Fried, A., Heikes, B. G., Huey, L. G., O'Sullivan, D. W., Sachse, G. W., Shetter, R. E., Singh, H. B., Campos, T. L., Cantrell, C. A., Flocke, F. M., Dunlea, E. J., Jimenez, J. L., Weinheimer, A. J., Crounse, J. D., Wennberg, P. O., Schauer, J. J., Stone, E. A., Jaffe, D. A., and Reidmiller, D. R.: A regional scale modeling analysis of aerosol and trace gas distributions over the eastern Pacific during the INTEX-B field campaign, Atmos. Chem. Phys., 10, 2091–2115, https://doi.org/10.5194/acp-10-2091-2010, 2010.

Andreae, M. O. and Gelencsér, A.: Black carbon or brown carbon? The nature of light-absorbing carbonaceous aerosols, Atmos. Chem. Phys., 6, 3131–3148, https://doi.org/10.5194/acp-6-3131-2006, 2006.

Antony Chen, L. W., Doddridge, B. G., Dickerson, R. R., Chow, J. C., Mueller, P. K., Quinn, J., and Butler, W. A.: Seasonal variations in elemental carbon aerosol, carbon monoxide and sulfur dioxide: Implications for sources, Geophys. Res. Lett., 28, 1711–1714, https://doi.org/10.1029/2000gl012354, 2001.

Badarinath, K. V. S., Latha, K. M., Chand, T. R. K., Reddy, R. R., Gopal, K. R., Reddy, L. S. S., Narasimhulu, K., and Kumar, K. R.: Black carbon aerosols and gaseous pollutants in an urban area in North India during a fog period, Atmos. Res., 85, 209–216, https://doi.org/10.1016/j.atmosres.2006.12.007, 2007.

Bergin, M. H., Tripathi, S. N., Jai Devi, J., Gupta, T., McKenzie, M., Rana, K., Shafer, M. M., Villalobos, A. M., and Schauer, J. J.: The discoloration of the Taj Mahal due to particulate carbon and dust deposition, Environ. Sci. Technol., 49, 808–812, https://doi.org/10.1021/es504005q, 2015.

Bergstrom, R. W., Pilewskie, P., Russell, P., Redemann, J., Bond, T., Quinn, P., and Sierau, B.: Spectral absorption properties of atmospheric aerosols, Atmos. Chem. Phys., 7, 5937–5943, https://doi.org/10.5194/acp-7-5937-2007, 2007.

Bisht, D. S., Dumka, U. C., Kaskaoutis, D. G., Pipal, A. S., Srivastava, A. K., Soni, V. K., Attri, S. D., Sateesh, M., and Tiwari, S.: Carbonaceous aerosols and pollutants over Delhi urban environment: temporal evolution, source apportionment and radiative forcing, Sci. Total Environ., 521, 431–445, https://doi.org/10.1016/j.scitotenv.2015.03.083, 2015.

Bonasoni, P., Laj, P., Marinoni, A., Sprenger, M., Angelini, F., Arduini, J., Bonafè, U., Calzolari, F., Colombo, T., Decesari, S., Di Biagio, C., di Sarra, A. G., Evangelisti, F., Duchi, R., Facchini, M. C., Fuzzi, S., Gobbi, G. P., Maione, M., Panday, A., Roccato, F., Sellegri, K., Venzac, H., Verza, G. P., Villani, P., Vuillermoz, E., and Cristofanelli, P.: Atmospheric Brown Clouds in the Himalayas: first two years of continuous observations at the Nepal Climate Observatory-Pyramid (5079 m), Atmos. Chem. Phys., 10, 7515–7531, https://doi.org/10.5194/acp-10-7515-2010, 2010.

Brimblecombe, P.: The effects of air pollution on the built environment, Imperial College Press, London, 2003.

Burney, J. and Ramanathan, V.: Recent climate and air pollution impacts on Indian agriculture, P. Natl. Acad. Sci. USA, 111, 16319–16324, https://doi.org/10.1073/pnas.1317275111, 2014.

Cao, J.-J., Zhu, C.-S., Chow, J. C., Watson, J. G., Han, Y.-M., Wang, G.-H., Shen, Z.-X., and An, Z.-S.: Black carbon relationships with emissions and meteorology in Xi'an, China, Atmos. Res., 94, 194–202, https://doi.org/10.1016/j.atmosres.2009.05.009, 2009.

Cong, Z., Kang, S., Kawamura, K., Liu, B., Wan, X., Wang, Z., Gao, S., and Fu, P.: Carbonaceous aerosols on the south edge of the Tibetan Plateau: concentrations, seasonality and sources, Atmos. Chem. Phys., 15, 1573–1584, https://doi.org/10.5194/acp-15-1573-2015, 2015a.

Cong, Z., Kawamura, K., Kang, S., and Fu, P.: Penetration of biomass-burning emissions from South Asia through the Himalayas: new insights from atmospheric organic acids, Sci. Rep.-UK, 5, 1–7, https://doi.org/10.1038/srep09580, 2015b.

Cristofanelli, P., Bracci, A., Sprenger, M., Marinoni, A., Bonafè, U., Calzolari, F., Duchi, R., Laj, P., Pichon, J. M., Roccato, F., Venzac, H., Vuillermoz, E., and Bonasoni, P.: Tro-

pospheric ozone variations at the Nepal Climate Observatory-Pyramid (Himalayas, 5079 m a.s.l.) and influence of deep stratospheric intrusion events, Atmos. Chem. Phys., 10, 6537–6549, https://doi.org/10.5194/acp-10-6537-2010, 2010.

Cristofanelli, P., Fierli, F., Marinoni, A., Calzolari, F., Duchi, R., Burkhart, J., Stohl, A., Maione, M., Arduini, J., and Bonasoni, P.: Influence of biomass burning and anthropogenic emissions on ozone, carbon monoxide and black carbon at the Mt. Cimone GAW-WMO global station (Italy, 2165 m a.s.l.), Atmos. Chem. Phys., 13, 15–30, https://doi.org/10.5194/acp-13-15-2013, 2013.

Das, S. K. and Jayaraman, A.: Role of black carbon in aerosol properties and radiative forcing over western India during premonsoon period, Atmos. Res., 102, 320–334, https://doi.org/10.1016/j.atmosres.2011.08.003, 2011.

Day, D., Hand, J., Carrico, C., Engling, G., and Malm, W.: Humidification factors from laboratory studies of fresh smoke from biomass fuels, J. Geophys. Res., 111, D22202, https://doi.org/10.1029/2006JD007221, 2006.

Dickerson, R. R., Andreae, M. O., Campos, T., Mayol-Bracero, O. L., Neusuess, C., and Streets, D. G.: Analysis of black carbon and carbon monoxide observed over the Indian Ocean: implications for emissions and photochemistry, J. Geophys. Res.-Atmos., 107, 8017, https://doi.org/10.1029/2001JD000501, 2002.

Duchi, R., Cristofanelli, P., Marinoni, A., Laj, P., Marcq, S., Villani, P., Sellegri, K., Angelini, F., Calzolari, F., Gobbi, G. P., Verza, G. P., Vuillermoz, E., Sapkota, A., and Bonasoni, P.: Continuous observations of synoptic-scale dust transport at the Nepal Climate Observatory-Pyramid (5079 m a.s.l.) in the Himalayas, Atmos. Chem. Phys. Discuss., 11, 4229–4261, https://doi.org/10.5194/acpd-11-4229-2011, 2011.

Forouzanfar, M. H., Alexander, L., Anderson, H. R., Bachman, V. F., Biryukov, S., Brauer, M., Burnett, R., Casey, D., Coates, M. M., and Cohen, A.: Global, regional, and national comparative risk assessment of 79 behavioural, environmental and occupational, and metabolic risks or clusters of risks in 188 countries, 1990–2013: a systematic analysis for the Global Burden of Disease Study 2013, Lancet, 386, 2287–2323, https://doi.org/10.1016/S0140-6736(15)00128-2, 2015.

Ganguly, D., Jayaraman, A., Rajesh, T. A., and Gadhavi, H.: Wintertime aerosol properties during foggy and nonfoggy days over urban center Delhi and their implications for shortwave radiative forcing, J. Geophys. Res., 111, 1–15, https://doi.org/10.1029/2005jd007029, 2006.

Gaur, A., Tripathi, S. N., Kanawade, V. P., Tare, V., and Shukla, S. P.: Four-year measurements of trace gases ($SO_2$, $NO_x$, CO, and $O_3$) at an urban location, Kanpur, in Northern India, J. Atmos. Chem., 71, 283–301, https://doi.org/10.1007/s10874-014-9295-8, 2014.

Gauri, K. L. and Holdren, G.: Pollutant effects on stone monuments, Environ. Sci. Technol., 15, 386–390, https://doi.org/10.1021/es00086a001, 1981.

Girach, I. A., Nair, V. S., Babu, S. S., and Nair, P. R.: Black carbon and carbon monoxide over Bay of Bengal during W_ICARB: source characteristics, Atmos. Environ., 94, 508–517, https://doi.org/10.1016/j.atmosenv.2014.05.054, 2014.

Hobbs, P. V., Sinha, P., Yokelson, R. J., Christian, T. J., Blake, D. R., Gao, S., Kirchstetter, T. W., Novakov, T., and Pilewskie, P.: Evolution of gases and particles from a savanna fire in South Africa, J. Geophys. Res., 108, 8485, https://doi.org/10.1029/2002JD002352, 2003.

Huang, X.-F., Sun, T.-L., Zeng, L.-W., Yu, G.-H., and Luan, S.-J.: Black carbon aerosol characterization in a coastal city in South China using a single particle soot photometer, Atmos. Environ., 51, 21–28, https://doi.org/10.1016/j.atmosenv.2012.01.056, 2012.

IPCC: Climate Change 2013: The physical science basis, in: Contribution of Working Group I to the Fifth Assessment Report of the Intergovernmental Panel on Climate Change, edited by: Stocker, T. F., Qin, D., Plattner, G.-K., Tignor, M., Allen, S. K., Boschung, J., Nauels, A., Xia, Y., Bex, V., and Midgley, P. M., Cambridge University Press, Cambridge, UK and New York, NY, USA, 1535 pp., 2013.

Jaffe, D., Mahura, A., Kelley, J., Atkins, J., Novelli, P. C., and Merrill, J.: Impact of Asian emissions on the remote North Pacific atmosphere: interpretation of CO data from Shemya, Guam, Midway and Mauna Loa, J. Geophys. Res., 102, 28627, https://doi.org/10.1029/96jd02750, 1997.

Joshi, H., Naja, M., Singh, K., Kumar, R., Bhardwaj, P., Babu, S. S., Satheesh, S., Moorthy, K. K., and Chandola, H.: Investigations of aerosol black carbon from a semi-urban site in the Indo-Gangetic Plain region, Atmos. Environ., 125, 346–359, https://doi.org/10.1016/j.atmosenv.2015.04.007, 2016.

Kirchstetter, T. W., Novakov, T., and Hobbs, P. V.: Evidence that the spectral dependence of light absorption by aerosols is affected by organic carbon, J. Geophys. Res., 109, 1–12, https://doi.org/10.1029/2004JD004999, 2004.

Kondo, Y., Matsui, H., Moteki, N., Sahu, L., Takegawa, N., Kajino, M., Zhao, Y., Cubison, M. J., Jimenez, J. L., Vay, S., Diskin, G. S., Anderson, B., Wisthaler, A., Mikoviny, T., Fuelberg, H. E., Blake, D. R., Huey, G., Weinheimer, A. J., Knapp, D. J., and Brune, W. H.: Emissions of black carbon, organic, and inorganic aerosols from biomass burning in North America and Asia in 2008, J. Geophys. Res., 116, D08204, https://doi.org/10.1029/2010JD015152, 2011.

Kulkarni, S., Sobhani, N., Miller-Schulze, J. P., Shafer, M. M., Schauer, J. J., Solomon, P. A., Saide, P. E., Spak, S. N., Cheng, Y. F., Denier van der Gon, H. A. C., Lu, Z., Streets, D. G., Janssens-Maenhout, G., Wiedinmyer, C., Lantz, J., Artamonova, M., Chen, B., Imashev, S., Sverdlik, L., Deminter, J. T., Adhikary, B., D'Allura, A., Wei, C., and Carmichael, G. R.: Source sector and region contributions to BC and $PM_{2.5}$ in Central Asia, Atmos. Chem. Phys., 15, 1683–1705, https://doi.org/10.5194/acp-15-1683-2015, 2015.

Kumar, B., Chakraborty, A., Tripathi, S. N., and Bhattu, D.: Highly time resolved chemical characterization of submicron organic aerosols at a polluted urban location, Environ. Sci.-Proc. Imp., 18, 1285–1296, https://doi.org/10.1039/C6EM00392C, 2016a.

Kumar, M., Tiwari, S., Murari, V., Singh, A. K., and Banerjee, T.: Wintertime characteristics of aerosols at middle Indo-Gangetic Plain: impacts of regional meteorology and long range transport, Atmos. Environ., 104, 162–175, https://doi.org/10.1016/j.atmosenv.2015.01.014, 2015.

Kumar, V., Sarkar, C., and Sinha, V.: Influence of post harvest crop residue fires on surface ozone mixing ratios in the NW IGP analyzed using two years of continuous in-situ trace gas measurements, J. Geophys. Res.-Atmos., 121, 3619–3633, https://doi.org/10.1002/2015JD024308, 2016b.

Latha, K. M. and Badarinath, K. V. S.: Correlation between black carbon aerosols, carbon monoxide and tropospheric ozone over a tropical urban site, Atmos. Res., 71, 265–274, https://doi.org/10.1016/j.atmosres.2004.06.004, 2004.

Lawrence, M. G. and Lelieveld, J.: Atmospheric pollutant outflow from southern Asia: a review, Atmos. Chem. Phys., 10, 11017–11096, https://doi.org/10.5194/acp-10-11017-2010, 2010.

Lüthi, Z. L., Škerlak, B., Kim, S. W., Lauer, A., Mues, A., Rupakheti, M., and Kang, S.: Atmospheric brown clouds reach the Tibetan Plateau by crossing the Himalayas, Atmos. Chem. Phys., 15, 6007–6021, https://doi.org/10.5194/acp-15-6007-2015, 2015.

Marinoni, A., Cristofanelli, P., Laj, P., Duchi, R., Putero, D., Calzolari, F., Landi, T. C., Vuillermoz, E., Maione, M., and Bonasoni, P.: High black carbon and ozone concentrations during pollution transport in the Himalayas: five years of continuous observations at NCO-P global GAW station, J. Environ. Sci., 25, 1618–1625, https://doi.org/10.1016/s1001-0742(12)60242-3, 2013.

Misra, A., Gaur, A., Bhattu, D., Ghosh, S., Dwivedi, A. K., Dalai, R., Paul, D., Gupta, T., Tare, V., Mishra, S. K., Singh, S., and Tripathi, S. N.: An overview of the physico-chemical characteristics of dust at Kanpur in the central Indo-Gangetic basin, Atmos. Environ., 97, 386–396, https://doi.org/10.1016/j.atmosenv.2014.08.043, 2014.

Mohnen, V. A., Goldstein, W., and Wang, W. C.: Tropospheric Ozone and Climate Change, Air and Waste, 43, 1332–1334, https://doi.org/10.1080/1073161x.1993.10467207, 1993.

Naja, M., Lal, S., and Chand, D.: Diurnal and seasonal variabilities in surface ozone at a high altitude site Mt Abu (24.6° N, 72.7° E, 1680 m a.s.l.) in India, Atmos. Environ., 37, 4205–4215, https://doi.org/10.1016/S1352-2310(03)00565-X, 2003.

Pan, X. L., Kanaya, Y., Wang, Z. F., Liu, Y., Pochanart, P., Akimoto, H., Sun, Y. L., Dong, H. B., Li, J., Irie, H., and Takigawa, M.: Correlation of black carbon aerosol and carbon monoxide in the high-altitude environment of Mt. Huang in Eastern China, Atmos. Chem. Phys., 11, 9735–9747, https://doi.org/10.5194/acp-11-9735-2011, 2011.

Panday, A. K. and Prinn, R. G.: Diurnal cycle of air pollution in the Kathmandu Valley, Nepal: observations, J. Geophys. Res., 114, 1–19, https://doi.org/10.1029/2008jd009777, 2009.

Park, S. S., Harrison, D., Pancras, J. P., and Ondov, J. M.: Highly time-resolved organic and elemental carbon measurements at the Baltimore Supersite in 2002, J. Geophys. Res., 110, D07S06, https://doi.org/10.1029/2004jd004610, 2005.

Pathak, B., Bhuyan, P. K., Biswas, J., and Takemura, T.: Long term climatology of particulate matter and associated microphysical and optical properties over Dibrugarh, North-East India and inter-comparison with SPRINTARS simulations, Atmos. Environ., 69, 334–344, https://doi.org/10.1016/j.atmosenv.2012.12.032, 2013.

Praveen, P. S., Ahmed, T., Kar, A., Rehman, I. H., and Ramanathan, V.: Link between local scale BC emissions in the Indo-Gangetic Plains and large scale atmospheric solar absorption, Atmos. Chem. Phys., 12, 1173–1187, https://doi.org/10.5194/acp-12-1173-2012, 2012.

Putero, D., Landi, T., Cristofanelli, P., Marinoni, A., Laj, P., Duchi, R., Calzolari, F., Verza, G., and Bonasoni, P.: Influence of open vegetation fires on black carbon and ozone variability in the southern Himalayas (NCO-P, 5079 m a.s.l.), Environ. Pollut., 184, 597–604, https://doi.org/10.1016/j.envpol.2013.09.035, 2014.

Putero, D., Cristofanelli, P., Marinoni, A., Adhikary, B., Duchi, R., Shrestha, S. D., Verza, G. P., Landi, T. C., Calzolari, F., Busetto, M., Agrillo, G., Biancofiore, F., Di Carlo, P., Panday, A. K., Rupakheti, M., and Bonasoni, P.: Seasonal variation of ozone and black carbon observed at Paknajol, an urban site in the Kathmandu Valley, Nepal, Atmos. Chem. Phys., 15, 13957–13971, https://doi.org/10.5194/acp-15-13957-2015, 2015.

Ram, K., Sarin, M., and Tripathi, S.: A 1 year record of carbonaceous aerosols from an urban site in the Indo-Gangetic Plain: characterization, sources, and temporal variability, J. Geophys. Res., 115, D24313, https://doi.org/10.1029/2010JD014188, 2010.

Ramachandran, S. and Rajesh, T. A.: Black carbon aerosol mass concentrations over Ahmedabad, an urban location in western India: comparison with urban sites in Asia, Europe, Canada, and the United States, J. Geophys. Res., 112, 1–19, https://doi.org/10.1029/2006JD007488, 2007.

Ramanathan, V. and Carmichael, G.: Global and regional climate changes due to black carbon, Nat. Geosci., 1, 221–227, 2008.

Ramanathan, V., Li, F., Ramana, M., Praveen, P., Kim, D., Corrigan, C., Nguyen, H., Stone, E. A., Schauer, J. J., and Carmichael, G.: Atmospheric brown clouds: hemispherical and regional variations in long-range transport, absorption, and radiative forcing, J. Geophys. Res., 112, 1–26, https://doi.org/10.1029/2006JD008124, 2007.

Rastogi, N., Singh, A., Sarin, M. M., and Singh, D.: Temporal variability of primary and secondary aerosols over northern India: impact of biomass burning emissions, Atmos. Environ., 125, 396–403, https://doi.org/10.1016/j.atmosenv.2015.06.010, 2016.

Rehman, I. H., Ahmed, T., Praveen, P. S., Kar, A., and Ramanathan, V.: Black carbon emissions from biomass and fossil fuels in rural India, Atmos. Chem. Phys., 11, 7289–7299, https://doi.org/10.5194/acp-11-7289-2011, 2011.

Retama, A., Baumgardner, D., Raga, G. B., McMeeking, G. R., and Walker, J. W.: Seasonal and diurnal trends in black carbon properties and co-pollutants in Mexico City, Atmos. Chem. Phys., 15, 9693-9709, https://doi.org/10.5194/acp-15-9693-2015, 2015.

Rupakheti, M., Panday, A. K., Lawrence, M. G., Kim, S. W., Sinha, V., Kang, S. C., Naja, M., Park, J. S., Hoor, P., Holben, B., Sharma, R. K., Mues, A., Mahata, K. S., Bhardwaj, P., Sarkar, C., Rupakheti, D., Regmi, R. P., and Gustafsson, Ö.: Air pollution in the Himalayan foothills: overview of the SusKat-ABC international air pollution measurement campaign in Nepal, Atmos. Chem. Phys. Discuss., in preparation, 2017.

Safai, P. D., Kewat, S., Pandithurai, G., Praveen, P. S., Ali, K., Tiwari, S., Rao, P. S. P., Budhawant, K. B., Saha, S. K., and Devara, P. C. S.: Aerosol characteristics during winter fog at Agra, North India, J. Atmos. Chem., 61, 101–118, https://doi.org/10.1007/s10874-009-9127-4, 2009.

Sahu, L. K., Kondo, Y., Moteki, N., Takegawa, N., Zhao, Y., Cubison, M. J., Jimenez, J. L., Vay, S., Diskin, G. S., Wisthaler, A., Mikoviny, T., Huey, L. G., Weinheimer, A. J., and Knapp, D. J.: Emission characteristics of black carbon in anthropogenic and biomass burning plumes over California during ARCTAS-CARB 2008, J. Geophys. Res., 117, D16302, https://doi.org/10.1029/2011jd017401, 2012.

Sandradewi, J., Prévôt, A. S. H., Szidat, S., Perron, N., Alfarra, M. R., Lanz, V. A., Weingartner, E., and Baltensperger, U.: Using aerosol light absorption measurements for the quantitative determination of wood burning and traffic emission contributions to particulate matter, Environ. Sci. Technol., 42, 3316–3323, https://doi.org/10.1021/es702253m, 2008.

Sarkar, C., Kumar, V., and Sinha, V.: Massive emissions of carcinogenic benzenoids from paddy residue burning in north India, Curr. Sci. India, 104, 1703–1709, 2013.

Schmid, O., Artaxo, P., Arnott, W. P., Chand, D., Gatti, L. V., Frank, G. P., Hoffer, A., Schnaiter, M., and Andreae, M. O.: Spectral light absorption by ambient aerosols influenced by biomass burning in the Amazon Basin. I: Comparison and field calibration of absorption measurement techniques, Atmos. Chem. Phys., 6, 3443–3462, https://doi.org/10.5194/acp-6-3443-2006, 2006.

Sharma, R. K., Bhattarai, B. K., Sapkota, B. K., Gewali, M. B., and Kjeldstad, B.: Black carbon aerosols variation in Kathmandu valley, Nepal, Atmos. Environ., 63, 282–288, https://doi.org/10.1016/j.atmosenv.2012.09.023, 2012.

Sharma, S., Brook, J. R., Cachier, H., Chow, J., Gaudenzi, A., and Lu, G.: Light absorption and thermal measurements of black carbon in different regions of Canada, J. Geophys. Res.-Atmos., 107, 1–11, https://doi.org/10.1029/2002JD002496, 2002.

Shindell, D., Kuylenstierna, J. C., Vignati, E., van Dingenen, R., Amann, M., Klimont, Z., Anenberg, S. C., Muller, N., Janssens-Maenhout, G., and Raes, F.: Simultaneously mitigating near-term climate change and improving human health and food security, Science, 335, 183–189, 2012.

Singh, N., Murari, V., Kumar, M., Barman, S. C., and Banerjee, T.: Fine particulates over South Asia: review and meta-analysis of $PM_{2.5}$ source apportionment through receptor model, Environ. Pollut., 223, 121–136, https://doi.org/10.1016/j.envpol.2016.12.071, 2017.

Sinha, V., Kumar, V., and Sarkar, C.: Chemical composition of pre-monsoon air in the Indo-Gangetic Plain measured using a new air quality facility and PTR-MS: high surface ozone and strong influence of biomass burning, Atmos. Chem. Phys., 14, 5921–5941, https://doi.org/10.5194/acp-14-5921-2014, 2014.

Skamarock, W., Klemp, J., Dudhia, J., Gill, D., and Barker, D.: A description of the Advanced Research WRF version 3, Technical Report NCAR/TN475 + STR, National Center for Atmospheric Research Technical Note, Boulder, Colorado, 2008.

Snider, G., Weagle, C. L., Murdymootoo, K. K., Ring, A., Ritchie, Y., Stone, E., Walsh, A., Akoshile, C., Anh, N. X., Balasubramanian, R., Brook, J., Qonitan, F. D., Dong, J., Griffith, D., He, K., Holben, B. N., Kahn, R., Lagrosas, N., Lestari, P., Ma, Z., Misra, A., Norford, L. K., Quel, E. J., Salam, A., Schichtel, B., Segev, L., Tripathi, S., Wang, C., Yu, C., Zhang, Q., Zhang, Y., Brauer, M., Cohen, A., Gibson, M. D., Liu, Y., Martins, J. V., Rudich, Y., and Martin, R. V.: Variation in global chemical composition of $PM_{2.5}$: emerging results from SPARTAN, Atmos. Chem. Phys., 16, 9629–9653, https://doi.org/10.5194/acp-16-9629-2016, 2016.

Stevenson, D., Dentener, F., Schultz, M., Ellingsen, K., Van Noije, T., Wild, O., Zeng, G., Amann, M., Atherton, C., and Bell, N.: Multimodel ensemble simulations of present-day and near-future tropospheric ozone, J. Geophys. Res., 111, 1–23, https://doi.org/10.1029/2005JD006338, 2006.

Subramanian, R., Kok, G. L., Baumgardner, D., Clarke, A., Shinozuka, Y., Campos, T. L., Heizer, C. G., Stephens, B. B., de Foy, B., Voss, P. B., and Zaveri, R. A.: Black carbon over Mexico: the effect of atmospheric transport on mixing state, mass absorption cross-section, and BC / CO ratios, Atmos. Chem. Phys., 10, 219–237, https://doi.org/10.5194/acp-10-219-2010, 2010.

Tiwari, S., Srivastava, A. K., Bisht, D. S., Parmita, P., Srivastava, M. K., and Attri, S. D.: Diurnal and seasonal variations of black carbon and $PM_{2.5}$ over New Delhi, India: influence of meteorology, Atmos. Res., 125, 50–62, https://doi.org/10.1016/j.atmosres.2013.01.011, 2013.

Tiwari, S., Pipal, A. S., Hopke, P. K., Bisht, D. S., Srivastava, A. K., Tiwari, S., Saxena, P. N., Khan, A. H., and Pervez, S.: Study of the carbonaceous aerosol and morphological analysis of fine particles along with their mixing state in Delhi, India: a case study, Environ. Sci. Pollut. R., 22, 10744–10757, https://doi.org/10.1007/s11356-015-4272-6, 2015.

Tiwari, S., Dumka, U. C., Kaskaoutis, D. G., Ram, K., Panicker, A. S., Srivastava, M. K., Tiwari, S., Attri, S. D., Soni, V. K., and Pandey, A. K.: Aerosol chemical characterization and role of carbonaceous aerosol on radiative effect over Varanasi in central Indo-Gangetic Plain, Atmos. Environ., 125, 437–449, https://doi.org/10.1016/j.atmosenv.2015.07.031, 2016.

Tiwari, S., Dumka, U. C., Gautam, A. S., Kaskaoutis, D. G., Srivastava, A. K., Bisht, D. S., Chakrabarty, R. K., Sumlin, B. J., and Solmon, F.: Assessment of $PM_{2.5}$ and $PM_{10}$ over Guwahati in Brahmaputra River Valley: temporal evolution, source apportionment and meteorological dependence, Atmos. Pollut. Res., 8, 13–28, https://doi.org/10.1016/j.apr.2016.07.008, 2017.

Val Martín, M., Honrath, R. E., Owen, R. C., Pfister, G., Fialho, P., and Barata, F.: Significant enhancements of nitrogen oxides, black carbon, and ozone in the North Atlantic lower free troposphere resulting from North American boreal wildfires, J. Geophys. Res., 111, D23S60, https://doi.org/10.1029/2006jd007530, 2006.

Verma, R. L., Sahu, L. K., Kondo, Y., Takegawa, N., Han, S., Jung, J. S., Kim, Y. J., Fan, S., Sugimoto, N., Shammaa, M. H., Zhang, Y. H., and Zhao, Y.: Temporal variations of black carbon in Guangzhou, China, in summer 2006, Atmos. Chem. Phys., 10, 6471–6485, https://doi.org/10.5194/acp-10-6471-2010, 2010.

Verma, R. L., Kondo, Y., Oshima, N., Matsui, H., Kita, K., Sahu, L. K., Kato, S., Kajii, Y., Takami, A., and Miyakawa, T.: Seasonal variations of the transport of black carbon and carbon monoxide from the Asian continent to the western Pacific in the boundary layer, J. Geophys. Res., 116, D21307, https://doi.org/10.1029/2011jd015830, 2011.

Verma, S., Worden, J., Payra, S., Jourdain, L., and Shim, C.: Characterizing the long-range transport of black carbon aerosols during Transport and Chemical Evolution over the Pacific (TRACE-P) experiment, Environ. Monit. Assess., 154, 85–92, https://doi.org/10.1007/s10661-008-0379-2, 2009.

Wan, X., Kang, S., Li, Q., Rupakheti, D., Zhang, Q., Guo, J., Chen, P., Tripathee, L., Rupakheti, M., Panday, A. K., Wang, W., Kawamura, K., Gao, S., Wu, G., and Cong, Z.: Organic molecular tracers in the atmospheric aerosols from Lumbini, Nepal, in the northern Indo-Gangetic Plain: influence of biomass burning, Atmos. Chem. Phys., 17, 8867–8885, https://doi.org/10.5194/acp-17-8867-2017, 2017.

WHO: Air quality guidelines: global update 2005: particulate matter, ozone, nitrogen dioxide, and sulfur dioxide, World Health Organization, Geneva, 22 pp., 2006.

Wiedinmyer, C., Akagi, S. K., Yokelson, R. J., Emmons, L. K., Al-Saadi, J. A., Orlando, J. J., and Soja, A. J.: The Fire INventory from NCAR (FINN): a high resolution global model to estimate the emissions from open burning, Geosci. Model Dev., 4, 625–641, https://doi.org/10.5194/gmd-4-625-2011, 2011.

Zhou, X., Gao, J., Wang, T., Wu, W., and Wang, W.: Measurement of black carbon aerosols near two Chinese megacities and the implications for improving emission inventories, Atmos. Environ., 43, 3918–3924, https://doi.org/10.1016/j.atmosenv.2009.04.062, 2009.

# Re-evaluating black carbon in the Himalayas and the Tibetan Plateau: concentrations and deposition

**Chaoliu Li**[1,3,4], **Fangping Yan**[3], **Shichang Kang**[2,4], **Pengfei Chen**[2], **Xiaowen Han**[1,5], **Zhaofu Hu**[2,5], **Guoshuai Zhang**[1], **Ye Hong**[6], **Shaopeng Gao**[1], **Bin Qu**[3], **Zhejing Zhu**[7], **Jiwei Li**[7], **Bing Chen**[7], and **Mika Sillanpää**[3,8]

[1]Key Laboratory of Tibetan Environment Changes and Land Surface Processes, Institute of Tibetan Plateau Research, Chinese Academy of Sciences, Beijing 100101, China
[2]State Key Laboratory of Cryospheric Sciences, Northwest Institute of Eco-Environment and Resources, Chinese Academy of Sciences, Lanzhou 730000, China
[3]Laboratory of Green Chemistry, Lappeenranta University of Technology, Sammonkatu 12, 50130 Mikkeli, Finland
[4]CAS Center for Excellence in Tibetan Plateau Earth Sciences, Beijing 100101, China
[5]University of Chinese Academy of Sciences, Beijing 100049, China
[6]Institute of Atmospheric Environment, China Meteorological Administration, Shenyang 110166, China
[7]Environmental Research Institute, Shandong University, Jinan 250100, China
[8]Department of Civil and Environmental Engineering, Florida International University, Miami, FL 33174, USA

*Correspondence to:* Chaoliu Li (lichaoliu@itpcas.ac.cn)

**Abstract.** Black carbon (BC) is the second most important warming component in the atmosphere after $CO_2$. The BC in the Himalayas and the Tibetan Plateau (HTP) has influenced the Indian monsoon and accelerated the retreat of glaciers, resulting in serious consequences for billions of Asian residents. Although a number of related studies have been conducted in this region, the BC concentrations and deposition rates remain poorly constrained. Because of the presence of arid environments and the potential influence of carbonates in mineral dust (MD), the reported BC concentrations in the HTP are overestimated. In addition, large discrepancies have been reported among the BC deposition derived from lake cores, ice cores, snow pits and models. Therefore, the actual BC concentration and deposition values in this sensitive region must be determined. A comparison between the BC concentrations in acid (HCl)-treated and untreated total suspected particle samples from the HTP showed that the BC concentrations previously reported for the Nam Co station (central part of the HTP) and the Everest station (northern slope of the central Himalayas) were overestimated by approximately $52 \pm 35$ and $39 \pm 24$ %, respectively, because of the influence of carbonates in MD. Additionally, the organic carbon (OC) levels were overestimated by approxi-

mately $22 \pm 10$ and $22 \pm 12$ % for the same reason. Based on previously reported values from the study region, we propose that the actual BC concentrations at the Nam Co and Everest stations are 61 and 154 $\mathrm{ng\,m^{-3}}$, respectively. Furthermore, a comprehensive comparison of the BC deposition rates obtained via different methods indicated that the deposition of BC in HTP lake cores was mainly related to river sediment transport from the lake basin as a result of climate change (e.g., increases in temperature and precipitation) and that relatively little BC deposition occurred via atmospheric deposition. Therefore, previously reported BC deposition rates from lake cores overestimated the atmospheric deposition of BC in the HTP. Correspondingly, BC deposition derived from snow pits and ice cores agreed well with that derived from models, implying that the BC depositions of these two methods reflect the actual values in the HTP. Therefore, based on reported values from snow pits and ice cores, we propose that the BC deposition in the HTP is $17.9 \pm 5.3 \mathrm{\,mg\,m^{-2}\,a^{-1}}$, with higher and lower values appearing along the fringes and central areas of the HTP, respectively. These adjusted BC concentrations and deposition values in the HTP are critical for performing accurate evaluations of other BC factors, such as atmospheric distribution, radiative forcing and chemical transport in the HTP.

# 1 Introduction

The Himalayas and the Tibetan Plateau (HTP) region is the highest mountain–plateau system in the world and is the source of approximately 10 large rivers in Asia. This region is also sensitive to climate change (Bolch et al., 2012; Kang et al., 2010; You et al., 2010). Black carbon (BC) in and around the HTP has been found to play key roles in climate change patterns in the HTP and Asia, including causing atmospheric warming (Xu et al., 2016; Ramanathan and Carmichael, 2008; Lau et al., 2010; Ji et al., 2015), promoting HTP glacial retreat (Xu et al., 2009; Qu et al., 2014; Li et al., 2017; Zhang et al., 2017b; Ming et al., 2009, 2013), altering monsoon system evolution (Bollasina et al., 2008) and affecting the fresh water supplies of billions of residents across Asia. To date, numerous studies have been conducted on the BC concentrations in the atmosphere (Zhao et al., 2013b; Ming et al., 2010; Cong et al., 2015; Marinoni et al., 2010; Wan et al., 2015) and atmospheric BC deposition as determined from lake core sediments (Han et al., 2015; Cong et al., 2013). However, all of these studies exhibit limitations because of certain special environmental factors in the HTP (e.g., high concentrations of mineral dust (MD) in aerosols and catchment inputs to lake core sediment). Therefore, the above studies should be reinvestigated to better define the actual BC values in the HTP. Therefore, in this article, we discussed the actual concentrations and deposition of BC in the HTP based on data of aerosols collected at two remote stations and previously reported BC deposition data.

At present, the thermal–optical method is a widely used method for measuring BC concentrations in aerosols from the HTP (Zhao et al., 2013b; Ming et al., 2010; Cong et al., 2015; Li et al., 2016d). An important factor influencing the accurate measurement of BC concentrations via this method is the presence of carbonates (inorganic carbon – IC) in MD. IC can also emit $CO_2$ in response to increasing temperature during measurements, thus causing an overestimation of the total carbon (TC) in carbonaceous aerosols (CAs) (Karanasiou et al., 2011). Hence, IC is generally excluded in CA studies (Bond et al., 2013). However, few studies of the HTP have considered the contributions of IC to TC and BC because one study concluded that IC can be neglected in studies of the TC and BC in midlatitude aerosols because the IC exists at far lower concentrations relative to TC and BC (Chow and Watson, 2002).

This conclusion cannot be blindly applied to other areas because of the complexities of midlatitude environments around the world (e.g., arid areas and deserts with intense dust storm events). For example, previous studies in Xi'an, midwestern and northeastern China showed that IC accounts for approximately 8 % (Cao et al., 2005) to 10 % (Ho et al., 2011) of the TC in particles with diameters less than 2.5 µm ($PM_{2.5}$) during dust storm events. Similar phenomena have also been found for both $PM_{2.5}$ and total suspended particle (TSP) samples in southern Europe (Sillanpää et al., 2005; Perrone et al., 2011). Because TSP samples contain more MD and carbonates than $PM_{2.5}$, they should have higher concentrations of IC.

The above phenomenon should also be taken into consideration in the study of CAs of the HTP. Similar to northern China, large sand dunes and deserts are widely distributed across the western HTP (Liu et al., 2005), and dust storms occur frequently in winter and spring (Wang et al., 2005). Thus, IC may account for a large portion of the CAs in the HTP. Unfortunately, the potential contributions of IC to the TC and BC in HTP aerosols have been overlooked (Cao et al., 2010; Cong et al., 2015; Li et al., 2016b; Ming et al., 2010; Wan et al., 2015; Zhao et al., 2013b). Additionally, IC contributions may be high because almost all of the reported data on CAs are based on the TSP content, which includes large volumes of coarse particles derived directly from MD. Therefore, the TC and BC concentrations in the HTP are likely overestimated. In fact, some published articles on aerosols collected from remote areas of the HTP have identified MD components (Cong et al., 2015; Zhao et al., 2013b), although neither of these two studies have directly discussed this issue or evaluated the effects of IC.

Because MD has lower influences on light than BC in the atmosphere (Clarke et al., 2004; Bond and Bergstrom, 2006) and on glacier surfaces (Qu et al., 2014), considering IC as BC will overestimate the BC-driven climate forcing. Organic carbon (OC) is generally considered to scatter sunlight. However, some components of OC also absorb sunlight and warm the atmosphere (Andreae and Gelencser, 2006). Therefore, the contributions of IC to the OC and BC values in HTP aerosols must be quantitatively evaluated. In this study, TSP samples from two remote stations in the HTP were collected to evaluate the contributions of IC to the TC and BC. Additionally, seasonal variations in the extent of the overestimations of TC and BC and possible causes were also examined. Finally, previously published TC and BC concentrations at these two stations were adjusted (Cong et al., 2015; Zhao et al., 2013a).

BC deposition is closely related to the BC transport processes, lifetime and radiative forcing. Depositional value can be measured from historical media, such as sediments (Gustafsson and Gschwend, 1998; Han et al., 2016) and ice cores (Ming et al., 2007; Ruppel et al., 2014), estimated from BC concentrations in the atmosphere (Jurado et al., 2008) or calculated using models (Zhang et al., 2015). At present, the BC deposition process remains poorly quantified in the HTP because of its complex terrain and dynamic regimes (Bond et al., 2013; Bauer et al., 2013). Thus far, only three studies have directly reported on BC deposition in the HTP. One model indicated that the BC deposition in the central HTP

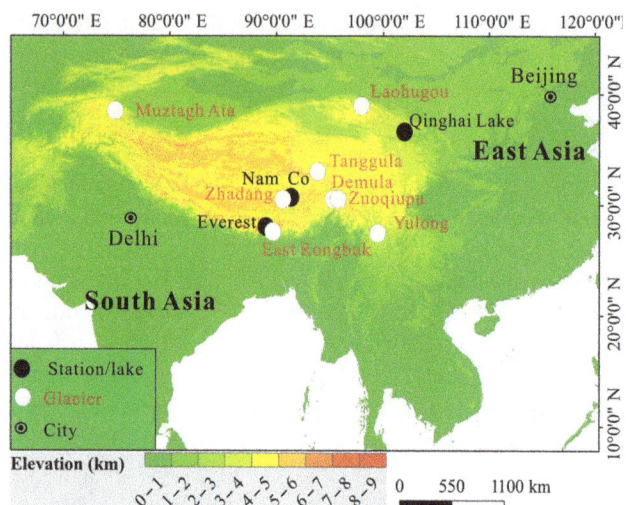

**Figure 1.** Selected study sites, including the HTP stations, lakes and glaciers.

was $9\,mg\,m^{-2}\,a^{-1}$ (Zhang et al., 2015), which is approximately 30 times lower than the values measured in lake cores at Nam Co and Qinghai lakes ($270–390\,mg\,m^{-2}\,a^{-1}$) (Fig. 1) (Cong et al., 2013; Han et al., 2011). Although considerable uncertainties exist in atmospheric BC deposition estimated from models (Koch et al., 2009; Bond et al., 2013) and lake core sediments (Yang, 2015; Cohen, 2003), these large differences need to be thoroughly investigated.

For instance, although the influence of sediment focusing on BC deposition in lake cores has been noted in other areas (Yang, 2015; Blais and Kalff, 1995), it has not been pointed out and evaluated in the HTP. Consequently, correcting for this process might result in incorrect data and explanations. Therefore, additional studies must be performed to provide more reliable BC deposition values. For instance, other researchers have reported BC concentrations and water accumulation rates in ice cores and snow pits from the HTP (Fig. 1) (Xu et al., 2009; Li et al., 2016a, c; Ming et al., 2008). Although these studies did not report BC deposition values directly, BC deposition rates could be easily calculated from the data reported in those articles. Because the cols of glaciers where the snow and ice samples were collected are generally located at the highest altitudes of a given region, BC is only deposited via wet and dry deposition from the atmosphere. Therefore, these data need to be comprehensively evaluated.

Notably, some uncertainties exist in the comparison of BC data among different studies. Despite recent technological achievements, accurately measuring BC concentrations in ambient samples remains a challenge in atmospheric chemistry research (Andreae and Gelencser, 2006; Bond et al., 2013; Lim et al., 2014). Because the methods used to measure BC concentrations and determine BC deposition levels are not the same, uncertainties will be introduced

when directly comparing the results from different studies. For instance, different thermal–optical methods with different temperature increase protocols (e.g., NIOSH vs. IMPROVE vs. EUSAAR_2) will produce different BC concentrations for the same sample (Karanasiou et al., 2015; Andreae and Gelencser, 2006). In general, BC concentrations derived from the IMPROVE method are 1.2–1.5 times higher than those derived from the NIOSH method (Chow et al., 2001; Reisinger et al., 2008), and BC concentrations from the EUSAAR_2 temperature protocol are approximately twice as high as those derived from the NIOSH protocol (Cavalli et al., 2010). Furthermore, lake core samples need to be pretreated with HCl and hydrofluoric acid (HF) several times prior to measurements with the thermal–optical methods (Han et al., 2015). However, because of the complex chemical properties of ambient samples, the "best" thermal–optical protocol has not been identified (Karanasiou et al., 2015), and an exact ratio for BC produced from different methods is difficult to determine. Therefore, although the direct comparison of BC concentrations and deposition levels across different studies presents certain uncertainties in this study, the comparison between data of lake core and snow pit is still reliable because BC deposition of the former was much higher (approximately 20 times) than that of the latter, up to 7 times more among different methods (Watson et al., 2005). For instance, although large uncertainties exist for BC concentrations within the same environmental matrix (Watson et al., 2005; Hammes et al., 2007; Han et al., 2011), the similarity of the BC deposition values among different glaciers (Table 1) in different studies implies that comparing BC deposition data is feasible for the glacial region in the HTP. In addition, because BC concentrations measured via the SP2 method are far lower than those measured via thermal–optical methods (Lim et al., 2014) (the former can only measure BC in grain sizes finer than 500 nm; Kaspari et al., 2011), SP2-based BC data were avoided in this study.

## 2 Methods

### 2.1 Collection of aerosols, surface soils and river sediments

TSP samples were collected from the Nam Co Monitoring and Research Station for Multisphere Interactions and the Qomolangma Atmospheric and Environmental Observation and Research Station (Everest station) (Fig. 1) from 2014 to 2016. The Nam Co station is located in the center of the HTP. The Everest station is located on the northern slopes of the Himalayas. Both of these two stations are generally considered to be located in remote areas of the HTP that receive BC transported over long distances from south Asia, and several BC studies have been conducted there (Chen et al., 2015; Cong et al., 2015; Ming et al., 2010; Li et al., 2016a). In detail, TSP samples were collected using 90 mm pre-combusted

**Table 1.** Monitored or recovered BC deposition ($mg\,m^{-2}\,a^{-1}$) from the HTP and other regions of the world.

| Region | Sites | Deposition | Period | References |
|--------|-------|------------|--------|------------|
| Tibet | Zuoqiupu glacier | 12 | 1970–2005 | 1 |
| | Muztagh Ata | 18 | 1970–2005 | 1 |
| | East Rongbuk ice core | 10.2 | 1995–2002 | 2 |
| | Laohugou glacier | 25 | 2013–2014 | 3, 4 |
| | Tanggula glacier | 21.2 | 2013–2014 | 3, 4 |
| | Zhangdang glacier | 22.8 | 2013–2014 | 3, 4 |
| | Demula glacier | 14.4 | 2013–2014 | 3, 4 |
| | Yulong glacier | 20.3 | 2013–2014 | 3, 4 |
| | Model results of central Tibetan Plateau | 9 | 2013–2014 | 5 |
| | Nam Co Lake core | 260 | 1960–2009 | 6 |
| | Qinghai Lake core | 270–390 | 1770 s–2011 | 7 |
| | Aerosol of Nam Co station | 10.5 | 2005–2007 | 8 |
| | Aerosol of Qinghai Lake | 92.7 | 2011–2012 | 8 |
| East China | Chaohu lake core, East China | 1160 | 1980–2012 | 9 |
| | Northern China | 1660 | Around 2010 | 10 |
| | North China Plain | 1500 | 2008–2009 | 11 |

Note: 1: Bauer et al. (2013); 2: BC concentration ($20.3\,ng\,g^{-1}$) and snow accumulation (500 mm) were adopted from Ming et al. (2008) and Li et al. (2016c), respectively; 3: Li et al. (2016c); 4: Li et al. (2016a); 5: Zhang et al. (2015); 6: Cong et al. (2013); 7: Han et al. (2015); 8: calculated in this study; 9: Han et al., (2016); 10: Fang et al. (2015); 11: Tang et al. (2014).

**Table 2.** Precipitation (mm) and BC concentration ($ng\,m^{-3}$) values used for the BC deposition calculations for Nam Co Lake and Qinghai Lake.

| | Nam Co Lake precipitation | BC concentration | Qinghai Lake precipitation | BC concentration |
|--------|------|------|------|------|
| Spring | 29.65 | 135.86 | 77.51 | 1000 |
| Summer | 190.05 | 90.97 | 244.02 | 530 |
| Autumn | 79.72 | 86.58 | 89.78 | 690 |
| Winter | 2.95 | 93.55 | 3.81 | 1050 |

(550 °C, 6 h) quartz fiber filters (Whatman Corp) with a vacuum pump (VT 4.8, Germany). Because the pump was not equipped with a flow meter, the air volumes passing through each filter could not be determined (Li et al., 2016d); however, this did not influence the objectives of this study (e.g., relative concentrations of TC and BC in the original and acid-treated samples). Four field blank filters were also collected from each station by exposing the filters in each sampler without pumping.

To compare the BC concentrations of the Nam Co Lake cores, two surface soil samples and four suspended particle samples from four rivers in the Nam Co Basin were collected during a period of peak river flow in 2015. The < 20 µm fraction of these samples was extracted (Li et al., 2009) and treated (Han et al., 2015) to measure the BC concentrations. In addition, 10 surface soil samples around the Everest station were collected to study the pH values.

## 2.2 Measurement of BC and elemental concentrations

The carbonates of the collected aerosol samples were removed via a fumigation process involving exposing a subset of samples to a vapor of 37 % hydrochloric acid (HCl) for 24 h. Then, the treated samples were held at 60 °C for over 1 h to remove any acid remaining on the filter (Li et al., 2016a; Pio et al., 2007; Chen et al., 2013; Bosch et al., 2014). The OC and elemental carbon (EC, the common chemical/mass definition of BC) concentrations of both the original and treated samples were measured using a Desert Research Institute (DRI) model 2001 thermal–optical carbon analyzer (Atmoslytic Inc., Calabasas, CA, USA) following the IMPROVE-A protocol (Chow and Watson, 2002). The OC and BC concentrations were determined based on varying transmission signals. To investigate the BC concentration measured by different methods, 16 acid-fumigated aerosol samples were measured following the EUSAAR_2 and NIOSH protocols for comparison with the results of the IMPROVE protocol. The results showed that the TC concentrations of three methods for the same sample were similar, as suggested by previous research (Chow et al., 2001). The ratios of $BC_{(IMPROVE)}/BC_{(NIOSH)}$ and $BC_{(EUSAAR_2)}/BC_{(NIOSH)}$ for the studied samples were $1.36 \pm 0.35$ and $1.88 \pm 0.60$, respectively, both of which agreed with the previously proposed ratios of 1.2–1.5 (Chow et al., 2001; Reisinger et al., 2008) and 2 (Cavalli et al., 2010), respectively. To evaluate the concentrations of MD, the concentrations of Ca, Fe, Al and Ti in the aerosol samples were measured by inductively coupled plasma optical emission spectroscopy (ICP-OES) following the method of Li et al. (2009). All the reported values in this study were corrected based on the values of the blanks. The contributions of MD (Maenhaut et al., 2002) and CA (Ram et al., 2010) of the collected samples were calculated using the following equations:

$$MD = (1.41 \times Ca + 2.09 \times Fe + 1.9 \times Al + 2.15 \times Si + 1.67 \times Ti) \times 1.16, \tag{1}$$

where Si is calculated from Al assuming an average ratio of Si/Al is 2.5 (Carrico et al., 2003), and

$$CA = OC \times 1.6 + BC. \tag{2}$$

### 2.3 Adoption and calculation of BC deposition data

To determine the actual BC deposition in the HTP, previously reported data were compiled and evaluated (Table 1). In addition, BC deposition rates from the Nam Co station and Qinghai Lake basin were estimated from the average BC concentrations in the atmosphere and average precipitation levels using the method described in detail in other studies (Jurado et al., 2008; Fang et al., 2015) (Table 2). In brief, the annual atmospheric deposition rate of BC ($\mu g\,m^{-2}\,a^{-1}$) was calculated as follows:

$$F_{DC} - F_{DD} + F_{WD} \tag{3}$$
$$F_{DD} = 7.78 \times 10^4 \cdot V_D \cdot C_{BC-TSP} \tag{4}$$
$$F_{WD} = 10^{-3} \cdot P_0 \cdot W_p \cdot C_{BC-TSP}, \tag{5}$$

where $F_{DD}$ and $F_{WD}$ are the seasonal dry and wet deposition ($\mu g\,m^{-2}$), respectively; $V_D$, $P_0$ and $W_p$ are the dry deposition velocity of aerosol ($0.15\,cm\,s^{-1}$), the precipitation amount (mm) in a given season and the particle washout ratio ($2.0 \times 10^5$), respectively (Fang et al., 2015); and $C_{BC-TSP}$ is the BC concentration of the TSPs ($\mu g\,m^{-3}$). The seasonal BC concentrations at the Nam Co station were monitored with an AE-31, and the average precipitation levels at the station were recorded from 2014–2015. The BC concentrations in Qinghai Lake are reported in Zhao et al. (2015), and the average 1961–2010 precipitation levels recorded by the China Meteorological Administration from the Huangyuan station in the lake basin were used. The values used in the BC deposition calculations for these two areas are shown in Table 2.

## 3 Results and discussion

### 3.1 Actual BC concentrations in the atmosphere over the HTP

#### 3.1.1 Contribution of carbonate carbon to both TC and BC

In this study, it was shown that carbonate carbon significantly contributes to the BC, TC and OC concentrations of the TSP samples of Nam Co and Everest stations after comparing BC and OC concentrations between original and acid-treated samples. The ratios of the TC, OC and BC levels of the aerosols treated with acid ($TC_A$, $OC_A$ and $BC_A$) to those of the original samples ($TC_O$, $OC_O$ and $BC_O$) were $0.81 \pm 0.13$, $0.78 \pm 0.10$ and $0.48 \pm 0.35$, respectively, for the Nam Co station and $0.76 \pm 0.12$, $0.78 \pm 0.12$ and $0.61 \pm 0.24$, respectively, for the Everest station. Meanwhile, because of heavy precipitation during monsoon period, influences of IC to both BC and TC during this time were lower than those of non-monsoon period at two studied stations (Fig. 2). As proposed in previous work (Chow and Watson, 2002), BC concentrations are more heavily influenced than OC and TC concentrations because carbonates are more prone to decompose at high temperatures along with BC during analyses. The OC concentrations in the treated samples used in this study also decreased, indicating that carbonates can also decompose at low temperatures (Karanasiou et al., 2011). Clear seasonal variations, i.e., low $TC_A/TC_O$ ratios during non-monsoon periods and high $TC_A/TC_O$ ratios during monsoon periods, were observed in the aerosols at the Nam Co station (Fig. 2). This pattern is consistent with the intense dust storms that occur during non-monsoon periods. However, clear seasonal patterns in the $TC_A/TC_O$ ratio at the Everest station were not observed, in accordance with the relatively stable seasonal variations in the $Ca^{2+}$ content in aerosols recorded at this station (Cong et al., 2015). To evaluate the relative ratio of MD and CA, MD/(MD + CA) values were calculated (Fig. 3). The MD/(MD + CA) levels recorded at the Nam Co station during non-monsoon periods were significantly higher than those recorded during monsoon periods ($p < 0.01$), whereas the corresponding values at the Everest station were not significantly different between the two periods ($p > 0.05$) (Fig. 3). Compared with those of other areas, the MD/(MD + CA) values recorded at the two stations were higher than those recorded at the NCO-P station ($27.95°$ N, $86.82°$ E; 5079 m.a.s.l) (70 and 73 % for the premonsoon and monsoon periods, respectively) located on the southern slope of the Himalayas (Decesari et al., 2010). This difference may be related to the serious levels of south Asian pollutants at the NCO-P station and the relative ease with which polluted clouds are transported to this station. However, because the measured particle size ($PM_{10}$) and the measurement methods of Ca, Mg and EC at the NCO-P station

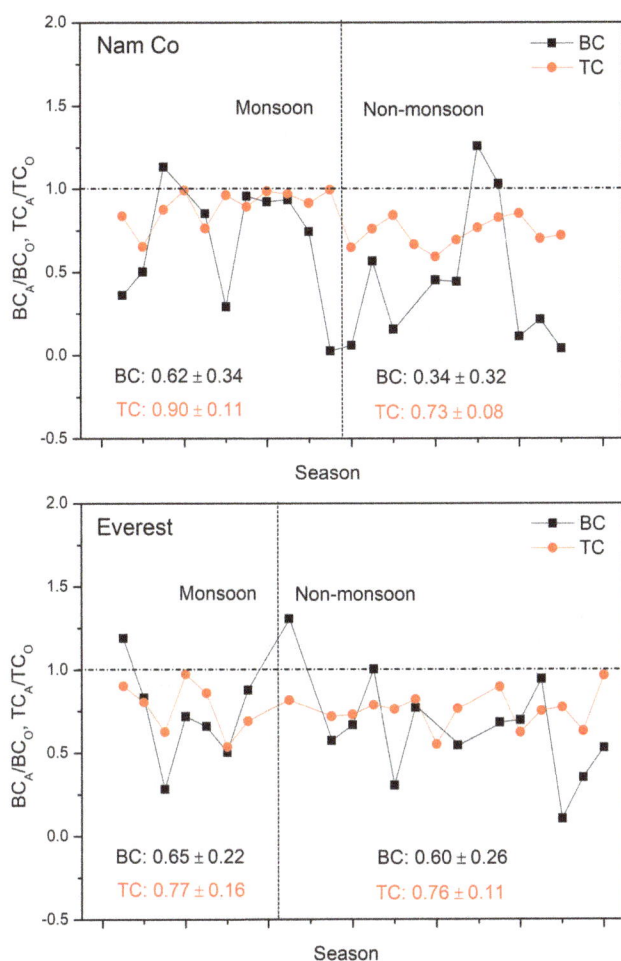

**Figure 2.** Seasonal variations in the BC and TC concentrations in the original and acid-treated aerosol samples collected at the Nam Co and Everest stations.

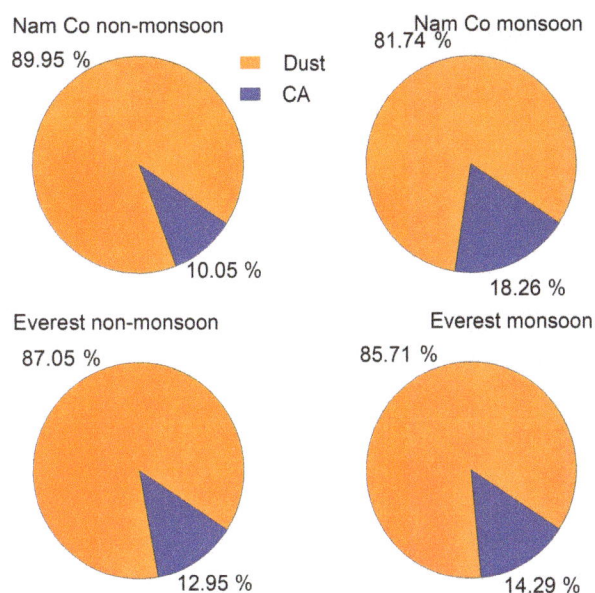

**Figure 3.** Percentage of MD and CA relative to their sum during both non-monsoon and monsoon periods at the Nam Co and Everest stations.

differed from those in this study, uncertainties exist in such a direct comparison.

The Everest station is located in a dry river valley with sparse vegetation cover (a typical barren site), and the MD derived from the local surface soil contributes considerably to aerosols collected during monsoon periods (Liu et al., 2017). However, the Nam Co station is located in a typical grassland region with limited amounts of locally sourced dust during monsoon periods. Additionally, the Everest station is located in the rain shadow of the Himalayas; thus, the precipitation level recorded at the Everest station (172 mm during the monsoon period between 2014 and 2015) is much lower than that at the Nam Co station (258 mm), causing high MD concentrations in the atmosphere of the Everest station during that period. Potential carbonate-induced biasing of aerosol samples has been proposed to occur in arid areas with alkaline soils (Chow and Watson, 2002). Because of the dry weather conditions, the pH values of the soil around the Nam Co and Everest stations are as high as 8 (Li et al.,

2008) and 8.3, respectively, implying considerable carbonate contributions. During non-monsoon periods, MD is mainly transported by westerlies from the arid western HTP, where MD is distributed across large deserts with sand dunes; thus, the aerosol samples were influenced by MD with high concentrations of carbonates. Finally, the significant positive relationship ($p < 0.01$) between Ca and IC ($TC_O$–$TC_A$) for the aerosols of these two stations further demonstrated the contributions of $CaCO_3$ to aerosol IC (Fig. 4). The ratio of Ca/IC was higher in the Everest station samples than that of Nam Co station, possibly reflecting different types of carbonate at these two stations.

### 3.1.2 Actual BC concentrations at the two stations and implications

In summary, we clearly showed that the presence of carbonates in MD led the TC levels in TSP samples in the HTP to be overestimated by approximately $19 \pm 13$ and $24 \pm 12\%$ at the Nam Co and Everest stations, respectively. These overestimates were higher than the corresponding value of 10 % found for coarse particles in the central Mediterranean region of Europe (Perrone et al., 2011). In addition, the related BC values were overestimated by approximately $52 \pm 35$ and $39 \pm 24\%$, respectively, thus implying that the actual BC concentrations at these two stations were lower than previously reported values. Although fumigation with HCl can cause the loss of volatile organic acids in treated samples (Chow et al., 1993), this potential influence is not important because of the significant relationship between $TC_O$–$TC_A$ and Ca (Fig. 4). Moreover, because of the large variations in the above val-

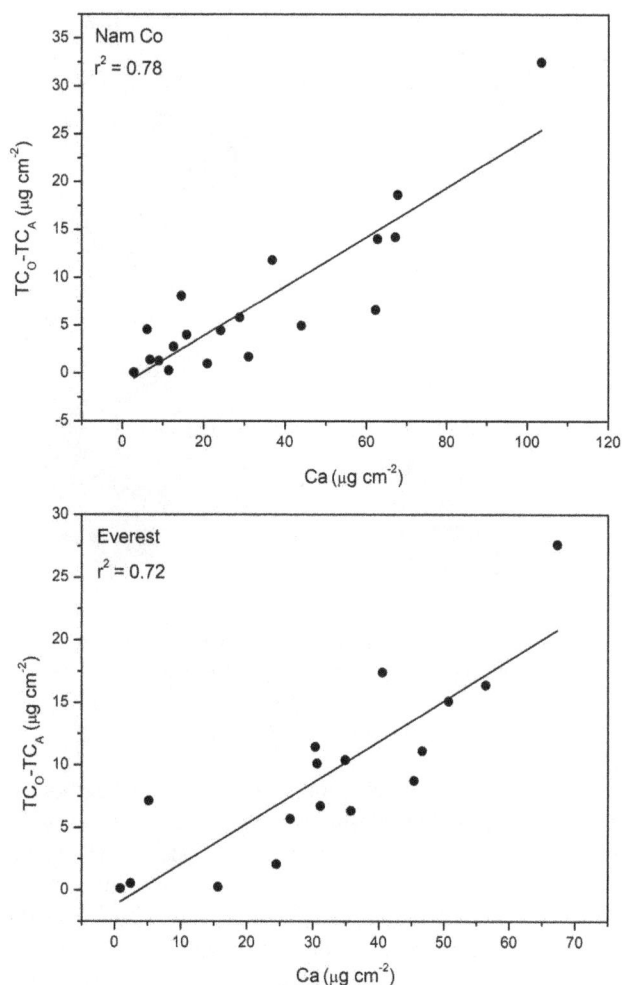

**Figure 4.** Relationship between aerosol IC and Ca at the Nam Co and Everest stations.

tion of OC was considered BC in the acid-treated samples (Jankowski et al., 2008).

Since the influence of carbonate carbon on TC has been observed in $PM_{2.5}$ samples from Qinghai Lake, northwest China (Zhao et al., 2015), this phenomenon should be obvious in the TSP samples in this study. Because dust storms in the northern and western parts of the HTP are more severe than those near the two studied stations during the non-monsoon periods, the effect of carbonates on the concentrations of OC and BC should be more pronounced in such areas and must be seriously considered in future studies. Therefore, the overestimation of BC values is likely greater in the northern and western parts of the HTP than near Nam Co, as we noted previously. MD concentrations have been shown to be much higher than BC concentrations in snow and ice core samples from the HTP (Qu et al., 2014; Li et al., 2017). However, numerous studies have measured BC concentrations without using an acid pretreatment step (Qu et al., 2014; Li et al., 2017; Zhang et al., 2017b). Therefore, the contribution of carbonates in MD to the BC concentrations in snow and ice core samples is likely considerable and needs to be quantitatively evaluated in a future study. Similarly, related HTP studies on other issues, such as BC radiative forcing and atmospheric transport models, based on in situ BC concentrations must be adjusted.

### 3.2 Actual BC deposition in the HTP

#### 3.2.1 Overestimated BC deposition in lake cores from the HTP

In general, the BC deposition levels measured via different methods should be consistent for a given region. For instance, in the severely polluted region of eastern China (Chen et al., 2013; Yan et al., 2015), the BC deposition rate recovered from a Chaohu lake core was $1660\,\mathrm{mg\,m^{-2}\,a^{-1}}$ (Han et al., 2016), which was close to the values of northern China calculated from the BC concentrations in aerosols (Fang et al., 2015) and determined via in situ monitoring on the North China Plain (Tang et al., 2014) (Table 1). However, this consistency was not the case in the HTP, where large discrepancies were found among the reported HTP BC deposition values. Catchment inputs have been shown to significantly influence the chemical deposition values reconstructed from lake cores (Yang, 2015). For instance, BC deposition rates derived from lake cores of Nam Co Lake (NMC09) and Qinghai Lake were 260 and $270–390\,\mathrm{mg\,m^{-2}\,a^{-1}}$, respectively, which were much higher than those derived from ice core and snow pit samples from the HTP (Table 1). We proposed that the BC deposition in the lake cores of Qinghai Lake mainly reflected atmospheric deposition followed by catchment inputs. However, the NMC09 value of Nam Co Lake was mainly influenced by catchment inputs.

Lake-core-derived BC deposition in Qinghai Lake was only 2–3 times higher than that estimated from the BC con-

ues, the corrected BC concentrations at the two stations have large uncertainties. Therefore, based on previously reported BC concentrations measured via the same method as in this study (Zhao et al., 2013a; Cong et al., 2015), the actual BC concentrations at the Nam Co and Everest stations were estimated to be 61 and $154\,\mathrm{ng\,m^{-3}}$, respectively.

Carbonates can decompose at relatively low temperatures during measurement, leading to overestimation of both BC and OC concentrations (Karanasiou et al., 2011). In addition, sometimes the acid-treated ambient samples transfer some components of OC to BC, leading to higher BC concentrations (Jankowski et al., 2008). However, this phenomenon was not common in the aerosol samples examined in this study, although several samples from both stations showed higher BC concentrations in the acid-treated samples (Fig. 2). Because $BC_A$ cannot be higher than $BC_O$, the samples with $BC_A / BC_O$ values greater than 1 were not included in the above calculations. Nevertheless, the ratio of $BC_A / BC_O$ was considered to be slightly overestimated, as some por-

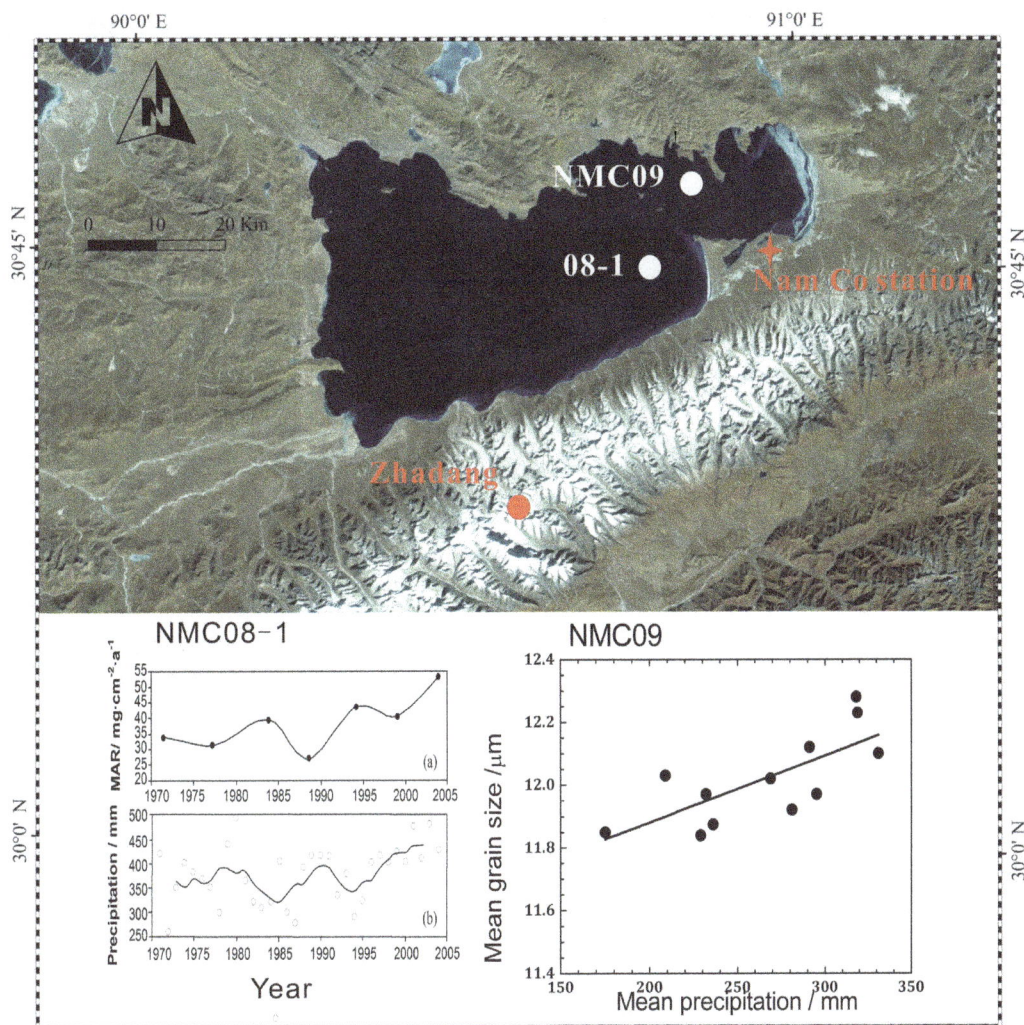

**Figure 5.** Similar variations in precipitation and mass accumulation rates **(a)** (Wang et al., 2011) and significant relationships between mean precipitation and mean grain size **(b)** (Li et al., 2014) in the Nam Co Lake cores.

centrations of PM$_{2.5}$ in the atmosphere (Zhao et al., 2015). Because PM$_{2.5}$ does not include all particles in the atmosphere, the actual BC concentration in the atmosphere should be higher than that of PM$_{2.5}$ (Li et al., 2016b; Viidanoja et al., 2002); therefore, the atmospheric BC deposition should be more similar to that of a lake core. In addition, a previous study showed that approximately 65 and 22 % of the surface sediments in Qinghai Lake resulted from atmospheric deposition and catchment inputs (Wan et al., 2012), respectively, further demonstrating the significant effects of atmospheric deposition on lake core sediments. Therefore, if the BC deposition from atmospheric particles and that of the lake core are the same, then the atmospheric BC deposition based on Qinghai Lake core data is overestimated by approximately 35 %.

Correspondingly, catchment inputs account for a large proportion of the NMC09 samples. BC is widely distributed throughout environmental materials (e.g., soil and river sediments) because of its inert characteristics (Cornelissen et al., 2005; Bucheli et al., 2004). Therefore, river inputs contribute sediments as well as BC to lakes. For instance, in the Nam Co Basin, BC concentrations within the < 20 μm fraction of surface soil and sediment reach 0.78 ± 0.48 mg g$^{-1}$, which is close to the Nam Co Lake core concentration of 0.74 mg g$^{-1}$ (Cong et al., 2013). In addition, several findings have demonstrated the contributions of catchment inputs to Nam Co Lake cores because of the focusing factor, which was shown in the following sections.

First, a large glacial area (141.88 km$^2$) is present within the Nam Co Basin (Fig. 5), and large volumes of glacier meltwater and sediment flow into the lake annually (Wu et al., 2007). Due to recent increasing temperatures and precipitation in the Nam Co Basin, glacier meltwater accounts for approximately 50.6 % of the lake's volume, which has

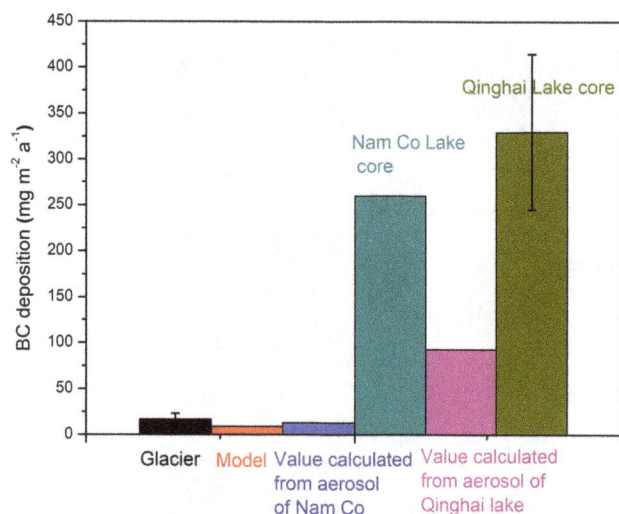

**Figure 6.** Comparison of atmospheric BC deposition rates derived from the glacial region, models, lake cores and values calculated from BC concentrations in the aerosols of the HTP.

increased over the last 30 years (Zhu et al., 2010). Originating at high-elevation glacier terminal, these rivers flow are at a steep angle, and large volumes of suspended allochthonous sediments are transported into Nam Co Lake annually (Doberschütz et al., 2014). A similar phenomenon was also observed in lake cores of a glacier-fed lake as a result of glacier meltwater effects (Bogdal et al., 2011). Second, previous studies on the accumulation rates in lake cores have revealed significant contributions of riverine particles. The accumulation rates in a Nam Co Lake core (NMC 08-1) are consistent with the precipitation variations recorded in the Nam Co Basin during the last 60 years (Fig. 5a) (Wang et al., 2011), indicating that heavy precipitation promotes the transport of large riverine particles to the lake, thus increasing the accumulation rates in the lake cores. Interestingly, the mean grain size of the lake core (NMC09) that reported BC atmospheric deposition showed a significant positive relationship with precipitation (Fig. 5b), thus reflecting the same relationship between catchment inputs and lake core accumulation rates (Li et al., 2014). Because these two lake cores were drilled from different sites (Fig. 5), their similar catchment input characteristics reflect a common feature of Nam Co sediment. As shown above, the BC concentrations in the fine fraction of the river sediments are nearly equivalent to those in the lake cores; thus, additional catchment inputs will increase the BC deposition rates within lake cores. Third, the atmospheric BC deposition rate calculated from BC concentrations in the atmosphere is much lower than the BC deposition rate recorded in the Nam Co Lake cores (Fig. 6), further reflecting the dominant contributions of catchment inputs relative to atmospheric inputs in lake cores.

The above evidence demonstrates that variations in the BC deposition in Nam Co Lake mainly reflect variations in catchment inputs rather than in atmospheric inputs; thus, atmospheric deposition plays a minor role relative to catchment inputs. Because most lakes in the HTP have increased in area over the last 20 years (Zhang et al., 2017a), this phenomenon likely occurs in many other lakes in the HTP.

### 3.2.2 Actual atmospheric BC deposition and potential uncertainties

BC deposition rates derived from ice cores and snow pits are proposed to be closer to the actual atmospheric values in the HTP. This hypothesis is supported by two lines of evidence. First, BC deposition levels in the snow pits of different glaciers are consistent. For example, the estimated BC deposition rates of Laohugou, Tanggula, Zhadang, Demula and Yulong are 25, 21.3, 20, 14.5 and 20.2 $mg\,m^{-2}\,a^{-1}$, respectively (Table 1), which reflects a homogeneous spatial distribution in BC deposition. The above values are also similar to those of ice cores described in other articles (e.g., 18, 12 and 10.1 $mg\,m^{-2}\,a^{-1}$ for the Muztagh Ata, Zuoqiupu and east Rongbuk glaciers, respectively (Xu et al., 2009; Bauer et al., 2013; Ming et al., 2008) (Table 1). Second, these values are nearly equivalent to those of atmospheric BC deposition rates derived from completely different methods (e.g., Community Atmosphere Model version 5 (Zhang et al., 2015) and other models (Bauer et al., 2013) (Table 1). In summary, despite some uncertainties associated with the remote study area, the atmospheric BC deposition rate of $17.9 \pm 5.3\,mg\,m^{-2}\,a^{-1}$ in the glacial region of the HTP is proposed.

### 4 Conclusions

The BC concentration and deposition in the HTP region, which features the largest glacial area in the middle latitudes, were investigated and re-evaluated in this article. Our findings indicated that carbonate carbon contributions from MD have led to overestimations of approximately $52 \pm 35$ and $39 \pm 24$ % in previously reported BC concentrations in TSP samples at the remote Nam Co and Everest stations, respectively, in the central and southern HTP. After omitting the contributions of carbonate carbon, the actual BC concentrations at the Nam Co and Everest stations should be 61 and 154 $ng\,m^{-3}$, respectively. In addition, the levels of OC and TC in TSP samples were also overestimated by $22 \pm 10$ and $19 \pm 13$ %, respectively, at the Nam Co station and by $22 \pm 12$ and $24 \pm 12$ %, respectively, at the Everest station. These values of TC were close to those of a study in the western HTP (Cao et al., 2009). Large arid areas that receive little precipitation are distributed across the western and northern HTP; thus, the effects of carbonates on BC measurements are expected to be greater in these areas and must be considered in future related studies. In addition, TSP samples must be treated with acid to eliminate the effects of carbonates prior to measuring BC. A comparison among BC deposition values based on different methods and materials showed that, because of catchment inputs, the BC deposition rates derived from HTP lake cores were higher than the actual atmospheric deposition values. Correspondingly, the BC deposition values measured from snow pits and ice cores in glacial regions were similar to those obtained via models; thus, these data reflect the actual atmospheric BC deposition values. Although the HTP is located adjacent to seriously polluted regions in south Asia and east China, the HTP BC deposition rates are relatively low because of the high elevation. Finally, our results indicate that the atmospheric BC deposition rate in the HTP is approximately $17.9 \pm 5.3 \, \mathrm{mg\,m^{-2}\,a^{-1}}$, with lower and higher values appearing in the central and peripheral areas of the HTP, respectively.

*Competing interests.* The authors declare that they have no conflict of interest.

*Special issue statement.* This article is part of the special issue "Atmospheric pollution in the Himalayan foothills: the SusKat-ABC international air pollution measurement campaign". It is not associated with a conference.

*Acknowledgements.* This study was supported by the NSFC (41630754, 41675130), the State Key Laboratory of Cryospheric Science (SKLCS-ZZ-2017) and the China Postdoctoral Science Foundation (2016M602897). This study is part of a framework across the HTP: atmospheric Pollution and Cryospheric Change (APCC). Additionally, the authors thank the staff of the Nam Co and Everest stations for collecting samples and providing precipitation data.

Edited by: Ernest Weingartner

# References

Andreae, M. O. and Gelencser, A.: Black carbon or brown carbon? The nature of light-absorbing carbonaceous aerosols, Atmos. Chem. Phys., 6, 3131–3148, https://doi.org/10.5194/acp-6-3131-2006, 2006.

Bauer, S. E., Bausch, A., Nazarenko, L., Tsigaridis, K., Xu, B., Edwards, R., Bisiaux, M., and McConnell, J.: Historical and future black carbon deposition on the three ice caps: ice core measurements and model simulations from 1850 to 2100, J. Geophys. Res.-Atmos., 118, 7948–7961, 2013.

Blais, J. M. and Kalff, J.: The influence of lake morphometry on sediment focusing, Limnol. Oceanogr., 40, 582–588, 1995.

Bogdal, C., Bucheli, T. D., Agarwal, T., Anselmetti, F. S., Blum, F., Hungerbühler, K., Kohler, M., Schmid, P., Scheringer, M., and Sobek, A.: Contrasting temporal trends and relationships of total organic carbon, black carbon, and polycyclic aromatic hydrocarbons in rural low-altitude and remote high-altitude lakes, J. Environ. Monitor., 13, 1316–1326, 2011.

Bolch, T., Kulkarni, A., Kääb, A., Huggel, C., Paul, F., Cogley, J. G., Frey, H., Kargel, J. S., Fujita, K., and Scheel, M.: The state and fate of Himalayan glaciers, Science, 336, 310–314, 2012.

Bollasina, M., Nigam, S., and Lau, K. M.: Absorbing aerosols and summer monsoon evolution over South Asia: an observational portrayal, J. Clim., 21, 3221–3239, https://doi.org/10.1175/2007jcli2094.1, 2008.

Bond, T. C. and Bergstrom, R. W.: Light absorption by carbonaceous particles: an investigative review, Aerosol Sci. Tech., 40, 27–67, 2006.

Bond, T. C., Doherty, S. J., Fahey, D. W., Forster, P. M., Berntsen, T., DeAngelo, B. J., Flanner, M. G., Ghan, S., Kaercher, B., Koch, D., Kinne, S., Kondo, Y., Quinn, P. K., Sarofim, M. C., Schultz, M. G., Schulz, M., Venkataraman, C., Zhang, H., Zhang, S., Bellouin, N., Guttikunda, S. K., Hopke, P. K., Jacobson, M. Z., Kaiser, J. W., Klimont, Z., Lohmann, U., Schwarz, J. P., Shindell, D., Storelvmo, T., Warren, S. G., and Zender, C. S.: Bounding the role of black carbon in the climate system: a scientific assessment, J. Geophys. Res.-Atmos., 118, 5380–5552, https://doi.org/10.1002/jgrd.50171, 2013.

Bosch, C., Andersson, A., Kirillova, E. N., Budhavant, K., Tiwari, S., Praveen, P., Russell, L. M., Beres, N. D., Ramanathan, V., and Gustafsson, Ö.: Source-diagnostic dual-isotope composition and optical properties of water-soluble organic carbon and elemental carbon in the South Asian outflow intercepted over the Indian Ocean, J. Geophys. Res.-Atmos., 119, 11743–711759, 2014.

Bucheli, T. D., Blum, F., Desaules, A., and Gustafsson, Ö.: Polycyclic aromatic hydrocarbons, black carbon, and molecular markers in soils of Switzerland, Chemosphere, 56, 1061–1076, 2004.

Cao, J. J., Lee, S. C., Zhang, X. Y., Chow, J. C., An, Z. S., Ho, K. F., Watson, J. G., Fung, K., Wang, Y. Q., and Shen, Z. X.: Characterization of airborne carbonate over a site near Asian dust source regions during spring 2002 and its climatic and environmental significance, J. Geophys. Res.-Atmos., 110, D03203, https://doi.org/10.1029/2004jd005244, 2005.

Cao, J.-J., Xu, B.-Q., He, J.-Q., Liu, X.-Q., Han, Y.-M., Wang, G.-H., and Zhu, C.-S.: Concentrations, seasonal variations, and transport of carbonaceous aerosols at a remote Mountainous region in western China, Atmos. Environ., 43, 4444–4452, https://doi.org/10.1016/j.atmosenv.2009.06.023, 2009.

Cao, J., Tie, X., Xu, B., Zhao, Z., Zhu, C., Li, G., and Liu, S.: Measuring and modeling black carbon (BC) contamination in the SE Tibetan Plateau, J. Atmos. Chem., 67, 45–60, https://doi.org/10.1007/s10874-011-9202-5, 2010.

Carrico, C. M., Bergin, M. H., Shrestha, A. B., Dibb, J. E., Gomes, L., and Harris, J. M.: The importance of carbon and mineral dust to seasonal aerosol properties in the Nepal Himalaya, Atmos. Environ., 37, 2811–2824, https://doi.org/10.1016/S1352-2310(03)00197-3, 2003.

Cavalli, F., Viana, M., Yttri, K. E., Genberg, J., and Putaud, J. P.: Toward a standardised thermal-optical protocol for measuring atmospheric organic and elemental carbon: the EUSAAR protocol, Atmos. Meas. Tech., 3, 79–89, https://doi.org/10.5194/amt-3-79-2010, 2010.

Chen, B., Andersson, A., Lee, M., Kirillova, E. N., Xiao, Q., Krusa, M., Shi, M., Hu, K., Lu, Z., Streets, D. G., Du, K., and Gustafsson, O.: Source forensics of black carbon aerosols from China, Environ. Sci. Technol., 47, 9102–9108, https://doi.org/10.1021/es401599r, 2013.

Chen, P., Kang, S., Bai, J., Sillanpää, M., and Li, C.: Yak dung combustion aerosols in the Tibetan Plateau: chemical characteristics and influence on the local atmospheric environment, Atmos. Res., 156, 58–66, https://doi.org/10.1016/j.atmosres.2015.01.001, 2015.

Chow, J. C. and Watson, J. G.: $PM_{2.5}$ carbonate concentrations at regionally representative interagency monitoring of protected visual environment sites, J. Geophys. Res.-Atmos., 107, 8344, https://doi.org/10.1029/2001jd000574, 2002.

Chow, J. C., Watson, J. G., Pritchett, L. C., Pierson, W. R., Frazier, C. A., and Purcell, R. G.: The dri thermal/optical reflectance carbon analysis system: description, evaluation and applications in US Air quality studies, Atmos. Environ. A.-Gen., 27, 1185–1201, https://doi.org/10.1016/0960-1686(93)90245-t, 1993.

Chow, J. C., Watson, J. G., Crow, D., Lowenthal, D. H., and Merrifield, T.: Comparison of IMPROVE and NIOSH carbon measurements, Aerosol Sci. Tech., 34, 23–34, https://doi.org/10.1080/027868201300081923, 2001.

Clarke, A. D., Shinozuka, Y., Kapustin, V. N., Howell, S., Huebert, B., Doherty, S., Anderson, T., Covert, D., Anderson, J., Hua, X., Moore, K. G., McNaughton, C., Carmichael, G., and Weber, R.: Size distributions and mixtures of dust and black carbon aerosol in Asian outflow: physiochemistry and optical properties, J. Geophys. Res.-Atmos., 109, D15S09, https://doi.org/10.1029/2003jd004378, 2004.

Cohen, A. S.: Paleolimnology: The History and Evolution of Lake Systems, Oxford University Press, 2003.

Cong, Z., Kang, S., Gao, S., Zhang, Y., Li, Q., and Kawamura, K.: Historical trends of atmospheric black carbon on Tibetan Plateau as reconstructed from a 150 year lake sediment record, Environ. Sci. Technol., 47, 2579–2586, 2013.

Cong, Z., Kang, S., Kawamura, K., Liu, B., Wan, X., Wang, Z., Gao, S., and Fu, P.: Carbonaceous aerosols on the south edge of the Tibetan Plateau: concentrations, seasonality and sources, Atmos. Chem. Phys., 15, 1573–1584, https://doi.org/10.5194/acp-15-1573-2015, 2015.

Cornelissen, G., Gustafsson, Ö., Bucheli, T. D., Jonker, M. T., Koelmans, A. A., and van Noort, P. C.: Extensive sorption of organic compounds to black carbon, coal, and kerogen in sediments and soils: mechanisms and consequences for distribution, bioaccumulation, and biodegradation, Environ. Sci. Technol., 39, 6881–6895, 2005.

Decesari, S., Facchini, M. C., Carbone, C., Giulianelli, L., Rinaldi, M., Finessi, E., Fuzzi, S., Marinoni, A., Cristofanelli, P., and Duchi, R.: Chemical composition of $PM_{10}$ and $PM_1$ at the high-altitude Himalayan station Nepal Climate Observatory-Pyramid (NCO-P) (5079 m a.s.l.), Atmos. Chem. Phys., 10, 4583–4596, https://doi.org/10.5194/acp-10-4583-2010, 2010.

Doberschütz, S., Frenzel, P., Haberzettl, T., Kasper, T., Wang, J., Zhu, L., Daut, G., Schwalb, A., and Mäusbacher, R.: Monsoonal forcing of Holocene paleoenvironmental change on the central Tibetan Plateau inferred using a sediment record from Lake Nam Co (Xizang, China), J. Paleolimnol., 51, 253–266, https://doi.org/10.1007/s10933-013-9702-1, 2014.

Fang, Y., Chen, Y., Tian, C., Lin, T., Hu, L., Huang, G., Tang, J., Li, J., and Zhang, G.: Flux and budget of BC in the continental shelf seas adjacent to Chinese high BC emission source regions, Global Biogeochem. Cy., 29, 957–972, https://doi.org/10.1002/2014gb004985, 2015.

Gustafsson, Ö. and Gschwend, P. M.: The flux of black carbon to surface sediments on the New England continental shelf, Geochim. Cosmochim. Ac., 62, 465–472, 1998.

Hammes, K., Schmidt, M. W. I., Smernik, R. J., Currie, L. A., Ball, W. P., Nguyen, T. H., Louchouarn, P., Houel, S., Gustafsson, O., Elmquist, M., Cornelissen, G., Skjemstad, J. O., Masiello, C. A., Song, J., Peng, P. a., Mitra, S., Dunn, J. C., Hatcher, P. G., Hockaday, W. C., Smith, D. M., Hartkopf-Froeder, C., Boehmer, A., Lueer, B., Huebert, B. J., Amelung, W., Brodowski, S., Huang, L., Zhang, W., Gschwend, P. M., Flores-Cervantes, D. X., largeau, C., Rouzaud, J.-N., Rumpel, C., Guggenberger, G., Kaiser, K., Rodionov, A., Gonzalez-Vila, F. J., Gonzalez-Perez, J. A., de la Rosa, J. M., Manning, D. A. C., Lopez-Capel, E., and Ding, L.: Comparison of quantification methods to measure fire-derived (black/elemental) carbon in soils and sediments using reference materials from soil, water, sediment and the atmosphere, Global Biogeochem. Cy., 21, GB3016, https://doi.org/10.1029/2006gb002914, 2007.

Han, Y. M., Cao, J. J., Yan, B. Z., Kenna, T. C., Jin, Z. D., Cheng, Y., Chow, J. C., and An, Z. S.: Comparison of elemental carbon in lake sediments measured by three different methods and 150 year pollution history in eastern China, Environ. Sci. Technol., 45, 5287–5293, 2011.

Han, Y., Wei, C., Bandowe, B., Wilcke, W., Cao, J., Xu, B., Gao, S., Tie, X., Li, G., and Jin, Z.: Elemental carbon and polycyclic aromatic compounds in a 150 year sediment core from Lake Qinghai, Tibetan Plateau, China: influence of regional and local sources and transport pathways, Environ. Sci. Technol., 49, 4176–4183, 2015.

Han, Y. M., Wei, C., Huang, R. J., Bandowe, B. A. M., Ho, S. S. H., Cao, J. J., Jin, Z. D., Xu, B. Q., Gao, S. P., Tie, X. X., An, Z. S., and Wilcke, W.: Reconstruction of atmospheric soot history in inland regions from lake sediments over the past 150 years, Sci. Rep.-UK, 6, 19151, https://doi.org/10.1038/srep19151, 2016.

Ho, K. F., Zhang, R. J., Lee, S. C., Ho, S. S. H., Liu, S. X., Fung, K., Cao, J. J., Shen, Z. X., and Xu, H. M.: Characteristics of carbonate carbon in $PM_{2.5}$ in a typical semi-arid area of Northeastern China, Atmos. Environ., 45, 1268–1274, https://doi.org/10.1016/j.atmosenv.2010.12.007, 2011.

Jankowski, N., Schmidl, C., Marr, I. L., Bauer, H., and Puxbaum, H.: Comparison of methods for the quantification of carbonate carbon in atmospheric $PM_{10}$ aerosol samples, Atmos. Environ., 42, 8055–8064, https://doi.org/10.1016/j.atmosenv.2008.06.012, 2008.

Ji, Z., Kang, S., Cong, Z., Zhang, Q., and Yao, T.: Simulation of carbonaceous aerosols over the Third Pole and adjacent regions: distribution, transportation, deposition, and climatic effects, Clim.

Dynam., 45, 2831–2846, https://doi.org/10.1007/s00382-015-2509-1, 2015.

Jurado, E., Dachs, J., Duarte, C. M., and Simo, R.: Atmospheric deposition of organic and black carbon to the global oceans, Atmos. Environ., 42, 7931–7939, 2008.

Kang, S., Xu, Y., You, Q., Flügel, W.-A., Pepin, N., and Yao, T.: Review of climate and cryospheric change in the Tibetan Plateau, Environ. Res. Lett., 5, 015101, https://doi.org/10.1088/1748-9326/5/1/015101, 2010.

Karanasiou, A., Diapouli, E., Cavalli, F., Eleftheriadis, K., Viana, M., Alastuey, A., Querol, X., and Reche, C.: On the quantification of atmospheric carbonate carbon by thermal/optical analysis protocols, Atmos. Meas. Tech., 4, 2409–2419, https://doi.org/10.5194/amt-4-2409-2011, 2011.

Karanasiou, A., Minguillón, M. C., Viana, M., Alastuey, A., Putaud, J. P., Maenhaut, W., Panteliadis, P., Močnik, G., Favez, O., and Kuhlbusch, T. A. J.: Thermal-optical analysis for the measurement of elemental carbon (EC) and organic carbon (OC) in ambient air a literature review, Atmos. Meas. Tech. Discuss., 8, 9649–9712, https://doi.org/10.5194/amtd-8-9649-2015, 2015.

Kaspari, S. D., Schwikowski, M., Gysel, M., Flanner, M. G., Kang, S., Hou, S., and Mayewski, P. A.: Recent increase in black carbon concentrations from a Mt. Everest ice core spanning 1860–2000 AD, Geophys. Res. Lett., 38, L04703, https://doi.org/10.1029/2010gl046096, 2011.

Koch, D., Schulz, M., Kinne, S., McNaughton, C., Spackman, J. R., Balkanski, Y., Bauer, S., Berntsen, T., Bond, T. C., Boucher, O., Chin, M., Clarke, A., De Luca, N., Dentener, F., Diehl, T., Dubovik, O., Easter, R., Fahey, D. W., Feichter, J., Fillmore, D., Freitag, S., Ghan, S., Ginoux, P., Gong, S., Horowitz, L., Iversen, T., Kirkevag, A., Klimont, Z., Kondo, Y., Krol, M., Liu, X., Miller, R., Montanaro, V., Moteki, N., Myhre, G., Penner, J. E., Perlwitz, J., Pitari, G., Reddy, S., Sahu, L., Sakamoto, H., Schuster, G., Schwarz, J. P., Seland, O., Stier, P., Takegawa, N., Takemura, T., Textor, C., van Aardenne, J. A., and Zhao, Y.: Evaluation of black carbon estimations in global aerosol models, Atmos. Chem. Phys., 9, 9001–9026, https://doi.org/10.5194/acp-9-9001-2009, 2009.

Lau, W. K. M., Kim, M.-K., Kim, K.-M., and Lee, W.-S.: Enhanced surface warming and accelerated snow melt in the Himalayas and Tibetan Plateau induced by absorbing aerosols, Environ. Res. Lett., 5, 025204, https://doi.org/10.1088/1748-9326/5/2/025204, 2010.

Li, C., Kang, S., Wang, X., Ajmone-Marsan, F., and Zhang, Q.: Heavy metals and rare earth elements (REEs) in soil from the Nam Co Basin, Tibetan Plateau, Environ. Geol., 53, 1433–1440, https://doi.org/10.1007/s00254-007-0752-4, 2008.

Li, C., Kang, S., Zhang, Q., and Wang, F.: Rare earth elements in the surface sediments of the Yarlung Tsangbo (Upper Brahmaputra River) sediments, southern Tibetan Plateau, Quatern. Int., 208, 151–157, 2009.

Li, Q., Kang, S., Zhang, Q., Huang, J., guo, J., Wang, K., and Wang, J.: A 150 year climate change history reconstructed by lake sediment of Nam Co, Tibetan Plateau (in Chinese), Acta Sedimentologica Sinica, 32, 669–676, 2014.

Li, C., Bosch, C., Kang, S., Andersson, A., Chen, P., Zhang, Q., Cong, Z., Chen, B., Qin, D., and Gustafsson, Ö.: Sources of black carbon to the Himalayan–Tibetan Plateau glaciers, Nat. Commun., 7, 12574, https://doi.org/10.1038/ncomms12574, 2016a.

Li, C., Chen, P., Kang, S., Yan, F., Hu, Z., Qu, B., and Sillanpää, M.: Concentrations and light absorption characteristics of carbonaceous aerosol in $PM_{2.5}$ and $PM_{10}$ of Lhasa city, the Tibetan Plateau, Atmos. Environ., 127, 340–346, https://doi.org/10.1016/j.atmosenv.2015.12.059, 2016b.

Li, C., Chen, P., Kang, S., Yan, F., Li, X., Qu, B., and Sillanpää, M.: Carbonaceous matter deposition in the high glacial regions of the Tibetan Plateau, Atmos. Environ., 141, 203–208, https://doi.org/10.1016/j.atmosenv.2016.06.064, 2016c.

Li, C., Yan, F., Kang, S., Chen, P., Hu, Z., Gao, S., Qu, B., and Sillanpää, M.: Light absorption characteristics of carbonaceous aerosols in two remote stations of the southern fringe of the Tibetan Plateau, China, Atmos. Environ., 143, 79–85, https://doi.org/10.1016/j.atmosenv.2016.08.042, 2016d.

Li, X., Kang, S., He, X., Qu, B., Tripathee, L., Jing, Z., Paudyal, R., Li, Y., Zhang, Y., Yan, F., Li, G., and Li, C.: Light-absorbing impurities accelerate glacier melt in the Central Tibetan Plateau, Sci. Total Environ., 587, 482–490, https://doi.org/10.1016/j.scitotenv.2017.02.169, 2017.

Lim, S., Faïn, X., Zanatta, M., Cozic, J., Jaffrezo, J. L., Ginot, P., and Laj, P.: Refractory black carbon mass concentrations in snow and ice: method evaluation and inter-comparison with elemental carbon measurement, Atmos. Meas. Tech., 7, 3307–3324, https://doi.org/10.5194/amt-7-3307-2014, 2014.

Liu, B., Cong, Z., Wang, Y., Xin, J., Wan, X., Pan, Y., Liu, Z., Wang, Y., Zhang, G., Wang, Z., Wang, Y., and Kang, S.: Background aerosol over the Himalayas and Tibetan Plateau: observed characteristics of aerosol mass loading, Atmos. Chem. Phys., 17, 449–463, https://doi.org/10.5194/acp-17-449-2017, 2017.

Liu, Y.-H., Dong, G.-R., Li, S., and Dong, Y.-X.: Status, causes and combating suggestions of sandy desertification in qinghai-tibet plateau, Chinese Geogr. Sci., 15, 289–296, https://doi.org/10.1007/s11769-005-0015-9, 2005.

Maenhaut, W., Schwarz, J., Cafmeyer, J., and Chi, X.: Aerosol chemical mass closure during the EUROTRAC-2 AEROSOL Intercomparison 2000, Nucl. Instrum. Meth. B, 189, 233–237, https://doi.org/10.1016/S0168-583X(01)01048-5, 2002.

Marinoni, A., Cristofanelli, P., Laj, P., Duchi, R., Calzolari, F., Decesari, S., Sellegri, K., Vuillermoz, E., Verza, G. P., Villani, P., and Bonasoni, P.: Aerosol mass and black carbon concentrations, a two year record at NCO-P (5079 m, Southern Himalayas), Atmos. Chem. Phys., 10, 8551–8562, https://doi.org/10.5194/acp-10-8551-2010, 2010.

Ming, J., Zhang, D., Kang, S., and Tian, W.: Aerosol and fresh snow chemistry in the East Rongbuk Glacier on the northern slope of Mt. Qomolangma (Everest), J. Geophys. Res., 112, D15307, https://doi.org/10.1029/2007JD008618, 2007.

Ming, J., Cachier, H., Xiao, C., Qin, D., Kang, S., Hou, S., and Xu, J.: Black carbon record based on a shallow Himalayan ice core and its climatic implications, Atmos. Chem. Phys., 8, 1343–1352, https://doi.org/10.5194/acp-8-1343-2008, 2008.

Ming, J., Xiao, C., Cachier, H., Qin, D., Qin, X., Li, Z., and Pu, J.: Black carbon (BC) in the snow of glaciers in west China and its potential effects on albedos, Atmos. Res., 92, 114–123, 2009.

Ming, J., Xiao, C., Sun, J., Kang, S., and Bonasoni, P.: Carbonaceous particles in the atmosphere and precipitation of the

Nam Co region, Central Tibet, J. Environ. Sci., 22, 1748–1756, 2010.

Ming, J., Xiao, C., Du, Z., and Yang, X.: An overview of black carbon deposition in High Asia glaciers and its impacts on radiation balance, Adv. Water Resour., 55, 80–87, https://doi.org/10.1016/j.advwatres.2012.05.015, 2013.

Perrone, M. R., Piazzalunga, A., Prato, M., and Carofalo, I.: Composition of fine and coarse particles in a coastal site of the central Mediterranean: carbonaceous species contributions, Atmos. Environ., 45, 7470–7477, 2011.

Pio, C. A., Legrand, M., Oliveira, T., Afonso, J., Santos, C., Caseiro, A., Fialho, P., Barata, F., Puxbaum, H., Sanchez-Ochoa, A., Kasper-Giebl, A., Gelencser, A., Preunkert, S., and Schock, M.: Climatology of aerosol composition (organic versus inorganic) at nonurban sites on a west–east transect across Europe, J. Geophys. Res.-Atmos., 112, D23S02, https://doi.org/10.1029/2006jd008038, 2007.

Qu, B., Ming, J., Kang, S. C., Zhang, G. S., Li, Y. W., Li, C. D., Zhao, S. Y., Ji, Z. M., and Cao, J. J.: The decreasing albedo of the Zhadang glacier on western Nyainqentanglha and the role of light-absorbing impurities, Atmos. Chem. Phys., 14, 11117–11128, https://doi.org/10.5194/acp-14-11117-2014, 2014.

Ram, K., Sarin, M. M., and Tripathi, S. N.: A 1 year record of carbonaceous aerosols from an urban site in the Indo-Gangetic Plain: characterization, sources, and temporal variability, J. Geophys. Res., 115, 9–12, 2010.

Ramanathan, V. and Carmichael, G.: Global and regional climate changes due to black carbon, Nat. Geosci., 1, 221–227, 2008.

Reisinger, P., Wonaschütz, A., Hitzenberger, R., Petzold, A., Bauer, H., Jankowski, N., Puxbaum, H., Chi, X., and Maenhaut, W.: Intercomparison of measurement techniques for black or elemental carbon under urban background conditions in wintertime: influence of biomass combustion, Environ. Sci. Technol., 42, 884–889, https://doi.org/10.1021/es0715041, 2008.

Ruppel, M. M., Isaksson, E., Ström, J., Beaudon, E., Svensson, J., Pedersen, C. A., and Korhola, A.: Increase in elemental carbon values between 1970 and 2004 observed in a 300 year ice core from Holtedahlfonna (Svalbard), Atmos. Chem. Phys., 14, 11447–11460, https://doi.org/10.5194/acp-14-11447-2014, 2014.

Sillanpää, M., Frey, A., Hillamo, R., Pennanen, A. S., and Salonen, R. O.: Organic, elemental and inorganic carbon in particulate matter of six urban environments in Europe, Atmos. Chem. Phys., 5, 2869–2879, https://doi.org/10.5194/acp-5-2869-2005, 2005.

Tang, Y., Han, G. L., and Xu, Z. F.: Atmospheric black carbon deposit in Beijing and Zhangbei, China, in: Geochemistry of the Earth's Surface Ges-10, edited by: Gaillardet, J., Procedia Earth and Planetary Science, 383–387, 2014.

Viidanoja, J., Sillanpaa, M., Laakia, J., Kerminen, V. M., Hillamo, R., Aarnio, P., and Koskentalo, T.: Organic and black carbon in $PM_{2.5}$ and $PM_{10}$: 1 year of data from an urban site in Helsinki, Finland, Atmos. Environ., 36, 3183–3193, https://doi.org/10.1016/s1352-2310(02)00205-4, 2002.

Wan, D. J., Jin, Z. D., and Wang, Y. X.: Geochemistry of eolian dust and its elemental contribution to Lake Qinghai sediment, Appl. Geochem., 27, 1546–1555, https://doi.org/10.1016/j.apgeochem.2012.03.009, 2012.

Wan, X., Kang, S., Wang, Y., Xin, J., Liu, B., Guo, Y., Wen, T., Zhang, G., and Cong, Z.: Size distribution of carbonaceous aerosols at a high-altitude site on the central Tibetan Plateau (Nam Co Station, 4730 m a.s.l.), Atmos. Res., 153, 155–164, https://doi.org/10.1016/j.atmosres.2014.08.008, 2015.

Wang, S., Wang, J., Zhou, Z., and Shang, K.: Regional characteristics of three kinds of dust storm events in China, Atmos. Environ., 39, 509–520, https://doi.org/10.1016/j.atmosenv.2004.09.033, 2005.

Wang, J., Zhu, L., Wang, Y., Gao, S., and Daut, G.: Spatial variability of recent sedimentation rate and variations in the past 60 years in Nam Co, Tibetan Palteau, China (in Chinese), Quaternary Sci. Rev., 31, 535–543, 2011.

Watson, J. G., Chow, J. C., and Chen, L. W. A.: Summary of organic and elemental carbon/black carbon analysis methods and intercomparisons, Aerosol Air Qual. Res., 5, 65–102, 2005.

Wu, Y., Zhu, L., Ye, Q., and Wang, L.: The response of lake-glacier area change to climate variations in Namco Basin, Central Tibetan Plateau, during the last three decades (in Chinese), Acta Geographica Sinica, 62, 301–311, 2007.

Xu, B., Cao, J., Hansen, J., Yao, T., Joswia, D. R., Wang, N., Wu, G., Wang, M., Zhao, H., and Yang, W.: Black soot and the survival of Tibetan glaciers, P. Natl. Acad. Sci. USA, 106, 22114–22118, 2009.

Xu, Y., Ramanathan, V., and Washington, W. M.: Observed high-altitude warming and snow cover retreat over Tibet and the Himalayas enhanced by black carbon aerosols, Atmos. Chem. Phys., 16, 1303–1315, https://doi.org/10.5194/acp-16-1303-2016, 2016.

Yan, C., Zheng, M., Sullivan, A. P., Bosch, C., Desyaterik, Y., Andersson, A., Li, X., Guo, X., Zhou, T., Gustafsson, Ö., and Collett Jr, J. L.: Chemical characteristics and light-absorbing property of water-soluble organic carbon in Beijing: biomass burning contributions, Atmos. Environ., 121, 4–12, https://doi.org/10.1016/j.atmosenv.2015.05.005, 2015.

Yang, H.: Lake sediments may not faithfully record decline of atmospheric pollutant deposition, Environ. Sci. Technol., 49, 12607–12608, https://doi.org/10.1021/acs.est.5b04386, 2015.

You, Q., Kang, S., Pepin, N., Fluegel, W.-A., Yan, Y., Behrawan, H., and Huang, J.: Relationship between temperature trend magnitude, elevation and mean temperature in the Tibetan Plateau from homogenized surface stations and reanalysis data, Glob. Planet. Change, 71, 124–133, https://doi.org/10.1016/j.gloplacha.2010.01.020, 2010.

Zhang, G., Yao, T., Piao, S., Bolch, T., Xie, H., Chen, D., Gao, Y., O'Reilly, C. M., Shum, C. K., Yang, K., Yi, S., Lei, Y., Wang, W., He, Y., Shang, K., Yang, X., and Zhang, H.: Extensive and drastically different alpine lake changes on Asia's high plateaus during the past four decades, Geophys. Res. Lett., 44, 252–260, https://doi.org/10.1002/2016gl072033, 2017a.

Zhang, R., Wang, H., Qian, Y., Rasch, P. J., Easter, R. C., Ma, P. L., Singh, B., Huang, J., and Fu, Q.: Quantifying sources, transport, deposition, and radiative forcing of black carbon over the Himalayas and Tibetan Plateau, Atmos. Chem. Phys., 15, 6205–6223, https://doi.org/10.5194/acp-15-6205-2015, 2015.

Zhang, Y., Kang, S., Cong, Z., Schmale, J., Sprenger, M., Li, C., Yang, W., Gao, T., Sillanpää, M., and Li, X.: Light-absorbing impurities enhance glacier albedo reduction in the southeastern Tibetan Plateau, J. Geophys. Res.-Atmos., 122, 6915–6933, 2017b.

Zhao, S., Ming, J., Sun, J., and Xiao, C.: Observation of carbonaceous aerosols during 2006–2009 in Nyainqntanglha Mountains and the implications for glaciers, Environ. Sci. Pollut. R., 20, 5827–5838, https://doi.org/10.1007/s11356-013-1548-6, 2013a.

Zhao, Z., Cao, J., Shen, Z., Xu, B., Zhu, C., Chen, L. W. A., Su, X., Liu, S., Han, Y., Wang, G., and Ho, K.: Aerosol particles at a high-altitude site on the Southeast Tibetan Plateau, China: implications for pollution transport from South Asia, J. Geophys. Res.-Atmos., 118, 11360–11375, https://doi.org/10.1002/jgrd.50599, 2013b.

Zhao, Z. Z., Cao, J. J., Shen, Z. X., Huang, R. J., Hu, T. F., Wang, P., Zhang, T., and Liu, S. X.: Chemical composition of PM$_{2.5}$ at a high-altitude regional background site over Northeast of Tibet Plateau, Atmos. Pollut. Res., 6, 815–823, https://doi.org/10.5094/apr.2015.090, 2015.

Zhu, L., Xie, M., and Wu, Y.: Quantitative analysis of lake area variations and the influence factors from 1971 to 2004 in the Nam Co basin of the Tibetan Plateau, Chinese Sci. Bull., 55, 1294–1303, 2010.

# Comparison of large-scale dynamical variability in the extratropical stratosphere among the JRA-55 family data sets: impacts of assimilation of observational data in JRA-55 reanalysis data

**Masakazu Taguchi**

Department of Earth Science, Aichi University of Education, Kariya, 448-8542, Japan

*Correspondence to:* Masakazu Taguchi (mtaguchi@auecc.aichi-edu.ac.jp)

**Abstract.** This study compares large-scale dynamical variability in the extratropical stratosphere, such as major stratospheric sudden warmings (MSSWs), among the Japanese 55-year Reanalysis (JRA-55) family data sets. The JRA-55 family consists of three products: a standard product (STDD) of the JRA-55 reanalysis data and two sub-products of JRA-55C (CONV) and JRA-55AMIP (AMIP). CONV assimilates only conventional surface and upper-air observations without assimilation of satellite observations, whereas AMIP runs the same numerical weather prediction model without assimilation of observational data. A comparison of the occurrence of MSSWs in Northern Hemisphere (NH) winter shows that, compared to STDD, CONV delays several MSSWs by 1 to 4 days and also misses a few MSSWs. CONV also misses the Southern Hemisphere (SH) MSSW in September 2002. AMIP shows significantly fewer MSSWs in Northern Hemisphere winter and especially lacks MSSWs of the high aspect ratio of the polar vortex in which the vortex is highly stretched or split. A further examination of daily geopotential height differences between STDD and CONV reveals occasional peaks in both hemispheres that are separated from MSSWs. The delayed and missed MSSW cases have smaller height differences in magnitude than such peaks. The height differences for those MSSWs include large contributions from the zonal component, which reflects underestimations in the weakening of the zonal mean polar night jet in CONV. We also explore strong planetary wave forcings and associated polar vortex weakenings for STDD and AMIP. We find a lower frequency of strong wave forcings and weaker vortex responses to such wave forcings in AMIP, consistent with the lower MSSW frequency.

## 1 Introduction

Large-scale dynamical variability is an important feature in the extratropical stratosphere, especially in the Northern Hemisphere (NH) winter stratosphere (e.g., Labitzke and van Loon, 1999; Yoden et al., 2002; Waugh and Polvani, 2010). The NH winter stratosphere exhibits large intraseasonal and interannual variations, reflecting anomalously strong and weak conditions of the polar vortex. The polar vortex in the Southern Hemisphere (SH) spring stratosphere also shows large variations, e.g., in its strength and distribution, while the SH winter stratosphere is more dynamically quiescent.

Some weak conditions of the polar vortex correspond to the occurrence of stratospheric sudden warmings (SSWs), during which the polar night jet weakens and polar stratospheric temperatures rise as the polar vortex largely distorts and/or breaks down (e.g., Limpasuvan et al., 2004; Charlton and Polvani, 2007). An SSW is classified as a major SSW (MSSW) when it accompanies a reversal of the zonal mean zonal wind often seen at 60° N, 10 hPa (e.g., Butler et al., 2015). An SSW without such a zonal wind reversal is classified as a minor SSW.

Previous studies investigate aspects of such dynamical variability in the extratropical stratosphere using multiple reanalysis data sets. Reanalysis data sets are a vital tool to understand atmospheric variability and relevant processes in the climate science including middle atmosphere dynamics, but different reanalyses sometimes yield different results for the same diagnostics. Martineau and Son (2010) examined time evolutions of the Northern Annular Mode index for stratospheric vortex weakening and intensification events among five reanalyses and found good agreement. Martineau et al. (2016) investigated dynamical consistency in the extratropical stratosphere among eight reanalyses as quantified by the residual of the zonal momentum equation. Applying a multiple linear regression analysis to nine reanalyses, Mitchell et al. (2015) studied signatures of interannual variability associated with natural forcings and found remarkable similarity among the data sets. Manney et al. (2005) conducted a diagnostic comparison of the SH MSSW in September 2002 among several meteorological data sets.

Some of these studies are part of a coordinated activity of the Stratosphere–troposphere Processes And their Role in Climate (SPARC) Reanalysis Intercomparison Project (S-RIP; Fujiwara et al., 2017). The climatology and interannual variability of monthly mean temperature and wind fields are surveyed in the S-RIP framework. Furthermore, various aspects of the dynamical coupling between the NH extratropical stratosphere and troposphere will also be investigated there.

The present study focuses on the Japanese 55-year Reanalysis (JRA-55; Kobayashi et al., 2015) data among others, which is one of newer reanalyses. A unique feature of JRA-55 is that, in addition to the standard product (STDD), two companion products, JRA-55C (CONV) and JRA-55AMIP (AMIP), are also available. CONV assimilates conventional surface and upper-air observations only, without assimilation of satellite observations. AMIP is an AMIP-type simulation using the same forecast model as in STDD and CONV, without assimilation of any observational data. The three products are called the "JRA-55 family data sets" as a whole (Kobayashi et al., 2014).

Previous studies investigated some aspects of the JRA-55 family data sets, especially in the stratosphere. Kobayashi and Iwasaki (2016) examined the Brewer–Dobson circulation in the lower stratosphere to show that the mass stream function at 100 hPa (vertically integrated northward mass flux above 100 hPa) is similar between STDD and CONV in annual and seasonal averages, but it is much weaker for AMIP. Kobayashi et al. (2014) and Kobayashi and Iwasaki (2016) also showed that the polar night jet in the winter stratosphere for each hemisphere is stronger for CONV than for STDD, which is associated with weaker upward wave propagation and driving. The differences of AMIP from STDD are qualitatively similar but are much larger in magnitude.

This study seeks to compare the climatology and large-scale dynamical variability, such as frequency and vortex geometry of SSWs, in the NH and SH extratropical stratosphere among the JRA-55 family data sets. A motivation for this study is that dynamical variability in the extratropical stratosphere remains relatively unexplored in the JRA-55 family data sets, while they provide a good opportunity for a clear comparison owing to their meticulous design. In order to better understand the differences of MSSWs, we further describe differences of daily geopotential height fields between STDD and CONV and relate them to the occurrence

of MSSWs. We also explore strong wave forcings and associated stratospheric vortex responses for STDD and AMIP.

The rest of the paper is organized as follows. Section 2 explains the data and analysis methods used in this study. Section 3 surveys the climatology and variability in the extratropical stratosphere in the JRA-55 family data sets. Section 4 further examines STDD and CONV, and Sect. 5 examines STDD and AMIP. Finally, Sect. 6 provides the summary and discussion.

## 2 Data and analysis method

### 2.1 Data

This study makes use of daily averages for the three products of the JRA-55 family data sets. The horizontal resolution is $2.5° \times 2.5°$, with 37 levels up to 1 hPa. The full period is from 1958 to 2012, but we use a shorter period in some comparisons (see below). We mainly use the zonal mean zonal wind, poleward eddy heat flux by wave components of zonal wave numbers 1–3 (waves 1–3), and geopotential height. The eddy heat flux in the extratropical lower stratosphere (e.g., 40–90° N, 100 hPa) is used as a proxy for planetary wave forcing from the troposphere, since it is proportional to the vertical component of the Eliassen–Palm (EP) flux under the quasi-geostrophic scaling (Andrews et al., 1987).

We regard STDD, or the standard JRA-55 reanalysis data (Kobayashi et al., 2015), as a good representation of the real world. A comparison of CONV to STDD elucidates the effects of assimilation of satellite data, since the inclusion or exclusion of assimilation of satellite data is the only difference between the two (Kobayashi et al., 2014). We also compare AMIP to STDD to examine model biases, since the AMIP data are obtained from an AMIP-type forecast simulation using the same numerical weather prediction model and same boundary conditions (sea surface temperatures, greenhouse gases, etc.; see Kobayashi et al., 2015) without assimilation of any observational data.

For the comparison between STDD and CONV, we use the data from 1972–1973 to 2012–2013 when CONV is available. For the comparison between STDD and AMIP, we use the data from 1957–1958 to 2012–2013 (i.e., full period). We sort the data in time for both hemispheres so that they begin from June in each year and end in May in the next year to facilitate our main focus on NH winter when the NH extratropical stratosphere is dynamically active. For both NH and SH, we refer to each year from June to May by the year to which the month of January belongs.

### 2.2 Analysis methods

#### 2.2.1 Identification of MSSW onset dates

The method outlined by Charlton and Polvani (2007) is basically followed to identify MSSWs. This method identifies

**Table 1.** Onset dates of DJF MSSWs for STDD and CONV, with differences in the onset dates (number of days) of CONV from STDD. Empty cells indicate the period when CONV is unavailable. The "M" letters mean that the MSSWs identified in STDD are missed in CONV.

| STDD | CONV | DIFF |
|------|------|------|
| 30 Jan 1958 | | |
| 17 Jan 1960 | | |
| 30 Jan 1963 | | |
| 18 Dec 1965 | | |
| 23 Feb 1966 | | |
| 7 Jan 1968 | | |
| 2 Jan 1970 | | |
| 18 Jan 1971 | | |
| 31 Jan 1973 | 1 Feb 1973 | 1 |
| 9 Jan 1977 | 13 Jan 1977 | 4 |
| 22 Feb 1979 | 22 Feb 1979 | 0 |
| 29 Feb 1980 | 29 Feb 1980 | 0 |
| 6 Feb 1981 | M | – |
| 4 Dec 1981 | M | – |
| 24 Feb 1984 | 24 Feb 1984 | 0 |
| 1 Jan 1985 | 4 Jan 1985 | 3 |
| 23 Jan 1987 | 23 Jan 1987 | 0 |
| 8 Dec 1987 | 8 Dec 1987 | 0 |
| 21 Feb 1989 | 21 Feb 1989 | 0 |
| 15 Dec 1998 | 16 Dec 1998 | 1 |
| 26 Feb 1999 | 26 Feb 1999 | 0 |
| 11 Feb 2001 | 11 Feb 2001 | 0 |
| 31 Dec 2001 | M | – |
| 18 Jan 2003 | 18 Jan 2003 | 0 |
| 5 Jan 2004 | 7 Jan 2004 | 2 |
| 21 Jan 2006 | 22 Jan 2006 | 1 |
| 24 Feb 2007 | 24 Feb 2007 | 0 |
| 22 Feb 2008 | 22 Feb 2008 | 0 |
| 24 Jan 2009 | 25 Jan 2009 | 1 |
| 9 Feb 2010 | 9 Feb 2010 | 0 |

the 10 hPa height according to Seviour et al. (2013). This method diagnoses where the center of the vortex is located (CL) and how stretched the vortex is (AR), where the vortex is defined as the region of the 10 hPa height lower than a threshold of Zb. The parameter Zb is taken as the climatological and zonal mean of the 10 hPa height at 60° N in each product and is different among the three products. The height fields are smoothed with a 5-day running mean in calculating CL and AR so as to filter out day-to-day fluctuations and capture dominant geometry features.

### 2.2.3 Calculation of RMSD of geopotential height fields

Differences in daily geopotential height fields, e.g., at 10 hPa, between STDD and CONV are evaluated with the root mean square difference (RMSD) as follows:

$$\text{RMSD} = \left[ \frac{\sum_{i=1}^{n} w_i \{Z_{\text{CONV}}(x_i) - Z_{\text{STDD}}(x_i)\}^2}{\sum_{i=1}^{n} w_i} \right]^{1/2}. \tag{1}$$

Here, $Z_{\text{STDD}}$ and $Z_{\text{CONV}}$ denote geopotential height fields at an arbitrary level on a day of interest for STDD and CONV, respectively; $x_i$ denotes spatial grid points (longitude and latitude); and $w_i$ is the cosine of latitude. The summations are taken for all extratropical grid points (indexed with $i$ from 1 to $n$) poleward of 30° N/S. Note that an arbitrary RMSD value is divided into contributions from the zonal and wave components as follows:

$$\text{RMSD}^2 = \text{RMSD}_{\text{zonal}}^2 + \text{RMSD}_{\text{wave}}^2. \tag{2}$$

Here, $\text{RMSD}_{\text{wave}}$ is calculated by applying Eq. (1) to wave fields of $Z_{\text{STDD}}$ and $Z_{\text{CONV}}$.

the onset date (denoted as lag $= 0$ day) of an MSSW as when the zonal mean zonal wind at 60° N, 10 hPa, reverses from a westerly wind to an easterly wind. We focus on MSSWs during the winter period of December–January–February (DJF). In order to identify two (or more) MSSWs in one season, the onset dates between two successive events must be separated more than 20 days, and the zonal wind must recover above $20 \, \text{m s}^{-1}$ between them. The latter condition is added to ensure that the polar vortex is sufficiently re-established after the first event. The resultant onset dates of the MSSWs identified for STDD and CONV are shown in Table 1. The onset dates for STDD are identical to those in Butler et al. (2017) as far as the DJF MSSWs are concerned.

### 2.2.2 Calculation of polar vortex geometry

In order to characterize the geometry of the polar vortex, centroid latitude (CL) and aspect ratio (AR) are calculated for

## 3 Survey of climatology and variability in the extratropical stratosphere during NH winter and SH spring

This section surveys the climatology (long-term mean) and variability in the extratropical stratosphere for NH winter and SH spring using the zonal mean zonal wind and 10 hPa geopotential height.

### 3.1 Climatology

Figure 1 shows the climatological zonal mean zonal wind in DJF and September–October–November (SON) for STDD, CONV, and AMIP. Color shades plot only differences of CONV or AMIP from STDD that are judged to be statistically significant according to a Student's $t$ test (two-sided test) at the 95 % confidence level. When taking differences between STDD and CONV, the STDD data after the 1972–1973 season are used as stated in Sect. 2.1. The degrees of

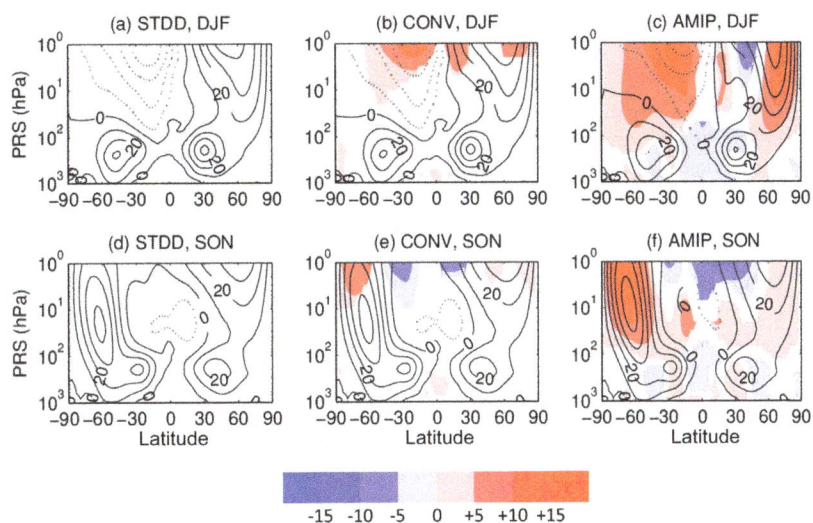

**Figure 1.** The climatological zonal mean zonal wind in the JRA-55 family data sets for **(a–c)** DJF and **(d–f)** SON in black contours. Panels **(a, d)** are for STDD, **(b, e)** for CONV, and **(c, f)** for AMIP. The contour interval is $10\,\mathrm{m\,s^{-1}}$. Panels **(b, c, e, f)** also plot differences from STDD using color shades (see the color bar) that are statistically significant at the 95 % level. When taking differences of CONV from STDD, the STDD data after the 1972–1973 season are used. PRS is pressure.

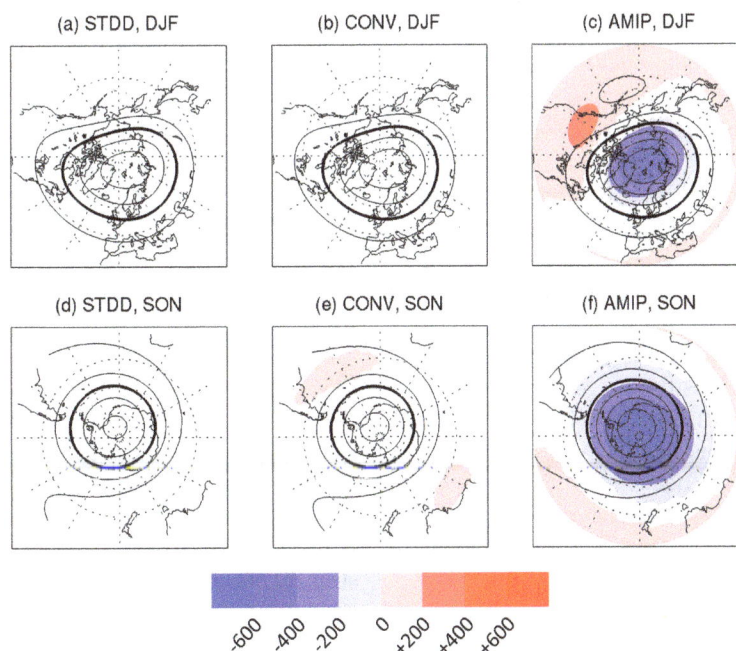

**Figure 2.** Same as in Fig. 1, but for the climatological mean 10 hPa height. The contour interval is 500 m. Thick contours denote 30 000 m.

freedom in the test are equated to the number of years, as the test uses DJF or SON means for the target years.

The climatological zonal wind is similar between STDD and CONV in a large part of the domain below the middle stratosphere for both seasons, although some differences are notable in the upper stratosphere. The differences are positive near the westerly jets in both hemispheres during the cold seasons, which indicate the stronger westerly jets for CONV.

The SH easterly winds for DJF also have positive differences. Negative differences appear in NH subtropical latitudes and SH midlatitudes for SON. Tropospheric wind differences are generally small in magnitude.

The wind differences of AMIP from STDD are roughly similar to those of CONV in spatial pattern but are larger in vertical extent and magnitude in the stratosphere. AMIP also shows significant differences in the troposphere, although

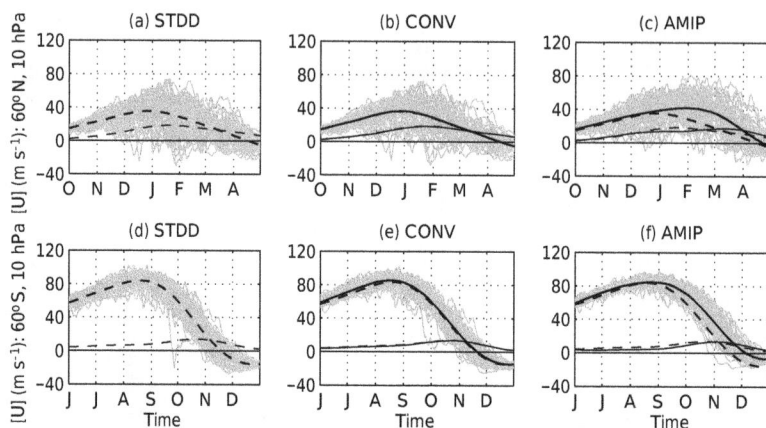

**Figure 3.** Daily time series of the zonal mean zonal wind [$U$] at 10 hPa (gray solid lines): **(a–c)** 60° N and **(d–f)** 60° S. The square brackets denote the zonal mean. Panels **(a, d)** are for STDD, **(b, e)** for CONV, and **(e, f)** for AMIP. The climatological seasonal cycle and standard deviation of interannual variability for each day are denoted by black broken lines in **(a, c)** and by black solid lines in the other panels. Broken lines in **(b, c, e, f)** denote the results from STDD. The STDD data after the 1972–1973 season are used in **(b, e)**. Month labels are placed at the 1st day of each month.

their magnitudes are smaller. In particular, the zonal wind has positive differences around 60° N for DJF, which extend to the stratosphere.

These differences in the polar night jet are consistently reflected in Fig. 2, which similarly shows maps of the climatological 10 hPa geopotential height. The polar vortex is very similar between STDD and CONV in strength and shape for both NH winter and SH spring (Fig. 2b, e). This feature corresponds to the absence of significant differences in the zonal wind at 10 hPa (Fig. 1b, e). On the other hand, AMIP simulates the vortex that is stronger than the STDD counterpart for both seasons (Fig. 2c, f), consistent with the positive wind differences around the polar night jet from the geostrophic wind relationship (Fig. 1c, f). Positive height differences are also notable in surrounding midlatitudes.

Some of these results about the climatological zonal wind are consistently seen in Kobayashi et al. (2014) and Kobayashi and Iwasaki (2016). The latter study further claimed that the stronger polar vortex in CONV and AMIP reflects weaker wave propagation and driving.

## 3.2 Variability

Next, we examine variability in the extratropical stratosphere by looking at daily time series of the zonal mean zonal wind at 60° N/S, 10 hPa (Fig. 3). The zonal mean zonal wind at this location is used as a measure of strength (and also flow direction) of the polar vortex for cold seasons in each hemisphere. The climatological seasonal cycle (long-term mean) and variability (standard deviation of interannual variability for each day) are also plotted in the figure. The climatology and standard deviation are smoothed in time so that they consist of low frequency components with periods longer than about 100 days.

The climatology and standard deviation of the zonal wind overlap between STDD and CONV for both NH and SH (Fig. 3b, e). A comparison of probability distribution functions (PDFs) of the daily zonal wind at 60° N for DJF and at 60° S for SON between STDD and CONV shows that they are very close to each other (Fig. 4b, e). It is difficult to notice frequency differences of easterly winds, or MSSWs, in NH during DJF in Figs. 3b and 4b, which are further examined in Sect. 3.3. In contrast, one can see that STDD shows SH easterly winds in late September, reflecting the occurrence of the MSSW in September 2002, whereas CONV underrepresents it as a minor SSW without a zonal wind reversal (Fig. 3d, e).

AMIP has stronger climatological winds from midwinter to spring and somewhat smaller variability around January in NH. One also sees in Fig. 3a, c that zonal wind reversals during DJF are less frequent for AMIP than for STDD. These features are reflected in PDFs of daily zonal wind data for STDD and AMIP (Fig. 4c). It is clear that the PDF for AMIP is biased toward the positive side, consistent with the stronger climatological westerly wind and less frequent zonal wind reversals. It is also notable that, apart from the climatological difference, the daily zonal wind data for AMIP has smaller variability. Another PDF is drawn in the thin solid line in Fig. 4c for zonal wind data for AMIP that are artificially decreased by the climatological wind difference between STDD and AMIP, so that this PDF has the same mean value as the STDD PDF. One sees that this PDF has somewhat lower frequencies of extreme (both strong and weak) wind values.

The AMIP zonal wind data are also biased toward stronger vortex states in the SH extratropical stratosphere for SH winter and spring (Figs. 3f and 4f). The stronger climatological wind is notable from September to November. No MSSW

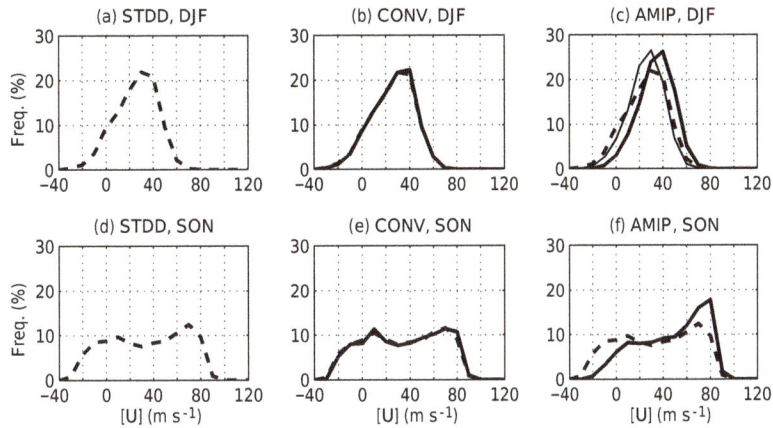

**Figure 4.** PDFs of the daily zonal mean zonal wind at 60° N, 10 hPa, for DJF **(a–c)** and at 60° S, 10 hPa, for SON **(d–f)** in broken lines **(a, d)** or in solid lines (the other panels). Panels **(a, d)** are for STDD, **(b, e)** for CONV, and **(e, f)** for AMIP. Broken lines in **(b, c, e, f)** denote the results from STDD. The STDD data after the 1972–1973 season are used in **(b, e)**. In panel **(c)**, an additional PDF is drawn in the thin solid line for the AMIP zonal wind data that are artificially decreased by the climatological difference from STDD, so that the decreased wind data have the same mean value as in STDD.

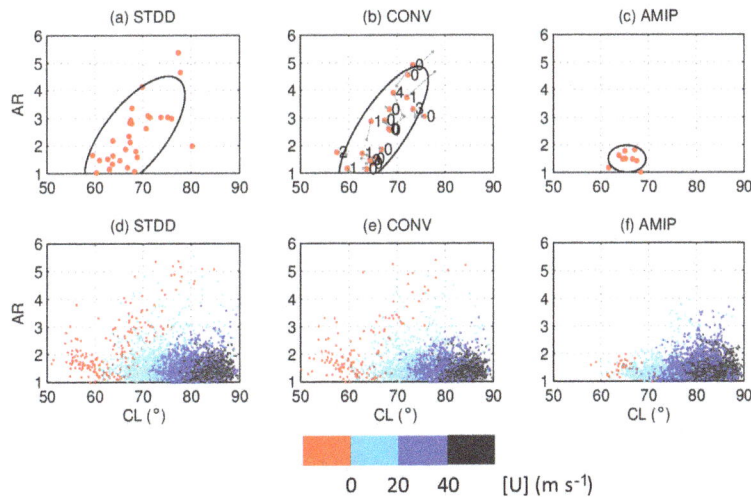

**Figure 5. (a–c)** Scatter plots between CL and AR of the 10 hPa height on the onset dates of the NH MSSWs: **(a)** STDD, **(b)** CONV, and **(c)** AMIP. Ellipses denote representative distributions of the data points extracted by an empirical orthogonal function analysis. Each data point in **(b)** is connected to the corresponding case in STDD (gray) and accompanied by a number denoting the time difference in the onset dates (Table 1). Panels **(d–f)** plot results for all DJF days available to each product. Each data point is colored by the zonal mean zonal wind at 60° N, 10 hPa (see the color bar).

(zonal wind reversal) is simulated from September to mid-October in AMIP.

### 3.3  Frequency and vortex geometry of MSSWs in NH

Figures 3 and 4 showed that the zonal wind variability in NH looks similar between STDD and CONV, suggesting that the occurrence of MSSWs is also similar. On the other hand, AMIP clearly lacks the zonal wind variability and easterly winds, which is suggestive of fewer MSSWs. In this subsection, we examine the frequency and also vortex geometry of NH MSSWs for the three products.

Table 1 lists the onset dates of DJF MSSWs identified for STDD and CONV. The frequency of DJF MSSWs for STDD is 30 events in the 56 seasons (53.6 % for each season) and 22 events in the 41 seasons (53.7 %) after 1972–1973. CONV shows 19 events in the 41 seasons (46.3 %). A comparison of the MSSW onset dates between STDD and CONV in the 41 seasons shows that CONV reproduces most MSSWs in STDD as inferred from Fig. 3 but delays seven cases by 1 to 4 days. CONV also misses three cases, underrepresenting them as minor SSWs that do not accompany a zonal wind reversal. The latter feature is similar to the SH MSSW in

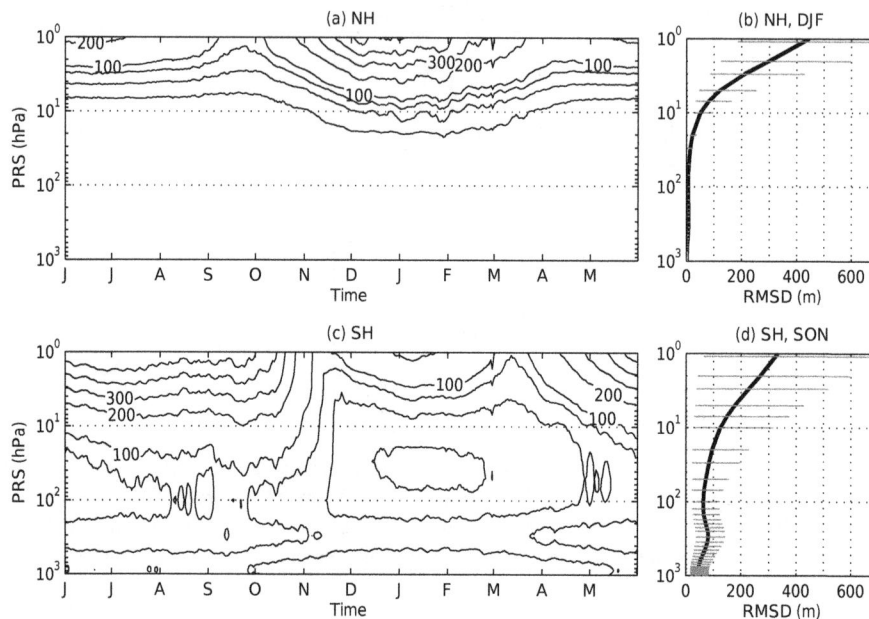

**Figure 6. (a, c)** Time–height sections of the climatology of RMSD: **(a)** NH and **(c)** SH. The contour interval is 100 m, with additional contours at 25, 50, and 75 m. Panels **(b, d)** plot seasonal means as a function of height: **(b)** NH for DJF and **(d)** SH for SON. The horizontal lines denote 5th and 95th percentiles.

September 2002, when CONV represents it as a minor SSW (Fig. 3). It is also noted that no opposite case exists where an onset date is represented earlier in CONV than in STDD or where an MSSW is represented only in CONV. The delayed and missed cases seem distributed randomly in time (year).

The frequency of DJF MSSWs for AMIP is 9 for the 56 seasons (16.1 %), which is much smaller than for STDD. The lower frequency of MSSWs in AMIP is consistent with the stronger westerly wind (Figs. 1–4).

The geometry of the polar vortex during the MSSWs for STDD is characterized in a scatter plot between CL and AR on lag $-0$ day (Fig. 5a). One sees that the data points for STDD roughly form a linear distribution, with a correlation coefficient of $+0.68$. Data points located near the lower-left end correspond to MSSWs of low CL (e.g., vortex displacement MSSWs), and those near the opposite end correspond to MSSWs of high AR (e.g., vortex-split MSSWs). Such a linear distribution was pointed out by Taguchi (2016). This distribution reflects that the zonal mean zonal wind at 60° N, 10 hPa, representing the vortex strength tends to weaken as CL decreases and/or AR increases (Fig. 5d).

CONV roughly reproduces a similar linear distribution of CL and AR on the MSSW onset dates. The differences in CL and AR vary from one MSSW to another.

In addition to the lower frequency, the MSSWs for AMIP show a notable feature about the vortex geometry – that all MSSWs have low CL and AR. No MSSW in AMIP shows high AR, e.g., over 2, on lag $=0$ day. This implies a lack of vortex-split MSSWs. The absence of MSSWs of high AR is

consistent with the scatter plot for all DJF data for AMIP (Fig. 5f): the data points for AMIP are biased toward high CL and low AR.

In the following, we further explore these differences in the frequency and vortex geometry of MSSWs. In Sect. 4, we relate the occurrence of MSSWs to general RMSD distributions of geopotential height fields between STDD and CONV. In Sect. 5, we examine strong planetary wave forcings and associated stratospheric vortex weakenings for STDD and AMIP.

## 4 Comparison of CONV to STDD

In Sect. 3, we showed that CONV is close to STDD in terms of the climatological zonal mean zonal wind and height in the NH extratropical stratosphere (Figs. 1 and 2). The daily time series and PDFs of the zonal wind also look similar between the two runs (Figs. 3 and 4), although the occurrence of MSSWs is slightly different. In this section, we describe general RMSD distributions of daily extratropical height fields between STDD and CONV at various levels and relate them to the occurrence of MSSWs.

### 4.1 Climatological RMSD distributions

First, we examine climatological distributions of RMSD for NH and SH. Figure 6a, c plot the long-term means of RMSD as a function of time (season) and height. It is common between the two hemispheres and all seasons for RMSD to in-

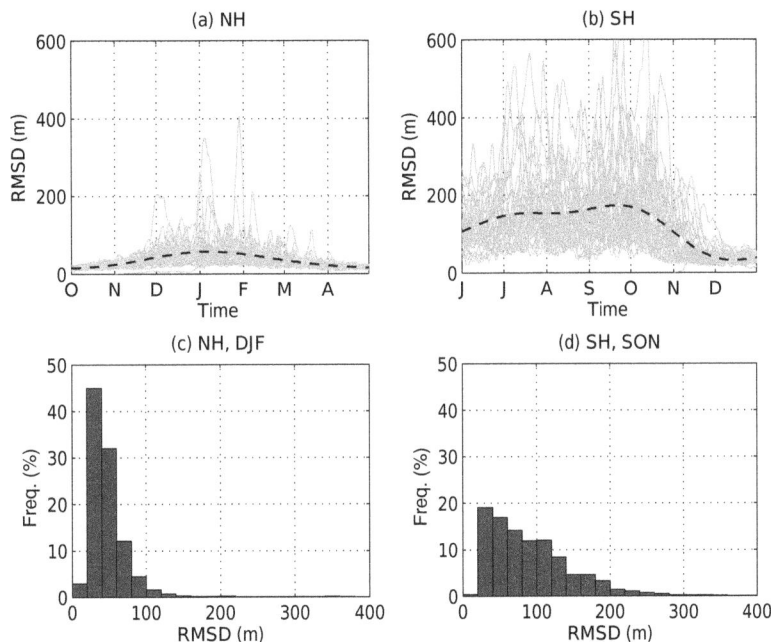

**Figure 7. (a, b)** Daily time series of RMSD at 10 hPa: **(a)** NH and **(b)** SH. Broken lines denote the climatological seasonal cycle. Panels **(c, d)** show PDFs of the RMSD values for DJF in NH and for SON in SH.

crease with height. This result will be expected, since the assimilation of conventional observations in CONV does not extend to the upper stratosphere. In the NH upper stratosphere, RMSD has a semiannual cycle, with two peaks in NH summer and winter. The winter peak is larger than the summer counterpart and extends deeper down to the middle stratosphere, such as 10 hPa. Such a semiannual cycle is also notable in the SH upper stratosphere. The winter peak also extends deeper than the summer peak in SH as in NH.

The climatological height distributions of RMSD for NH winter and SH spring are extracted in Fig. 6b, d to emphasize interhemispheric differences during the dynamically active seasons. It confirms the increase in RMSD with height for both hemispheres. It also reveals that RMSD is larger in SH than in NH in a large part of the domain, except for the upper stratosphere. The SH has considerable magnitudes of RMSD even in the troposphere. It may be that conventional observations are insufficient in SH, and assimilation of satellite observations lead to larger differences.

An examination of year-to-year changes in RMSD suggests a trend in SH, which is discussed in Sect. 6.2.

### 4.2 Case-to-case variability in RMSD and MSSWs in NH

We further examine day-to-day variability in RMSD, particularly for cases of extreme RMSD values, and compare it to the occurrence of MSSWs. Figure 7a, b show daily time series of RMSD at 10 hPa in both NH and SH during dynamically active seasons. PDFs of the RMSD values are also

plotted in Fig. 6c, d: DJF and SON data are used for NH and SH, respectively. In addition to the climatological difference, notable minimum and maximum RMSD values are also larger in SH than in NH. SH RMSD often attains large local maxima, e.g., over 400 m, and even NH RMSD sometimes exhibits sharp peaks.

In order to examine specific geopotential height distributions, we identify cases of extremely large RMSD values for NH and SH as when RMSD attains local maxima over a threshold. The threshold is defined as the 95th percentile of all RMSD values during DJF in NH or SON in SH for the 41 seasons. The local maxima must be separated by more than 30 days when they are identified in the same season. RMSD values around the MSSW onset dates (from lag $= -10$ to $+10$ days) are excluded from this procedure and are used separately.

Figure 8 presents scatter plots of the zonal and wave components contributing to the total RMSD values for these cases. Note that the distance of each data point from the origin gives the total RMSD value (Eq. 2). The figure also includes the results of all MSSW onset dates identified in STDD. The data points use different colors according to the zonal mean zonal wind at 60° N/S, 10 hPa, in STDD on the target day of each case (see the color bar). Note that data points plotted in red corresponds to MSSWs in STDD.

One sees in Fig. 8a that several data points are located far from the origin and hence have the largest total RMSD values. These cases, except for one MSSW, occur separately from the MSSWs. Some of the data points are located relatively close to the $y$ axis, implying large contributions from

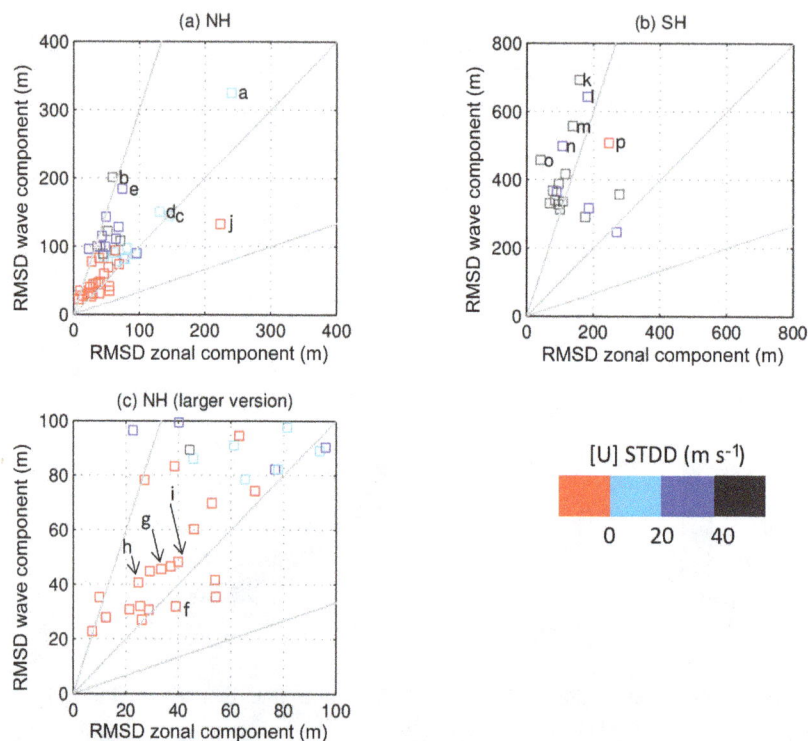

**Figure 8. (a, b)** Scatter plots between zonal and wave contributions to total RMSD for large RMSD peaks exceeding the 95th percentile of all RMSD values: **(a)** NH for DJF and **(b)** SH for SON. Results for all MSSW onset dates in STDD are also shown. Panel **(c)** is a larger version of **(a)**. The data points are plotted in different colors, which denote the zonal mean zonal wind in STDD at 60° N/S, 10 hPa, on the target dates (see the color bar). Red data points correspond to the MSSW cases. Letters a–j in **(a, c)** and k–p in **(b)** denote the cases used in Fig. 9. Gray lines denote $y = x/3$, $y = x$, and $y = 3x$.

the wave component. The zonal mean zonal wind in STDD is relatively strong (over $20\,\mathrm{m\,s^{-1}}$) for these cases when they have large values of the $y$ component (100 m).

The opposite also holds for other data points, when the zonal component makes large contributions and the zonal wind is weaker ($20\,\mathrm{m\,s^{-1}}$). We also examined zonal wind differences of CONV from STDD for these cases but did not notice any systematic relationship, as both overestimations and underestimations by CONV occur (not shown). The five largest values of total RMSD are labeled as a–e in Fig. 8a for Fig. 9.

Regarding the MSSWs, one may expect that the missed and delayed cases have larger RMSD values than the other MSSWs, but it turns out that this is not the case (see also Fig. 8c). All of the three missed cases (labeled as f–h) and one of the delayed cases (with a time difference of 4 days, labeled as i) appear in the intermediate data points near the coordinate point of (40, 40). The exceptional MSSW, located near (200, 100), is a delayed case with a time difference of 3 days (labeled as j).

As for SH, a large part of the data points are located near the $y$ axis, implying an important role of the wave component. For these cases, STDD has strong westerly winds over $20\,\mathrm{m\,s^{-1}}$, and both overestimations and underestimations in

the zonal wind by CONV occur. The missed MSSW case (September 2002 MSSW in STDD, labeled as p) also occurs with a large contribution from the wave component.

Figure 9a–e plot 10 hPa height maps for the five largest RMSD values in NH selected from Fig. 8a. In these figure panels, we see large differences in the 10 hPa height occurring in different (or various) locations depending on the target cases, which often occur near the edge of the polar vortex rather than near the center of the vortex. These differences do not have strong zonal symmetry but vary in the zonal direction, consistent with the large contributions from the wave component (Fig. 8a). These differences are reflected in different distributions of contour lines of 30 000 m representative of the vortex edge, implying different shapes of the vortex between STDD and CONV.

The height differences during the four MSSWs (three missed cases and one delayed case, Fig. 9f–i) are generally smaller in magnitude. Both STDD and CONV show very similar vortex shapes in terms of the 30 000 m contour lines. The difference fields have negative, albeit small in magnitude, values over polar latitudes, suggesting that the extratropical zonal wind is stronger in CONV. This is consistent with the fact that these MSSWs are missed or delayed in CONV.

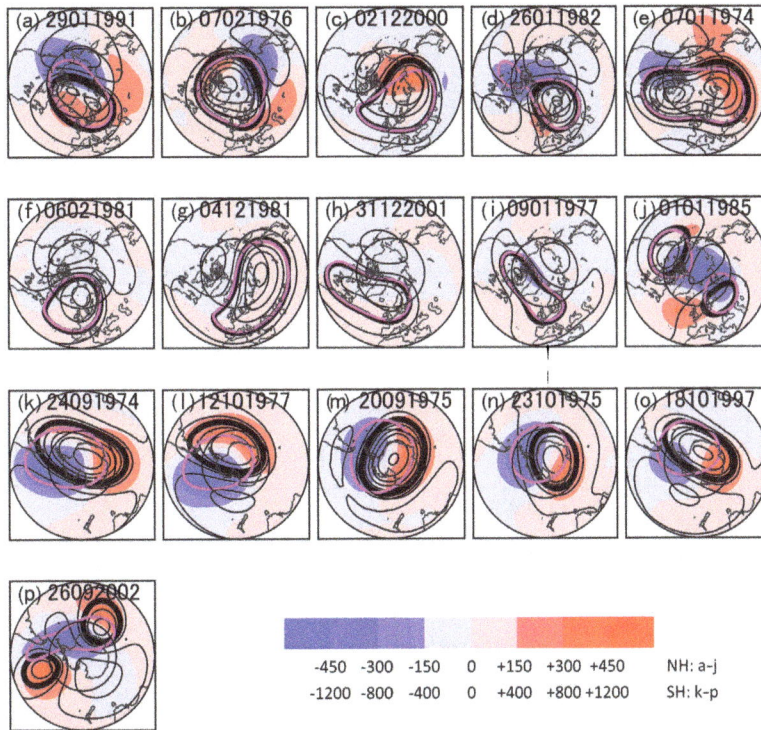

**Figure 9.** (Black contours in **(a–e)** and **(k–o)**.) Maps of the 10 hPa height in STDD for the five largest RMSD values: **(a–e)** NH and **(k–o)** SH. These labels are the same as in Fig. 8a, b. The target dates are also denoted in the ddmmyyyy format. The contour interval is 500 m, and thick black contours denote 30 000 m in STDD. Magenta contours denote 30 000 m for the 10 hPa height in CONV on the same dates. Color shades denote differences of CONV from STDD (see the color bar). Panels **(f–j)** are similar, but for the three missed MSSW cases and the two delayed cases in NH. These labels are also denoted in Fig. 8a, c. Panel **(p)** is similar, but for the 2002 MSSW in SH (see Fig. 8b).

One exception is the MSSW with the time difference of 3 days (Fig. 9j). In this case, the polar vortex similarly splits into two cyclones for both STDD and CONV. Nonetheless, the height differences have largely negative values at high latitudes.

Figure 9k–o are similar, but for the SH cases of the five largest RMSD values (Fig. 8b). The differences in SH are generally larger in magnitude than the NH counterparts and are dominated by the wave 1 component. The dominance of wave 1 reflects that the location of the polar vortex (or distribution of the 30 000 m contour) in CONV is notably deviated from the STDD counterpart.

As for the SH MSSW, STDD shows a split of the vortex into two cyclones, whereas CONV shows that the vortex (30 000 m contour line) shifts away from the South Pole and highly stretches, without a split (Fig. 9p). The height differences are comparable in magnitude to the preceding five cases. Negative values prevail in the height difference over polar latitudes, consistent with the underrepresentation of this case as a minor SSW in CONV.

## 5 Comparison of AMIP to STDD

We showed in Sect. 3 that the MSSW frequency for AMIP is much lower than that for STDD. In particular, AMIP lacks MSSWs of high AR. In order to better understand these differences (especially the former), in this section we examine extreme planetary wave forcings (poleward eddy heat fluxes) and stratospheric vortex weakening responses to them. Since it is well known that MSSWs are a response of the polar vortex to anomalously strong planetary wave forcings from the troposphere (Matsuno, 1971; Limpasuvan et al., 2004), the differences may be explained in terms of wave forcing from the troposphere and/or the vortex response in the stratosphere.

### 5.1 Lower MSSW frequency for AMIP

The first issue is the lower frequency of MSSWs in AMIP. One may hypothesize that it is due to a lower frequency of strong wave forcings and/or weaker vortex responses to such wave forcings.

In Fig. 10a, we examine how the poleward eddy heat flux in the extratropical lower stratosphere (40–90° N, 100 hPa) and zonal wind deceleration are related for all MSSWs in

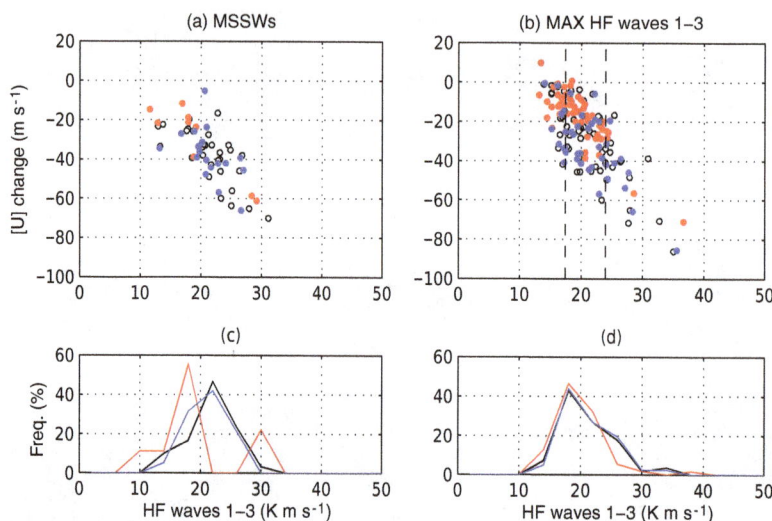

**Figure 10. (a)** Scatter plot between 21-day mean heat flux (lag = −20 to 0 day) of waves 1–3 and associated zonal wind deceleration for the 21 days for all MSSWs in STDD (black), CONV (blue), and AMIP (red). Panel **(b)** is similar but uses the maximum of the 21-day mean heat flux of waves 1–3 in each DJF season as the key. The vertical broken lines denote the 25th and 75th percentile values of the heat flux in STDD, which is used for the composite analysis in Fig. 11. Panels **(c, d)** show frequency distributions of the 21-day mean heat flux for the respective cases. The full period data are used for STDD.

STDD and AMIP. The heat flux is averaged in time from lag = −20 to 0 day, and the zonal wind deceleration is also calculated for the 21 days. We focus on the 21 day period, since the correlation between the two quantities for the MSSWs in STDD maximizes when we average the heat flux for about 10 to 30 days (not shown). This is consistent with the fact that polar vortex strength is highly correlated to the heat flux when the latter is averaged for a few weeks or longer (Polvani and Waugh, 2004). This method follows Taguchi (2017), who examined the Coupled Model Intercomparison Project Phase 5 (CMIP5) historical simulations with 30 models. It is common between STDD and AMIP that, as expected, the zonal wind decelerates more strongly when the wave forcing is stronger. The distribution for STDD has a correlation coefficient of −0.70.

Since this plot is based on the data around the MSSWs, it does not explain how or why the MSSWs are much lower for AMIP. In order to obtain a clue for the difference in the MSSW frequency, we look at extreme planetary wave forcings by identifying a maximum of the poleward heat flux (averaged for 21 days, from 20 day before each day) of waves 1–3 for each winter season in STDD and AMIP. We also extract zonal wind decelerations associated with the maximum wave forcings (Fig. 10b).

The plot shows two features in addition to similar linear distributions between the wave forcing and wind deceleration for both STDD and AMIP as in the MSSWs. First, AMIP lacks relatively strong forcings. For example, STDD has 10 samples over a threshold of $25 \, \mathrm{K \, m \, s^{-1}}$, whereas AMIP has only 3. Second, the zonal wind decelerations for wave forc-

ings around $20 \, \mathrm{K \, m \, s^{-1}}$ seem stronger for STDD than for AMIP.

The second point is further examined in a composite analysis with respect to the days of the maximum heat flux that ranges from the 25th percentile ($17.4 \, \mathrm{K \, m \, s^{-1}}$) to the 75th percentile ($23.9 \, \mathrm{K \, m \, s^{-1}}$) of all maximum heat flux values in STDD (Fig. 11). The two percentile values are similar for AMIP: 16.9 and $22.0 \, \mathrm{K \, m \, s^{-1}}$. The range limitation intends to subsample wave forcings of similar magnitudes for both STDD and AMIP.

As a result, the composite daily heat flux increases from lag ≈ −20 to 0 day for both STDD and AMIP (Fig. 11c). The lag = 0 day here denotes when the 21-day mean heat flux maximizes. This feature is quite similar between the two products by construction. The 21-day means from lag = −20 to 0 day are not significantly different at the 95 % level. The composite residual mean meridional wind (in the transformed Eulerian-mean equations) and wave driving (EP flux divergence–convergence) are also similar between the two (Fig. 11b, d), and their 21-day means are not significantly different at the 95 % level. In contrast, the composite zonal wind evolution is different, as the zonal wind decelerates more strongly for STDD. It is around $40 \, \mathrm{m \, s^{-1}}$ before lag ≈ −20 days in both STDD and AMIP, before decreasing to lag ≈ 0 day. The wind deceleration from lag = −20 to 0 day is $28.5 \, \mathrm{m \, s^{-1}}$ for STDD and $17.6 \, \mathrm{m \, s^{-1}}$ for AMIP, and this difference is judged to be significantly different at the 95 % level. This feature is consistent with Fig. 10b.

Thus, these results suggest that the lower frequency of MSSWs for AMIP can be attributed to the lower frequency of extreme wave forcings and the weaker vortex response in

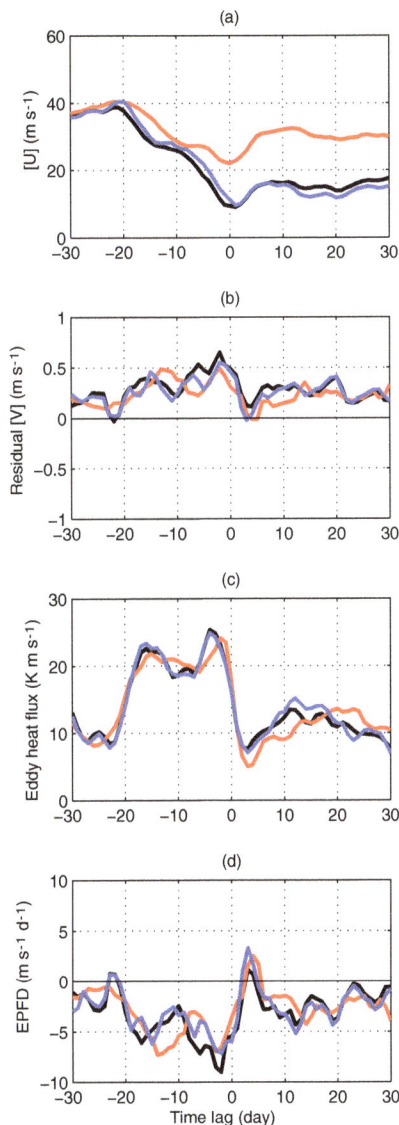

**Figure 11.** Composite time series with respect to the maximum of the 21-day mean heat flux of waves 1–3 in each DJF season for STDD (black), CONV (blue), and AMIP (red): **(a)** zonal mean zonal wind at 60° N, 10 hPa, **(b)** residual mean meridional wind in the transformed Eulerian-mean equations at 60° N, 10 hPa, **(c)** heat flux of waves 1–3 at 40–90° N, 100 hPa, and **(d)** EP flux divergence–convergence (EPFD) of waves 1–3 at 60° N, 10 hPa. The lag = 0 day denotes when the 21-day mean heat flux maximizes. The full period data are used for STDD.

the stratosphere. These differences may reflect model biases in AMIP (or effects of data assimilation in STDD), which will be discussed in Sect. 6.

Figures 10 and 11 additionally include results from CONV for reference. It is shown that the results from CONV are similar to those from STDD in terms of planetary wave forcing from the troposphere and vortex response in the stratosphere.

## 5.2  Lack of MSSWs of high AR in AMIP

The second issue is the lack of MSSWs of high AR in AMIP. Again, this may be due to a lack of appropriate wave forcings leading to such MSSWs and/or a difference in vortex responses to such forcings.

An analysis of AR changes associated with the maximum 21-day mean heat flux of waves 1–3 (defined for Figs. 10b, d and 11) in STDD and AMIP indicates that the lack of MSSWs of high AR in AMIP can be explained by a different vortex response in the stratosphere to strong wave forcings from the troposphere. The heat flux maxima of waves 1–3 include similar contributions from wave 1 and wave 2 between STDD and AMIP, but extreme AR values occur only for STDD (not shown).

It is noted that the climatological, or background, difference in the 10 hPa height will play a role in explaining the different vortex response. The climatological difference is negative and increases in magnitude toward polar latitudes (Fig. 2). Therefore, it is likely to follow that when the polar vortex in STDD largely stretches for a high AR value in response to a wave forcing, it will be more difficult for the deeper AMIP vortex to exhibit similar stretching even for the same forcing. This idea is related to Seviour et al. (2016), who pointed out a close connection between the frequency of vortex-split MSSWs and climatological ARs for 13 CMIP5 models.

## 6  Summary and discussion

### 6.1  Summary

This study has compared large-scale dynamical variability in the NH and SH extratropical stratosphere, such as MSSWs, among the JRA-55 family data sets. In spite of the importance of dynamical variability in the stratosphere, such an aspect was relatively unexplored in the JRA-55 family data sets and also in other reanalyses. This study owes the comparison to the meticulous design of the JRA-55 family data sets.

First, a survey of the climatological states confirms the stronger polar night jet in both hemispheres for CONV and AMIP than for STDD. This difference is more notable for AMIP. A comparison of MSSWs reveals that CONV reproduces a large part of MSSWs identified in STDD. However, CONV delays several cases by 1 to 4 days and also misses three cases, underrepresenting them as minor SSWs. The SH MSSW in September 2002 is also underrepresented as a minor SSW in CONV. AMIP lacks MSSWs, especially those of high AR values. The differences in CONV could be understood by the bias of the numerical weather prediction model (as seen in AMIP) and the paucity of data assimilation as hypothesized by Kobayashi et al. (2014). It is also suggested that, due to the bias of the model, even STDD (or CONV)

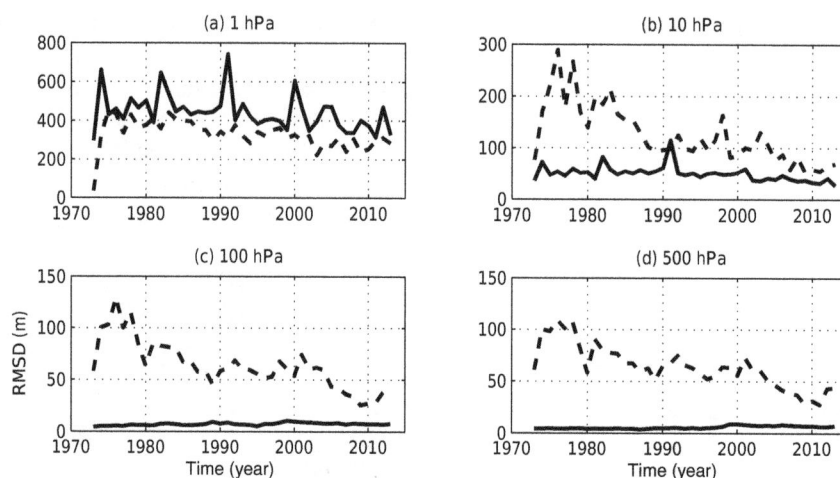

**Figure 12.** Year-to-year variations of RMSD at four levels as indicated. Solid lines denote results in NH for DJF, and broken lines denote results in SH for SON.

may underestimate stratospheric variability compared to the real world.

Next, we examined daily RMSD distributions of geopotential height fields between STDD and CONV and relates them to the occurrence of MSSWs. In spite of the slight climatological differences in the zonal mean zonal wind and 10 hPa geopotential height, the examination shows that CONV is sometimes significantly different from STDD in terms of daily height fields. The differences are especially large in the upper stratosphere in both hemispheres during dynamically active seasons, with larger values in SH. The RMSD values in NH winter sometimes have sharp peaks, apart from the MSSWs. The RMSD peaks include contributions from the wave component that are larger than or comparable to those from the zonal component. RMSD values for some of the delayed MSSWs and all of the missed MSSWs are smaller than such peaks. In such MSSW cases, CONV overestimates the zonal mean zonal wind in the extratropical stratosphere by definition, and both of the zonal and wave components make important contributions.

Furthermore, we examined strong planetary wave forcings and associated zonal wind decelerations (polar vortex weakenings) in STDD and AMIP. It turns out that in AMIP, extreme wave forcings are fewer and zonal wind decelerations to such wave forcings are weaker. The stronger polar vortex in AMIP may also contribute to the lack of MSSWs since it will act to prohibit "preconditioning" of the vortex for MSSWs. These features are commonly seen in some of the CMIP5 historical simulations (Taguchi, 2017). These model biases may be reflected in CONV, where they are likely to contribute to the delay and miss several MSSWs. The latter factor (different vortex responses in AMIP from STDD) is also suggested to be relevant to the lack of MSSWs of high AR values in AMIP.

## 6.2 Discussion

Finally, we discuss several aspects unanswered in this study.

This study reveals large RMSD values for stratospheric height fields between STDD and CONV (Figs. 7–9), but it does not fully answer how or why such differences occur. It does not seem simple to understand what patterns such large differences take and when they occur. They could be explained by the effects of the assimilation of satellite data, but future study is needed to understand the effects more specifically.

Whereas we mostly focus on the middle stratosphere, such as 10 hPa height distributions, the RMSD increase with height suggests that upper stratospheric heights have larger differences between STDD and CONV. This will invite a further study about differences in vertical vortex structures between the two products. For example, it is possible that even MSSWs identified on the same onset dates in STDD and CONV have notable differences in the upper stratosphere between them.

A further examination of year-to-year variations in RMSD at several levels suggests a decreasing trend in SH (Fig. 12). The SH decreasing trend is statistically significant at the 95 % level for all four levels. Such a trend is absent or much weaker in NH. This suggests that the impacts of assimilation of satellite data decrease over the decadal timescale especially in SH. Further details and mechanisms will be worthy of investigation.

The differences between STDD and AMIP, which are relevant to model biases, are characterized by the stronger climatological polar vortex and lack of MSSWs for AMIP. The lack of MSSWs in AMIP is not fully explained by the climatological difference, although it is consistent. This suggests that some additional factors or processes, such as tropospheric planetary wave forcing and stratospheric vortex response, may play a role.

Detailed analyses of these aspects will therefore be useful. A decomposition technique of eddy heat flux (e.g., Nishii et al., 2009; Fletcher and Kushner, 2011) could be applied so as to better understand the lack of extreme wave forcings from the troposphere in AMIP. A zonal momentum budget analysis as in Martineau et al. (2016) could also be conducted to diagnose stratospheric vortex responses to strong wave forcings in the three products, including how or why several MSSWs are delayed or missed in CONV compared to STDD.

*Competing interests.* The author declares that he has no conflict of interest.

*Special issue statement.* This article is part of the special issue "The SPARC Reanalysis Intercomparison Project (S-RIP) (ACP/ESSD inter-journal SI)". It is not associated with a conference.

*Acknowledgements.* The author thanks the Japan Meteorological Agency for making the JRA-55 family data sets available. The data were obtained from the Research Data Archive at the National Center for Atmospheric Research, Computational and Information Systems Laboratory. The author acknowledges that comments from three anonymous reviewers and Patrick Martineau improved the manuscript. This study is supported by the Grant-in-Aid for Scientific Research (S) 24224011 and (C) 15K05286.

Edited by: Peter Haynes

# References

Andrews, D. G., Holton, J. R., and Leovy, C. B.: Middle Atmosphere Dynamics, Academic Press, San Diego, California, 1987.

Butler, A. H., Seidel, D. J., Hardiman, S. C., Butchart, N., Birner, T., and Match, A.: Defining sudden stratospheric warmings, B. Am. Meteorol. Soc., 96, 1913–1928, https://doi.org/10.1175/BAMS-D-13-00173.1, 2015.

Butler, A. H., Sjoberg, J. P., Seidel, D. J., and Rosenlof, K. H.: A sudden stratospheric warming compendium, Earth Syst. Sci. Data, 9, 63–76, https://doi.org/10.5194/essd-9-63-2017, 2017.

Charlton, A. J. and Polvani, L. M.: A new look at stratospheric sudden warmings. Part I: Climatology and modeling benchmarks, J. Climate, 20, 449–469, https://doi.org/10.1175/JCLI3996.1, 2007.

Fletcher, C. G. and Kushner, P. J.: The role of linear interference in the annular mode response to tropical SST forcing, J. Climate, 24, 778–794, https://doi.org/10.1175/2010JCLI3735.1, 2011.

Fujiwara, M., Wright, J. S., Manney, G. L., Gray, L. J., Anstey, J., Birner, T., Davis, S., Gerber, E. P., Harvey, V. L., Hegglin, M. I., Homeyer, C. R., Knox, J. A., Krüger, K., Lambert, A.,

Long, C. S., Martineau, P., Molod, A., Monge-Sanz, B. M., Santee, M. L., Tegtmeier, S., Chabrillat, S., Tan, D. G. H., Jackson, D. R., Polavarapu, S., Compo, G. P., Dragani, R., Ebisuzaki, W., Harada, Y., Kobayashi, C., McCarty, W., Onogi, K., Pawson, S., Simmons, A., Wargan, K., Whitaker, J. S., and Zou, C.-Z.: Introduction to the SPARC Reanalysis Intercomparison Project (S-RIP) and overview of the reanalysis systems, Atmos. Chem. Phys., 17, 1417–1452, https://doi.org/10.5194/acp-17-1417-2017, 2017.

Kobayashi, C. and Iwasaki, T.: Brewer-Dobson circulation diagnosed from JRA-55, J. Geophys. Res.-Atmos., 121, 1493–1510, https://doi.org/10.1002/2015JD023476, 2016.

Kobayashi, C., Endo, H., Ota, Y., Kobayashi, S., Onoda, H., Harada, Y., Onogi, K., and Kamahori, H.: Preliminary results of the JRA-55C, an atmospheric reanalysis assimilating conventional observations only, SOLA, 10, 78–82, https://doi.org/10.2151/sola.2014-016, 2014.

Kobayashi, S., Ota, Y., Harada, Y., Ebita, A., Moriya, M., Onoda, H., Onogi, K., Kamahori, H., Kobayashi, C., Endo, H., Miyaoka, K., and Takahashi, K.: The JRA-55 Reanalysis: General specifications and basic characteristics, J. Meteorol. Soc. Jpn. Ser. II, 93, 5–48, https://doi.org/10.2151/jmsj.2015-001, 2015.

Labitzke, K. G. and van Loon, H.: The Stratosphere: Phenomena, History, and Relevance, Springer, Berlin, 1999.

Limpasuvan, V., Thompson, D. W. J., and Hartmann, D. L.: The life cycle of the Northern Hemisphere sudden stratospheric warmings, J. Climate, 17, 2584–2596, https://doi.org/10.1175/1520-0442(2004)017<2584:TLCOTN>2.0.CO;2, 2004.

Manney, G. L., Allen, D. R., Krüger, K., Naujokat, B., Santee, M. L., Sabutis, J. L., Pawson, S., Swinbank, R., Randall, C. E., Simmons, A. J., and Long, C.: Diagnostic comparison of meteorological analyses during the 2002 Antarctic winter, Mon. Weather Rev., 133, 1261–1278, https://doi.org/10.1175/MWR2926.1, 2005.

Martineau, P. and Son, S.-W.: Quality of reanalysis data during stratospheric vortex weakening and intensification events, Geophys. Res. Lett., 37, 1–5, https://doi.org/10.1029/2010GL045237, 2010.

Martineau, P., Son, S.-W., and Taguchi, M.: Dynamical consistency of reanalysis data sets in the extratropical stratosphere, J. Climate, 29, 3057–3074, https://doi.org/10.1175/JCLI-D-15-0469.1, 2016.

Matsuno, T.: A dynamical model of the stratospheric sudden warming, J. Atmos. Sci., 28, 1479–1494, https://doi.org/10.1175/1520-0469(1971)028<1479:ADMOTS>2.0.CO;2, 1971.

Mitchell, D. M., Gray, L. J., Fujiwara, M., Hibino, T., Anstey, J. A., Ebisuzaki, W., Harada, Y., Long, C., Misios, S., Stott, P. A., and Tan, D.: Signatures of naturally induced variability in the atmosphere using multiple reanalysis datasets, Q. J. Roy. Meteor. Soc., 141, 2011–2031, https://doi.org/10.1002/qj.2492, 2015.

Nishii, K., Nakamura, H., and Miyasaka, T.: Modulations in the planetary wave field induced by upward-propagating Rossby wave packets prior to stratospheric sudden warming events: A case-study, Q. J. Roy. Meteor. Soc., 135, 39–52, https://doi.org/10.1002/qj.359, 2009.

Polvani, L. M. and Waugh, D. W.: Upward wave activity flux as a precursor to extreme stratospheric events and subsequent anomalous surface weather regimes, J.

Climate, 17, 3548–3554, https://doi.org/10.1175/1520-0442(2004)017<3548:UWAFAA>2.0.CO;2, 2004.

Seviour, W. J. M., Mitchell, D. M., and Gray, L. J.: A practical method to identify displaced and split stratospheric polar vortex events, Geophys. Res. Lett., 40, 5268–5273, https://doi.org/10.1002/grl.50927, 2013.

Seviour, W. J. M., Gray, L. J., and Mitchell, D. M.: Stratospheric polar vortex splits and displacements in the high-top CMIP5 climate models, J. Geophys. Res.-Atmos., 121, 1400–1413, https://doi.org/10.1002/2015JD024178, 2016.

Taguchi, M.: Connection of predictability of major stratospheric sudden warmings to polar vortex geometry, Atmos. Sci. Lett., 17, 33–38, https://doi.org/10.1002/asl.595, 2016.

Taguchi, M.: A study of different frequencies of major stratospheric sudden warmings in CMIP5 historical simulations, J. Geophys. Res., 122, 5144–5156, https://doi.org/10.1002/2016JD025826, 2017.

Waugh, D. W. and Polvani, L. M.: Stratospheric polar vortices, in: The Stratosphere: Dynamics, Transport, and Chemistry, edited by: Polvani, L. M., Sobel, A. H., and Waugh, D. W., American Geophysical Union, Washington, D.C., 43–57, 2010.

Yoden, S., Taguchi, M., and Naito, Y.: Numerical Studies on Time Variations of the Troposphere-Stratosphere Coupled System, J. Meteorol. Soc. Jpn., 80, 811–830, 2002.

# Cross-polar transport and scavenging of Siberian aerosols containing black carbon during the 2012 ACCESS summer campaign

Jean-Christophe Raut[1], Louis Marelle[1,a], Jerome D. Fast[2], Jennie L. Thomas[1], Bernadett Weinzierl[3,4,5], Katharine S. Law[1], Larry K. Berg[2], Anke Roiger[3], Richard C. Easter[2], Katharina Heimerl[3,4,5], Tatsuo Onishi[1], Julien Delanoë[1], and Hans Schlager[3]

[1]LATMOS/IPSL, UPMC Univ. Paris 06 Sorbonne Universités, UVSQ, CNRS, Paris, France
[2]Pacific Northwest National Laboratory, Richland, WA, USA
[3]Deutsches Zentrum für Luft- und Raumfahrt (DLR), Institut für Physik der Atmosphäre, Oberpfaffenhofen, Germany
[4]Ludwig-Maximilians-Universität, Meteorologisches Institut, Munich, Germany
[5]University of Vienna, Aerosol Physics and Environmental Physics, Vienna, Austria
[a]now at: Center for International Climate and Environmental Research, Oslo, Norway

*Correspondence to:* Jean-Christophe Raut (jean-christophe.raut@latmos.ipsl.fr)

**Abstract.** During the ACCESS airborne campaign in July 2012, extensive boreal forest fires resulted in significant aerosol transport to the Arctic. A 10-day episode combining intense biomass burning over Siberia and low-pressure systems over the Arctic Ocean resulted in efficient transport of plumes containing black carbon (BC) towards the Arctic, mostly in the upper troposphere (6–8 km). A combination of in situ observations (DLR Falcon aircraft), satellite analysis and WRF-Chem simulations is used to understand the vertical and horizontal transport mechanisms of BC with a focus on the role of wet removal. Between the northwestern Norwegian coast and the Svalbard archipelago, the Falcon aircraft sampled plumes with enhanced CO concentrations up to 200 ppbv and BC mixing ratios up to 25 ng kg$^{-1}$. During transport to the Arctic region, a large fraction of BC particles are scavenged by two wet deposition processes, namely wet removal by large-scale precipitation and removal in wet convective updrafts, with both processes contributing almost equally to the total accumulated deposition of BC. Our results underline that applying a finer horizontal resolution (40 instead of 100 km) improves the model performance, as it significantly reduces the overestimation of BC levels observed at a coarser resolution in the mid-troposphere. According to the simulations at 40 km, the transport efficiency of BC (TE$_{BC}$) in biomass burning plumes was larger (60 %), because it was impacted by small accumulated precipita-

tion along trajectory (1 mm). In contrast TE$_{BC}$ was small (< 30 %) and accumulated precipitation amounts were larger (5–10 mm) in plumes influenced by urban anthropogenic sources and flaring activities in northern Russia, resulting in transport to lower altitudes. TE$_{BC}$ due to large-scale precipitation is responsible for a sharp meridional gradient in the distribution of BC concentrations. Wet removal in cumulus clouds is the cause of modeled vertical gradient of TE$_{BC}$, especially in the mid-latitudes, reflecting the distribution of convective precipitation, but is dominated in the Arctic region by the large-scale wet removal associated with the formation of stratocumulus clouds in the planetary boundary layer (PBL) that produce frequent drizzle.

## 1 Introduction

The Arctic region is particularly sensitive to environmental change, as it is predicted to warm faster than any other region (Manabe et al., 1992). In the past decades, a significant loss of the summertime sea-ice extent has been detected (e.g., Lindsay et al., 2009). A substantial increase in oil and gas extraction activities as well as in shipping may result from a warmer Arctic (Corbett et al., 2010; IPCC, 2013). Recently, studies have highlighted the role of shorter-lived climate forcers (SLCFs), e.g., aerosols and ozone in this warming (Tomasi et al., 2007; Law and Stohl, 2007; Quinn et al., 2008; AMAP, 2015), perturbing the polar radiative balance directly by absorbing and scattering radiation (Shindell and Faluvegi, 2009), and indirectly due to aerosol effects on clouds properties leading to increases in shortwave scattering efficiency and IR emissivity of Arctic clouds (Garrett and Zhao, 2006; Lubin and Vogelmann, 2006; Zhao and Garrett, 2015). Black carbon (BC) is only a minor contributor to aerosol mass but is clearly a significant SLCF in the atmosphere and can be deposited onto snow and ice surfaces, reducing their albedo due to multiple scattering in the snowpack and the much larger absorption coefficient of BC than ice (Warren and Wiscombe, 1980; Hansen and Nazarenko, 2004; Jacobson, 2004), modifying snow grain size and facilitating snow melt, which drives changes in surface temperature (Flanner et al., 2007, 2009; Jacobson, 2010).

Winter and spring have been the most favored seasons for Arctic aerosol research, while fewer summertime measurement studies have been carried out when aerosol mixing ratios are generally smaller (Law and Stohl, 2007). More attention has been given to surface measurements, whereas only very few airborne campaigns (e.g., POLARCAT, ARCTAS) of aerosol chemical composition have been performed in summer (Schmale et al., 2011). During summer, inefficient long-range transport reveals that summertime aerosol originates from the marginal Arctic (sea-ice boundary and boreal forest regions; Warneke et al., 2010) and from episodic and rapid transport events from heavily areas, and leading to dense haze layers aloft (Browse et al., 2012; Di Pierro et al., 2013; Ancellet et al., 2014; Law et al., 2014). Some studies (Brock et al., 1989; Stohl, 2006; Paris et al., 2009) have identified boreal and temperate forest fires, especially Siberian fires, as a dominant source of BC at this season, especially above the planetary boundary layer (PBL) at altitudes between 3 and 10 km, and even into the lower stratosphere by injection in pyrocumulus clouds (Lavoué et al., 2000). Air masses can also be uplifted to the polar tropopause during transport in warm conveyor belts (WCBs) of synoptic low-pressure systems. Eckhardt et al. (2004) and Madonna et al. (2014) have highlighted frequent WCBs over northeastern China and Russia based on climatological analyses. During the IPY (International Polar Year) in 2008, different Lagrangian studies (e.g., Roiger et al., 2011; Sodemann et al., 2011; Matsui et al., 2011) identified several Asian pollution plumes transported to the Arctic after a strong uplift in WCBs located over the Russian east coast and subsequently transported to the Arctic. Franklin et al. (2014) and Taylor et al. (2014) have documented the impact of wet removal in Canadian biomass burning plumes and confirmed that wet deposition is the dominant mechanism for BC removal from the atmosphere and consequently determining its lifetime and atmospheric burden and affecting vertical profiles of number and mass concentration.

State-of-the-art global chemical transport models have large discrepancies in modeling Arctic BC mixing ratios and seasonal variability (Shindell et al., 2008; Koch et al., 2009; Lee et al., 2013; Schwarz et al., 2013; AMAP, 2015; Schwarz et al., 2017). Vignati et al. (2010), Bourgeois and Bey (2011) and Browse et al. (2012) identified the representation of wet removal of BC as one of the main reasons of these differences. More recently, Breider et al. (2014) compared GEOS-Chem modeled BC with observations from the ARCTAS summer campaign. The mean vertical profile of BC measurements in July was always lower than $25 \, \text{ng} \, \text{m}^{-3}$, but the model strongly overestimated the concentrations at high altitudes influenced by long-range transport of deep convective outflow from Asian pollution (summer monsoon) and Siberian fire plumes. This result agrees with the overestimation of BC concentrations in the upper Arctic troposphere found by Koch and Hansen (2005). Despite an enhanced scavenging efficiency of BC in deep convection, Wang et al. (2014) noticed a strong overestimation of BC in the midtroposphere. Sharma et al. (2013) explained this overprediction of BC by an incorrect production of convective precipitation in summer in the Arctic. These results reinforce the fact that aerosol and cloud physical and chemical processing (removal, oxidation and microphysics) is the primary source of uncertainty in modeling aerosol distributions in the Arctic (Eckhardt et al., 2015).

During the summer of 2012, extensive boreal forest fires occurred both in western and in eastern Siberia (in the Yakutsk region). These fires were associated with a hottest summer on record over Siberia, leading to elevated emissions of trace gases and particles (Ponomarev, 2013). In this paper, we focus on a case of pollutant transport from Asia and Russia across the Arctic because they may ultimately lead to the transport of pollution to Europe and North America. During the ACCESS (Arctic Climate Change, Economy and Society) campaign over northern Norway in July 2012, the DLR Falcon 20 aircraft sampled plumes representative of cross-polar transport reaching the Norwegian coasts. Combining an analysis of aircraft measurements, satellite observations, Lagrangian trajectories and simulations from a regional online chemistry transport model (CTM), we investigate the horizontal spatial and vertical structure of BC-containing air masses. We focus on the factors affecting this transport, including the meteorological situation, the dry and wet deposition processes and we assess the relative contribution of anthropogenic, flaring and biomass burning sources to the BC concentrations. Section 2 describes the ACCESS campaign, as well as the modeling tools used in this study. Model output is validated against ACCESS observations in terms of meteorological and chemistry variables, aerosol optical depth (AOD), precipitation and cloud structures (Sect. 3). The export processes, the identification of plumes and the transport pathways are described in Sect. 4. Finally, Sect. 5 investigates the contribution of the various pathway-dependent aerosols removal processes and the transport efficiency of BC reaching the Norwegian coasts.

## 2  Methodology

### 2.1  ACCESS campaign

Roiger et al. (2015) have recently given a detailed overview of the ACCESS aircraft campaign. Therefore we only describe it briefly here. This campaign was based out of Andenes, Norway (69.3° N, 16.1° E), in July 2012 and included 14 flights over the northwestern coast of Norway using the DLR Falcon 20 aircraft (Fig. 1). Most of the flights were devoted to studying the impacts of local pollution sources (ships, oil and gas extraction) on Arctic atmospheric composition (Marelle et al., 2016), but pollution from remote sources was also measured during some specific flights in the middle and upper troposphere (Roiger et al., 2015). In this study, we focus on all flights but describe in particular the two flights that both occurred on 17 July (namely flights 17a and 17b), which were specifically dedicated to probe father into the Arctic in order to study plumes transported from boreal and Asian sources (Fig. 1). Infrared Atmospheric Sounding Interferometer (IASI)-retrieved CO total columns as well as global trace gas forecasts from the Monitoring Atmospheric Composition and Climate (MACC) were used during the campaign to target such flights in order to identify polluted air masses in the mid- and upper troposphere (Roiger et al., 2015). In addition to the suite of meteorological instruments, the Falcon 20 was equipped with trace gas ($NO_x$, $SO_2$, $O_3$, and CO) and aerosol instrumentation including BC. CO was measured every second from VUV fluorescence with an accuracy of 10 %.

BC mass mixing ratios were measured using a single-particle soot photometer (SP2, Droplet Measurement Technologies, Boulder, CO, USA) based on laser-induced incandescence. The instrument measures the refractory black carbon (rBC; hereafter referred to as BC) on a single-particle basis (Schwarz et al., 2006). Particles are drawn through a high-intensity 1064 nm Nd:YAG laser which heats BC-containing particles to incandescence. Schwarz et al. (2017) detailed how the instrument had been calibrated and how the measurements had been corrected for the ACCESS campaign. The instrument has been calibrated using the recommended calibration material (fullerene soot). The individual particle masses of the BC cores of BC-containing particles were averaged over 10 s intervals to obtain mass mixing ratios of BC. Since the SP2 instrument detected BC cores only in the size range of 80–470 nm, assuming a density of $1.8\,g\,cm^{-3}$ (Moteki et al., 2010), the derived mass mixing ratios were scaled by a factor to account for particles outside the detection range (as in, for example, Schwarz et al., 2006) to obtain the total mass mixing ratio. The factor was derived from the average mass size distribution of BC cores separately for

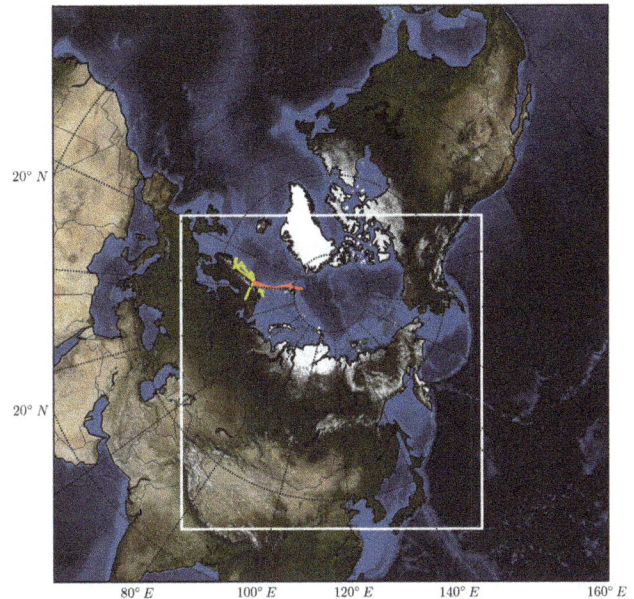

**Figure 1.** Map of the WRF-Chem domain (white) and the flights conducted as part of ACCESS used to evaluate the model. The Falcon 20 flights are in yellow. Flights of 17 July are highlighted in red.

each flight and ranged from 1.1 to 1.55, depending on the encountered aerosol type. Total relative uncertainty of the derived total BC mass mixing ratio is about 30 % (Laborde et al., 2012; Schwarz et al., 2017).

### 2.2  Modeling strategy

#### 2.2.1  WRF-Chem model setup

In order to study the transport and processing of aerosols, regional chemical transport simulations are performed using the fully compressible mesoscale meteorological Weather Research and Forecasting model (Skamarock and Klemp, 2008) with chemistry modules (WRF-Chem Version 3.5.1) (Grell et al., 2005). The chemistry component is fully consistent with the meteorological part (Fast et al., 2006). The different schemes and parameterizations used in this work are summarized in Table 1. Land surface processes are resolved using the Noah Land Surface model scheme with four soil layers (Ek et al., 2003). PBL processes are parameterized according to the Mellor–Yamada–Janjic local scheme (Janjic, 1994). It is coupled to a surface layer based on Monin–Obukhov with Zilitinkevich thermal roughness length and standard similarity functions from look-up tables. We use the Morrison two-moment microphysics scheme (Morrison et al., 2009). To represent the effects of subgrid-scale convective processes on the grid variables, we use the Kain–Fritsch convective implicit parameterization (Kain, 2004), which has been recently improved by Berg et al. (2013) (details below). The shortwave and longwave radiation schemes are from the

**Table 1.** Schemes and parameterizations used for the WRF-Chem simulations.

| Physical process | WRF-Chem option |
|---|---|
| Planetary boundary layer | Mellor–Yamada–Janjic (Janjic, 1994) |
| Surface layer | Monin–Obukhov |
| Land surface | Noah Land Surface model (Ek et al., 2003) |
| Microphysics | Morrison two-moment (Morrison et al., 2009) |
| Shortwave radiation | RRTMG (Iacono et al., 2008) |
| Longwave radiation | RRTMG (Iacono et al., 2008) |
| Cumulus parameterization | Kain-Fritsch-CuP (KFCuP) (Berg et al., 2013, 2015) |
| Photolysis | Fast-J (Wild et al., 2000) |
| Gas-phase chemistry | CBM-Z (Zaveri and Peters, 1999) |
| Aerosol model | MOSAIC eight bins with aqueous chemistry (Zaveri et al., 2008) |

RRTMG model (Iacono et al., 2008). The Fast-J photolysis scheme (Wild et al., 2000) is coupled with hydrometeors as well as aerosols.

The model domain is set up on a polar stereographic grid of 8800 km ×8800 km (Fig. 1). It has a horizontal resolution of 40 km and 50 vertical levels up to 50 hPa. The lowest vertical grid spacing is about 25 m. The simulation uses a model time step of 180 s, begins at 00:00 UTC on 4 July 2012 (after 3 days of spinup) and ends at 00:00 UTC on 21 July 2012. Initial conditions and lateral boundaries of the meteorological fields are provided by National Centers for Environmental Prediction (NCEP) Global Forecast System (GFS) Final Analysis (FNL) 1° resolution data, which has 26 pressure levels and is updated every 6 h. The simulations are also initialized with NCEP-archived 0.5° sea-surface temperatures fields (updated every 6 h) and sea-ice data (updated every day). The modeled horizontal wind components, atmospheric temperature and humidity are nudged towards the 6-hourly NCEP/FNL reanalysis data (Stauffer and Seaman, 1990) above the PBL. The offline global chemical transport model MOZART-4 (the Model for Ozone and Related Chemical Tracers) provides initial and spatially and temporally (6-hourly) varying boundary conditions for chemistry and aerosol species (Emmons et al., 2010).

In this study, the MOSAIC (Model for Simulating Aerosol Interactions and Chemistry) aerosol model (Zaveri et al., 2008) coupled with the CBM-Z (Carbon Bond Mechanism) photochemical mechanism (Zaveri and Peters, 1999) is used. MOSAIC employs a sectional approach to simulate size distributions of a wide number of internally mixed aerosol components (BC, organics, sea salt, sulfate, nitrate, ammonium, dust, and others). The size distribution for each species is divided into eight size bins between 39 nm and 10 μm. The MOSAIC aerosol scheme includes physical and chemical processes of nucleation, condensation, coagulation, aqueous-phase chemistry, and water uptake by aerosols. It is also coupled to the microphysics and cumulus schemes, as well as to the shortwave and longwave radiation schemes.

Aerosol dry deposition is modeled using the "Resistance in Series Parameterization" (Wesely and Hicks,

2000). Wet removal of aerosols by grid-resolved stratiform clouds/precipitation includes in-cloud and below-cloud removal by rain, snow, and graupel by Brownian diffusion, interception, and impaction mechanisms, following Easter et al. (2004) and Chapman et al. (2009). Gustafson et al. (2007) included aerosol–cloud interactions in the model to calculate the activation and resuspension between dry aerosols and cloud droplets. Furthermore, WRF-Chem model is now able to represent aerosol and trace gases within subgrid-scale convective clouds. Berg et al. (2015) indeed included fractional coverage of active and passive clouds, vertical transport, activation and resuspension, wet removal, and aqueous chemistry for cloud-borne particles. This new treatment, coupled to MOSAIC aerosol scheme, is important in our Arctic-scale modeling study to investigate the cloud effects on aerosol within parameterized shallow and deep convective clouds.

### 2.2.2 Emissions

Our simulations use daily, 1 km horizontal resolution fire emissions from the Fire INventory from NCAR (FINN-v1) (Wiedinmyer et al., 2011), based on fire count information from the MODIS (Moderate Resolution Imaging Spectrometer) instrument. Appropriate fire properties are obtained from a synergy between remote sensing observations, land use and carbon fuel datasets to determine in which columns the fires are located and the plume rise is simulated explicitly, accounting for thermal buoyancy associated with fires and local atmospheric stability (Freitas et al., 2007; Grell et al., 2011; Sessions et al., 2011). Anthropogenic emissions are taken from the HTAPv2 (Hemispheric Transport of Air Pollution version 2) 0.1° × 0.1° inventory (http://edgar.jrc.ec.europa. eu/htap_v2/). Volatile organic compounds (VOCs) are given as a bulk VOC mass and distributed into CBM-Z emission categories assuming the speciation reported by Murrells et al. (2010). Hourly and weekly variability of anthropogenic emissions is applied for each emission sector (van der Gon et al., 2011). Flaring emissions are from the ECLIPSE (Evaluating the CLimate and Air Quality ImPacts of Short-livEd

Pollutants) inventory (Stohl et al., 2015). Sea salt and soil-derived (dust) aerosol emissions are incorporated into WRF-Chem (Shaw et al., 2008). Biogenic emissions are calculated online following MEGAN (Model of Emissions of Gases and Aerosols from Nature) (Guenther et al., 2006).

### 2.2.3　Model simulations

The fine-scale structure of the Arctic atmosphere (e.g., due to the thermal stratification) poses a major problem for atmospheric transport model simulations. Here we run the WRF-Chem on a polar-stereographic projection at relatively high horizontal (40 km) and vertical (50 levels up to 50 hPa) resolution to adequately represent the actual structure of the Arctic atmosphere, avoiding an overly rapid diffusion and decay along the boundaries of advected plumes (Rastigejev et al., 2010). A total of nine simulations are conducted: a baseline WRF-Chem simulation (BASE) and eight sensitivity tests to investigate the effect of several processes and emission sources on BC transport to the Arctic: a run without biomass burning emissions (NoFire), a run without anthropogenic emissions (NoAnthro), a run without emissions from flaring activities (NoFlr), a run without dry deposition (NoDry), a run without wet deposition in grid-scale clouds (NoWet), a run without wet deposition in parameterized cumulus clouds (NoWetCu), and a run without wet removal in any clouds (NoWetAll). An additional simulation (Run100) similar to the BASE simulation, but at a coarser resolution (100 km) is also performed to study the influence of horizontal resolution on the transport and deposition of carbonaceous aerosols. The list of the WRF-Chem runs performed is summarized in Table 2.

### 2.2.4　FLEXPART-WRF

We use the Lagrangian particle dispersion model FLEXPART-WRF (Fast and Easter, 2006; Brioude et al., 2013) to assess pollution transport and dispersion from individual sources and to identify the origins of measured pollution plumes during the ACCESS campaign. FLEXPART-WRF is derived from the FLEXPART dispersion model (Stohl et al., 2005), driven by meteorological fields simulated by WRF. We use FLEXPART-WRF in backward mode to study source–receptor relationships: 10 000 particles are released at a receptor point and are transported backward in time using the meteorological fields from the BASE run. Source–receptor relationships are quantified by calculating the potential emission sensitivities (PES), which represent the amount of time spent by the particles in every grid cell. In this paper, meteorological variables are interpolated in time and space over trajectories from the WRF runs. FLEXPART-WRF is therefore useful to identify the origins and transport pathways of pollution plumes observed during the ACCESS campaign and to

derive precipitation and deposition efficiencies along those transport pathways (Sect. 5).

## 3　Model validation

### 3.1　Meteorological parameters

To evaluate the skill in the modeled weather and transport patterns predictions, model results are first compared to ACCESS observations in terms of potential temperature, relative humidity (RH), wind speed and wind direction. The model outputs available at hourly intervals are linearly interpolated in latitude, longitude, altitude and time, every minute along the flight track. The ensemble of key observed and simulated meteorological parameters interpolated along the 14 flight tracks are then gathered to compute vertical profiles. The comparison with Falcon observations is shown in Fig. 2. The predicted temperature is in very close agreement with airborne measurements, with only a small negative bias (about $-1$ K) at high altitudes. The modeled wind direction also matches the observations closely, with a slight positive normalized mean bias (NMB $< 6$ %) in the boundary layer. We note however that statistical performances for wind direction should be taken with caution as they are derived from a vector quantity. The simulated wind speed shows good agreement with a correlation coefficient of $R^2 \simeq 0.866$, although a slightly negative bias (lower than 3 %) at altitudes above 6 km. The water vapor mixing ratio and the resulting RH are also well reproduced by the model with high correlation coefficients (Table 3). Overall WRF-Chem shows appreciable skill in capturing the variability seen in the observations, suggesting that the large-scale transport patterns and major aspects of flow conditions are well represented in the WRF simulations for this campaign.

### 3.2　Trace gases and aerosols

Figure 3 shows the vertical profiles of the CO and BC mixing ratios measured during the ACCESS campaign and simulated by WRF-Chem (BASE and Run100 simulations). On average, the ACCESS measurements show enhanced BC and CO concentrations between 6 and 9 km altitude. The two measured profiles are well correlated with maximum CO values of 200 ppbv at 7–8 km, associated with elevated BC values reaching 25 ng kg$^{-1}$. The corresponding mean values for CO and BC in this altitude range are 135 ppbv and 12 ng kg$^{-1}$, respectively. The small underestimation in CO between 6 and 9 km is a common feature observed by most models (Emmons et al., 2015; AMAP, 2015). Variability in models, run with the same emissions, appears to be driven by differences in chemical schemes influencing modeled OH and/or differences in modeled vertical export efficiency of CO (via frontal or convective transport) from mid-latitude source regions to the Arctic (Monks et al., 2015). A noticeable increase in ozone concentrations is also clearly seen at 7 km,

**Table 2.** List of WRF-Chem simulations performed.

| Name | Description |
|------|-------------|
| BASE | Baseline simulation at 40 km |
| NoFire | Run at 40 km without biomass burning emissions |
| NoAnthro | Run at 40 km without anthropogenic emissions |
| NoFlr | Run at 40 km without flaring emissions |
| NoDry | Run at 40 km without dry deposition |
| NoWet | Run at 40 km without wet deposition in grid-scale clouds |
| NoWetCu | Run at 40 km without wet deposition in parameterized cumulus clouds |
| NoWetAll | Run at 40 km without wet removal in any clouds |
| Run100 | Run at 100 km with the same vertical resolution as in the BASE run |

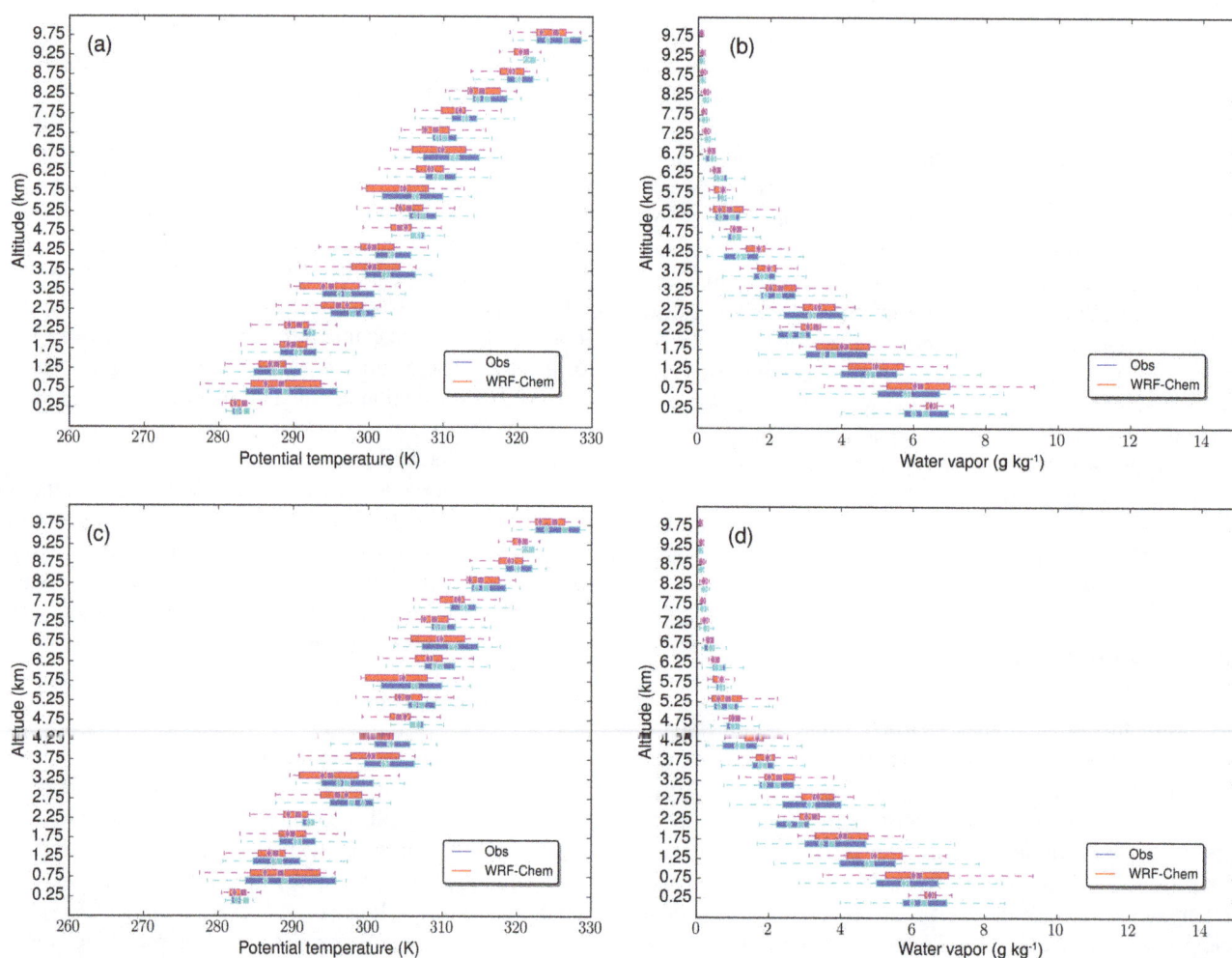

**Figure 2.** Vertical profiles of observed (blue) and modeled (red) **(a)** potential temperature (K), **(b)** water vapor mixing ratio (g kg$^{-1}$), **(c)** wind speed (m s$^{-1}$) and **(d)** wind direction (°) interpolated along all the 14 Falcon flight tracks. Boxes represent the interquartile range, diamonds are the median values, squares show the means and whiskers show the minimum and maximum of the data.

with a mean value of 135 ppbv (not shown). The vertical profiles of CO, BC and ozone suggest that, during summer 2012, pollution transported to the Arctic increased trace gases and aerosol mixing ratios in the middle and upper troposphere compared to typical clean background values (Roiger et al.,

2015). The statistical performance of the model in regards to the representation of key chemical species extracted along flight tracks are given in Table 3. The comparison of simulated CO (NMB < 1.5 %) against the Falcon flight measurements proves that the model predictions are able to capture

**Table 3.** Performance statistics for meteorological variables and chemical species (BC, CO). For all flights, $R^2$, MB, RMSE and NMB represent the Pearson correlation coefficient, the mean bias, the root-mean-square error, and the normalized mean bias, respectively.

| Variable (unit) | $R^2$ | MB | RMSE | NMB (%) |
|---|---|---|---|---|
| Pressure (hPa) | 0.999 | 10.14 | 13.61 | 1.37 |
| Potential temperature (K) | 0.997 | −0.96 | 1.60 | −0.32 |
| Relative humidity (%) | 0.709 | 1.29 | 14.60 | 1.73 |
| Wind speed ($m\,s^{-1}$) | 0.866 | −0.27 | 2.39 | −3.33 |
| Water vapor mixing ratio ($g\,kg^{-1}$) | 0.924 | 0.15 | 0.61 | 3.67 |
| BC ($ng\,kg^{-1}$) | 0.468 | 1.58 | 7.44 | 27.30 |
| CO (ppbv) | 0.524 | 1.47 | 17.63 | 1.49 |

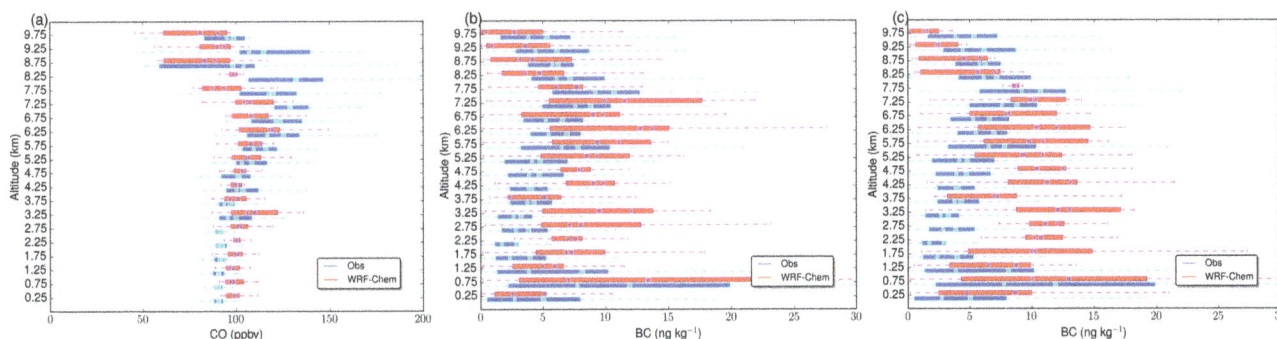

**Figure 3.** Vertical profiles of observed (blue) and modeled (red) mixing ratios of **(a)** CO (ppbv), **(b)** BC ($ng\,kg^{-1}$) at a horizontal resolution 40 km (BASE), and **(c)** BC ($ng\,kg^{-1}$) at a horizontal resolution 100 km (Run100) interpolated along all the 14 Falcon flight tracks. Boxes represent the interquartile range, diamonds are the median values, squares show the means and whiskers show the minimum and maximum of the data.

the magnitude and temporal variability seen in observed values. The model demonstrates reasonable skill in capturing the general structure of the vertical profile of BC, but overestimates the BC mixing ratio in the mid-troposphere. The resulting mean bias on BC is $1.5\,ng\,kg^{-1}$ and the corresponding normalized mean bias (27.3 %) is lower than the error on the SP2 instrument (30 %) and much lower than biases reported for most models in the Arctic region. Eckhardt et al. (2015) and AMAP (2015) indeed reported that BC concentrations in July–September are overestimated in the mean of intercompared models by 88 %. The mean profile of BC sampled during the ACCESS campaign is of the same order of magnitude as those reported by Breider et al. (2014) ($25\,ng\,m^{-3}$ in July 2008) and by Schwarz et al. (2013) during the HIPPO (HIAPER Pole-to-Pole Observations) campaign at similar latitudes (60–80° N) in the remote Pacific region (10–20 $ng\,kg^{-1}$). In Sect. 4, we discuss the origin and transport of plumes leading to this noticeable increases of CO and BC between 6 and 9 km altitude, associated with higher ozone mixing ratios.

### 3.3 Model resolution

The result of the Run100 simulation is also shown in Fig. 3. The corresponding BC vertical profile shows the same en-

hancement between 6 and 9 km as the BASE run, but it also clearly highlights a strong overestimation (×3) in the mid-troposphere (1.5–5 km). The CO concentrations are also enhanced in the Run100 simulation by 4–5 ppbv as compared to the BASE run. This suggests that, at a coarser resolution, the model is unable to resolve the fine structure of plumes transported aloft, illustrates the difficulty to represent the cloud and precipitation structures and points to the need for improved representation of BC processing in global models. More generally, global models are commonly run at horizontal and vertical resolutions that are inadequate for representing the actual structure of the Arctic atmosphere, mostly because of computational limited resources. Schwarz et al. (2013, 2017) studied the bias between BC measurements and the AeroCOM model suite in remote regions. Global models generally overestimated BC mass concentration, especially in the Arctic upper troposphere, where the overestimation was about a factor of 13, suggesting that the aerosol lifetime and removal was not correctly in models. Model features that govern the vertical distribution and lifetimes of SLCFs in the Arctic atmosphere must still be improved (AMAP, 2015). Ma et al. (2013) showed that BC was better simulated with higher spatial resolution, and described that there is likely less wet removal at higher spatial resolution since aerosols and clouds do not overlap as much. In our study, at a hor-

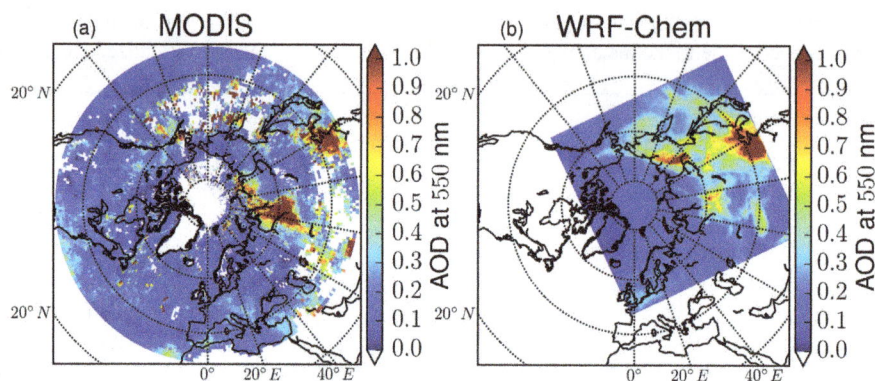

**Figure 4.** Aerosol optical depth (AOD) at 550 nm **(a)** measured by MODIS instrument aboard Aqua and **(b)** simulated by a Mie code in the WRF-Chem model, averaged between 4 and 21 July 2012. White areas in panel **(a)** indicate the presence of clouds, preventing the AOD retrieval.

izontal resolution of 40 km, the model-to-observation comparison indicates that the WRF-Chem model correctly represents the transport pathways of pollution into the Arctic in summer from mid-latitudes and wildfire/biomass burning. At a finer resolution, the model is also able to better represent mean vertical motions associated with large-scale synoptic systems, i.e., conveyor belts, that lift PBL aerosols into the free troposphere. Resolution can therefore affect long-range transport, since it affects how high aerosols are lifted into the free troposphere.

### 3.4   Aerosol optical depth

To evaluate the total aerosol column simulated by WRF-Chem during the ACCESS campaign, we calculate the aerosol optical depth (AOD) at 550 nm, defined as the integrated extinction coefficient over a vertical column of unit cross section, taking into account the attenuation of the radiance by aerosol scattering and absorption. This is compared to the AOD retrieved at the same wavelength by the MODIS instrument aboard Aqua satellite passing over the Equator in the afternoon. We average, over the period of the ACCESS campaign (4–21 July), daily observations of AOD obtained from MODIS level 3 products (Hubanks et al., 2008), accounting for missing observations primarily due to the presence of clouds within the column (Fig. 4). Note that the AOD retrieved from MODIS above Terra satellite is similar to that obtained from Aqua (not shown). In general, the model correctly represents the main features of the spatial distribution of AOD, including the large values over northeastern China and western Siberia. In China, high AODs reaching 2.3 (1.8 in the model) are linked to high pollution levels in the PBL combined with strong dust episodes. Over western Siberia, the extended area with substantially enhanced values of AOD (up to 2.5) is due to large fires (Ponomarev, 2013) and with a potential contribution from intense flaring activities where oil and gas resources are exploited (Stohl et al., 2013). The AOD is, however, underestimated by the model by 25 % over this

latter region. The model also highlights a strong signal over the Yakutsk region (maximum AOD of 2.2), which is less visible in the MODIS images due to the continuous presence of clouds in July 2012. This is due to intense biomass burning plumes at this period (Ponomarev, 2013). The plumes extend towards the east and above the Bering Strait. Other significant enhancements of modeled AOD (0.4–0.5) are also seen above northern India and the Pacific Ocean, corresponding to the outflow area of China. The good performance in representing the main features of AOD distribution, especially in eastern Asia and western Siberia, gives good confidence in the model to transport plumes to the Arctic. The underestimation above northern Russia indicates that the transport of plumes from that area should be considered with caution. Nevertheless, due to the large amount of precipitation in this area (Sect. 3.5), the potential impact of this source area is reduced.

### 3.5   Precipitation

The lofting of BC can occur isentropically at low RH or can be caused by rapid ascent and heavy precipitation (Matsui et al., 2011). Studying precipitation during transport is therefore crucial and it needs to be modeled correctly to understand the role of wet removal of aerosols. Daily GPCP (Global Precipitation Climatology Project) precipitation data are used to assess the reproducibility of precipitation in the BASE simulation. Here we use GPCP data (version 1.2) at 1° resolution, retrieved from a combination of satellite observations and rain gauge measurements (Huffman et al., 2001). Model estimates of rainfall are interpolated every day to the GPCP 1° grid and then averaged between 4 and 21 July to give an assessment of the mean precipitation during the ACCESS campaign. The WRF-Chem simulation overall reproduces the spatial distribution of the observed precipitation reasonably well (Fig. 5), with highest precipitation intensity over Southeast Asia (South Korea and Japan) with about 17 mm day$^{-1}$, high values over northern China and Mongolia

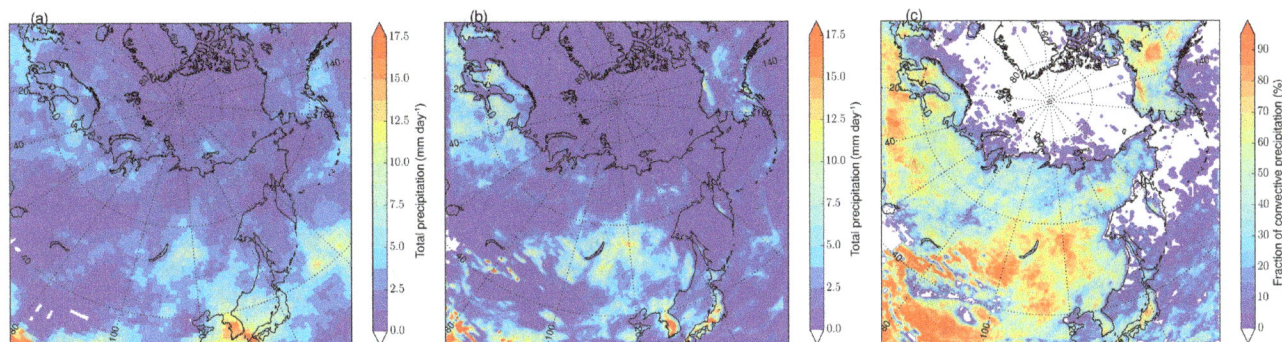

**Figure 5.** Average precipitation from period 4–21 July **(a)** obtained from 1° daily GPCP data and **(b)** simulated by WRF-Chem. The fraction of convective precipitation is shown in panel **(c)**.

and in the vicinity of Lake Baikal (9–11 mm day$^{-1}$) or over the Pacific Ocean due to storm tracks and associated WCBs (10–13 mm day$^{-1}$). Modest values are detected over Europe with 5–8 mm day$^{-1}$. The model nevertheless overestimates the precipitation intensity over southern Russia close to Lake Baikal and over Europe (by 20 %) and underestimates values over the Pacific Ocean. Over Siberia and central Arctic the precipitation intensity is low (1–3 mm day$^{-1}$). The highest values of rainfall are correlated with high convective precipitation fractions (Fig 5), reaching 60 to 90 % over land. Over the oceans and northern Russia the main contribution is from non-convective precipitation (e.g., drizzle).

## 3.6 Cloud vertical distributions

Prediction of Arctic clouds (regional extent and heights) remains a challenge but is also crucial, since the vertical distribution of clouds as well as the state of cloud microphysics is important to understand how models assess the wet removal of aerosols. To evaluate the vertical distributions of clouds in our BASE simulation, we use the DARDAR-MASK v1 data set, which employs a combination of the CloudSat and CALIPSO (Cloud-Aerosol Lidar and Infrared Pathfinder Satellite Observations) products (Delanoë and Hogan, 2010). To retrieve cloud-phase properties, the algorithm uses the 94 GHz radar reflectivity from CloudSat, the lidar backscatter coefficient at 532 nm and vertical feature mask from CALIPSO, as well as ECMWF (European Centre for Medium-Range Weather Forecasts) thermodynamic variables. A range of categories is returned from the DARDAR-MASK: clear, ground, stratospheric features, aerosols, rain, supercooled liquid water, liquid warm, mixed-phase and ice. The algorithm also includes an uncertain classification, in regions where the lidar signal is heavily attenuated or is missing. In this study, we use a simple DARDAR-MASK simulator for WRF-Chem enabling a clear comparison between the model and satellite observations. Mass mixing ratios of liquid water, ice crystals, rain, graupel and snow, as well as the temperature, are interpolated in time and space from

the model along each CALIPSO or CloudSat track passing through the WRF-Chem domain during the simulation (4–21 July 2012). The observed and simulated masks are then temporally averaged at each altitude bin for each cloud-phase category: uncertain, ice, mixed-phase (ice and supercooled water), liquid warm, supercooled water and rain. The result is a mean vertical profile for each category and is represented in Fig. 6. Not surprisingly, the cloud fraction is more important in the lowest layers than at higher altitudes, as the source of water in clouds is the evaporation from the Arctic Ocean and the humid soil. In the upper troposphere, the fraction of occurrence is 20 % both in DARDAR and WRF-Chem, and clouds are composed only of ice crystals. This is directly liked to the temperature of clouds, below −40 °C at those altitudes. The ability of the model to represent those high clouds is excellent. In the PBL, the cloud fraction is larger (between 20 and 35 % if the uncertain category is not taken into account) and is dominated by liquid warm and rain. The main discrepancy between the model and the satellite observations can be ascribed to the uncertain cloud type. When the cloud type is classified as uncertain because the lidar signal is extinguished or too attenuated below optically thick clouds, the model generally predicts no cloud. The model slightly overestimates the fraction of clouds containing supercooled water in the PBL (by 2 %), whereas this fraction is underestimated in the mid-troposphere. The model also slightly underestimates the fraction of ice and supercooled droplets in the upper troposphere. If we do not consider the uncertain class, the result of the WRF-Chem model shows appreciable skill in capturing the average vertical distribution of cloud phases. In particular, the vertical distribution of the liquid warm and rain categories is well reproduced. This is very promising in simulating the wet deposition efficiency of aerosols in plumes transported to the Arctic region.

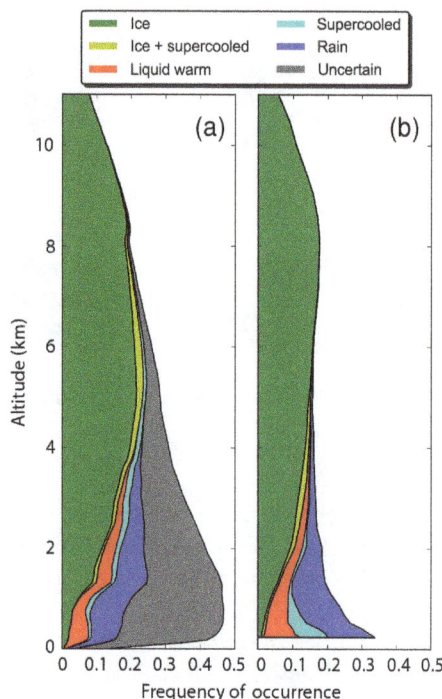

**Figure 6.** Mean vertical profiles of the different cloud-phase categories **(a)** returned from the DARDAR-MASK and **(b)** simulated by the WRF-Chem model between 4 and 21 July.

## 4 Source attribution of BC particles

### 4.1 Export processes

To identify the sources and mechanisms of pollution BC export to the Arctic, we compute the horizontal poleward eddy heat flux at the 850 hPa level using the model (Fig. 7). Regions with large values of the horizontal poleward eddy heat flux generally illustrate the activity of migratory cyclones and may reflect strongly ascending moist airstreams (WCBs) in extratropical cyclones that rise from the PBL to the free troposphere (FT) on short timescales (1 or 2 days) (Eckhardt et al., 2004). The horizontal poleward eddy heat flux is calculated as $\overline{v'\theta'}$, where the overbar denotes time-averaging over the ACCESS period (4–21 July 2012). Because the timescale of the migratory cyclones is about 1 week, $v'$ and $\theta'$ are the instantaneous deviations from the 5-day running means of the meridian wind component and the potential temperature, respectively. This eddy heat flux is computed at each grid point of the WRF-Chem simulation. Large values of eddy heat transport are found over the West Siberian Plain (60–70° N, 40–80° E) extending above the Barents Sea, over the Sakha (Yakutia) Republic (65–70° N, 110–150° E) and finally over the Pacific storm track around northern Japan. If those regions are located in the vicinity of BC sources, they are likely to contribute to export pollution to the Norwegian Arctic, including areas sampled by the Falcon.

In Fig. 7, the horizontal distribution of the mean upward BC mass fluxes averaged over the ACCESS period ($\overline{[BC]w}$), where $w$ denotes the vertical velocity) is also represented at 700 hPa. Three regions with strong upward BC fluxes can be identified: central and northeastern China (30–45° N, 110–120° E), south of the Krasnoyarsk region (50–60° N, 80–100° E) and Yakutia (60–68° N, 125–150° E). Those regions characterized by a strong ascent of BC mass fluxes are also co-located with areas presenting large values for the convergence of horizontal BC flux in the PBL, defined here as the 700–1000 hPa layer and large values for the divergence of horizontal BC flux in the FT, defined here as the 700–200 hPa layer (Fig. 7). They illustrate the locations of source regions for BC emissions and the divergences represent the horizontal transport of BC emitted from those source regions to the surrounding regions (Oshima et al., 2013). Convergence of BC flux in the lowest layers can indeed uplift air parcels from the surface to the FT, which are then exported to outflow regions. A good spatial correlation is also observed with the spatial distribution of AOD (Fig. 4), illustrating that the main emission areas present a significant fraction of carbonaceous aerosols. BC emitted from central and northeastern China is mostly of anthropogenic origin, whereas the two other regions (south of the Krasnoyarsk region and Yakutia) are co-located with intense biomass burning sources, explaining the strong values of the upward BC flux. Aerosols and CO emitted from those source regions may potentially be exported towards the Arctic through the areas presenting large horizontal poleward eddy heat flux: the Sakha (Yakutia) Republic (65–70° N, 110–150° E) and the Pacific storm track. We also note that the source of BC identified in western Siberia is also partly located in the region where the Russian oil and gas flaring emissions are very high and along the main low-level pathway of air masses entering the Arctic (Stohl et al., 2013). They may be exported towards the Arctic through the region with high values of $\overline{v'\theta'}$, extending above the Barents Sea. Anthropogenic emissions in Europe may also be exported to the Arctic (large values of $\overline{v'\theta'}$ in northern Norway) but to a lower extent since the corresponding horizontal and vertical mass fluxes of BC are low in summer.

### 4.2 Plume identification

As indicated in Sect. 2.1, the two flights of 17 July (namely flight 17a in the morning and flight 17b in the afternoon) were specifically dedicated to plumes transported in the upper troposphere from boreal and Asian sources to the Arctic region. Data obtained during these flights are thus the most relevant for the study of BC transport to the Arctic. The vertical cross sections of CO and BC interpolated from the WRF-Chem BASE simulation onto the two flight tracks of 17 July are shown in Fig. 8. In situ measurements of these variables are also represented. The horizontal and vertical variabilities of the CO and BC enhancements ascribed to different source regions can be distinguished from the modeled time–

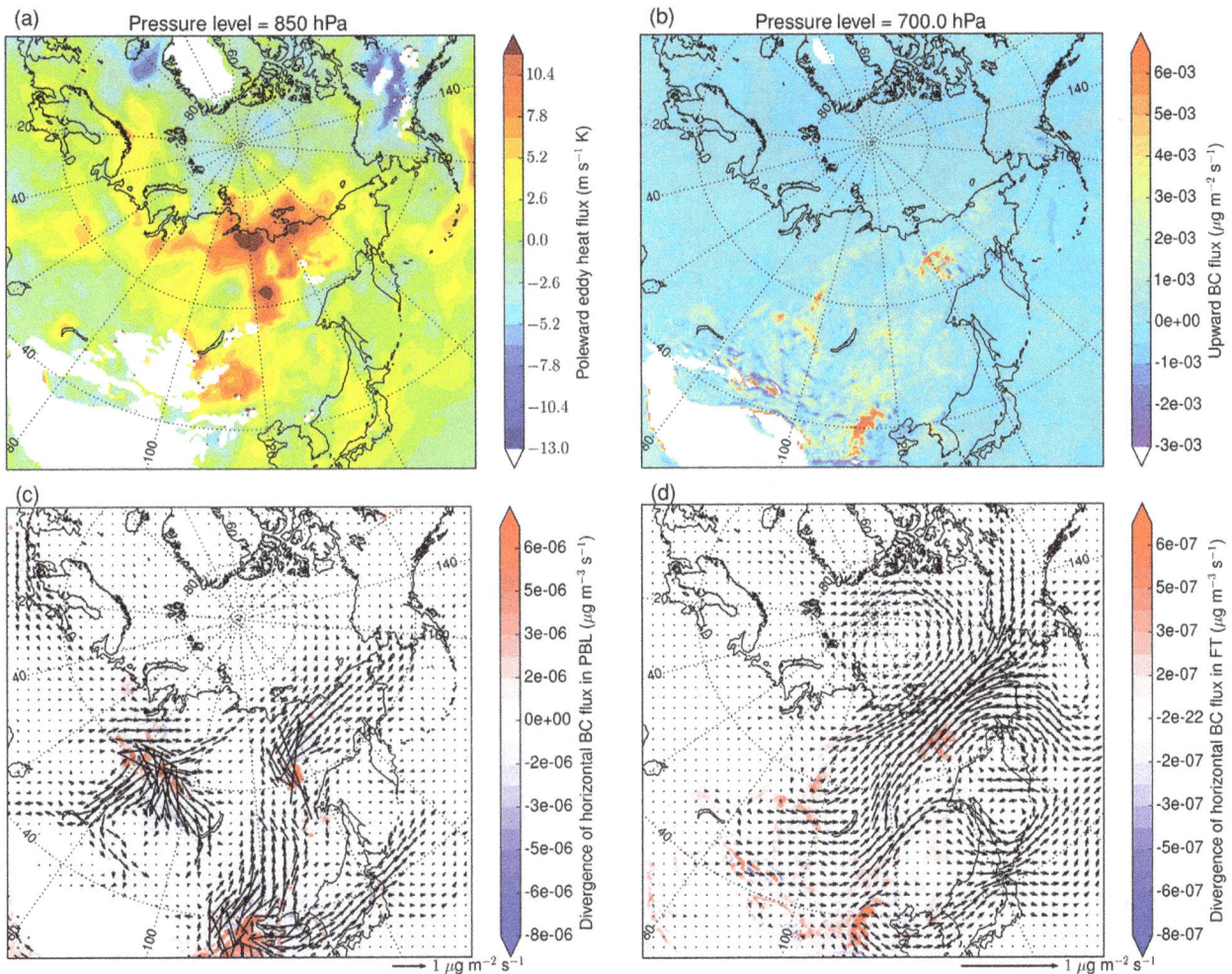

**Figure 7.** Time-averaged (**a**) horizontal poleward eddy heat flux $\overline{v'\theta'}$ at the 850 hPa level, (**b**) upward mass flux of BC $\overline{[BC]w}$ at the 700 hPa level, and divergence of the horizontal mass flux of BC integrated over (**c**) the PBL and (**d**) the free troposphere (FT) during the ACCESS period in the WRF-Chem BASE simulation. Regions without data in white have a high orography (surface pressure below the plotted level). Vectors representing the time-averaged horizontal mass flux of BC integrated over the PBL and FT during the ACCESS period are shown in panels (**c**) and (**d**) respectively, as well as the scaling near the lower right corner.

altitude cross sections along the selected flight tracks. The Falcon transected pollution plumes located between Andøya (69.1° N, 15.7° E) and Spitsbergen (78.9° N, 18.0° E) during flight 17a and sampled the same plumes again during flight 17b before returning to Andøya. The model simulates various CO enhancements during each flight period. Increases in modeled BC mixing ratios are well co-located with those CO enhancements. The PV (potential vorticity) vertical cross section suggests that all those enhancements are in the troposphere: at the time of the flights, the stratospheric air masses are confined in regions with potential temperature larger than 310 K, with a fold between 09:30 and 13:30 UTC, bringing stratospheric intrusions down to 310 K. In the upper troposphere, the model predicts three periods of enhanced CO and BC during each flight, near the 310 K isentrope (altitude between 6 and 8 km). In this altitude

range, when the Falcon crosses the plumes predicted by the model, the instruments also detect higher concentrations of CO and BC. WRF-Chem however underestimates the magnitude of the mixing ratios of carbon monoxide in the peaks, reaching 210 ppbv. This is mostly due to numerical diffusion caused by the finite-difference method applied to the advection equation on the model grid. In the model, the gradients are simply taken along coordinate surfaces and hence are imperfectly described. A sixth-order numerical diffusion is used in our simulations. The CO underestimation can be also due to the fact that the Falcon sampled the southern edge of a larger and more intense plume located farther north before returning to Andenes (Roiger et al., 2015). The underestimation of CO can also partly be ascribed to too low emissions or problems in modeling OH. This plume is shifted towards the north by 40 km in the model simulation, explaining the strong

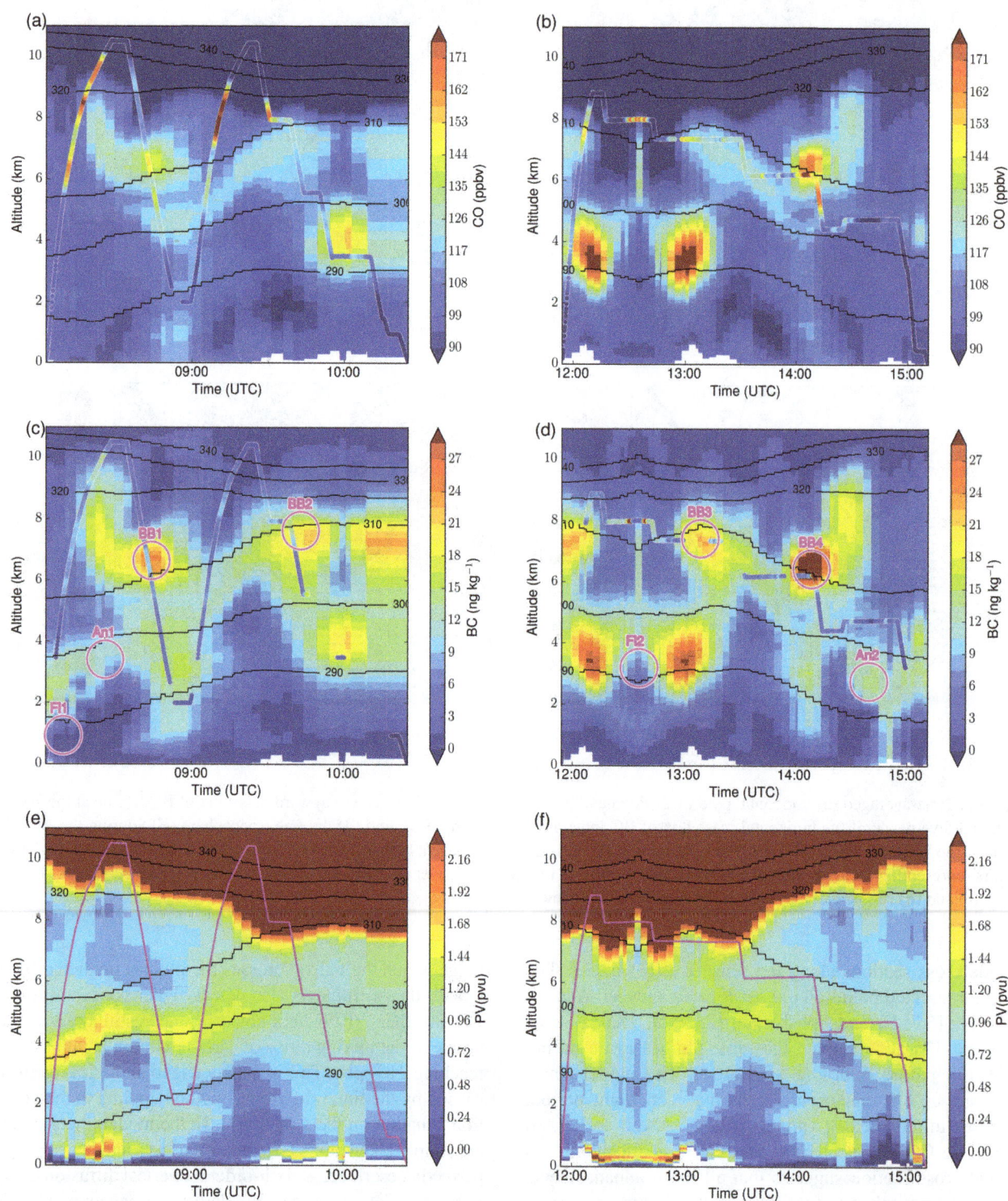

**Figure 8.** Vertical cross sections of (**a, b**) CO mixing ratio, (**c, d**) BC mixing ratio, and (**e, f**) PV along Falcon flight tracks on 17 July 2012 extracted from WRF-Chem between the island of Andøya and the Svalbard archipelago. In situ measurements aboard the Falcon averaged using a 2 min running mean are shown in each panel with colored dots, using the same color scale. Only BC values outside of clouds are reported. Black solid lines represent the dry isentropes between 290 and 340 K. Magenta circles highlight the air masses discussed in Sect. 4.4.

underestimation of the model at 09:15 UTC. In terms of BC, the agreement between WRF-Chem and the SP2 instrument is reasonable. Inside the plumes, values of 20–25 ng kg$^{-1}$ are reported, with peak values larger than 30 ng kg$^{-1}$. In the mid-troposphere, CO and BC concentrations are also significantly increased sporadically in the band delimited by the 290 and 300 K isentropes (between 2 and 4 km) with values reaching 180 ppbv and 30 ng kg$^{-1}$, respectively.

### 4.3   Contribution from the different sources

To understand the contributions of the different source emissions (anthropogenic, biomass burning and flaring) to the BC transported to the region of the flights, we use the difference between the BASE run and the NoAnthro, NoFire, and NoFlr simulations. Model results from each run are interpolated in time and space to the position of the Falcon during the two flights on 17 July. The relative contribution of each source to total BC is obtained by dividing by the results derived from the BASE simulation. Figure 9 shows the time–altitude cross sections of the relative contributions of BC concentrations due to the different emission sources. The two vertical cross sections clearly underline that the mass mixing ratio of BC is strongly dominated by the fire contribution, which is generally larger than 80 % at all altitudes and times, and often larger than 90 %. We note two exceptions to the dominance of fires. In the vicinity of the Norwegian coast, i.e., just after take-off from Andenes before 09:00 and on the way back to Norway after 14:00 UTC, a strong influence of anthropogenic sources (about 85 %) is clearly identified between 2 and 4 km, and to a lesser extent in the PBL, where a weaker contribution of Norwegian flares is predicted (40–70 %). A second zone where the influence of fires is low (below 15 %) is detected in the mid-troposphere (2–5 km) between 12:30 and 13:00 UTC. In this region, the contribution of flaring emissions from Siberian oil exploration is dominant, about 80 %.

In Sect. 4.2, we identified significant enhancements in CO and BC between 6 and 8 km. According to Fig. 9, these plumes originate from large biomass burning sources in Siberia. In WRF-Chem, the fire emission heights are determined by the online plume rise injection model (Grell et al., 2011). These emissions may then be lofted to the faster winds of the free atmosphere. We should note that in this altitude range (5–10 km), the contribution of anthropogenic sources is weak (maximum 15 %) but not zero. There is likely a small influence of the vertical transport by continental convective lifting over Asia in summer and frontal lifting due to mid-latitude cyclones. The transport pathways from the different source regions to the Arctic is discussed in Sect. 4.4. In the mid-troposphere (2–4 km), the CO and BC concentrations have also increased due to the influence of fires, except in the vicinity of the northern Norwegian coast, where the impact of anthropogenic sources is larger. The flaring emissions

play a very localized role in the area of the flights and do not lead to any noticeable increase in CO or BC.

### 4.4   Transport pathways

To investigate the origin and transport pathways of air masses and provide insight into the WRF-Chem representation of BC, the FLEXPART-WRF model is run in backward mode. We identify four plumes originating from boreal fires on 17 July from Fig. 8 in the upper troposphere between 6 and 8 km and confirm them by in situ measurements from the aircraft. The center of those plumes has been found at 08:42 (BB1), 09:39 (BB2), 13:06 (BB3) and 14:10 (BB4) UTC. Their dominant biomass burning origin has been suggested by Fig. 9. We also consider two air masses predicted by the model to be influenced by flaring emissions, in the PBL (0.7 km) at 08:05 UTC (Fl1) and in the mid-troposphere (3.7 km) at 12:35 UTC (Fl2). Finally, two anthropogenic air masses have been selected in the mid-troposphere (3.2–3.5 km), in the proximity of Andøya at 08:30 (An1) and 14:45 UTC (An2). For each selected air mass, 10 000 FLEXPART-WRF particles are released in a volume of 40 km × 40 km (horizontally) and 1 km (vertically). Since transport timescales are typically lower than 10 days, each simulation is run backwards for 10 days to track the origin of air masses reported in Fig. 8.

Figure 10 shows the 0–20 km column of FLEXPART-WRF PES integrated for 10 days for the first fire plume (released at 08:42 UTC on 17 July), together with the altitude and BC concentration retrieved from the model and interpolated along each of the 10 000 trajectories. A vertical cross section of BC interpolated along the plume centroid location is also shown. We note that the results are very similar for the four backward simulations in biomass burning plumes (not shown). Figure 10 highlights a cross-polar transport of biomass burning emissions from Siberian fires into the Arctic, which reach the Svalbard archipelago thanks to a pronounced northerly flow. These plumes have been probed in detail by the aircraft on the two flights on 17 July. During the ACCESS campaign in July 2012, this cross-polar transport of biomass burning pollution was caused by Arctic low-pressure systems, regular phenomena of the summertime Arctic circulation (Orsolini and Sorteberg, 2009). WRF-Chem meteorological analyses suggest that these Arctic transport events in the upper troposphere are due to WCBs linked to two wave cyclones over extreme northern Russia and eastern Siberia. The first one was present at 72° N, 80° E between 11 and 12 July and reached 970 hPa. The second cyclone, responsible of the plume curl over the North Pole (Fig. 10), moved northward and extended from 80° N, 140° E to 88° N, 180° E, progressively deepening to reach a minimum surface pressure of 975 hPa.

Two large plumes enriched in CO and BC form over the south of the Krasnoyarsk region on 8–9 July and over Yakutia on 11–12 July and merge over eastern Siberia into a

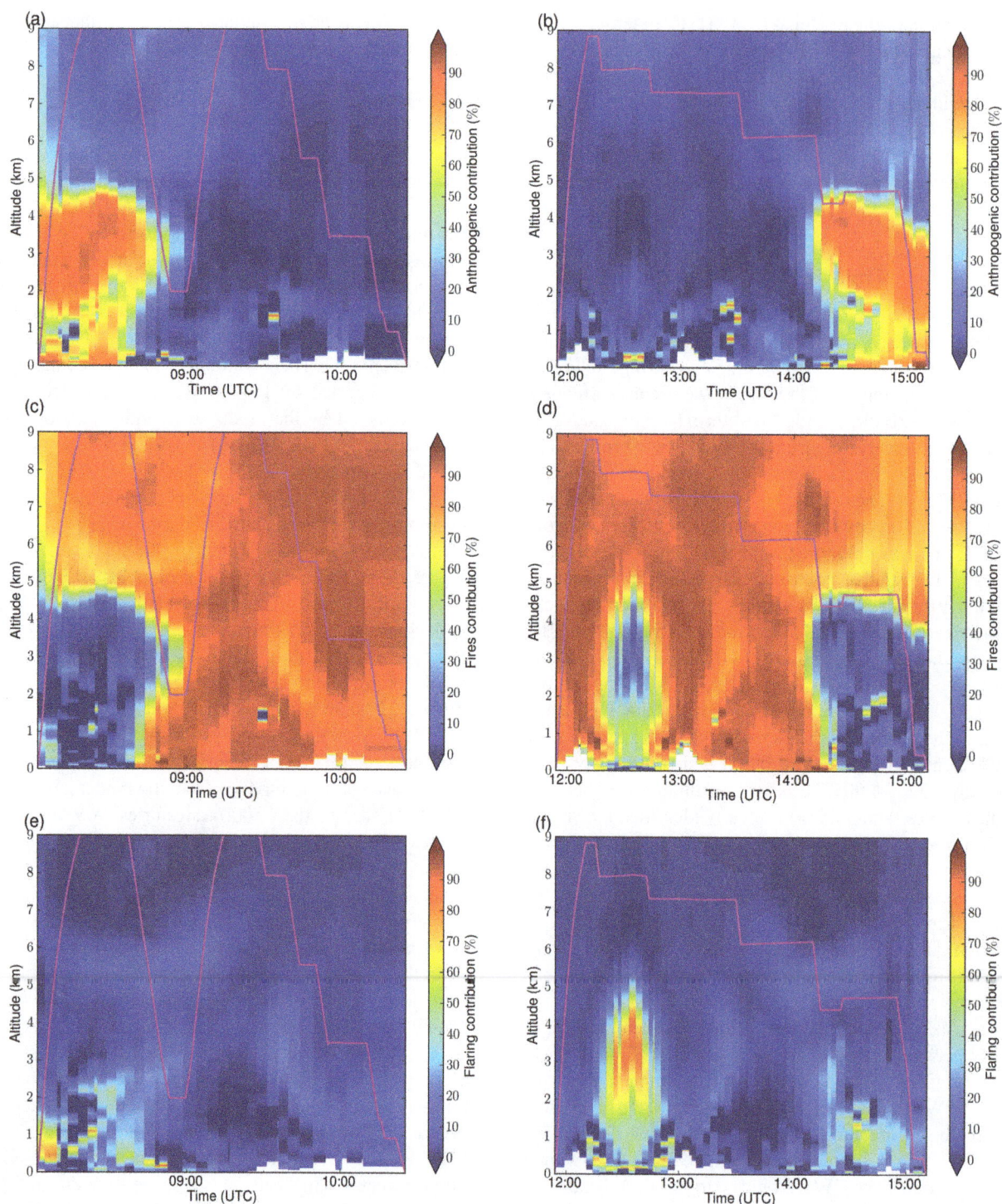

**Figure 9.** Time–altitude cross sections of the relative contributions (%) of BC concentrations due to (**a, b**) anthropogenic, (**c, d**) biomass burning and (**e, f**) flaring emissions along flight tracks 17a (in panels **a, c** and **e**) and 17b (in panels **b, d** and **f**). The altitude of the two flights is indicated in magenta in each panel.

large plume advected to higher latitudes. When the first low-pressure system forms at the surface, the plume intrudes into the Arctic atmosphere, where it elongates and becomes narrower (12 July). The second low-pressure system forms at 80° N, 140° E, which facilitates the progression of the eastern part of the plume further north aloft. Due to the progression of the cyclone to the northeast (88° N, 180° E), the plume elongates and moves almost over the North Pole (13 July).

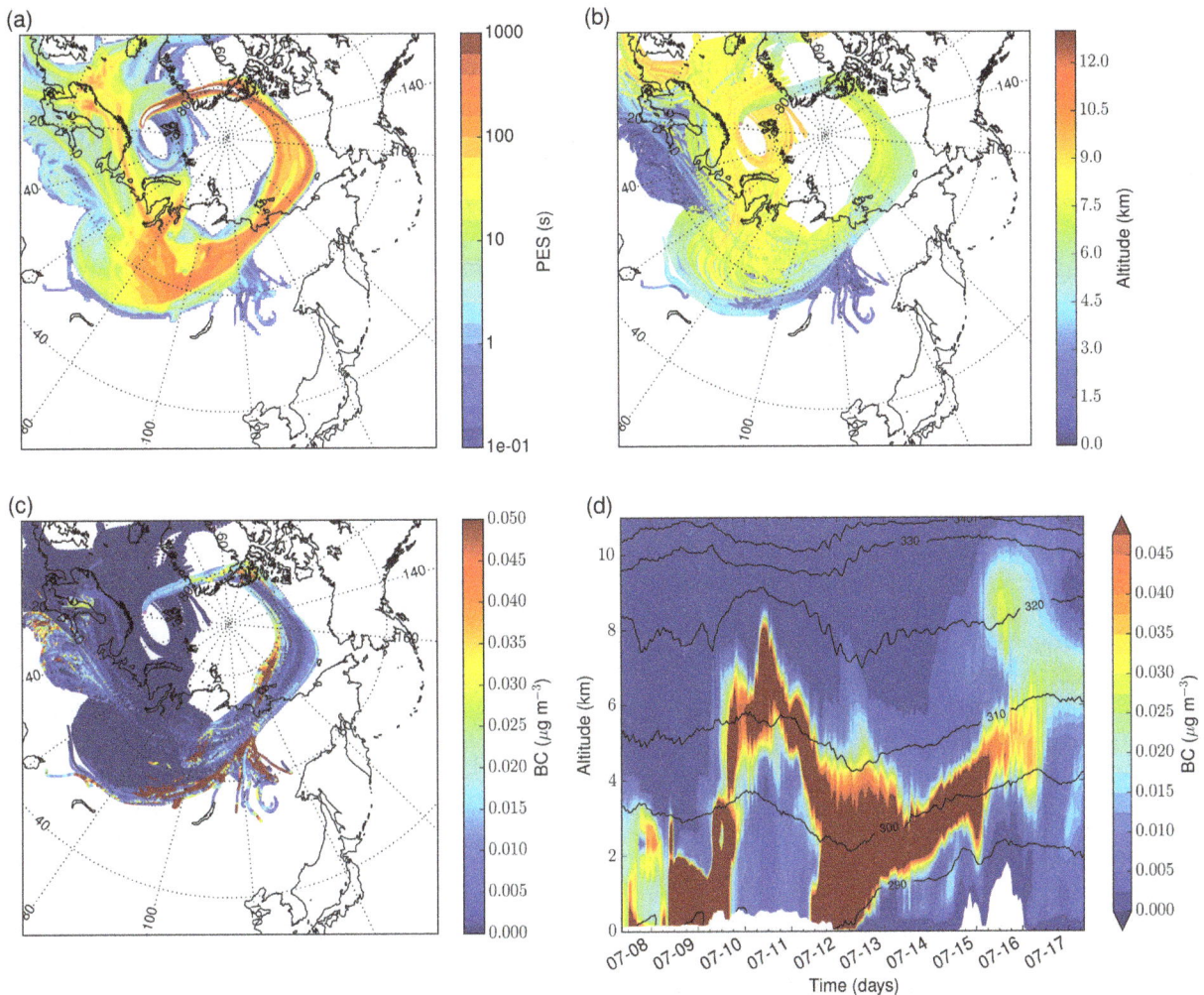

**Figure 10.** FLEXPART-WRF backward simulation from the biomass burning plume observed by the Falcon aircraft at 73.1° N, 18.2° E and 6.5 km between 08:37 and 08:47 UTC. **(a)** Column-integrated PES, **(b)** altitude and **(c)** BC mass concentration interpolated along the 10 000 trajectories for 10 days prior to the release. In panel **(d)**, the vertical cross section of BC interpolated along the plume centroid location is shown, together with the dry isentropes (black lines) between 290 and 340 K.

Along the plume boundaries, some mixing processes with the surrounding air occur, partly eroding its outer parts and leading to a filamentary structure. The aging plume is split into two branches reaching northern Canada on 14 July and moving towards Svalbard. The relatively fine resolution of our simulations (40 km) can resolve the increased filamentation of the pollution plumes, an advantage over most global models with coarser grids (Sodemann et al., 2011; Ma et al., 2014). The concentrations of CO and BC interpolated along trajectories generally decrease with time due to mixing with the surrounding clean atmosphere and deposition processes along transport (Fig. 10), except on 14–15 July, when two plume branches of the same initial plume merge together.

Figure 9 also demonstrates a small contribution (15 %) of anthropogenic sources to the enhanced CO and BC detected by the Falcon observations. As suggested by Fig. 10, the an-thropogenic emissions that arrive near Norway may originate from fossil-fuel burning in northern China, where high levels of BC have been modeled 10 days before the flights on 17 July (Fig. 7). These emissions undergo a strong and efficient uplift from the PBL to the middle and upper troposphere within WCBs and therefore enter the Arctic at high altitudes. Those Asian anthropogenic air masses are characterized by high water vapor mixing ratios ($14\,g\,kg^{-1}$) and RH values larger than 80 %, facilitating their uplift over cold, dense Arctic air masses. The fact that the contribution of fossil-fuel emissions is found to be weak over Scandinavia suggests that significant deposition of aerosols occurs during transport. Wet removal is due to cloud formation and a high amount of precipitation during uplift, mostly convective rain (Fig. 5). It is also characterized by an enhancement of potential temperature from 300 to 315 K caused by latent heat release from water vapor condensation.

According to Serreze and Barrett (2008) and Orsolini and Sorteberg (2009), such pollution transport events characterized by the presence of low-pressure systems in summer could be a common and effective transport mechanism to the Arctic atmosphere for Siberian forest fires and Asian industrial emissions. Such transport pathways reaching the Arctic have been observed by Sodemann et al. (2011) and Roiger et al. (2011) during July 2008. In our study, the low-pressure systems are more intense (surface pressures ranging from 970 to 975 hPa), suggesting an effective mechanism for intrusion of pollutants into northern Norway.

The pollution plumes exported from the flaring emission sources follow different pathways as a function of their altitude. The plume reaching Scandinavia in the PBL is exported above the West Siberian Plain, where large values of the poleward eddy heat flux are predicted (Fig. 7), and passes over the Barents Sea before progressing poleward at low level. However, another plume emitted from the flares in western Siberia is transported eastward along the Siberian coast, moves to the north over Yakutia, curls over the North Pole before reaching Svalbard in the mid-troposphere. Finally, the two anthropogenic plumes identified in the mid-troposphere close to Andøya island are exported from Europe in about 6 days.

## 5 Deposition during transport

### 5.1 Contribution of deposition processes

To understand the contributions of the different deposition processes (dry deposition, wet removal in grid-resolved clouds and in parameterized convective clouds) on the mass of BC transported to the region of the flights, we use the normalized differences between the NoX simulations and the BASE run, where NoX represents the NoDry, NoWet or NoWetCu simulations: $100 \times \left( \frac{\mathrm{NoX} - \mathrm{BASE}}{\mathrm{NoX}} \right)$. Model results from those four runs are interpolated in time and latitude and longitude at the position of the Falcon during the two flights on 17 July. Figure 11 shows the time–altitude cross sections of the relative contributions of the different deposition processes.

The dry deposition is only significant (more than 50 %) in the lowest layers, and close to the Norwegian coast. Due to the fact that BC particles predominantly occupy the Aitken and accumulation mode (90 % of BC particles have a diameter lower than 2.5 μm in our simulation), the impact of the sedimentation process is very weak. Dry deposition of BC-containing particles therefore arises from the aerodynamic transport down through the atmospheric layer to a very thin layer of stagnant air just adjacent to the surface (by turbulent diffusion) and from the Brownian transport across this quasi-laminar sublayer to the surface itself. As a consequence, the role of dry deposition is limited to the lower troposphere

(PBL). More important are the impacts of wet deposition processes in grid-scale and subgrid parameterized clouds. Grid-scale wet removal has the largest effect (80 % BC removal) in the mid-troposphere, where European anthropogenic influence is also large (Fig. 9). Altitudes influenced by biomass burning plumes (Fig. 9) experience similar BC removal rates, 40–60 %, from both grid-scale and subgrid-scale clouds. In the areas where the biomass burning plumes were more intense, as identified in Sect. 4.2, the impact of the deposition processes is generally smaller, almost negligible for dry deposition, about 30 % for wet removal in grid-scale clouds and 10 % in convective clouds. The fact that the maximum in BC concentrations corresponds to zones less affected by deposition processes illustrates the heterogeneous vertical cross sections of BC (Fig. 8). The vertical profile of BC (Fig. 3) would have therefore peaked more intensively in the mid- and upper troposphere if wet removal had been less efficient. Between 4 and 6 km, the BC mixing ratio would have been multiplied by a factor of 3 without wet removal in subgrid convective clouds, or by a factor of 4 without wet removal in grid-scale clouds.

BC particles coated with non-refractory and water-soluble secondary compounds can be CCN (cloud condensation nuclei) active in liquid clouds. BC-containing particles can also act as IN (ice nuclei). In the MOSAIC aerosol module used in WRF-Chem, aerosol components are assumed to undergo a instantaneous internal mixing in each size bin. Petters et al. (2009) showed that internally mixed BC particles can be more hygroscopic and easily removed by wet scavenging compared to externally mixed BC. This should not have a large effect on our results as BC particles sampled by the Falcon have principally been transported in biomass burning plumes from Siberia (Sect. 4.4). Kondo et al. (2011) and Sahu et al. (2012) observed that BC in biomass burning plumes is often more internally mixed than in fossil-fuel emissions, frequently shows very thick coatings (Dahlkötter et al., 2014) and is thought to occur in the first few hours after emission (Abel et al., 2003; Akagi et al., 2012).

During the long-range transport of plumes towards the Arctic region, the WCBs observed in this study loft aerosols from the PBL to the free troposphere, or rapidly inject them into the upper troposphere, in the warm section of mid-latitude cyclones. Clouds form and a significant portion of particles becoming efficient CCN is scavenged by cloud droplets or rain drops and potentially removed from the atmosphere via wet deposition in those vertical transport pathways. A fraction of aerosols may survive this deposition associated with the cyclones, reach higher altitudes and travel to polar latitudes. However, a small portion of the remaining interstitial BC particles can be removed through impaction scavenging by collection of cloud or rain droplets. This process is said to be more effective for BC removal, both in mixed-phase (Twohy et al., 2010) and in ice clouds (Baumgardner et al., 2008; Stith et al., 2011). This is particularly important in this study as Fig. 6 indicated that ice, super-

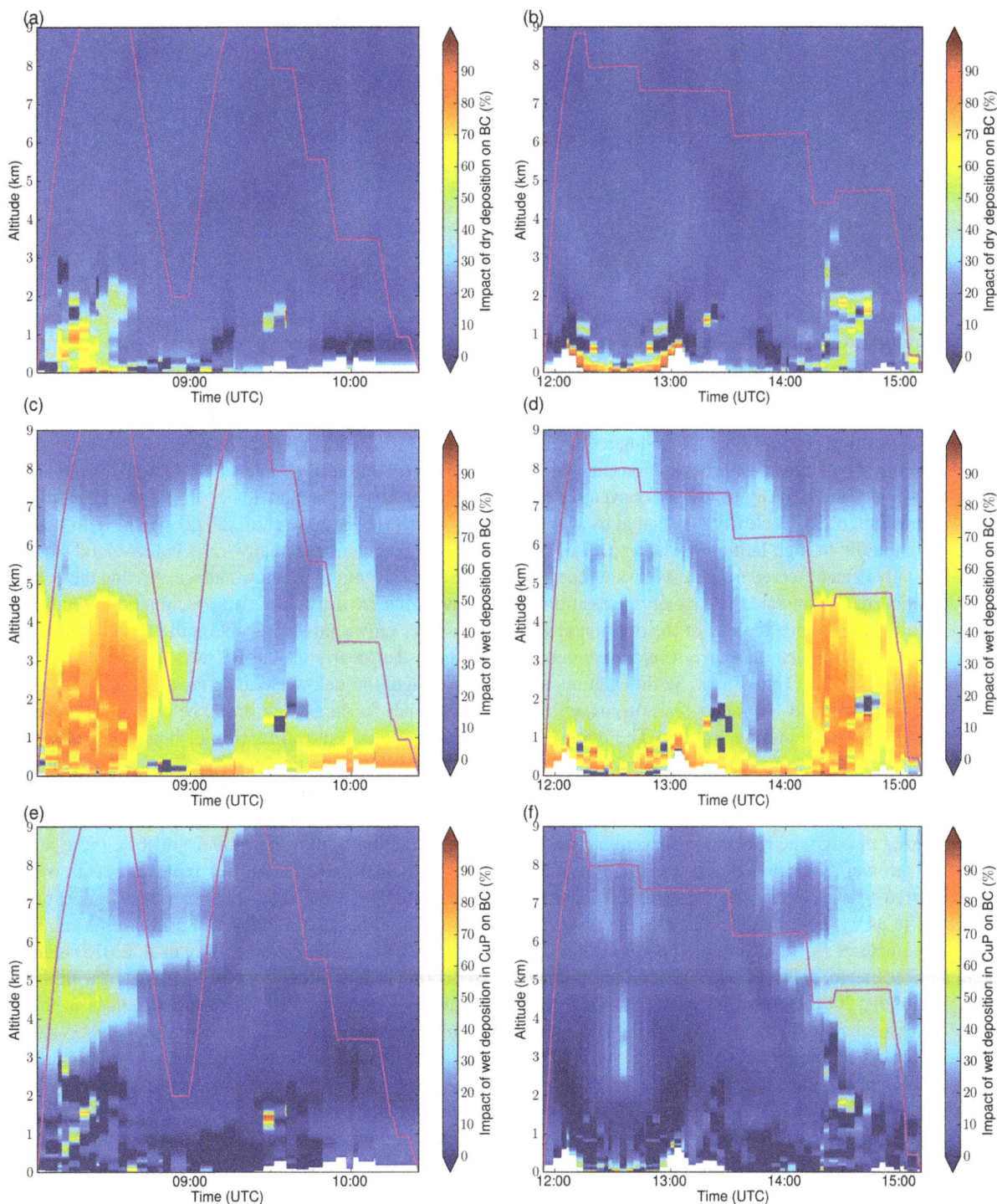

**Figure 11.** Time–altitude cross sections of the relative contributions (%) of (**a, b**) dry deposition, (**c, d**) wet deposition in grid-resolved clouds and (**e, f**) wet deposition in parameterized cumulus clouds to the BC mass mixing ratio along flight tracks 17a (in panels **a**, **c** and **e**) and 17b (in panels **b**, **d** and **f**). The altitude of the two flights is indicated in magenta in each panel.

cooled and mixed-phase clouds are predominant in the mid-troposphere during the ACCESS period.

## 5.2  Impact of deposition

Figure 12 shows the accumulated deposition of BC between 7 and 21 July, which is the period during which aerosols

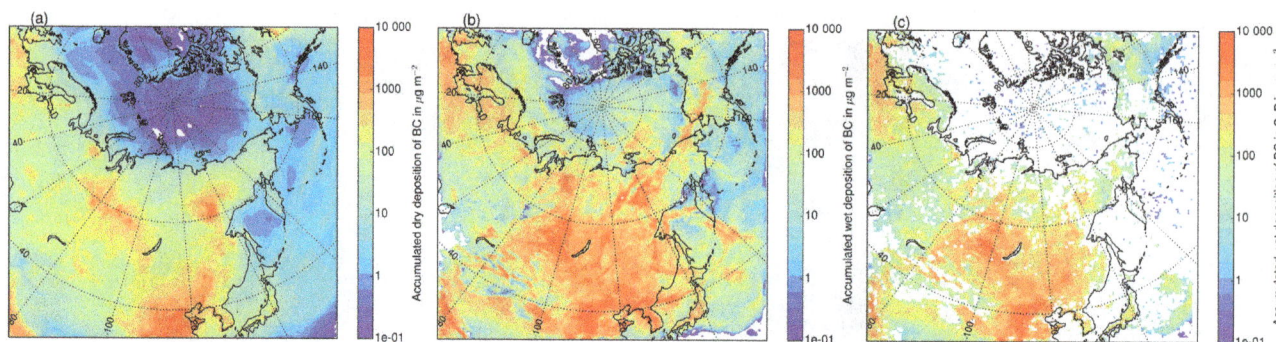

**Figure 12.** Accumulated deposition of BC due the **(a)** dry deposition, **(b)** wet removal in all clouds (grid-resolved and cumuli) and **(c)** wet removal in parameterized cumulus clouds between 7 and 21 July. Note the logarithmic color scale.

were transported from mid-latitudes to the Scandinavia. The three types of deposition processes are studied here: dry deposition, wet removal by large-scale precipitation (first-order rainout and washout) and scavenging in wet convective updrafts. The deposition of BC is clearly dominated by wet removal from the middle to high latitudes. The impact of dry deposition can only been observed near emission sources: the spatial distribution of accumulated BC by dry deposition is co-located with the divergence regions of the horizontal BC flux in the PBL (Fig.7), which may therefore correspond to the spatial distribution of BC emissions. Sedimentation and turbulent mix-out are indeed the major sink of coarse particles containing BC near emission sources. The dry deposition flux of BC particles from the troposphere to the surface is moreover almost proportional to the local BC mass mixing ratio near the surface. In large-scale clouds, BC particles are removed through the nucleation scavenging mechanism, or through wet deposition processes (in-cloud scavenging is referred to as rainout, and below-cloud scavenging is referred to as washout), where particles are collected by falling hydrometeors (in stratiform precipitation) through Brownian motion, electrostatic forces, collision or impaction (Seinfeld and Pandis, 2006). Interstitial and cloud-borne BC-containing particles can finally be removed in the strong updraft core of deep convective clouds parameterized in KFCuP (Kain-Fritsch-Cumulus Parameterization).

The horizontal distribution of BC wet removal results from a combined effect of the precipitation amounts (Fig. 5) and the upward BC flux (Fig. 7). During the vertical transport of plumes containing carbonaceous aerosols, the wet deposition is indeed directly linked to the distribution of precipitation. The spatial distribution of BC scavenged by wet deposition in parameterized cumulus clouds matches that of convective precipitation well (Fig. 5). Aerosol wet removal from predicted deep and shallow convective clouds is, however, lower than the removal from grid-scale clouds. This is the case even where the cumulus scheme predicts greater amounts of precipitation than the microphysical scheme (Fig. 5). This KFCuP scheme triggers almost everywhere in the mid-

latitudes, but not so much over the oceans and in the Arctic or sub-Arctic (beyond 60° N), where it is mostly dominated by the formation of low-level stratocumulus cloud decks in the boundary layer and lower troposphere, producing frequent drizzle (Browse et al., 2012). We have to note that the KFCuP mechanism actually lacks below-cloud wet scavenging and scavenging by snow and ice. Adding the below-cloud scavenging by the convective precipitation would nevertheless have only a minor impact on the BC concentrations: assuming that nearly all the BC was in the 80–470 nm diameter range, the below-cloud removal efficiencies are indeed very small for large rain drops below cumuli. Some of the enhanced wet removal in cumuli must be compensated for by the enhanced vertical transport of aerosols from the PBL into the free troposphere during deep convection (Berg et al., 2015). It is also important to bear in mind that the cumulus cloud fraction within the model grid box (here 40 km) can be quite small, so that the wet removal occurs over a relatively small area and does not have a large impact on the total aerosol loading within the model grid box. While validating the KFCuP cumulus scheme, Berg et al. (2015) highlighted a reduction in low altitude aerosol associated with the venting of aerosol from the PBL as well as changes in the vertical structure associated with the cumulus induced subsidence, but little change in the average aerosol aloft. This may also be consistent with the small amount of secondary activation in the simulations.

During the summer of 2012, the wet removal is the dominant process at all latitudes, which illustrates that a substantial fraction of BC mass concentrations is removed in the regions south of 70° N before reaching the Arctic.

The influence of wet removal of BC scavenging is also predicted in the high latitudes in summer, underlining the role of drizzle from low-level clouds and fogs. The combination of low-level scavenging in the Arctic region and transport decrease from mid-latitudes is the cause of the low summertime BC concentrations. We note the fact that the WRF-Chem performs worse at 100 km in reproducing the mean BC profile observed by the Falcon, particularly in the mid-troposphere

(Fig. 3) suggests that the coarse resolution and the increased role of convective clouds result in less efficient BC removal during transport. This is due to the presence of weakly precipitating shallow clouds at high latitudes, which do not scavenge aerosols adequately.

### 5.3 Transport efficiency of BC particles

#### 5.3.1 Role of accumulated precipitation and condensed water along FLEXPART-WRF trajectories

To investigate the transport efficiency of BC particles and evaluate the effects of the wet removal of BC from the atmosphere by precipitation during transport, we use the NoWetAll simulation, which does not consider aerosol wet deposition both in grid-scale and in parameterized convective clouds. We define the transport efficiency of BC as

$$TE_{BC} = \frac{[BC]_{BASE}}{[BC]_{NoWetAll}}. \tag{1}$$

Similar quantities have been used in previous studies (e.g., Koike et al., 2003; Matsui et al., 2011; Oshima et al., 2012), but the denominator was often chosen as the observed BC-to-CO ratios for the source regions rather than its value extracted from a simulation where wet removal has been turned off, as in Oshima et al. (2013).

Following the methodology of Oshima et al. (2012), who investigated the role of wet removal of BC in Asian outflow towards the Pacific Ocean, we estimate the accumulated precipitation along trajectory (APT) and use it as a proxy for wet deposition. APT represents the amount of precipitation that an air parcel had experienced during transport. For every sampled air parcel by the Falcon and identified as a plume on 17 July in Sect. 4.4, the $TE_{BC}$ and APT values are computed along each of the 10 000 FLEXPART-WRF trajectories released at the time and location of the plume. Specifically, the APT is derived by integrating the WRF-Chem precipitation amounts from the BASE run along each Lagrangian 10-day backward trajectory of the corresponding uplifted air parcel. For the plume originating from Russian flaring sources (rapidly transported at a low altitude; Sect. 4.4), the trajectory runs backwards for 6 days. At hourly intervals, the rain, ice, snow and graupel precipitation rates (including values in cumuli) are interpolated at the closest WRF-Chem grid box of the FLEXPART-WRF trajectory, summed and then integrated in time. This method is preferred to the one based on the WRF-Chem surface precipitation amounts that would not take into account the relative vertical distribution of precipitation occurrence and air parcels. Similarly, we also extract the accumulated condensed water along trajectory (ACWT) to account for the fact that BC particles are also removed from the plumes during activation. This process, called the nucleation scavenging mechanism, refers to the transfer of particles from interstitial aerosol to cloud-borne aerosol. The ACWT is calculated using the method applied for APT but based on the mixing ratios of cloud liquid water, ice, snow, rain and graupel contents both in grid-scale and convective parameterized clouds. The mixing ratios are summed and integrated vertically in each model grid box crossed by the trajectory and finally along this trajectory. The nucleation scavenging mechanism can be considered as aerosol removal from the plume when cloud droplets containing aerosols reach the sizes of precipitating rain drops in other cells, e.g., those above or below the grid box crossed by the trajectory. $TE_{BC}$, APT and ACWT values extracted along each of the trajectories are then averaged to give a mean trajectory corresponding to an uplifted air parcel.

#### 5.3.2 Results along trajectories

In Sect. 4.4, eight distinct plumes have been identified: four plumes originating from boreal fire sources in Russia (associated with a small influence of anthropogenic Asian air masses; Fig 9), two plumes from flaring activities in northern Russia and two anthropogenic plumes from Europe. Figure 13 shows the dependence of $TE_{BC}$ for these eight plumes on the APT and ACWT as a function of the FLEXPART-WRF transport time. $TE_{BC}$ systematically decreases with increasing APT or ACWT, reflecting the crucial role of precipitation and nucleation scavenging in removing particles containing BC from advected plumes along transport, and thereby in controlling the amount of BC reaching the Arctic. An exception occurs on 14–15 July, when $TE_{BC}$ slightly increases in some plumes, because two branches of the same initial plume merge over northern Canada (Sect. 4.4). $TE_{BC}$ decline is not linear with APT or ACWT (logarithmic scale in Fig. 13) but starts to significantly decrease 6–7 days (144–168 h) before reaching Scandinavia, corresponding to APT values of 0.2–0.4 mm and ACWT between 200 and 400 mm.

When reaching Scandinavia (retroplume age very young), plumes present various values of $TE_{BC}$ as a function of their origins. $TE_{BC}$ is the greatest for Siberian biomass-burning-impacted air masses. As shown in Fig. 9, 85 % of the BC sampled by the aircraft in such plumes originated from boreal fires and 15 % from Asian anthropogenic masses. In such plumes, the $TE_{BC}$ was as high as 56–68 % and was caused by low APT and ACWT values (1 mm and 300–1000 mm, respectively). In contrast, $TE_{BC}$ was low (21–28 %) and APT and ACWT amounts were high for European anthropogenic air parcels (5–10 and 2000–3000 mm, respectively). Air parcels influenced by flaring emissions exhibit very different behavior as a function of their altitude: moderate $TE_{BC}$ values (50 %) were calculated in plumes reaching the middle Arctic troposphere (3.7 km), whereas $TE_{BC}$ decreased to 20 % in plumes transported in the PBL. The fact that the relation between $TE_{BC}$ and ACWT is different from the one between $TE_{BC}$ and APT, especially in flaring plumes, indicates a role of cloud liquid water and ice crystals in clouds to remove BC during transport. This is caused by the nucleation scavenging mechanism. ACWT appears to be a better proxy

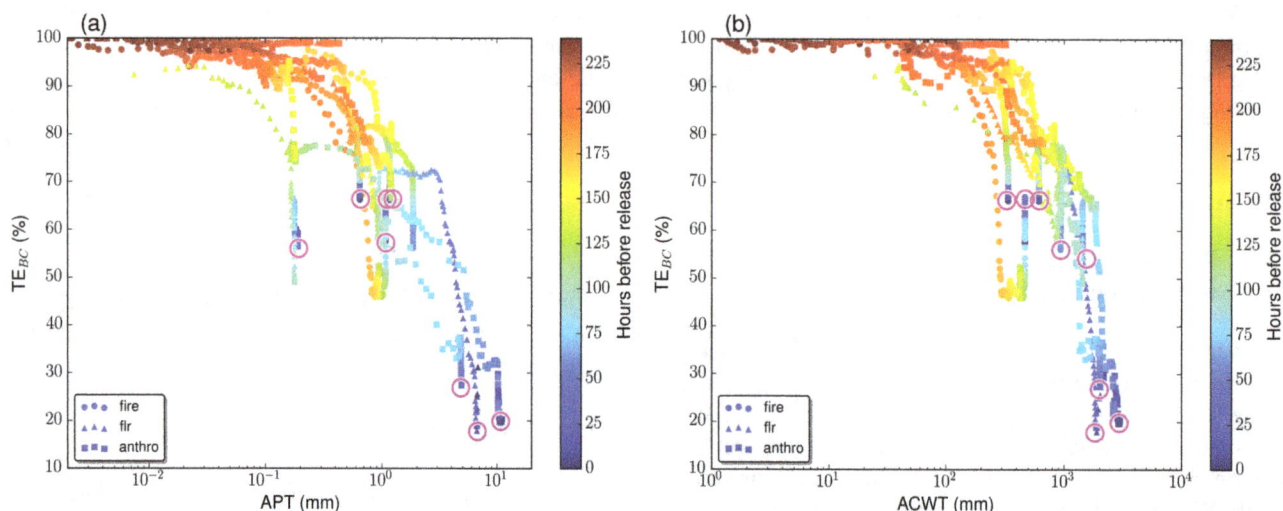

**Figure 13.** Transport efficiency of BC ($TE_{BC}$) for eight plumes identified from sampled air parcels by the Falcon (four boreal fire plumes, two plumes from flaring activities in northern Russia and two anthropogenic plumes from Europe) as a function of **(a)** the accumulated precipitation along trajectory (APT) and **(b)** the accumulated condensed water along trajectory (ACWT). The points are colored by the time (in hours) before the release of the trajectories in FLEXPART-WRF. Magenta circles emphasize the starting points (release time = 0) of the eight air masses discussed in Sect. 4.4.

for characterizing the efficiency of aerosol removal in such air parcels, since the values are of the same order of magnitude in both flaring-impacted air masses (1000–2000 mm), while APT varies from 0.2 mm (in mid-tropospheric plumes) to 8 mm (in PBL plumes).

The magnitude of the differences in $TE_{BC}$ values obtained during the same period due to various origins is particularly noticeable. European anthropogenic air parcels and flaring emissions are generally transported at lower altitudes than biomass burning air masses, according to Fig. 8. They encountered more non-convective precipitation caused by frequent drizzle in low-level stratocumulus clouds, scavenging aerosols during transport. In contrast, Russian and to a lesser extent Asian air masses have exported BC more efficiently towards the Arctic. Most air parcels from those sources have experienced transport in WCBs, removing part of the BC loadings. However, the transport of aerosols has been facilitated thanks to the injection of fire plumes at high altitudes and to the fact that biomass burning sources in Siberia are located far north in summer (45–65° N). These results are in agreement with previous studies from Matsui et al. (2011) and Sharma et al. (2013). They underlined a significant inhibition of emissions from eastern Asia (90 %) compared to biomass burning sources in Russia on the total Arctic BC, because of larger precipitation rates and stronger scavenging.

### 5.3.3 Zonal mean of transport efficiency

The vertical distributions of the mean BC mass concentrations zonally averaged during the ACCESS period in the BASE run and the corresponding mean $TE_{BC}$ values due to grid-scale clouds and all clouds are shown in Fig. 14. The

WRF-Chem domain (Fig. 1) is appropriate for this calculation because all sources influencing the Arctic BC during the ACCESS period are in this domain: the sum of anthropogenic, flaring and fire contributions was found to be larger than 98 % (Fig. 9). The BC mass concentrations are greatest below the altitude of 4 km on the mid-latitude region (40–66° N) with values larger than $50 \, ng \, m^{-3}$. The highest concentrations ($> 100 \, ng \, m^{-3}$) are found close to emission sources and they continuously decrease with height. A high contrast is observed between the mid-latitudes and the Arctic. In the mid-troposphere (3–6 km), the contrast is less pronounced as moderate BC concentrations are spread over a wide latitude range. As a consequence, above 80° N, lowest BC concentrations are found at the surface, and the maximum is detected between 2 and 4 km ($15 \, ng \, m^{-3}$).

In contrast, $TE_{BC}$ due to only grid-scale precipitation exhibits a very distinct dependency on latitude: above 70° N, $TE_{BC}$ decreases from 70–85 to 44–69 %. This sharp meridional gradient is due to the fact that $TE_{BC}$ is indeed smaller for air experiencing wet removal over Asia or Siberia and that is subsequently transported to the outflow regions (high latitudes). This explains the contrast between the BC concentrations in the mid-latitudes and in the Arctic region. Within the Arctic, the maximum of $TE_{BC}$ ($\simeq 69 \%$) is found at higher altitudes (4 and 6 km) than the maximum in BC concentrations. This provides evidence that the transport in WCBs partly, but not completely, removes carbonaceous aerosols by wet deposition. The low BC concentrations in the lowest polar layers are correlated with a weak transport efficiency ($\simeq 45 \%$). This is due to frequent drizzle produced by the formation of low-level stratocumulus clouds in the Arctic PBL and lower

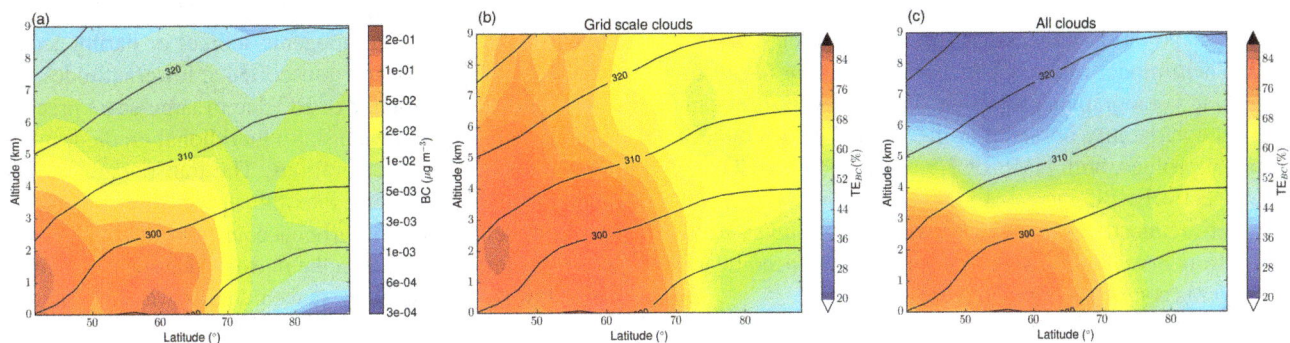

**Figure 14.** Latitude–altitude cross sections of **(a)** the BC mass concentrations, and the corresponding $TE_{BC}$ values due to **(b)** grid-scale clouds and **(c)** all clouds, averaged zonally and temporally over the ACCESS campaign period in the BASE run. On each panel, black solid lines represent the dry isentropes between 290 and 330 K.

troposphere, together with a slower transport in the lowest altitudes. It is also interesting to note that in the mid-latitude regions, higher $TE_{BC}$ are decoupled from the surface, reflecting the strong uplift of Siberian fire emissions above boreal forest. The associated strong ascent generally occurs at subgrid scale ($< 40$ km), which explains the inability of wet removal in large-scale precipitation to clean the atmosphere above sources.

$TE_{BC}$ due to wet removal in all clouds exhibits a totally different pattern. $TE_{BC}$ values are correlated more closely with BC mass concentrations, suggesting that wet removal in subgrid parameterized clouds (cumuli) is also fundamental to understand the vertical distribution of BC mass. It also suggests that cumuli in the mid-latitudes removed more of the low-level aerosol by scavenging or washout than by uplifting it. Aerosol removal in cumulus clouds is responsible of the vertical gradient of $TE_{BC}$, especially in the mid-latitudes. Two noticeable differences between the zonal cross sections of BC and $TE_{BC}$ are observed. First, in mid-latitudes (40–70° N), $TE_{BC}$ is low above 6 km ($TE_{BC} < 20$ %), whereas BC concentrations reach moderate values in this altitude range: this reflects the efficient BC wet removal due to convection and precipitation during the rapid ascent over the source regions. The BC mass concentrations observed above 6 km mainly arise from the boundaries. Second, in the Arctic, the maximum of $TE_{BC}$ is located between 3.5 and 5.5 km (as high as 64 %): it illustrates the fact that the transport of Siberian fire plumes closely followed isotropes, as also noticed by Matsui et al. (2011). Wet removal of aerosols tightly modulates long-range transport from Asia and Siberia to the Arctic, with a marked zonal component and transport at higher altitudes. It is important to note that the transport efficiency is probably slightly overestimated in our study because the model slightly overpredicts BC mixing ratios in the mid-troposphere (Fig. 3).

The Siberian plume has been sampled by the Falcon in the upper troposphere (6–8 km). In Fig. 14, we find that $TE_{BC}$ is about 60 %, which is in agreement with the value found by

along trajectories (Fig. 13). The altitude dependence of BC in the ACCESS aircraft measurements (Fig. 3 and Fig. 8) is consistent with the zonally and temporally averaged $TE_{BC}$ distribution simulated in the BASE simulation during the ACCESS campaign period. This suggests that the general features of wet removal of BC transported in plumes from Siberia and Asia during summer 2012 were well captured in the ACCESS aircraft measurements. We finally note that the $TE_{BC}$ values in this study are much larger than the ones reported by Matsui et al. (2011). That discrepancy is partially due to the different year (2008 vs. 2012) with different emissions and precipitation patterns (APT values much lower in our study). In addition, the definition of $TE_{BC}$, used by Matsui et al. (2011), is the transport efficiency of BC uplifted air masses sampled by the aircraft relative to their corresponding sources. In our study, $TE_{BC}$ represents the transport efficiency of BC due only to wet removal in grid-scale and subgrid parameterized convective clouds. If a plume emitted from a specific source diverges before reaching the receptor area, the transport efficiency calculated by the method of Matsui et al. (2011) will decrease. In our modeling study, the transport efficiencies are only influenced by precipitation and are therefore higher. They illustrate the role of aerosol removal during transport rather than the contribution of the transport itself.

## 6 Summary and conclusions

During the ACCESS campaign in summer 2012, extensive boreal forest fires associated with temperatures record in both western and eastern Siberia strongly impacted the atmospheric composition of the Arctic atmosphere. The Falcon aircraft regularly sampled transported pollution. During two dedicated flights on 17 July, it measured plumes in the middle and upper troposphere that were representative of cross-polar transport from Siberia and Asia towards Scandinavia and the Svalbard archipelago. Specifically, enhanced CO concentrations and BC mixing ratios up to 200 ppbv and 25 ng kg$^{-1}$

are reported between 6 and 8 km altitude, reflecting the influence of pollution transported from remote sources. The WRF-Chem model run at a horizontal resolution of 40 km on a polar stereographic projection shows good skill in capturing the variability observed in the vertical distribution of chemical profiles, with NMB of 1.5 and 27.3 % for CO and BC respectively. BC increases in the middle and upper troposphere are mostly linked to biomass burning plumes transported from boreal sources (85 %), weakly influenced by anthropogenic sources in northern China (15 %).

Our results underline that a coarser horizontal resolution (100 km instead of 40 km) deteriorates the model performance, with a significant overestimation of BC levels in the mid-troposphere. This suggests that, at a coarser resolution, the model is unable to resolve the fine structure of plumes transported in altitude across the North Pole, illustrates the difficulty to represent the cloud and precipitation structures and scavenging processes at subgrid scale and points to the need for improved representation of BC processing in global models in the Arctic.

With a horizontal resolution of 40 km, a good spatial correlation is found between the horizontal distribution of AOD, the strong divergence regions of the BC horizontal fluxes in the FT and the spatial distribution of the mean upward BC mass fluxes, underlying ideal conditions for BC poleward transport. This highlights two major biomass burning regions in Siberia, south of Krasnoyarsk region and Yakutia, and one source of anthropogenic pollution in central and northeastern China. Aerosols and CO emitted from the two boreal forest fire regions are exported towards the Arctic through the north of Yakutia in the region presenting a large horizontal poleward eddy heat flux. Their transport towards the North Pole is caused by low-pressure systems controlling the progression and path of the plumes. In the lowest Arctic atmospheric layers, weak CO and BC enhancements are linked to the inefficient transport of European anthropogenic emissions and to a lesser extent to emissions from flaring activities in northern Russia. These emissions are exported over the Barents Sea and transported at low altitudes into the Arctic.

During the cross-polar transport of Siberian and Asian plumes to the Arctic region, a large fraction of BC particles mixed with sufficient water-soluble compounds become CCN active and are scavenged by cloud droplets or rain drops via wet deposition associated with the warm section of mid-latitude cyclones. The two wet deposition processes, namely the wet removal by large-scale precipitation and the scavenging in wet convective updrafts, contribute almost similarly to the total accumulated deposition of BC. The weak role of dry deposition is limited to the lower troposphere. The simulated transport efficiency of BC ($TE_{BC}$), accumulated precipitation along trajectory (APT) and accumulated condensed water along trajectory (ACWT) values provide insights for understanding the wet removal of aerosols transported to the Arctic. The $TE_{BC}$ in biomass burning plumes is about 60 % and is found to be due to low APT (1 mm). In contrast, $TE_{BC}$

is small (< 30 %) and APTs are larger (5–10 mm) in plumes influenced by urban anthropogenic sources or flaring activities in northern Russia and transported at lower altitudes.

The transport efficiency of BC due to large-scale precipitation is responsible for the sharp meridional gradient in the distribution of BC concentrations. The transport in WCBs, rapidly injecting aerosols into the upper troposphere, removes some but not all carbonaceous aerosols by wet deposition. Cumulus clouds in the mid-latitudes removed more of the low-level aerosol by scavenging or washout than by uplifting it and are responsible for the vertical gradient of $TE_{BC}$, especially in the mid-latitudes. Wet deposition in convective updrafts is essentially active below 60° N, reflecting the distribution of convective precipitation, but is dominated in the Arctic region by the large-scale wet removal associated with the formation of stratocumulus clouds in the PBL producing frequent drizzle. An understanding of the respective contributions of the different wet deposition processes is therefore critically important since it drives the vertical profiles of BC and tightly modulates the long-range transport of BC aerosol from Asia and Siberia to the Arctic, with a marked zonal component and transport at higher altitudes.

*Code availability.* We used WRF-Chem version 3.5.1 modified to include the KFCuP scheme. After version 3.8, the KFCuP scheme is included in the released code, available at http://www2.mmm.ucar.edu/wrf/users/download/get_source.html.

*Author contributions.* The first author performed the WRF-Chem simulations and analyses and drafted the paper; the co-authors contributed text and ideas and discussed the results. Most of the authors participated in the ACCESS campaign in July 2012.

*Competing interests.* The authors declare that they have no conflict of interest.

*Acknowledgements.* The research leading to these results received funding from the European Union under grant agreement 265863-ACCESS (http://www.access-eu.org) within the Ocean of Tomorrow call of the European Commission Seventh Framework Programme. The French co-authors acknowledge support from the national Chantier Arctique project, PARCS (Pollution in the Arctic System), funded by CNRS-INSU. We thank the pilots, engineers, and scientists from the DLR flight department for their excellent support during the field campaign. Computer modeling performed by Jean-Christophe Raut benefited from access to IDRIS HPC resources (GENCI allocations 2015-017141 and 2016-017141) and the IPSL mesoscale computing center (CICLAD: Calcul Intensif pour le Climat, l'Atmosphère et la Dynamique). We thank the EDGAR team for compiling the HTAPv2 emissions. We acknowledge use of the WRF-Chem preprocessor tool (mozbc, fire_emiss, bio_emiss) provided by the Atmospheric Chemistry Observations and Modeling Lab (ACOM) of NCAR. We acknowledge use of MOZART-4 global model output available at http://www.acom.ucar.edu/wrf-chem/mozart.shtml. Jerome D. Fast, Larry K. Berg and Richard C. Easter were supported by the Office of Science of the US Department of Energy as part of the Atmospheric System Research (ASR) program. The Pacific Northwest National Laboratory (PNNL) is operated by DOE by the Battelle Memorial Institute under contract DE-A06-76RLO976 1830. B. Weinzierl and K. Heimerl were supported by the Helmholtz Association (grant number VH-NG-606) and by the European Research Council under the European Community's Horizon 2020 research and innovation framework program/ERC grant agreement 640458 (A-LIFE). We thank the AERIS-ICARE Data and Services Center for providing access to the DARDAR data used in this study.

Edited by: Federico Fierli

# References

Abel, S. J., Haywood, J. M., Highwood, E. J., Li, J., and Buseck, P. R.: Evolution of biomass burning aerosol properties from an agricultural fire in southern Africa, Geophys. Res. Lett., 30, 1783, https://doi.org/10.1029/2003GL017342, 2003.

Akagi, S. K., Craven, J. S., Taylor, J. W., McMeeking, G. R., Yokelson, R. J., Burling, I. R., Urbanski, S. P., Wold, C. E., Seinfeld, J. H., Coe, H., Alvarado, M. J., and Weise, D. R.: Evolution of trace gases and particles emitted by a chaparral fire in California, Atmos. Chem. Phys., 12, 1397–1421, https://doi.org/10.5194/acp-12-1397-2012, 2012.

AMAP: Black Carbon and Ozone as Arctic Climate Forcers, Arctic Monitoring and Assessment Programme (AMAP), Oslo, Norway vii þ, 2015.

Ancellet, G., Pelon, J., Blanchard, Y., Quennehen, B., Bazureau, A., Law, K. S., and Schwarzenboeck, A.: Transport of aerosol to the Arctic: analysis of CALIOP and French aircraft data during the spring 2008 POLARCAT campaign, Atmos. Chem. Phys., 14, 8235–8254, https://doi.org/10.5194/acp-14-8235-2014, 2014.

Baumgardner, D., Subramanian, R., Twohy, C., Stith, J., and Kok, G.: Scavenging of black carbon by ice crystals over the northern Pacific, Geophys. Res. Lett., 35, L22815, https://doi.org/10.1029/2008GL035764, 2008.

Berg, L. K., Shrivastava, M., Easter, R. C., Fast, J. D., Chapman, E. G., Liu, Y., and Ferrare, R. A.: A new WRF-Chem treatment for studying regional-scale impacts of cloud processes on aerosol and trace gases in parameterized cumuli, Geosci. Model Dev., 8, 409–429, https://doi.org/10.5194/gmd-8-409-2015, 2015.

Berg Jr., L. K., Gustafson Jr., W., Kassianov, E. I., and Deng, L.: Evaluation of a Modified Scheme for Shallow Convection: Implementation of CuP and Case Studies, Mon. Weather Rev., 141, 134–147, https://doi.org/10.1175/MWR-D-12-00136.1, 2013.

Bourgeois, Q. and Bey, I.: Pollution transport efficiency toward the Arctic: Sensitivity to aerosol scavenging and source regions, J. Geophys. Res.-Atmos., 116, D08213, https://doi.org/10.1029/2010JD015096, 2011.

Breider, T. J., Mickley, L. J., Jacob, D. J., Wang, Q., Fisher, J. A., Chang, R., and Alexander, B.: Annual distributions and sources of Arctic aerosol components, aerosol optical depth, and aerosol

absorption, J. Geophys. Res.-Atmos., 119, 4107–4124, 2014.

Brioude, J., Arnold, D., Stohl, A., Cassiani, M., Morton, D., Seibert, P., Angevine, W., Evan, S., Dingwell, A., Fast, J. D., Easter, R. C., Pisso, I., Burkhart, J., and Wotawa, G.: The Lagrangian particle dispersion model FLEXPART-WRF version 3.1, Geosci. Model Dev., 6, 1889–1904, https://doi.org/10.5194/gmd-6-1889-2013, 2013.

Brock, C. A., Radke, L. F., Lyons, J. H., and Hobbs, P. V.: Arctic hazes in summer over Greenland and the North American Arctic. I: Incidence and origins, J. Atmos. Chem., 9, 129–148, 1989.

Browse, J., Carslaw, K. S., Arnold, S. R., Pringle, K., and Boucher, O.: The scavenging processes controlling the seasonal cycle in Arctic sulphate and black carbon aerosol, Atmos. Chem. Phys., 12, 6775–6798, https://doi.org/10.5194/acp-12-6775-2012, 2012.

Chapman, E. G., Gustafson Jr., W. I., Easter, R. C., Barnard, J. C., Ghan, S. J., Pekour, M. S., and Fast, J. D.: Coupling aerosol-cloud-radiative processes in the WRF-Chem model: Investigating the radiative impact of elevated point sources, Atmos. Chem. Phys., 9, 945–964, https://doi.org/10.5194/acp-9-945-2009, 2009.

Corbett, J. J., Lack, D. A., Winebrake, J. J., Harder, S., Silberman, J. A., and Gold, M.: Arctic shipping emissions inventories and future scenarios, Atmos. Chem. Phys., 10, 9689–9704, https://doi.org/10.5194/acp-10-9689-2010, 2010.

Dahlkötter, F., Gysel, M., Sauer, D., Minikin, A., Baumann, R., Seifert, P., Ansmann, A., Fromm, M., Voigt, C., and Weinzierl, B.: The Pagami Creek smoke plume after long-range transport to the upper troposphere over Europe – aerosol properties and black carbon mixing state, Atmos. Chem. Phys., 14, 6111–6137, https://doi.org/10.5194/acp-14-6111-2014, 2014.

Delanoë, J. and Hogan, R. J.: Combined CloudSat-CALIPSO-MODIS retrievals of the properties of ice clouds, J. Geophys. Res.-Atmos., 115, D00H29, https://doi.org/10.1029/2009JD012346, 2010.

Di Pierro, M., Jaeglé, L., Eloranta, E. W., and Sharma, S.: Spatial and seasonal distribution of Arctic aerosols observed by the CALIOP satellite instrument (2006–2012), Atmos. Chem. Phys., 13, 7075–7095, https://doi.org/10.5194/acp-13-7075-2013, 2013.

Easter, R. C., Ghan, S. J., Zhang, Y., Saylor, R. D., Chapman, E. G., Laulainen, N. S., Abdul-Razzak, H., Leung, L. R., Bian, X., and Zaveri, R. A.: MIRAGE: Model description and evaluation of aerosols and trace gases, J. Geophys. Res.-Atmos., 109, D20210, https://doi.org/10.1029/2004JD004571, 2004.

Eckhardt, S., Stohl, A., Wernli, H., James, P., Forster, C., and Spichtinger, N.: A 15-year climatology of warm conveyor belts, J. Climate, 17, 218–237, 2004.

Eckhardt, S., Quennehen, B., Olivié, D. J. L., Berntsen, T. K., Cherian, R., Christensen, J. H., Collins, W., Crepinsek, S., Daskalakis, N., Flanner, M., Herber, A., Heyes, C., Hodnebrog, Ø., Huang, L., Kanakidou, M., Klimont, Z., Langner, J., Law, K. S., Lund, M. T., Mahmood, R., Massling, A., Myriokefalitakis, S., Nielsen, I. E., Nøjgaard, J. K., Quaas, J., Quinn, P. K., Raut, J.-C., Rumbold, S. T., Schulz, M., Sharma, S., Skeie, R. B., Skov, H., Uttal, T., von Salzen, K., and Stohl, A.: Current model capabilities for simulating black carbon and sulfate concentrations in the Arctic atmosphere: a multi-model evaluation using a comprehensive measurement data set, Atmos. Chem. Phys., 15,

9413–9433, https://doi.org/10.5194/acp-15-9413-2015, 2015.

Ek, M., Mitchell, K., Lin, Y., Rogers, E., Grunmann, P., Koren, V., Gayno, G., and Tarpley, J.: Implementation of Noah land surface model advances in the National Centers for Environmental Prediction operational mesoscale Eta model, J. Geophys. Res.-Atmos., 108, 8851, https://doi.org/10.1029/2002JD003296, 2003.

Emmons, L. K., Walters, S., Hess, P. G., Lamarque, J.-F., Pfister, G. G., Fillmore, D., Granier, C., Guenther, A., Kinnison, D., Laepple, T., Orlando, J., Tie, X., Tyndall, G., Wiedinmyer, C., Baughcum, S. L., and Kloster, S.: Description and evaluation of the Model for Ozone and Related chemical Tracers, version 4 (MOZART-4), Geosci. Model Dev., 3, 43–67, https://doi.org/10.5194/gmd-3-43-2010, 2010.

Emmons, L. K., Arnold, S. R., Monks, S. A., Huijnen, V., Tilmes, S., Law, K. S., Thomas, J. L., Raut, J.-C., Bouarar, I., Turquety, S., Long, Y., Duncan, B., Steenrod, S., Strode, S., Flemming, J., Mao, J., Langner, J., Thompson, A. M., Tarasick, D., Apel, E. C., Blake, D. R., Cohen, R. C., Dibb, J., Diskin, G. S., Fried, A., Hall, S. R., Huey, L. G., Weinheimer, A. J., Wisthaler, A., Mikoviny, T., Nowak, J., Peischl, J., Roberts, J. M., Ryerson, T., Warneke, C., and Helmig, D.: The POLARCAT Model Intercomparison Project (POLMIP): overview and evaluation with observations, Atmos. Chem. Phys., 15, 6721–6744, https://doi.org/10.5194/acp-15-6721-2015, 2015.

Fast, J. D. and Easter, R. C.: A Lagrangian particle dispersion model compatible with WRF, in: 7th WRF User's Workshop, Boulder, CO, USA, Citeseer, 19–22 June 2006, 2006.

Fast, J. D., Gustafson, W. I., Easter, R. C., Zaveri, R. A., Barnard, J. C., Chapman, E. G., Grell, G. A., and Peckham, S. E.: Evolution of ozone, particulates, and aerosol direct radiative forcing in the vicinity of Houston using a fully coupled meteorology-chemistry-aerosol model, J. Geophys. Res.-Atmos., 111, D21305, https://doi.org/10.1029/2005JD006721, 2006.

Flanner, M. G., Zender, C. S., Randerson, J. T., and Rasch, P. J.: Present-day climate forcing and response from black carbon in snow, J. Geophys. Res.-Atmos., 112, 2007.

Flanner, M. G., Zender, C. S., Hess, P. G., Mahowald, N. M., Painter, T. H., Ramanathan, V., and Rasch, P. J.: Springtime warming and reduced snow cover from carbonaceous particles, Atmos. Chem. Phys., 9, 2481–2497, https://doi.org/10.5194/acp-9-2481-2009, 2009.

Franklin, J. E., Drummond, J. R., Griffin, D., Pierce, J. R., Waugh, D. L., Palmer, P. I., Parrington, M., Lee, J. D., Lewis, A. C., Rickard, A. R., Taylor, J. W., Allan, J. D., Coe, H., Walker, K. A., Chisholm, L., Duck, T. J., Hopper, J. T., Blanchard, Y., Gibson, M. D., Curry, K. R., Sakamoto, K. M., Lesins, G., Dan, L., Kliever, J., and Saha, A.: A case study of aerosol scavenging in a biomass burning plume over eastern Canada during the 2011 BORTAS field experiment, Atmos. Chem. Phys., 14, 8449–8460, https://doi.org/10.5194/acp-14-8449-2014, 2014.

Freitas, S. R., Longo, K. M., Chatfield, R., Latham, D., Silva Dias, M. A. F., Andreae, M. O., Prins, E., Santos, J. C., Gielow, R., and Carvalho Jr., J. A.: Including the sub-grid scale plume rise of vegetation fires in low resolution atmospheric transport models, Atmos. Chem. Phys., 7, 3385–3398, https://doi.org/10.5194/acp-7-3385-2007, 2007.

Garrett, T. J. and Zhao, C.: Increased Arctic cloud longwave emissivity associated with pollution from mid-latitudes, Nature, 440, 787–789, https://doi.org/10.1038/nature04636, 2006.

Grell, G., Freitas, S. R., Stuefer, M., and Fast, J.: Inclusion of biomass burning in WRF-Chem: impact of wildfires on weather forecasts, Atmos. Chem. Phys., 11, 5289–5303, https://doi.org/10.5194/acp-11-5289-2011, 2011.

Grell, G. A., Peckham, S. E., Schmitz, R., McKeen, S. A., Frost, G., Skamarock, W. C., and Eder, B.: Fully coupled "online" chemistry within the WRF model, Atmos. Environ., 39, 6957–6975, 2005.

Guenther, A., Karl, T., Harley, P., Wiedinmyer, C., Palmer, P. I., and Geron, C.: Estimates of global terrestrial isoprene emissions using MEGAN (Model of Emissions of Gases and Aerosols from Nature), Atmos. Chem. Phys., 6, 3181–3210, https://doi.org/10.5194/acp-6-3181-2006, 2006.

Gustafson, W. I., Chapman, E. G., Ghan, S. J., Easter, R. C., and Fast, J. D.: Impact on modeled cloud characteristics due to simplified treatment of uniform cloud condensation nuclei during NEAQS 2004, Geophys. Res. Lett., 34, L19809, https://doi.org/10.1029/2007GL030021, 2007.

Hansen, J. and Nazarenko, L.: Soot climate forcing via snow and ice albedos, P. Natl. Acad. Sci. USA, 101, 423–428, 2004.

Hubanks, P. A., King, M. D., Platnick, S., and Pincus, R.: MODIS atmosphere L3 gridded product algorithm theoretical basis document, ATBD Reference Number: ATBD-MOD-30, 30, 96 pp., 2008.

Huffman, G. J., Adler, R. F., Morrissey, M. M., Bolvin, D. T., Curtis, S., Joyce, R., McGavock, B., and Susskind, J.: Global precipitation at one-degree daily resolution from multisatellite observations, J. Hydrometeorol., 2, 36–50, 2001.

Iacono, M. J., Delamere, J. S., Mlawer, E. J., Shephard, M. W., Clough, S. A., and Collins, W. D.: Radiative forcing by long-lived greenhouse gases: Calculations with the AER radiative transfer models, J. Geophys. Res.-Atmos., 113, D13103, https://doi.org/10.1029/2008JD009944, 2008.

IPCC: Climate Change 2013: The Physical Science Basis. Contribution of Working Group I to the Fifth Assessment Report of the Intergovernmental Panel on Climate Change, Cambridge University Press, Cambridge, UK and New York, NY, USA, https://doi.org/10.1017/CBO9781107415324, 2013.

Jacobson, M. Z.: Climate response of fossil fuel and biofuel soot, accounting for soot's feedback to snow and sea ice albedo and emissivity, J. Geophys. Res.-Atmos., 109, D21201, https://doi.org/10.1029/2004JD004945, 2004.

Jacobson, M. Z.: Short-term effects of controlling fossil-fuel soot, biofuel soot and gases, and methane on climate, Arctic ice, and air pollution health, J. Geophys. Res.-Atmos., 115, D14209, https://doi.org/10.1029/2009JD013795, 2010.

Janjic, Z. I.: The step-mountain eta coordinate model: Further developments of the convection, viscous sublayer, and turbulence closure schemes, Mon. Weather Rev., 122, 927–945, 1994.

Kain, J. S.: The Kain–Fritsch Convective Parameterization: An Update, J. Appl. Meteorol., 43, 170–181, https://doi.org/10.1175/1520-0450(2004)043<0170:TKCPAU>2.0.CO;2, 2004.

Koch, D. and Hansen, J.: Distant origins of Arctic black carbon: a Goddard Institute for Space Studies ModelE experiment, J. Geophys. Res.-Atmos., 110, D04204, https://doi.org/10.1029/2004JD005296, 2005.

Koch, D., Schulz, M., Kinne, S., McNaughton, C., Spackman, J. R., Balkanski, Y., Bauer, S., Berntsen, T., Bond, T. C., Boucher, O., Chin, M., Clarke, A., De Luca, N., Dentener, F., Diehl, T., Dubovik, O., Easter, R., Fahey, D. W., Feichter, J., Fillmore, D., Freitag, S., Ghan, S., Ginoux, P., Gong, S., Horowitz, L., Iversen, T., Kirkevåg, A., Klimont, Z., Kondo, Y., Krol, M., Liu, X., Miller, R., Montanaro, V., Moteki, N., Myhre, G., Penner, J. E., Perlwitz, J., Pitari, G., Reddy, S., Sahu, L., Sakamoto, H., Schuster, G., Schwarz, J. P., Seland, Ø., Stier, P., Takegawa, N., Takemura, T., Textor, C., van Aardenne, J. A., and Zhao, Y.: Evaluation of black carbon estimations in global aerosol models, Atmos. Chem. Phys., 9, 9001–9026, https://doi.org/10.5194/acp-9-9001-2009, 2009.

Koike, M., Kondo, Y., Kita, K., Takegawa, N., Masui, Y., Miyazaki, Y., Ko, M., Weinheimer, A., Flocke, F., Weber, R., Thornton, D. C., Sachse, G. W., Vay, S. A., Blake, D. R., Streets, D. G., Eisele, F. L., Sandholm, S. T., Singh, H. B., and Talbot, R. W.: Export of anthropogenic reactive nitrogen and sulfur compounds from the East Asia region in spring, J. Geophys. Res.-Atmos., 108, 8789, https://doi.org/10.1029/2002JD003284, 2003.

Kondo, Y., Matsui, H., Moteki, N., Sahu, L., Takegawa, N., Kajino, M., Zhao, Y., Cubison, M., Jimenez, J., Vay, S., Diskin, G. S., Anderson, B., Wisthaler, A., Mikoviny, T., Fuelberg, H. E., Blake, D. R., Huey, G., Weinheimer, A. J., Knapp, D. J., and Brune, W. H.: Emissions of black carbon, organic, and inorganic aerosols from biomass burning in North America and Asia in 2008, J. Geophys. Res.-Atmos., 116, D08204, https://doi.org/10.1029/2010JD015152, 2011.

Laborde, M., Schnaiter, M., Linke, C., Saathoff, H., Naumann, K.-H., Möhler, O., Berlenz, S., Wagner, U., Taylor, J. W., Liu, D., Flynn, M., Allan, J. D., Coe, H., Heimerl, K., Dahlkötter, F., Weinzierl, B., Wollny, A. G., Zanatta, M., Cozic, J., Laj, P., Hitzenberger, R., Schwarz, J. P., and Gysel, M.: Single Particle Soot Photometer intercomparison at the AIDA chamber, Atmos. Meas. Tech., 5, 3077–3097, https://doi.org/10.5194/amt-5-3077-2012, 2012.

Lavoué, D., Liousse, C., Cachier, H., Stocks, B. J., and Goldammer, J. G.: Modeling of carbonaceous particles emitted by boreal and temperate wildfires at northern latitudes, J. Geophys. Res.-Atmos., 105, 26871–26890, 2000.

Law, K. S. and Stohl, A.: Arctic Air Pollution: Origins and Impacts, Science, 315, 1537–1540, https://doi.org/10.1126/science.1137695, 2007.

Law, K. S., Stohl, A., Quinn, P. K., Brock, C. A., Burkhart, J. F., Paris, J.-D., Ancellet, G., Singh, H. B., Roiger, A., Schlager, H., Dibb, J., Jacob, D. J., Arnold, S. R., Pelon, J., and Thomas, J. L.: Arctic air pollution: new insights from POLARCAT-IPY, B. Am. Meteorol. Soc., 95, 1873–1895, 2014.

Lee, Y. H., Lamarque, J.-F., Flanner, M. G., Jiao, C., Shindell, D. T., Berntsen, T., Bisiaux, M. M., Cao, J., Collins, W. J., Curran, M., Edwards, R., Faluvegi, G., Ghan, S., Horowitz, L. W., McConnell, J. R., Ming, J., Myhre, G., Nagashima, T., Naik, V., Rumbold, S. T., Skeie, R. B., Sudo, K., Takemura, T., Thevenon, F., Xu, B., and Yoon, J.-H.: Evaluation of preindustrial to present-day black carbon and its albedo forcing from Atmospheric Chemistry and Climate Model Intercomparison Project (ACCMIP), Atmos. Chem. Phys., 13, 2607–2634, https://doi.org/10.5194/acp-13-2607-2013, 2013.

Lindsay, R. W., Zhang, J., Schweiger, A., Steele, M., and Stern, H.: Arctic Sea Ice Retreat in 2007 Follows Thinning Trend, J. Climate, 22, 165–176, https://doi.org/10.1175/2008JCLI2521.1, 2009.

Lubin, D. and Vogelmann, A. M.: A climatologically significant aerosol longwave indirect effect in the Arctic, Nature, 439, 453–456, 2006.

Ma, P.-L., Rasch, P. J., Wang, H., Zhang, K., Easter, R. C., Tilmes, S., Fast, J. D., Liu, X., Yoon, J.-H., and Lamarque, J.-F.: The role of circulation features on black carbon transport into the Arctic in the Community Atmosphere Model version 5 (CAM5), J. Geophys. Res.-Atmos., 118, 4657–4669, https://doi.org/10.1002/jgrd.50411, 2013.

Ma, P.-L., Rasch, P. J., Fast, J. D., Easter, R. C., Gustafson Jr., W. I., Liu, X., Ghan, S. J., and Singh, B.: Assessing the CAM5 physics suite in the WRF-Chem model: implementation, resolution sensitivity, and a first evaluation for a regional case study, Geosci. Model Dev., 7, 755–778, https://doi.org/10.5194/gmd-7-755-2014, 2014.

Madonna, E., Wernli, H., Joos, H., and Martius, O.: Warm Conveyor Belts in the ERA-Interim Dataset (1979–2010). Part I: Climatology and Potential Vorticity Evolution, J. Climate, 27, 3–26, https://doi.org/10.1175/JCLI-D-12-00720.1, 2014.

Manabe, S., Spelman, M., and Stouffer, R.: Transient responses of a coupled ocean-atmosphere model to gradual changes of atmospheric $CO_2$. Part II: Seasonal response, J. Climate, 5, 105–126, 1992.

Marelle, L., Thomas, J. L., Raut, J.-C., Law, K. S., Jalkanen, J.-P., Johansson, L., Roiger, A., Schlager, H., Kim, J., Reiter, A., and Weinzierl, B.: Air quality and radiative impacts of Arctic shipping emissions in the summertime in northern Norway: from the local to the regional scale, Atmos. Chem. Phys., 16, 2359–2379, https://doi.org/10.5194/acp-16-2359-2016, 2016.

Matsui, H., Kondo, Y., Moteki, N., Takegawa, N., Sahu, L., Zhao, Y., Fuelberg, H., Sessions, W., Diskin, G., Blake, D., Wisthaler, A., and Koike, M.: Seasonal variation of the transport of black carbon aerosol from the Asian continent to the Arctic during the ARCTAS aircraft campaign, J. Geophys. Res.-Atmos., 116, D05202, https://doi.org/10.1029/2010JD015067, 2011.

Monks, S. A., Arnold, S. R., Emmons, L. K., Law, K. S., Turquety, S., Duncan, B. N., Flemming, J., Huijnen, V., Tilmes, S., Langner, J., Mao, J., Long, Y., Thomas, J. L., Steenrod, S. D., Raut, J. C., Wilson, C., Chipperfield, M. P., Diskin, G. S., Weinheimer, A., Schlager, H., and Ancellet, G.: Multi-model study of chemical and physical controls on transport of anthropogenic and biomass burning pollution to the Arctic, Atmos. Chem. Phys., 15, 3575–3603, https://doi.org/10.5194/acp-15-3575-2015, 2015.

Morrison, H., Thompson, G., and Tatarskii, V.: Impact of cloud microphysics on the development of trailing stratiform precipitation in a simulated squall line: Comparison of one-and two-moment schemes, Mon. Weather Rev., 137, 991–1007, 2009.

Moteki, N., Kondo, Y., and Nakamura, S.-I.: Method to measure refractive indices of small nonspherical particles: Application to black carbon particles, J. Aerosol Sci., 41, 513–521, 2010.

Murrells, T., Passant, N., Thistlethwaite, G., Wagner, A., Li, Y., Bush, T., Norris, J., Walker, C., Stewart, R., Tsagatakis, I., Whiting, R., Conolly, C., Okamura, S., Peirce, M., Sneddon, S., Webb, J., Thomas, J., MacCarthy, J., Choudrie, S., and Brophy, N.: UK emissions of air pollutants 1970 to 2008, AEA Energy and Environment, 2010.

Orsolini, Y. J. and Sorteberg, A.: Projected changes in Eurasian and Arctic summer cyclones under global warming in the Bergen Climate Model, Atmos. Oceanic Sci. Lett., 2, 62–67, 2009.

Oshima, N., Kondo, Y., Moteki, N., Takegawa, N., Koike, M., Kita, K., Matsui, H., Kajino, M., Nakamura, H., Jung, J., and Kim, Y. J.: Wet removal of black carbon in Asian outflow: Aerosol Radiative Forcing in East Asia (A-FORCE) aircraft campaign, J. Geophys. Res.-Atmos., 117, D03204, https://doi.org/10.1029/2011JD016552, 2012.

Oshima, N., Koike, M., Kondo, Y., Nakamura, H., Moteki, N., Matsui, H., Takegawa, N., and Kita, K.: Vertical transport mechanisms of black carbon over East Asia in spring during the A-FORCE aircraft campaign, J. Geophys. Res.-Atmos., 118, 13175–13198, https://doi.org/10.1002/2013JD020262, 2013.

Paris, J.-D., Stohl, A., Nédélec, P., Arshinov, M. Yu., Panchenko, M. V., Shmargunov, V. P., Law, K. S., Belan, B. D., and Ciais, P.: Wildfire smoke in the Siberian Arctic in summer: source characterization and plume evolution from airborne measurements, Atmos. Chem. Phys., 9, 9315–9327, https://doi.org/10.5194/acp-9-9315-2009, 2009.

Petters, M. D., Carrico, C. M., Kreidenweis, S. M., Prenni, A. J., DeMott, P. J., Collett, J. L., and Moosmüller, H.: Cloud condensation nucleation activity of biomass burning aerosol, J. Geophys. Res.-Atmos., 114, D22205, https://doi.org/10.1029/2009JD012353, 2009.

Ponomarev, E. I.: Radiative power of wildfires in Siberia on the basis of TERRA/Modis imagery processing, Folia Forestalia Polonica, Seria A-Forestry, 55, 102–110, 2013.

Quinn, P. K., Bates, T. S., Baum, E., Doubleday, N., Fiore, A. M., Flanner, M., Fridlind, A., Garrett, T. J., Koch, D., Menon, S., Shindell, D., Stohl, A., and Warren, S. G.: Short-lived pollutants in the Arctic: their climate impact and possible mitigation strategies, Atmos. Chem. Phys., 8, 1723–1735, https://doi.org/10.5194/acp-8-1723-2008, 2008.

Rastigejev, Y., Park, R., Brenner, M. P., and Jacob, D. J.: Resolving intercontinental pollution plumes in global models of atmospheric transport, J. Geophys. Res.-Atmos., 115, D02302, https://doi.org/10.1029/2009JD012568, 2010.

Roiger, A., Schlager, H., Schäfler, A., Huntrieser, H., Scheibe, M., Aufmhoff, H., Cooper, O. R., Sodemann, H., Stohl, A., Burkhart, J., Lazzara, M., Schiller, C., Law, K. S., and Arnold, F.: In-situ observation of Asian pollution transported into the Arctic lowermost stratosphere, Atmos. Chem. Phys., 11, 10975–10994, https://doi.org/10.5194/acp-11-10975-2011, 2011.

Roiger, A., Thomas, J.-L., Schlager, H., Law, K. S., Kim, J., Schäfler, A., Weinzierl, B., Dahlkötter, F., Krisch, I., and Marelle, L.: Quantifying emerging local anthropogenic emissions in the Arctic region: The ACCESS aircraft campaign experiment, B. Am. Meteorol. Soc., 96, 441–460, 2015.

Sahu, L., Kondo, Y., Moteki, N., Takegawa, N., Zhao, Y., Cubison, M., Jimenez, J., Vay, S., Diskin, G., Wisthaler, A., Mikoviny, T., Huey, L. G., Weinheimer, A. J., and Knapp, D. J.: Emission characteristics of black carbon in anthropogenic and biomass burning plumes over California during ARCTAS-CARB 2008, J. Geophys. Res.-Atmos., 117, 2156–2202, https://doi.org/10.1029/2011JD017401, 2012.

Schmale, J., Schneider, J., Ancellet, G., Quennehen, B., Stohl, A., Sodemann, H., Burkhart, J. F., Hamburger, T., Arnold, S. R., Schwarzenboeck, A., Borrmann, S., and Law, K. S.: Source identification and airborne chemical characterisation of aerosol pollution from long-range transport over Greenland during POLAR-CAT summer campaign 2008, Atmos. Chem. Phys., 11, 10097–10123, https://doi.org/10.5194/acp-11-10097-2011, 2011.

Schwarz, J., Gao, R., Fahey, D., Thomson, D., Watts, L., Wilson, J., Reeves, J., Darbeheshti, M., Baumgardner, D., Kok, G., Chung, S. H., Schulz, M., Hendricks, J. Lauer, A., Kärcher, B., Slowik, J. G., Rosenlof, K. H., Thompson, T. L., Langford, A. O., Loewenstein, M., and Aikin, K. C.: Single-particle measurements of midlatitude black carbon and light-scattering aerosols from the boundary layer to the lower stratosphere, J. Geophys. Res.-Atmos., 111, D16207, https://doi.org/10.1029/2006JD007076, 2006.

Schwarz, J., Samset, B., Perring, A., Spackman, J., Gao, R., Stier, P., Schulz, M., Moore, F., Ray, E. A., and Fahey, D.: Global-scale seasonally resolved black carbon vertical profiles over the Pacific, Geophys. Res. Lett., 40, 5542–5547, 2013.

Schwarz, J. P., Weinzierl, B., Samset, B. H., Dollner, M., Heimerl, K., Markovic, M. Z., Perring, A. E., and Ziemba, L.: Aircraft measurements of black carbon vertical profiles show upper tropospheric variability and stability, Geophys. Res. Lett., 44, 1132–1140, https://doi.org/10.1002/2016GL071241, 2017.

Seinfeld, J. H. and Pandis, S. N.: Atmospheric chemistry and physics, vol. 450, Wiley, Hoboken, USA, 2006.

Serreze, M. C. and Barrett, A. P.: The summer cyclone maximum over the central Arctic Ocean, J. Climate, 21, 1048–1065, 2008.

Sessions, W. R., Fuelberg, H. E., Kahn, R. A., and Winker, D. M.: An investigation of methods for injecting emissions from boreal wildfires using WRF-Chem during ARCTAS, Atmos. Chem. Phys., 11, 5719–5744, https://doi.org/10.5194/acp-11-5719-2011, 2011.

Sharma, S., Ishizawa, M., Chan, D., Lavoué, D., Andrews, E., Eleftheriadis, K., and Maksyutov, S.: 16-year simulation of Arctic black carbon: Transport, source contribution, and sensitivity analysis on deposition, J. Geophys. Res.-Atmos., 118, 943–964, 2013.

Shaw, W. J., Allwine, K. J., Fritz, B. G., Rutz, F. C., Rishel, J. P., and Chapman, E. G.: An evaluation of the wind erosion module in DUSTRAN, Atmos. Environ., 42, 1907–1921, 2008.

Shindell, D. and Faluvegi, G.: Climate response to regional radiative forcing during the twentieth century, Nat. Geosci., 2, 294–300, https://doi.org/10.1038/ngeo473, 2009.

Shindell, D. T., Chin, M., Dentener, F., Doherty, R. M., Faluvegi, G., Fiore, A. M., Hess, P., Koch, D. M., MacKenzie, I. A., Sanderson, M. G., Schultz, M. G., Schulz, M., Stevenson, D. S., Teich, H., Textor, C., Wild, O., Bergmann, D. J., Bey, I., Bian, H., Cuvelier, C., Duncan, B. N., Folberth, G., Horowitz, L. W., Jonson, J., Kaminski, J. W., Marmer, E., Park, R., Pringle, K. J., Schroeder, S., Szopa, S., Takemura, T., Zeng, G., Keating, T. J., and Zuber, A.: A multi-model assessment of pollution transport to the Arctic, Atmos. Chem. Phys., 8, 5353–5372, https://doi.org/10.5194/acp-8-5353-2008, 2008.

Skamarock, W. C. and Klemp, J. B.: A time-split nonhydrostatic atmospheric model for weather research and forecasting applications, J. Comput. Phys., 227, 3465–3485, 2008.

Sodemann, H., Pommier, M., Arnold, S. R., Monks, S. A., Stebel,

K., Burkhart, J. F., Hair, J. W., Diskin, G. S., Clerbaux, C., Coheur, P.-F., Hurtmans, D., Schlager, H., Blechschmidt, A.-M., Kristjánsson, J. E., and Stohl, A.: Episodes of cross-polar transport in the Arctic troposphere during July 2008 as seen from models, satellite, and aircraft observations, Atmos. Chem. Phys., 11, 3631–3651, https://doi.org/10.5194/acp-11-3631-2011, 2011.

Stauffer, D. R. and Seaman, N. L.: Use of four-dimensional data assimilation in a limited-area mesoscale model. Part I: Experiments with synoptic-scale data, Mon. Weather Rev., 118, 1250–1277, 1990.

Stith, J. L., Twohy, C. H., DeMott, P. J., Baumgardner, D., Campos, T., Gao, R., and Anderson, J.: Observations of ice nuclei and heterogeneous freezing in a Western Pacific extratropical storm, Atmos. Chem. Phys., 11, 6229–6243, https://doi.org/10.5194/acp-11-6229-2011, 2011.

Stohl, A.: Characteristics of atmospheric transport into the Arctic troposphere, J. Geophys. Res.-Atmos., 111, D11306, https://doi.org/10.1029/2005JD006888, 2006.

Stohl, A., Forster, C., Frank, A., Seibert, P., and Wotawa, G.: Technical note: The Lagrangian particle dispersion model FLEXPART version 6.2, Atmos. Chem. Phys., 5, 2461–2474, https://doi.org/10.5194/acp-5-2461-2005, 2005.

Stohl, A., Klimont, Z., Eckhardt, S., Kupiainen, K., Shevchenko, V. P., Kopeikin, V. M., and Novigatsky, A. N.: Black carbon in the Arctic: the underestimated role of gas flaring and residential combustion emissions, Atmos. Chem. Phys., 13, 8833–8855, https://doi.org/10.5194/acp-13-8833-2013, 2013.

Stohl, A., Aamaas, B., Amann, M., Baker, L. H., Bellouin, N., Berntsen, T. K., Boucher, O., Cherian, R., Collins, W., Daskalakis, N., Dusinska, M., Eckhardt, S., Fuglestvedt, J. S., Harju, M., Heyes, C., Hodnebrog, Ø., Hao, J., Im, U., Kanakidou, M., Klimont, Z., Kupiainen, K., Law, K. S., Lund, M. T., Maas, R., MacIntosh, C. R., Myhre, G., Myriokefalitakis, S., Olivié, D., Quaas, J., Quennehen, B., Raut, J.-C., Rumbold, S. T., Samset, B. H., Schulz, M., Seland, Ø., Shine, K. P., Skeie, R. B., Wang, S., Yttri, K. E., and Zhu, T.: Evaluating the climate and air quality impacts of short-lived pollutants, Atmos. Chem. Phys., 15, 10529–10566, https://doi.org/10.5194/acp-15-10529-2015, 2015.

Taylor, J. W., Allan, J. D., Allen, G., Coe, H., Williams, P. I., Flynn, M. J., Le Breton, M., Muller, J. B. A., Percival, C. J., Oram, D., Forster, G., Lee, J. D., Rickard, A. R., Parrington, M., and Palmer, P. I.: Size-dependent wet removal of black carbon in Canadian biomass burning plumes, Atmos. Chem. Phys., 14, 13755–13771, https://doi.org/10.5194/acp-14-13755-2014, 2014.

Tomasi, C., Vitale, V., Lupi, A., Di Carmine, C., Campanelli, M., Herber, A., Treffeisen, R., Stone, R. S., Andrews, E., Sharma, S., Radionov, V., von Hoyningen-Huene, W., Stebel, K., Hansen, G. H., Myhre, C. L., Wehrli, C., Aaltonen, V., Lihavainen, H., Virkkula, A., Hillamo, R., Ström, J., Toledano, C., Ca-chorro, V. E., Ortiz, P., de Frutos, A. M., Blindheim, S., Frioud, M., Gausa, M., Zielinski, T., Petelski, T., and Yamanouchi, T.: Aerosols in polar regions: A historical overview based on optical depth and in situ observations, J. Geophys. Res.-Atmos., 112, D16205, https://doi.org/10.1029/2007JD008432, 2007.

Twohy, C. H., DeMott, P. J., Pratt, K. A., Subramanian, R., Kok, G. L., Murphy, S. M., Lersch, T., Heymsfield, A. J., Wang, Z., Prather, K. A., and Seinfeld, J. H.: Relationships of biomass-burning aerosols to ice in orographic wave clouds, J. Atmos. Sci., 67, 2437–2450, 2010.

van der Gon, H. D., Hendriks, C., Kuenen, J., Segers, A., and Visschedijk, A.: Description of Current Temporal Emission Patterns and Sensitivity of Predicted AQ for Temporal Emission Patterns, EU FP7 MACC deliverable report, 2011.

Vignati, E., Karl, M., Krol, M., Wilson, J., Stier, P., and Cavalli, F.: Sources of uncertainties in modelling black carbon at the global scale, Atmos. Chem. Phys., 10, 2595–2611, https://doi.org/10.5194/acp-10-2595-2010, 2010.

Wang, H., Rasch, P. J., Easter, R. C., Singh, B., Zhang, R., Ma, P.-L., Qian, Y., Ghan, S. J., and Beagley, N.: Using an explicit emission tagging method in global modeling of source-receptor relationships for black carbon in the Arctic: Variations, sources, and transport pathways, J. Geophys. Res.-Atmos., 119, 12888–12909, https://doi.org/10.1002/2014JD022297, 2014.

Warneke, C., Froyd, K., Brioude, J., Bahreini, R., Brock, C., Cozic, J., De Gouw, J., Fahey, D., Ferrare, R., Holloway, J., Middlebrook, A. M., Miller, L., Montzka, S., Schwarz, J. P., Sodemann, H., Spackman, J. R., and Stohl, A.: An important contribution to springtime Arctic aerosol from biomass burning in Russia, Geophys. Res. Lett., 37, L01801, https://doi.org/10.1029/2009GL041816, 2010.

Warren, S. G. and Wiscombe, W. J.: A model for the spectral albedo of snow. II: Snow containing atmospheric aerosols, J. Atmos. Sci., 37, 2734–2745, 1980.

Wesely, M. and Hicks, B.: A review of the current status of knowledge on dry deposition, Atmos. Environ., 34, 2261–2282, 2000.

Wiedinmyer, C., Akagi, S. K., Yokelson, R. J., Emmons, L. K., Al-Saadi, J. A., Orlando, J. J., and Soja, A. J.: The Fire INventory from NCAR (FINN): a high resolution global model to estimate the emissions from open burning, Geosci. Model Dev., 4, 625–641, https://doi.org/10.5194/gmd-4-625-2011, 2011.

Wild, O., Zhu, X., and Prather, M. J.: Fast-J: Accurate simulation of in-and below-cloud photolysis in tropospheric chemical models, J. Atmos. Chem., 37, 245–282, 2000.

Zaveri, R. A. and Peters, L. K.: A new lumped structure photochemical mechanism for large-scale applications, J. Geophys. Res.-Atmos., 104, 30387–30415, 1999.

Zaveri, R. A., Easter, R. C., Fast, J. D., and Peters, L. K.: Model for simulating aerosol interactions and chemistry (MOSAIC), J. Geophys. Res.-Atmos., 113, L01801, https://doi.org/10.1029/2009GL041816, 2008.

Zhao, C. and Garrett, T. J.: Effects of Arctic haze on surface cloud radiative forcing, Geophys. Res. Lett., 42, 557–564, 2015.

# Impacts of large-scale atmospheric circulation changes in winter on black carbon transport and deposition to the Arctic

**Luca Pozzoli[1], Srdan Dobricic[1], Simone Russo[2], and Elisabetta Vignati[1]**

[1]European Commission, Joint Research Centre (JRC), Directorate for Energy, Transport and Climate, Air and Climate Unit, Ispra (VA), 21027, Italy
[2]European Commission, Joint Research Centre (JRC), Directorate for Competences, Modelling, Indicators and Impact Evaluation Unit, Ispra (VA), 21027, Italy

*Correspondence to:* Luca Pozzoli (luca.pozzoli@ec.europa.eu)

**Abstract.** Winter warming and sea-ice retreat observed in the Arctic in the last decades may be related to changes of large-scale atmospheric circulation pattern, which may impact the transport of black carbon (BC) to the Arctic and its deposition on the sea ice, with possible feedbacks on the regional and global climate forcing. In this study we developed and applied a statistical algorithm, based on the maximum likelihood estimate approach, to determine how the changes of three large-scale weather patterns associated with increasing temperatures in winter and sea-ice retreat in the Arctic impact the transport of BC to the Arctic and its deposition. We found that two atmospheric patterns together determine a decreasing winter deposition trend of BC between 1980 and 2015 in the eastern Arctic while they increase BC deposition in the western Arctic. The increasing BC trend is mainly due to a pattern characterized by a high-pressure anomaly near Scandinavia favouring the transport in the lower troposphere of BC from Europe and North Atlantic directly into to the Arctic. Another pattern with a high-pressure anomaly over the Arctic and low-pressure anomaly over the North Atlantic Ocean has a smaller impact on BC deposition but determines an increasing BC atmospheric load over the entire Arctic Ocean with increasing BC concentrations in the upper troposphere. The results show that changes in atmospheric circulation due to polar atmospheric warming and reduced winter sea ice significantly impacted BC transport and deposition. The anthropogenic emission reductions applied in the last decades were, therefore, crucial to counterbalance the most likely trend of increasing BC pollution in the Arctic.

## 1 Introduction

The Arctic has warmed during the recent decades more strongly than other regions due to the polar amplification of the global warming signal (Serreze and Barry, 2011). Satellite observations show a decline in summer sea-ice extent since the late 1970s with a dramatic acceleration in the last decade. The sea ice is also characterized by increased fraction of thinner and younger sea ice, which is melting during the polar summer (Comiso, 2012; Kwok and Rothrock, 2009; Kwok, 2009; Maslanik et al., 2011; Rothrock et al., 2008). The rate of change of Arctic sea ice has increased so quickly over the last decades that the expression "new Arctic" has recently been coined (e.g. Döscher et al., 2014). During winter the sea-ice cover and concentration have changed significantly: a decreasing trend of sea-ice concentration in winter months, December–January–February (DJF), for the period 1979–2014 was observed by satellites over the Barents Sea at a rate of up to 20 % per decade. Several studies investigated whether the varying winter sea-ice cover over the Barents Sea may further produce hemispheric-scale impacts in the atmosphere with changes in the large-scale atmospheric circulation (e.g. Barnes et al., 2014; Cohen et al., 2014; Deser et al., 2007; Mori et al., 2014; Overland and Wang, 2010; Petoukhov and Semenov, 2010; Screen et al., 2013; Semenov and Latif, 2015). Dobricic et al. (2016) analysed the atmospheric reanalysis of the past decades with a statistical approach, the independent component analysis (ICA), which has not been applied frequently in climate studies. ICA can be an efficient methodology to extract independent components (ICs), i.e. share the minimum information without the

Gaussianity assumption (Hyvärinen and Oja, 2000). Dobricic et al. (2016) found that three independent atmospheric patterns, connected to the North Atlantic Oscillation (NAO), the Scandinavian blocking (SB) and the El Niño–Southern Oscillation (ENSO), are closely related to the ongoing hemispheric increase of near-surface temperature over the Barents Sea and may be related to the sea-ice cover shrinkage in this region. Thus, the changes in the ocean–sea ice–atmosphere energy fluxes may also impact the large-scale atmospheric circulation in the Northern Hemisphere, as well as the transport of atmospheric pollutants with possible feedbacks on the regional and global climate.

The aim of this study is to estimate how these changes in the large-scale circulation patterns of the Northern Hemisphere in winter could affect the transport of pollutants to the Arctic and in particular black carbon (BC). BC is a short-lived climate pollutant and thus one of the key targets for emission reduction strategies to mitigate the effects of climate change (Shindell et al., 2012). BC plays a key role in the Arctic climate system as it affects the Earth's radiative balance through different mechanisms. BC particles directly absorb solar radiation, and above a surface with high albedo, such as snow and sea ice, they warm the atmosphere inside the haze layer and at higher altitudes (Shaw and Stamnes, 1980). However, BC may also cool the Arctic climate due to surface dimming and decreased poleward heat flux caused by weakened latitudinal temperature gradient from BC heating of the upper troposphere (Flanner, 2013; Sand et al., 2013a, b; Shindell and Faluvegi, 2009). An indirect BC radiative forcing process is determined by its deposition on snow and ice, which can accelerate their melting. Hadley and Kirchstetter (2012) measured in laboratory how BC snow contamination can reduce snow albedo, with amplified BC radiative perturbation in larger snow grains size, which is also consistent with the parameterizations of BC and soot concentrations in snow included in climate models (Aoki et al., 2000; Flanner and Zender, 2006; Yasunari et al., 2011). Due to lower insolation in the Arctic during winter and early spring, BC exerts a negligible radiative forcing, but particles that deposit to snow and ice surfaces can re-emerge at the surface when melt commences in the summer (Conway et al., 1996), meaning that winter transport and deposition of BC is also affecting the Arctic climate.

Measurements show that the equivalent BC (EBC; filter-based absorption measurements of aerosol particles) surface concentrations in the Arctic, as well as those of other atmospheric pollutants, such as sulfate, are largest in winter and early spring, when the transport of pollutants from lower latitudes is more efficient and the removal processes slower (e.g. Eleftheriadis et al., 2009; Gong et al., 2010; Hirdman et al., 2010; Sharma et al., 2006, 2013). The extreme cold surface temperatures in the Arctic determine strong inversions that built the so-called polar dome (Klonecki, 2003; Stohl, 2006), characterized by stable air masses inside the dome and limited exchange between the boundary layer and the free tro-

posphere, meaning that in the Arctic local sources of pollution are transported more efficiently. In contrast, the long-range transport of BC from lower latitudes may contribute more to the Arctic haze due to the significantly larger mid-latitudes emission levels compared to those in the Arctic region (Sand et al., 2015). Another barrier which isolate the Arctic from the lower latitudes is the Arctic front (Barrie, 1986). Polluted air masses can reach the Arctic lower troposphere only if they are emitted in a sufficiently cold region located north of Arctic front, a situation which mainly occurs during winter in northern Eurasia. Pollution emitted south of the Arctic front is lifted above it, generally with cloud formation and scavenging of aerosol particles and deposition to the surface by precipitation. Pollutants emitted in North America and East Asia can reach the Arctic mainly through this pathway (Stohl, 2006). As a result, the main contribution to winter BC surface concentrations and deposition in the Arctic originated from northern Eurasia (Hirdman et al., 2010). The BC concentration and deposition vary considerably on interannual timescale, and large-scale atmospheric circulation processes can favour or reduce the transport of pollutants from the main source regions (North America and Eurasia) toward the Arctic. Hirdman et al. (2010) analysed the long-term trends of EBC and sulfate measured at three stations in the Arctic. They found that EBC surface concentrations decreased at two Arctic stations, Alert (Canada) and Zeppelin (Svalbard islands, Norway), while there is no trend detectable at Barrow (Alaska). Analysing the atmospheric back trajectories at the three stations, they concluded that the observed decreasing trends are mainly driven by changes in the emissions, while the impact of atmospheric circulation can only explain a minor fraction of the downward trend. They found significant correlations between the North Atlantic Oscillation index (NAOI) and air masses from North America and northern Eurasia at both Alert and Barrow but not for the station at Zeppelin. However, full understanding of how the transport changes related to large-scale circulation patterns, such as the NAO, impact the BC concentrations and deposition to the Arctic has not yet been established.

In this study we apply a statistical methodology, based on a Bayesian approach, to investigate the relationships between large-scale atmospheric circulation patterns and the long-range atmospheric transport of air pollutants to the Arctic in recent decades. We estimate the most likely BC distribution associated with large-scale atmospheric patterns which can approximate the near-surface temperature trend in the Arctic, both spatially and temporally, of two atmospheric reanalysis of the past decades (Dobricic et al., 2016). The distributions of BC surface concentration, total column and deposition fluxes were taken from a hindcast simulation of tropospheric chemistry composition for the period 1980–2005. Three different BC simulations are analysed, one with changing anthropogenic emissions for the entire period and two with fixed anthropogenic emissions from the years 1980 and 2000. With this methodology we aim to quantify the trends and interannual variability of BC surface concentrations, load and deposition in the Arctic associated with winter large-scale atmospheric circulation changes occurred in the last decades.

## 2  Methods and data

In the following sections we will first describe the statistical method (Sect. 2.1) and the data used for the analysis. Section 2.2 provides a short introduction to the large-scale atmospheric patterns identified by Dobricic et al. (2016), which mainly contribute to the winter polar warming of the last decades. Section 2.3 describes the data used to estimate the maximum likelihood distribution of BC concentration and deposition associated with the specific atmospheric patterns.

### 2.1  Maximum likelihood estimate (MLE) of atmospheric pollutants

The MLE of the distribution of a pollutant in the atmosphere associated with a specific atmospheric pattern may be derived starting from Bayes' theorem. The probability, $p(c|a)$, of an unknown joint distribution of an atmospheric pollutant and corresponding atmospheric state, $c$, given a specific atmospheric pattern, $a$, may be expressed as

$$p(c|a) \cong p(a|c)\, p(c), \qquad (1)$$

where $p(c)$ is the priori probability of coupled pollutant distribution (e.g. atmospheric concentrations, surface deposition or atmospheric burden) coupled with corresponding atmospheric conditions, and $p(a|c)$ is the probability that a certain atmospheric pattern (e.g. surface pressure, geopotential height, wind speed) is associated with the atmospheric state coupled with pollutant distributions. The prior probability, $p(c)$, can be estimated, for example, by using chemistry–climate model simulations of the past or future atmospheric physical parameters and the corresponding coupled chemical compositions. In order to simplify the mathematical solution of the problem we assume that the two distributions, $p(a|c)$ and $p(c)$, are approximately Gaussian, meaning that their product, $p(c|a)$, will also be Gaussian. This assumption is supported by the central limit theorem, which tells us that the distribution produced by several processes with non-Gaussian distributions should appear closer to a Gaussian distribution (e.g. Hyvärinen and Oja, 2000). Thus it is possible to estimate the probability of a pollutant distribution for any atmospheric pattern $a_k$, which can be taken from an independent atmospheric model simulation or a specific large-scale atmospheric pattern, like the ENSO or the NAO. Equation (1) becomes

$$p(c|a_k) = \mathrm{const} \times$$

$$\exp\left\{ -\frac{1}{2}[a_k - H(c)]^T D^{-1}[a_k - H(c)] - \frac{1}{2}c^T C^{-1}c \right\}, \qquad (2)$$

where $H(c)$ is an operator which interpolates the coupled atmospheric states, $c$, on the model grid of the atmospheric pattern, $a_k$.

The MLE of $p(c|a_k)$ is the one with the minimum absolute value of the argument of the exponential function in Eq. (2), which is the minimum of the cost function $J$:

$$J = \frac{1}{2}[a_k - H(c)]^T \mathbf{D}^{-1}[a_k - H(c)] + \frac{1}{2}c^T \mathbf{C}^{-1}c. \qquad (3)$$

Assuming that the mapping $H$ is linear (e.g. a bilinear interpolation, $H(c) = \mathbf{H}c$), $J$ becomes a quadratic function with a single minimum which may be estimated by a linear minimizer. The numerical stability may be further increased by defining a control subspace $z = \mathbf{Z}^+ c$, where $\mathbf{Z}$ is a square root of $\mathbf{C}$ and the superscript $+$ indicates the generalized inverse. The cost function becomes

$$J = \frac{1}{2}[a_k - \mathbf{HZ}z]^T \mathbf{D}^{-1}[a_k - \mathbf{HZ}z] + \frac{1}{2}z^T z. \qquad (4)$$

After finding $z$, the most likelihood coupled anomaly is obtained from

$$c = \mathbf{Z}z. \qquad (5)$$

The cost function $J$ is minimized using the quasi-Newtonian limited memory Broyden–Fletcher–Goldfarb–Shanno (L-BFGS) minimizer (Byrd et al., 1995). It is assumed that $\mathbf{D}$ is a diagonal matrix and elements along the diagonal have equal variance divided by the area of each grid point. The atmospheric patterns $a_k$, for which we estimate the maximum likelihood distribution of atmospheric pollutants, are those previously described by Dobricic et al. (2016) and briefly described in Sect. 2.2. Matrix $\mathbf{C}$ is estimated from coupled anomalies of a multi-year simulation with a coupled atmosphere–chemistry model, described in more details in Sect. 2.3. Coupled anomalies contain both the atmospheric physical parameters and pollution concentrations. In order to filter out statistically insignificant relationships, the covariance matrix of anomalies is approximated by forming the empirical orthogonal function (EOF) decomposition and by maintaining only EOFs with major eigenvalues. The minimum of the cost function $J$ is found for each winter month (DJF) and for each atmospheric pattern $a_k$, separately.

### 2.2  Atmospheric patterns

Dobricic et al. (2016) performed an ICA of atmospheric reanalysis data finding a link between the increasing trend of near-surface temperature in the Arctic during winter (DJF) and three atmospheric patterns. Hannachi et al. (2009) first proposed applying the ICA instead of the commonly used EOF in climate studies, showing that ICA may be explained by a rotation of EOFs. The major difference between the two estimates is that ICA does not assume the Gaussian distributions of event probabilities. This property ensures that components are truly independent and not just uncorrelated;

however, it is not possible to determine the order of the ICs. In Dobricic et al. (2016) the FastICA algorithm by Hyvärinen and Oja (2000) was applied to extract independent atmospheric components in winter during the period 1980–2015. The ICA algorithm finds the best approximation of a matrix $\mathbf{X}$ containing rows of temporal anomalies in the physical space:

$$\mathbf{X} \cong \mathbf{AS}, \tag{6}$$

where the columns of matrix $\mathbf{A}$ represent spatially varying intensities and the rows of the orthogonal matrix $\mathbf{S}$ are the temporally varying independent components. Matrix $\mathbf{A}$ is full rank, but its columns are not orthogonal. This means that there may be overlapping spatial features in different columns of A. Details on implementing ICA for large-scale atmospheric processes may be found in Dobricic et al. (2016); here we introduce the main results of their study, which are used as input for our analysis. A set of three independent large-scale atmospheric structures were found, with significant linear trends and together approximating the spatial variability of near-surface temperature trend during winter, as well as atmospheric anomalies of wind intensity, temperature and geopotential height at different pressure levels from surface up to 10 mbar. Figure 1 shows the spatial distribution of the trends from 1980 to 2015 averaged over the three winter months for near-surface temperature at 1000 mbar (T1000) and geopotential height at 850 mbar (H850). The spatial distribution of the three ICs with statistically significant trends were named by visually recognizing their similarity to well-known large-scale weather patterns: the NAO, SB and ENSO. In this paper we will refer to the IC patterns as $IC_{NAO}$, $IC_{SB}$ and $IC_{ENSO}$ to avoid confusion with the NAO, SB and ENSO indices. The mean winter H850 trend of $IC_{NAO}$ (Fig. 1a) with increasing geopotential height over the Arctic and decreasing in the Atlantic Ocean near the Azores clearly appears as a tendency toward the negative phase of the NAO. T1000 increases over the Arctic with maximum over western Greenland, the Canadian archipelago and the Barents Sea, while at the same time it decreases over northern Europe and Siberia (Fig. 1d). The $IC_{SB}$ trend shows an increasing geopotential height over Scandinavia and north-western Siberia, which indicates a prevailing anticyclonic anomaly over the area, bringing warm air from Europe to the Arctic and cold air from the Arctic to Eurasia (T1000 in Fig. 1e). As shown in Fig. 1c and f, $IC_{ENSO}$ has a small T1000 trend over the Arctic, compared to the other two IC patterns. The reanalysis trends of T1000 and H850 are approximated well by summing the three ICs; all dominant features are captured both spatially and temporally, in particular the prominent dipole between the strong warming in the Arctic and cooling over Siberia (Dobricic et al., 2016). Consistent results were obtained using two different atmospheric reanalysis (Dobricic et al. 2016), the National Center for Environmental Prediction (NCEP; Kalnay et al., 1996)

and ERA-Interim (Dee et al., 2011) from the European Centre for Medium-Range Weather Forecasts (ECMWF). In our study we will estimate the maximum likelihood distribution of BC atmospheric concentration, load and deposition associated with the $IC_{NAO}$ and $IC_{SB}$ atmospheric patterns found in Dobricic et al. (2016). We do not discuss the third atmospheric pattern, the $IC_{ENSO}$, which showed weaker connection to the changes seen in near-surface temperature and geopotential height in the Arctic region.

## 2.3 Atmospheric chemical composition

We used winter monthly mean anomalies (DJF) of BC concentrations and surface deposition fluxes of the period 1980–2005 from hindcast simulations of the fully coupled aerosol–chemistry–climate model, ECHAM5–HAMMOZ (Pozzoli et al., 2011), which is composed of the general circulation model ECHAM5 (Roeckner et al., 2003), the tropospheric chemistry and aerosol module HAMMOZ (Pozzoli et al., 2008a). The horizontal resolution of the simulations is about $2.8° \times 2.8°$, with 31 vertical levels from the surface up to 10 hPa. The simulation was forced by nudging the reanalysis meteorological fields from the ECMWF ERA-40 reanalysis (Uppala et al., 2005) from year 1980 to 2000 and the operational analyses (IFS cycle-32r2) was used until year 2005. The dry deposition scheme of aerosol particles follows Ganzeveld and Lelieveld (1995), while in-cloud and below-cloud scavenging follows the scheme described by Stier et al. (2005). The anthropogenic emissions of CO, $NO_x$ and volatile organic carbons for the period 1980–2000 are taken from the RETRO inventory (Endresen et al., 2003; Schultz et al., 2007, 2008). The AeroCom hindcast aerosol emission inventory (Diehl et al., 2012) was used for the annual total anthropogenic emissions of primary BC, organic carbon aerosols and sulfur dioxide ($SO_2$). The BC anthropogenic emissions remained almost constant globally during the simulated period (1980–2005), at $4.9 \, Tg \, yr^{-1}$, but large changes occurred in North America, Europe, Former Soviet Union (FSU) and East Asia (Supplement Fig. S1a, c). Most of the anthropogenic BC is emitted between 30 and 60° N and decreased after the 1990s from about 3 to about $2.6 \, Tg \, yr^{-1}$ after 2000. Above 60° N BC anthropogenic emissions are a small fraction of the total and decreased from 100 to $30 \, Gg \, yr^{-1}$. The simulation also includes interannual varying biomass burning emissions, from tropical savannah burning, deforestation fires and mid- and high-latitude forest fires published by Schultz et al. (2008). BC biomass burning emissions range between 10 and $170 \, Gg \, yr^{-1}$ above 60° N and between 35 and $460 \, Gg \, yr^{-1}$ at mid-latitudes, with peak years connected to interannual meteorological variability (Fig. S1b, d).

The model has been extensively evaluated in previous studies by comparing simulated chemical concentrations and physical parameters to observations (Auvray et al., 2007; Pausata et al., 2012, 2013; Pozzoli et al., 2008a, b; Rast et

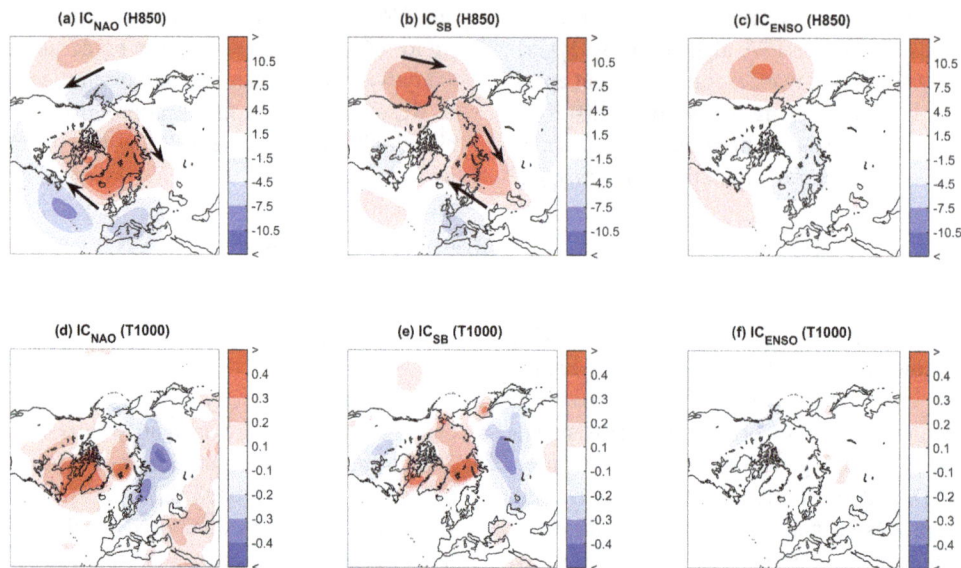

**Figure 1.** Winter trends (1980–2015) of geopotential height at 850 mbar (H850, $m\,yr^{-1}$, **a–c**) and near-surface temperature at 1000 mbar (T1000 $K\,yr^{-1}$, **d–f**) of the independent atmospheric patterns related to the North Atlantic Oscillation ($IC_{NAO}$), Scandinavian Blocking ($IC_{SB}$) and El Niño–Southern Oscillation ($IC_{ENSO}$). The arrows provide a qualitative representation of the circulation paths tendencies associated with the **(a)** $IC_{NAO}$ and **(b)** $IC_{SB}$ independent component patterns.

al., 2014; Stier et al., 2005; Bourgeois and Bey, 2011) and within model intercomparison studies (Kim et al., 2014; Pan et al., 2015; Tsigaridis et al., 2014). As shown by Bourgeois and Bey (2011), ECHAM5–HAMMOZ largely underestimates BC concentrations over the Arctic, both near the surface and in the atmospheric column. Compared to the BC measurements from SP2 instrument (Moteki and Kondo, 2007; Schwarz et al., 2006) of the Arctic Research of the Composition of the Troposphere from Aircraft and Satellites (ARCTAS; Jacob et al., 2010), the simulated BC concentrations show a mean absolute bias in the troposphere of 95 % in spring, while in summer the BC is well simulated in the upper troposphere and overestimated by 50 % near the surface. Bourgeois and Bey (2011) identified the wet scavenging of aerosol particles as one of the main processes responsible for model bias in winter; a model simulation with revisited wet scavenging coefficients from Henning et al. (2004) considerably improved the simulated BC concentrations in the troposphere in winter, reducing the mean absolute bias to 38 %. The large bias of simulated BC and EBC concentrations in the Arctic is a known issue, shared with several global climate and chemical transport models (Eckhardt et al., 2015; Sand et al., 2017). Qi et al. (2017) estimated that the Wegener–Bergeron–Findeisen process in mixed-phase clouds increases BC in the atmosphere by 25 to 70 % by reducing wet scavenging efficiency. Other factors which may improve the simulated BC distribution in the Arctic are dry deposition velocities calculated with resistance-in-series method over all surfaces (ocean, snow/ice) and improved BC flaring emissions (Qi et al., 2017; Stohl et al.,

2013). Jiao et al. (2014) show that BC concentrations in snow are poorly correlated with measurements, and a large spread is found among 25 model simulations, with BC lifetime in the Arctic ranging from about 4 to 23 days, implying large differences in local BC deposition efficiency. In this study we will focus on the impacts of large-scale atmospheric circulation trends on the transport of BC to the Arctic through a novel statistical methodology, assuming that underestimating BC concentrations does not significantly affect the spatial distribution of the trends. Three different simulations of the period 1980–2005 are available for our analysis: a reference simulation with annually varying anthropogenic emissions (thereinafter named REF) and two additional simulations with anthropogenic emissions kept constant for the entire period (1980–2005) at the levels of the years 1980 and 2000 (thereinafter named FIX1980 and FIX2000). These three simulations will provide a preliminary estimate of the uncertainty of our findings due to the different levels and their geographical distribution of anthropogenic emissions used in the different simulations. Other model simulations may be analysed in the future to quantify the uncertainty associated with different chemical mechanisms and physical parameterizations.

## 3 Results

In this study we will focus on the trends of BC transport patterns from northern hemispheric mid-latitudes towards the Arctic during winter months. Trends are calculated by the Sen–Kendall method (Sen, 1968). The statistical significance for all trends is set to the 0.05 level and is estimated by the Mann–Kendall test (Mann, 1945). Areas with significant trends are marked with small grey dots in the figures except for the trends of atmospheric patterns and maximum likelihood estimates of BC that have spatially uniform slopes. We first analyse the trends as simulated by a coupled chemistry–climate model under three different anthropogenic emission scenarios (Sect. 3.1). This first analysis will provide an estimate of the relative contribution of anthropogenic emissions and natural variability to the transport of BC toward the Arctic. In the second part, through the statistical method described in Sect. 2, the trend associated with the natural variability are further decoupled into the contributions from the northern hemispheric large-scale atmospheric weather patterns (Sect. 3.2), which were found to be associated with the observed near-surface Arctic warming and sea-ice retreat (Dobricic et al., 2016). In Sect. 3.3 we will illustrate the total trends due to the three independent atmospheric components and their impact on the interannual variability of BC wet deposition and load over the Arctic. The statistical method was applied to a combination of different datasets in order to analyse the spread of different solutions obtained using two different atmospheric reanalysis and three different chemistry–climate model simulations and for different time periods (Sect. 3.4).

### 3.1 Total trends of BC simulated by ECHAM5–HAMMOZ

Figure 2 shows the 26-year (1980–2005) trends of BC dry and wet deposition, surface concentration and vertically integrated atmospheric load over to the Arctic for the REF simulation, which includes the effects of both annually varying anthropogenic emissions and interannual meteorological variability. Decreasing and statistically significant trends are simulated over almost the entire Arctic for dry and wet deposition, as well as surface concentrations. The main reductions occurred in the eastern part of the Arctic, from the Scandinavian Peninsula along almost the entire Russian northern coastline, indicating the important role of the emission reductions occurred in the last decades in Europe and Russia. Previous studies also showed that the main transport pathway of air pollution to the Arctic in winter originates from the Eurasian continent (Sharma et al., 2013; Stohl, 2006). Nevertheless, it is interesting to note that an increasing trend of BC burden is simulated over the entire western Arctic and also part of Eastern Siberia. A possible explanation for this result can be found in the increasing anthropogenic emissions in East Asia in the last decades (Fig. S1). The long-range transport of aerosol particles from East Asia to the Arctic occurs in the middle and upper Arctic troposphere, as the warm air masses originated south of the Arctic front are lifted above and affect the BC burden more than surface concentration and deposition due to slow mixing with surface atmospheric layers. In contrast, the emissions in Eurasia can be often lo-

cated north of the Arctic front during winter and therefore BC is transported to the lower troposphere, affecting more surface concentrations and deposition (AMAP, 2015). Similar results were found by Sharma et al. (2013), who estimated a contribution of East Asia that was 3 times larger than the BC burden in the Arctic between 1990 and 2005; in the same period the contribution to BC burden and surface concentrations from the FSU declined by 50 and 70 %, respectively. The REF, FIX2000 and FIX1980 simulations are driven by the same atmospheric reanalysis and annually varying biomass burning emissions. The interannual meteorological variability of the last decades combined with constant anthropogenic emissions (FIX2000 in Fig. 3 and FIX1980 in Fig. S2) determined a significant increasing trend of BC dry deposition and surface concentrations over the Arctic, while BC wet deposition increased over the Canadian Archipelago and Greenland. Both FIX2000 and FIX1980 simulations show similar patterns, with differences due to the changing geographical distribution of anthropogenic emissions, larger in Europe and North America in 1980 and predominant in East Asia in 2000. Neither FIX2000 nor FIX1980 shows a significant trend in BC burden over the Arctic (Figs. 2d and S2d), which further confirms the role of anthropogenic emissions in East Asia that affect the transport of air pollution at higher altitudes to the Arctic. In some regions, like the Canadian Archipelago and the western Arctic Ocean, the changes in BC trends due to meteorology and natural variability contribute as much as the changes due to anthropogenic emissions. In the next sections we will further decompose the transport pathways of air pollution to the Arctic associated with the main large-scale atmospheric circulation patterns, which describe the trend of the warming amplification in the polar region and which have characterized the changing climate of the Northern Hemisphere of the last decades.

### 3.2 Trends of BC due to atmospheric circulation changes

Three large-scale atmospheric patterns were identified by Dobricic et al. (2016) as closely related to the near-surface warming trend in the polar region during winter months (DJF). Two of them, $IC_{NAO}$ and $IC_{SB}$, showed statistically significant trends of surface temperature and geopotential height over the Arctic during the last 36 years (1980–2015), and together could reproduce well the spatial and temporal distribution of the trends found in two atmospheric reanalysis (NCEP and ECMWF ERA-Interim; Sect. 2.2) In this section we limit our analysis to trends of MLE of BC wet deposition, concentration and load from the FIX2000 simulation and for the longest period available for the construction of coupled anomalies, 26 years (1980–2005). The trends of BC deposition and load are obtained by multiplying the MLE of the BC field by the trend of the associated atmospheric pattern, from the ICA previously performed by Dobricic et al. (2016). We will not discuss the trends of BC dry deposition as in the

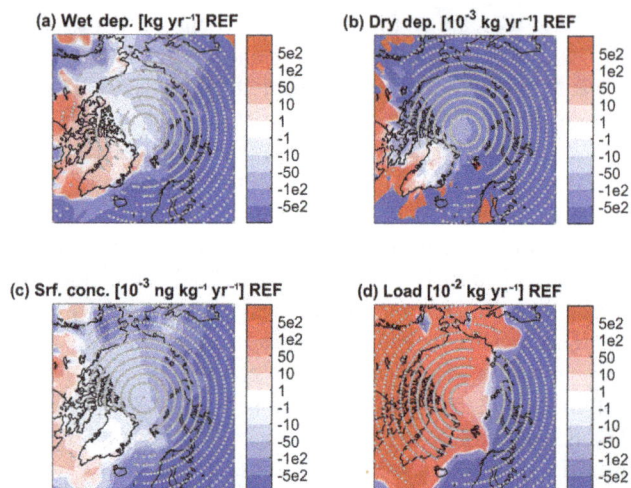

**Figure 2.** Winter (DJF) trends (1980–2005) of BC wet deposition (a), dry deposition (b), surface concentration (d) and total load (d) for the ECHAM5–HAMMOZ REF simulation. Grey dots represent the grid points with trend significant at 5 % level.

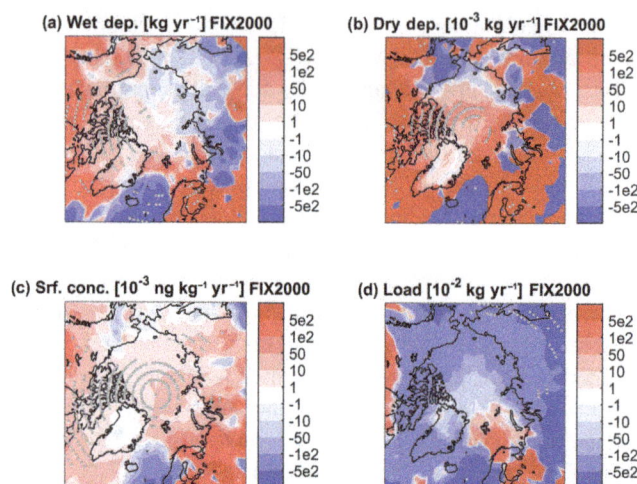

**Figure 3.** Winter (DJF) trends (1980–2005) of BC wet deposition (a), dry deposition (b), surface concentration (c) and total load (d) for the ECHAM5–HAMMOZ FIX2000 simulation. Grey dots represent the grid points with trend significant at 5 % level.

ECHAM5–HAMMOZ simulations it is only a small fraction of the total trend of BC deposition in the Arctic (Figs. 1b, 2b and S2b).

Figure 1 shows the trends of geopotential height at 850 mbar associated with $IC_{NAO}$ and $IC_{SB}$ patterns estimated by the ICA of NCEP atmospheric reanalysis data by Dobricic et al. (2016), while the arrows visually indicate how these trends may favour certain pollution transport pathways. We can expect that due to the different dynamical structures of $IC_{NAO}$ and $IC_{SB}$ they will differently impact the transport and deposition of pollutants from emission areas in the middle latitudes and their deposition rate once they reach the Arctic atmosphere. The tendency of $IC_{NAO}$ toward the negative phase of the NAO (Fig. 1a) forms an anticyclonic anomaly over the large part of the Arctic Ocean and a cyclonic anomaly in the North Atlantic Ocean. The intensity of westerly winds is decreased in the lower troposphere, with lower transport of pollution from North America across the Atlantic Ocean. In contrast, the $IC_{NAO}$ slightly increases the transport of pollution from north-western America towards the Arctic Ocean. This is also consistent with the results of Hirdman et al. (2010), which found significant correlations between the NAO index and EBC surface concentrations in Alert and Barrow, both in the western Arctic, with decreasing impact from northern Eurasia and increasing impact from North America. Consistent with the circulation pathways described in Fig. 1a, the MLEs of BC wet deposition trends related to $IC_{NAO}$ (Fig. 4) show a decreasing trend north of the Eurasian coast and an increasing trend north of America and Greenland. A correlation between the negative phase of the NAO and increasing precipitations and snow accumulation over western Greenland was also found by previous studies (e.g. Appenzeller et al., 1998; Mosley-Thompson et

al., 2005). The BC load has a positive trend over most of the Arctic Ocean, Greenland and the Canadian Archipelago, which may be associated with the dipole of pressure anomalies over the Pacific Ocean that also favours the export of polluted air masses from East Asia into North America and the Arctic (Fig. 1a). Sharma et al. (2013) previously showed that the contribution of East Asian BC emissions in the Arctic above 200 mb is the largest. The trends of average BC concentrations at different altitudes above 60° N at four different longitudinal portions (quadrants) of the Arctic (0–90° E; 90–180° E; 180–90° W; 90–0° W) are shown in Fig. 5. In all four Arctic quadrants an increasing trend of BC concentrations is estimated at an altitude of about 10 km, near the tropopause. Above the tropopause a shift in the vertical profile of BC concentration is observed, with a decreasing trend between the tropopause and 15 km and an increasing trend above 15 km and 70° N. As shown in Dobricic et al. (2016), the evolution of the $IC_{NAO}$ pattern from December to February indicates a weakening of the stratospheric vortex linking the tropospheric perturbation with the stratosphere, in agreement with findings by Feldstein and Lee (2014). There is an enhanced mixing anomaly between the troposphere and the stratosphere indicated by increased BC concentrations above and decreased below the tropopause. Near the surface the $IC_{NAO}$ has a small impact on BC concentrations with few differences in the four Arctic quadrants. Only between 0 and 90° E is there a small increasing trend above 70° N, while between 90 and 0° E the BC concentrations increase only below 70° N.

The $IC_{SB}$ pattern (Fig. 1b), similar to Scandinavia blocking, consists of an anticyclonic centre near Scandinavia and weaker centres of opposite sign over south-western Europe and Siberia/Mongolia. In this case advection changes in the

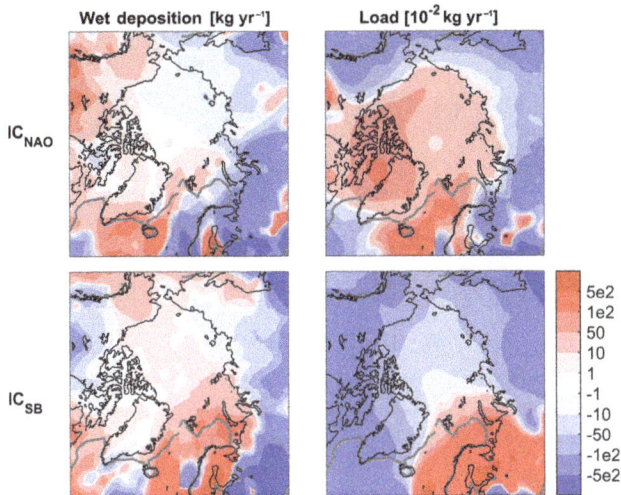

**Figure 4.** Winter (DJF) trends (1980–2015) of maximum likelihood estimates (MLEs) of BC wet deposition ($kg\,yr^{-1}$) and load ($kg \times 10^{-2}\,yr^{-1}$) associated with $IC_{NAO}$ and $IC_{SB}$. The grey line represents the mean winter sea ice and snow cover larger than 50 % from 1980 to now. NAO and SB trends are both significant at 5 % level.

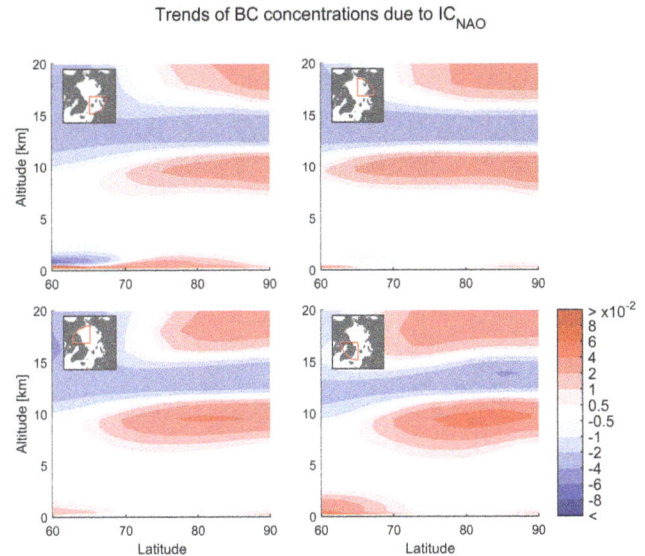

**Figure 5.** Winter (DJF) trends (1980–2015) of maximum likelihood estimates of BC concentrations ($ng\,kg^{-1}\,yr^{-1}$) associated with $IC_{NAO}$ circulation pattern. BC concentrations at each model vertical level and latitudes above 60° N are averaged over 90° longitudes segments of the Arctic, indicated on the map on the left upper corner of each plot.

lower troposphere indicated by the arrows favour the larger transport of pollutants from Europe directly to the sea-ice-covered areas of the Arctic Ocean, while due to the southward position of the anticyclonic anomaly the vertical stability over the Arctic is not changed significantly. The large high-pressure anomaly over the north Pacific Ocean decreases the Asian influence on the Arctic. In a dynamically consistent way it advects the warm air from the south to the Barents Sea and the cold air from the Arctic into Siberia. This circulation pattern results in an increased transport of pollution from central Europe directly to the Barents Sea and the Arctic, while the transport from Siberia to the Arctic is suppressed. All estimates of the deposition of BC related to $IC_{SB}$ show a strong positive trend that extends from the Fram Strait, the Barents and Kara seas to the central part of the Arctic Ocean (Fig. 4). The positive trend of the load spreads over the Norwegian Sea and Scandinavia and indicates the northward advection of the pollution from Europe towards the Arctic Ocean. The relationship between $IC_{SB}$ deposition and load trends and the trend of the anticyclonic circulation over Scandinavia is especially evident in February (not shown) when the Scandinavian anticyclone trend is the most pronounced (Dobricic et al., 2016). The vertical structure of the estimated BC concentration trend (Fig. 6) shows that in these conditions the maximum concentration trend is located in the lower troposphere below 3 km and mainly in the first quadrant of the Arctic (0–90° E). A strong increasing trend extends from mid-latitudes up to 90° N near the surface. In contrast, a negative trend is observed over the entire Arctic near the tropopause and above. The $IC_{SB}$ pattern is associated with a negative phase of the ENSO (La Niña), with

colder surface temperature in the equatorial Pacific. The opposite signs of perturbations in the equatorial Pacific and in Scandinavia maintain a planetary wave trapped in the troposphere and, compared to the $IC_{NAO}$ pattern, a lower transport to the stratosphere (Dobricic et al., 2016).

### 3.3 Total trends of BC maximum likelihood estimates

Figure 7 shows the reconstructed total trends of BC wet deposition and load, as the sum of the winter seasonal mean of maximum likelihood estimates associated with three independent atmospheric patterns ($IC_{NAO}$, $IC_{SB}$ and $IC_{ENSO}$) multiplied by their temporal trends. The BC wet deposition in winter increased in the last decades due to changing atmospheric circulation over a large part of the polar region. The largest increases were estimated over the Barents Sea, the Kara Sea and the Fram Strait, southern Greenland, Alaska and the Canadian Archipelago, but the increasing deposition trend extends over most of the Arctic Ocean covered by sea ice in winter (the grey line in Fig. 7 represents the limit of sea-ice- and snow-covered areas averaged since the 1980s). Decreasing BC wet deposition is estimated over most of Siberia and the Laptev Sea and East Siberian Sea. The trend of atmospheric load of BC reconstructed by the three ICs shows a net increasing trend over the Arctic north of about 75° N and the decreasing trend over the mid-latitudes, except over western Russia, the Scandinavian Peninsula and the Norwegian Sea. The trends reconstructed by the MLE analysis are similar to the trends simulated by the FIX2000

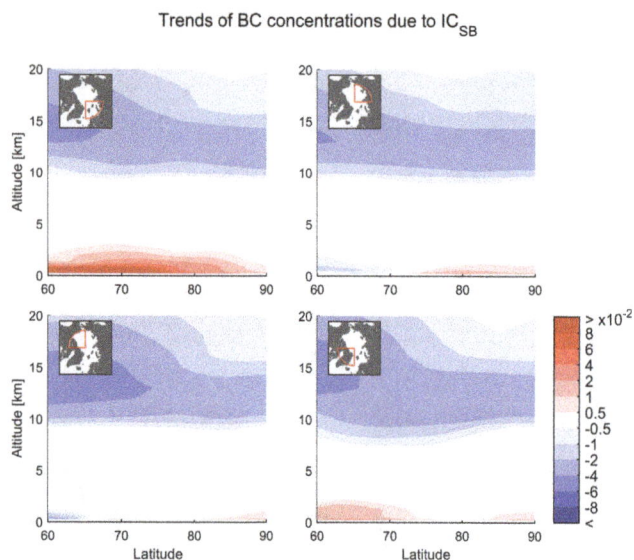

Trends of BC concentrations due to IC$_{SB}$

**Figure 6.** Winter (DJF) trends (1980–2005) of maximum likelihood estimates of BC concentrations (ng kg$^{-1}$ yr$^{-1}$) associated with IC$_{SB}$ circulation pattern. BC concentrations at each model vertical level and latitudes above 60° N are averaged over 90° longitudes segments of the Arctic, indicated on the map on the left upper corner of each plot.

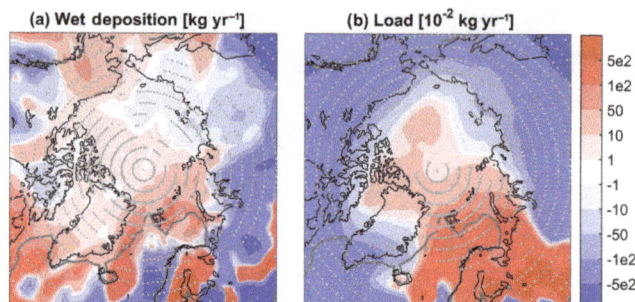

**Figure 7.** Total winter (DJF) trends (1980–2015) of maximum likelihood estimates (MLEs) of BC wet deposition (kg yr$^{-1}$) and load (kg $\times 10^{-2}$ yr$^{-1}$) associated with three atmospheric circulation patterns (Total $=$ IC$_{NAO}$ $+$ IC$_{SB}$ $+$ IC$_{ENSO}$). The grey line represents the mean winter sea ice and snow cover larger than 50 % from 1980 to now. Grey dots represent the grid points with trend significant at 5 % level.

chemistry–climate model simulation (Fig. 3a, d). In particular wet deposition trends are well represented in terms of both magnitude and spatial distribution. The trend of BC load in the FIX2000 simulation is not significant over the polar region; however, part of the noise was removed from the coupled atmospheric chemistry anomalies by performing and EOF transformation and retaining only the components with largest eigenvalues.

The temporal variability of the BC deposition anomalies in the period of 1980–2015 may be estimated for any area and for each pattern. In particular it is interesting to evaluate the combined and individual impacts of IC$_{NAO}$, IC$_{SB}$ and IC$_{ENSO}$ patterns on the total BC wet deposition and load over the Arctic Ocean, between 80 and 90° N. Figure 8 shows that the IC$_{NAO}$ pattern produces deposition anomalies with a smaller interannual variability than those driven by the IC$_{SB}$ pattern. The contribution of IC$_{ENSO}$ to the total deposition is relevant only in the third quadrant of the Arctic (180–90° W), north of the Canadian Archipelago. The largest variability is found in the first quadrant, north of Barents and Kara seas, up to a factor of 10 larger than the other sectors of the Arctic. Deposition anomalies from the IC$_{NAO}$ pattern are mostly negative until the year 2000 and after become positive. Those from the IC$_{SB}$ pattern show a strong increase after the year 2000, which dominates the whole trend from 1980 to 2015, corresponding well to an acceleration in the winter sea-ice decrease rate over the Barents Sea that is often coupled with the formation of the anticyclonic anomaly over Scandinavia (Cohen et al., 2014; Dobricic et al., 2016; Sato et al., 2014;

Screen et al., 2013). Figure 9 shows the temporal contribution of IC$_{NAO}$, IC$_{SB}$ and IC$_{ENSO}$ patterns to the total BC load over the Arctic quadrants. In this case, the IC$_{NAO}$ pattern drives the interannual variability, with negative values until the late 1990s and positive in the last 15 years. The IC$_{NAO}$ pattern affects BC load over the entire Arctic in a similar way, while the contribution of the IC$_{SB}$ pattern has an opposite trend except for the first quadrant, where it is only a small fraction of the total variability. As discussed in the previous section, the IC$_{SB}$ pattern increase the transport of BC directly from central Europe to the Barents and Kara Sea, while it decreases the transport from Siberia. The IC$_{SB}$ pattern has a negative contribution in the last 10 years in the other parts of the Arctic, and in this regions it can be as large as the IC$_{NAO}$ contribution. In Figs. 8 and 9 we can see that also the temporal variability is partially reconstructed by the applied Bayesian statistical approach. The correlation coefficients between the interannual variability reconstructed by the MLE and the FIX2000 model simulation for the common period 1980–2005 range between 0.35 and 0.4 for BC wet deposition and between 0.14 and 0.41 for BC load over the different sectors of the Arctic.

## 3.4 Uncertainty of the BC maximum likelihood estimates

In order to test the robustness of the statistical method, we have estimated the MLE of pollutant distributions for a set of eight combinations of coupled atmospheric reanalysis and BC simulations (Table 1). Dobricic et al. (2016) applied the ICA to two different atmospheric reanalysis of the period 1980–2015, NCEP and ERA-Interim, and similar atmospheric patterns and trends of the ICs were found. A first set of MLEs combined the atmospheric patterns computed from NCEP and ERA-Interim with two global chemistry–climate simulations, which used constant anthropogenic emissions

**Table 1.** Combination of atmospheric reanalysis, global chemistry–climate model simulations and time periods used to generate MLE of BC deposition and load.

| Estimate name | Atmospheric reanalysis | BC simulation | Period |
|---|---|---|---|
| NCEP-2000-A | NCEP | FIX2000 | 1980–2005 |
| ERA-2000-A | ERA-Interim | FIX2000 | 1980–2005 |
| NCEP-1980-A | NCEP | FIX1980 | 1980–2005 |
| ERA-1980-A | ERA-Interim | FIX1980 | 1980–2005 |
| NCEP-REF-B | NCEP | REF | 1993–2005 |
| ERA-REF-B | ERA-Interim | REF | 1993–2005 |
| NCEP-2000-B | NCEP | FIX2000 | 1993–2005 |
| ERA-2000-B | ERA-Interim | FIX2000 | 1993–2005 |

(FIX2000 and FIX1980; see Sect. 2.3). The MLEs were computed using the simulated pollutant fields for the longest period available: 26 years from 1980 to 2005. The results from NCEP-2000-A have been explained in details in the previous sections.

Figure 10 shows the total trends $(IC_{NAO} + IC_{SB} + IC_{ENSO})$ of BC wet deposition estimated using the FIX2000 and FIX1980 simulations to build the pollutants and atmospheric coupled anomalies. All four estimates are consistent in terms of geographical distribution, magnitude of the estimated trends and their statistical significance. Consistent results are obtained also for the total trends of BC load (Fig. S3). Only small differences are due to the two atmospheric reanalysis and the two atmospheric composition simulations, which have a different geographical distribution of anthropogenic emissions (Fig. S1).

In a second set of MLEs, we used the REF model simulation, which includes annually varying anthropogenic emissions. In this case, the sudden drop of anthropogenic emissions in Eastern Europe and Russia and the increase in East Asia in the early 1990s (Fig. S1d) introduced a strong discontinuous change in the transport variability that was impossible to resolve by the statistical method. Thus we chose a shorter period to form the simulated chemistry and atmospheric coupled anomalies, 13 years (1993–2005) after the drop of anthropogenic emissions. Figure 11 shows the estimated trends of BC wet deposition (trends of BC load are shown in Fig. S4) using only 13 years of coupled anomalies, from the REF simulation and also from the FIX2000 simulation for comparison. Also in this case the results shown in Figs. 10 and 11 are consistent, using a shorter period to form the simulated coupled anomalies and with both constant and varying anthropogenic emissions (without large emission changes in a short time). In the most remote Arctic region covered by sea ice, far from the main anthropogenic sources, the entire set of estimated trends of BC transport and deposition are very close to each other, meaning that the applied methodology can robustly approximate the transport

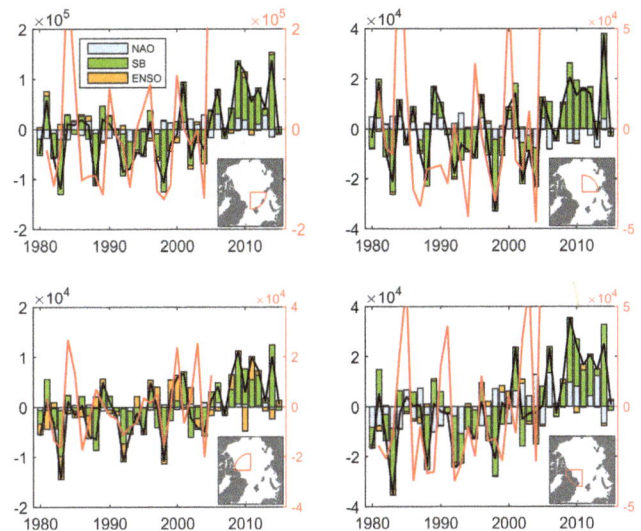

**Figure 8.** Temporal variability of total BC wet deposition anomaly over 90° longitude segments of the Arctic and between 80 and 90° N (kg yr$^{-1}$) related to the circulation patterns $IC_{NAO}$, $IC_{SB}$ and $IC_{ENSO}$. The black line is the sum of the three components; the red line (right $y$ axis) is the total BC wet deposition anomaly from the FIX2000 simulation.

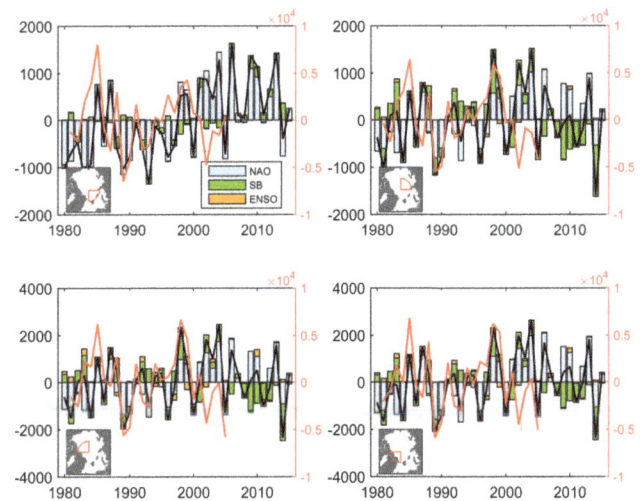

**Figure 9.** Temporal variability of total BC load anomaly over 90° longitude segments of the Arctic and between 80 and 90° N (kg yr$^{-1}$) related to the circulation patterns $IC_{NAO}$, $IC_{SB}$ and $IC_{ENSO}$. The black line is the sum of the three components; the red line (right $y$ axis) is the total BC load anomaly from the FIX2000 simulation.

of pollutants to this region and the trends associated with different large-scale circulation patterns.

## 4 Conclusions

The feedbacks between the global warming and Arctic amplification with sea-ice retreat and impacts on large-scale at-

BC wet deposition [kg yr$^{-1}$]

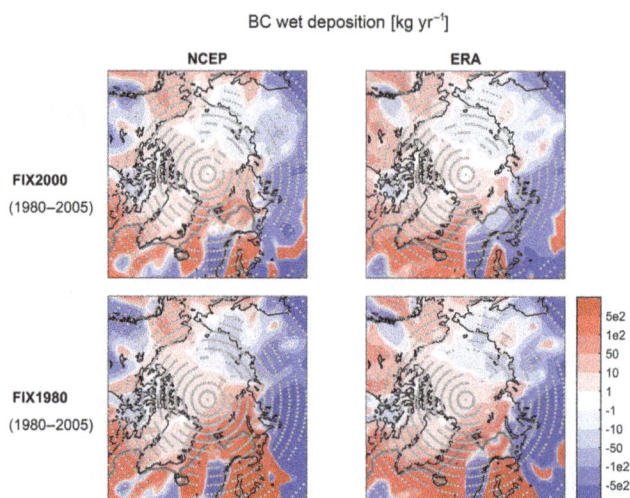

**Figure 10.** Four different estimates (see also Table 1) of total trends of maximum likelihood BC wet deposition (kg yr$^{-1}$) associated with three atmospheric circulation patterns (Total = NAO + SB + ENSO). The grey line represents the mean winter sea ice and snow cover larger than 50 % from 1980 to now. Grey dots represent the grid points with trend significant at 5 % level.

BC wet deposition [kg yr$^{-1}$]

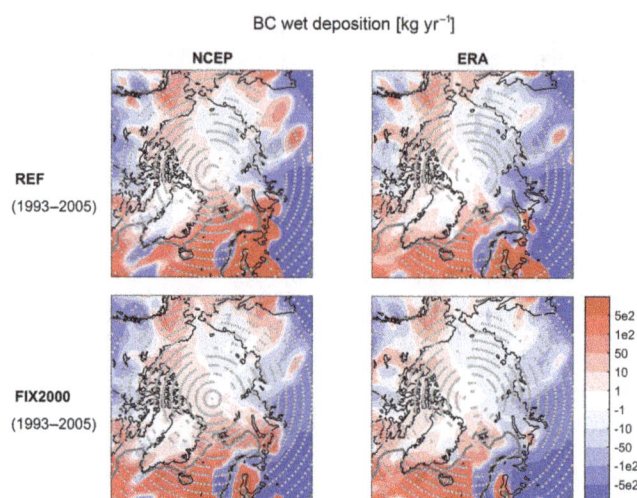

**Figure 11.** Four different estimates (see also Table 1) of total trends of maximum likelihood BC wet deposition (kg yr$^{-1}$) associated with three atmospheric circulation patterns (Total = NAO + SB + ENSO). The grey line represents the mean winter sea ice and snow cover larger than 50 % from 1980 to now. Grey dots represent the grid points with trend significant at 5 % level.

mospheric circulation are still contradictory. The response of mid-latitude weather to the Arctic warming and sea-ice cover changes of the last decades is highly uncertain due to non-linear processes involved in the Arctic and sub-Arctic climate system (Overland et al., 2016). Some studies find only weak or non-existent relationships between mid-latitude weather structures and Arctic warming (e.g. Screen and Simmonds, 2013; Barnes et al., 2014), while others found correlations between sea-ice retreat in winter over the Barents and Kara seas and hemispheric-scale impacts (e.g. Deser et al., 2007; Petoukhov and Semenov, 2010; Screen et al., 2013; Mori et al., 2014; Di Capua and Coumou, 2016). Recently, the trends from two atmospheric reanalysis (Figs. 2 and S3 of Dobricic et al., 2016) of winter near-surface warming of the Northern Hemisphere are approximated by three independent components which are also very similar to well-known atmospheric oscillations, the NAO, SB and ENSO. We developed a statistical algorithm to combine the estimates of these independent components (IC$_{NAO}$, IC$_{SB}$ and IC$_{ENSO}$) with chemistry–climate model simulations of transport and deposition of BC over the Arctic. In this way we estimate how changes in the large-scale atmospheric circulation impact the BC transport and deposition independently of the changes in emissions. For a given independent component of the atmospheric variability our algorithm estimates the most likelihood deposition and load of BC consistent with the variability of the atmosphere–chemistry coupled anomalies. To test the robustness of the method and the spread of the possible solutions, the method is applied with two atmospheric reanalysis, NCEP and ERA-Interim, and three realizations of coupled anomalies from ECHAM5–HAMMOZ

global model simulations, with constant and annually varying anthropogenic emissions.

The main results are summarized in the following points.

– The trends of BC concentrations and deposition fluxes simulated by a global chemistry–climate model (ECHAM5–HAMMOZ) indicated a strong reduction of near-surface transport and deposition to the Arctic due to anthropogenic emission reductions in Eurasia, in agreement with observations and other modelling studies (Gong et al., 2010; Hirdman et al., 2010), and an increasing BC load mainly due to the emissions from East Asia (Sharma et al., 2013). In contrast, model simulations with constant anthropogenic emissions indicated how the effect of changing meteorology in the last decades may have also determined significant trends of BC transport to the Arctic.

– The increasing trend of the anticyclonic circulation anomaly over the Arctic (IC$_{NAO}$) determines a decrease of the pollution transport pathway from Eurasia to the Arctic, with a decrease in BC surface concentrations and deposition particularly in the eastern Arctic. However, the IC$_{NAO}$ trend favours a stable vertical stratification which determines an increase of BC transport in the upper troposphere from East Asia and North America with increasing trend of total column BC concentrations over the entire Arctic Ocean.

– A second component, named IC$_{SB}$, directly transports BC in the Arctic lower troposphere from central Europe. The increasing trend of an anticyclonic anomaly over

Scandinavia and Western Siberia favours the transport of air pollution directly from Europe to the Barents Sea with a significant increase of BC wet deposition over the entire Arctic Ocean. Also, this pattern determines a decrease of pollution transport from East Eurasia to the Arctic and an overall decrease of BC load over the entire Arctic upper troposphere, except over the Barents Sea where the surface concentration trends is strongly increasing.

- A third component, which is more related to anomalies in the tropical Pacific Ocean and named $IC_{ENSO}$, has a smaller and statistically less significant impact on the transport and deposition of BC to the Arctic.

- The combined impact of all three atmospheric patterns on BC wet deposition determines a significant increasing trend between 1980 and 2015 over almost the entire Arctic Ocean, the Barents Sea and Fram Strait, Greenland, Alaska and the Canadian Archipelago. A small, significant decreasing trend is estimated in the East Siberian and Laptev seas. The BC load increases over the whole Arctic Ocean, with a significant trend only north of Greenland and the Barents Sea.

- The estimated interannual variability and trend of BC wet deposition over the Arctic Ocean is mainly driven by the $IC_{SB}$ pattern, with positive anomalies persisting for the last 10 years, 2005–2015. Smaller contributions are estimated for $IC_{ENSO}$ between 180 and 90° W and for NAO between 90 and 0° W. The interannual variability and trend of BC load over the Arctic is mainly driven by the $IC_{NAO}$ atmospheric pattern, in particular north of the Barents Sea. In the other sectors of the Arctic, $IC_{SB}$ also contributes to the load total variability.

- We have shown the robustness of a Bayesian approach to estimate the maximum likelihood distribution and trends of air pollutants associate with specific atmospheric patterns. The estimates obtained from different atmospheric reanalysis and global chemistry–climate model simulations are consistent. Problems may arise to resolve the total variability in presence of strong emission discontinuous changes. Although it was possible to separate the effects of changing anthropogenic emissions from the meteorological variability using only global chemistry–climate simulations, MLE further provided information on the feedback impacts of single atmospheric processes related to global warming and sea ice retreat on pollution transport. By using outputs from existing coupled model simulations the method may be applied with other atmospheric processes related to the pollution transport and to other temporal intervals.

Our results show how winter atmospheric processes linked to enhanced Arctic warming and sea-ice retreat may impact the BC deposition on the sea ice itself. BC deposition is mainly decreasing over the area that in the recent years is characterized by the thin first-year sea ice, while it is increasing over the area with the thicker multi-year sea ice, which has become a smaller fraction of the total sea-ice cover in the summers of recent years. During the winter polar night the deposited BC has no impact on sea-ice melting. When the sun appears in the spring the shortwave radiation starts to melt the fresh snow deposited on the sea ice. The BC impact on the albedo may be the most important at the end of the winter season, just before the snow starts to melt and surface becomes darker. Therefore, the BC deposition in winter may accelerate the initial melting of the snow and by an integrated effect it may result in the increased melting at the end of summer. The first-year sea ice may be completely melted every summer and the decrease of the BC deposition might not be the most important factor controlling its coverage in the late summer. However, with the increased deposition of BC the multi-year sea ice may melt more and its thickness may become reduced with respect to the previous year. If this process happens consistently over several decades it might significantly contribute to reduction of the thickness of the multi-year sea ice. Our estimate of the temporal variability of the BC deposition over the multi-year sea ice indicates that the deposition in the winter due to the sea-ice retreat feedback increased especially in the last 10 years.

The increasing trend of the BC deposition due to changes in atmospheric circulation emphasizes the importance of the reduction of BC anthropogenic emissions in the mid-latitudes. In particular the $IC_{SB}$ pattern favours the transport to the Arctic Ocean of polluted air masses from Europe and increased deposition on the multi-year sea ice. Policies that reduced anthropogenic emissions in the last decades, therefore, reduced the risk of a further acceleration of the melting of the multi-year sea ice. A strong shift of the anthropogenic emissions from North America and Europe to East and South Asia also contributed to reduce the transport of BC to the Arctic from Eurasia, which represents the main transport pathway. We may have an indication of the impact of this geographical emissions shift by comparing the trends estimated using the FIX2000 and FIX1980 BC simulations (Fig. S5). The BC deposition trend increased over the East Siberian and Laptev Sea considering the anthropogenic emissions of year 2000 compared to 1980, indicating that the contribution from the Asian emissions may play an important role in the future (e.g. Sand et al., 2015). This increase is affecting deeper into the Arctic Ocean in particular in association with the $IC_{NAO}$ pattern. A similar impact of the increasing emissions in Asia was estimated for the BC burden, in particular associated with $IC_{NAO}$ and extending from Siberia, across the Arctic Ocean up to the Canadian Archipelago and Greenland.

In order to further understand the link between global warming and trends of BC transport and deposition, it will be necessary to extend the study to other seasons. In particular the major precipitation and wet deposition of BC happening in the summer when the reduction of the ice cover in the recent years is the largest. Furthermore, although the use of different estimates for atmospheric patterns and coupled atmosphere–chemistry relationships provides some information on their uncertainty, in the future the same method may be applied with several coupled models and multi-year observations in the Arctic to further assess the uncertainties. The same methodology may be also applied to estimate trends of BC transport to the Arctic associated with atmospheric circulation changes in the future, for example using the global climate simulations instead of atmospheric reanalysis.

*Competing interests.*  The authors declare that they have no conflict of interest.

*Acknowledgements.*  We thank the anonymous reviewers for the helpful comments and suggestions during the review phase.

Edited by: Timothy J. Dunkerton

# References

AMAP: AMAP Assessment 2015: Black carbon and ozone as Arctic climate forcers, available from: http://www.amap.no/documents/download/2506 (last access: 25 September 2017), 2015.

Aoki, T., Aoki, T., Fukabori, M., Hachikubo, A., Tachibana, Y., and Nishio, F.: Effects of snow physical parameters on spectral albedo and bidirectional reflectance of snow surface, J. Geophys. Res., 105, 10219, https://doi.org/10.1029/1999JD901122, 2000.

Appenzeller, C., Schwander, J., Sommer, S., and Stocker, T. F.: The North Atlantic Oscillation and its imprint on precipitation and ice accumulation in Greenland, Geophys. Res. Lett., 25, 1939–1942, 1998.

Auvray, M., Bey, I., Llull, E., Schultz, M. G., and Rast, S.: A model investigation of tropospheric ozone chemical tendencies in long-range transported pollution plumes, J. Geophys. Res.-Atmos., 112, 1–17, https://doi.org/10.1029/2006JD007137, 2007.

Barnes, E. A., Dunn-sigouin, E., Masato, G., and Woollings, T.: Exploring recent trends in Northern Hemisphere blocking, Geophys. Res. Lett., 2, 638–644, https://doi.org/10.1002/2013GL058745, 2014.

Barrie, L. A.: Arctic air pollution: An overview of current knowledge, Atmos. Environ., 20, 643–663, https://doi.org/10.1016/0004-6981(86)90180-0, 1986.

Bourgeois, Q. and Bey, I.: Pollution transport efficiency toward the Arctic: Sensitivity to aerosol scavenging and source regions, 116, J. Geophys. Res.-Atmos., 1–18, https://doi.org/10.1029/2010JD015096, 2011.

Byrd, R., Lu, P., Nocedal, J., and Zhu, C.: A Limited Memory Algorithm for Bound Constrained Optimization, SIAM J. Sci. Comput., 16, 1190–1208, https://doi.org/10.1137/0916069, 1995.

Di Capua, G. and Coumou, D.: Changes in meandering of the Northern Hemisphere circulation, Environ. Res. Lett., 11,

094028, https://doi.org/10.1088/1748-9326/11/9/094028, 2016.

Cohen, J., Screen, J. A., Furtado, J. C., Barlow, M., Whittleston, D., Coumou, D., Francis, J., Dethloff, K., Entekhabi, D., Overland, J., and Jones, J.: Recent Arctic amplification and extreme mid-latitude weather, Nat. Geosci., 7, 627–637, https://doi.org/10.1038/ngeo2234, 2014.

Comiso, J. C.: Large decadal decline of the arctic multiyear ice cover, J. Climate, 25, 1176–1193, https://doi.org/10.1175/JCLI-D-11-00113.1, 2012.

Conway, H., Gades, A., Raymond, C. F.: Albedo of dirty snow during conditions of melt, Water Resour. Res., 32, 1713–1718, https://doi.org/10.1029/96WR00712, 1996.

Dee, D. P., Uppala, S. M., Simmons, A. J., Berrisford, P., Poli, P., Kobayashi, S., Andrae, U., Balmaseda, M. A., Balsamo, G., Bauer, P., Bechtold, P., Beljaars, A. C. M., van de Berg, L., Bidlot, J., Bormann, N., Delsol, C., Dragani, R., Fuentes, M., Geer, A. J., Haimberger, L., Healy, S. B., Hersbach, H., Holm, E. V., Isaksen, L., Kallberg, P., Kohler, M., Matricardi, M., Mcnally, A. P., Monge-Sanz, B. M., Morcrette, J. J., Park, B. K., Peubey, C., de Rosnay, P., Tavolato, C., Thepaut, J. N., and Vitart, F.: The ERA-Interim reanalysis: Configuration and performance of the data assimilation system, Q. J. Roy. Meteor. Soc., 137, 553–597, https://doi.org/10.1002/qj.828, 2011.

Deser, C., Tomas, R. A., and Peng, S.: The transient atmospheric circulation response to North Atlantic SST and sea ice anomalies, J. Climate, 20, 4751–4767, https://doi.org/10.1175/JCLI4278.1, 2007.

Diehl, T., Heil, A., Chin, M., Pan, X., Streets, D., Schultz, M., and Kinne, S.: Anthropogenic, biomass burning, and volcanic emissions of black carbon, organic carbon, and $SO_2$ from 1980 to 2010 for hindcast model experiments, Atmos. Chem. Phys. Discuss., https://doi.org/10.5194/acpd-12-24895-2012, 2012.

Dobricic, S., Vignati, E., and Russo, S.: Large-scale atmospheric warming in winter and the Arctic sea ice retreat, J. Climate, 29, 2869–2888, https://doi.org/10.1175/JCLI-D-15-0417.1, 2016.

Döscher, R., Vihma, T., and Maksimovich, E.: Recent advances in understanding the Arctic climate system state and change from a sea ice perspective: a review, Atmos. Chem. Phys., 14, 13571–13600, https://doi.org/10.5194/acp-14-13571-2014, 2014.

Eckhardt, S., Quennehen, B., Olivié, D. J. L., Berntsen, T. K., Cherian, R., Christensen, J. H., Collins, W., Crepinsek, S., Daskalakis, N., Flanner, M., Herber, A., Heyes, C., Hodnebrog, Ø., Huang, L., Kanakidou, M., Klimont, Z., Langner, J., Law, K. S., Lund, M. T., Mahmood, R., Massling, A., Myriokefalitakis, S., Nielsen, I. E., Nøjgaard, J. K., Quaas, J., Quinn, P. K., Raut, J.-C., Rumbold, S. T., Schulz, M., Sharma, S., Skeie, R. B., Skov, H., Uttal, T., von Salzen, K., and Stohl, A.: Current model capabilities for simulating black carbon and sulfate concentrations in the Arctic atmosphere: a multi-model evaluation using a comprehensive measurement data set, Atmos. Chem. Phys., 15, 9413–9433, https://doi.org/10.5194/acp-15-9413-2015, 2015.

Eleftheriadis, K., Vratolis, S., and Nyeki, S.: Aerosol black carbon in the European Arctic?: Measurements at Ny-Ålesund, Svalbard from 1998–2007, Geophys. Res. Lett., 36, L02809, https://doi.org/10.1029/2008GL035741, 2009.

Endresen, Ø., Sørgard, E., Sundet, J., Dalsøren, S., Isaksen, I., and Berglen, T.: Emission from international sea transportation and environmental impact, J. Geophys. Res., 108, 14.1–14.22, https://doi.org/10.1029/2002JD002898, 2003.

Feldstein, S. B. and Lee, S.: Intraseasonal and interdecadal jet shifts in the Northern Hemisphere: The role of warm pool tropical convection and sea ice, J. Climate, 27, 6497–6518, https://doi.org/10.1175/JCLI-D-14-00057.1, 2014.

Flanner, M. G.: Arctic climate sensitivity to local black carbon, J. Geophys. Res.-Atmos., 118, 1840–1851, https://doi.org/10.1002/jgrd.50176, 2013.

Flanner, M. G. and Zender, C. S.: Linking snowpack microphysics and albedo evolution, J. Geophys. Res.-Atmos., 111, 1–12, https://doi.org/10.1029/2005JD006834, 2006.

Ganzeveld, L. and Lelieveld, J.: Dry deposition parameterization in a chemistry general circulation model and its influence on the distribution of reactive trace gases, J. Geophys. Res.-Atmos., 100, 20999–21012, https://doi.org/10.1029/95JD02266, 1995.

Gong, S. L., Zhao, T. L., Sharma, S., Sauntry, D. T., Lavoué, D., Zhang, X. B., Leaitch, W. R., and Barrie, L. A.: Identification of trends and interannual variability of sulfate and black carbon in the Canadian High Arctic?: 1981–2007, J. Geophys. Res.-Atmos., 115, 1–9, https://doi.org/10.1029/2009JD012943, 2010.

Hadley, O. L. and Kirchstetter, T. W.: Black-carbon reduction of snow albedo, Nat. Clim. Chang., 2, 437–440, https://doi.org/10.1038/nclimate1433, 2012.

Hannachi, A., Unkel, S., Trendafilov, N. T., and Jolliffe, I. T.: Independent component analysis of climate data: A new look at EOF rotation, J. Climate, 22, 2797–2812, https://doi.org/10.1175/2008JCLI2571.1, 2009.

Henning, S., Bojinski, S., Diehl, K., Ghan, S., Nyeki, S., Weingartner, E., Wurzler, S., and Baltensperger, U.: Aerosol partitioning in natural mixed-phase clouds, Geophys. Res. Lett., 31, L06101, https://doi.org/10.1029/2003GL019025, 2004.

Hirdman, D., Burkhart, J. F., Sodemann, H., Eckhardt, S., Jefferson, A., Quinn, P. K., Sharma, S., Ström, J., and Stohl, A.: Long-term trends of black carbon and sulphate aerosol in the Arctic: changes in atmospheric transport and source region emissions, Atmos. Chem. Phys., 10, 9351–9368, https://doi.org/10.5194/acp-10-9351-2010, 2010.

Hyvärinen, A. and Oja, E.: Independent component analysis: Algorithms and applications, Neural Networks, 13, 411–430, https://doi.org/10.1016/S0893-6080(00)00026-5, 2000.

Jacob, D. J., Crawford, J. H., Maring, H., Clarke, A. D., Dibb, J. E., Emmons, L. K., Ferrare, R. A., Hostetler, C. A., Russell, P. B., Singh, H. B., Thompson, A. M., Shaw, G. E., McCauley, E., Pederson, J. R., and Fisher, J. A.: The Arctic Research of the Composition of the Troposphere from Aircraft and Satellites (ARCTAS) mission: design, execution, and first results, Atmos. Chem. Phys., 10, 5191–5212, https://doi.org/10.5194/acp-10-5191-2010, 2010.

Jiao, C., Flanner, M. G., Balkanski, Y., Bauer, S. E., Bellouin, N., Berntsen, T. K., Bian, H., Carslaw, K. S., Chin, M., De Luca, N., Diehl, T., Ghan, S. J., Iversen, T., Kirkevåg, A., Koch, D., Liu, X., Mann, G. W., Penner, J. E., Pitari, G., Schulz, M., Seland, Ø., Skeie, R. B., Steenrod, S. D., Stier, P., Takemura, T., Tsigaridis, K., van Noije, T., Yun, Y., and Zhang, K.: An AeroCom assessment of black carbon in Arctic snow and sea ice, Atmos. Chem. Phys., 14, 2399–2417, https://doi.org/10.5194/acp-14-2399-2014, 2014.

Kalnay, E., Kanamitsu, M., Kistler, R., Collins, W., Deaven, D., Gandin, L., Iredell, M., Saha, S., White, G., Woollen, J., Zhu, Y., Chelliah, M., Ebisuzaki, W., Higgins, W.,

Janowiak, J., Mo, K. C., Ropelewski, C., Wang, J., Leetmaa, A., Reynolds, R., Jenne, R., and Joseph, D.: The NCEP/NCAR 40-year reanalysis project, B. Am. Meteorol. Soc., 77, 437–471, https://doi.org/10.1175/1520-0477(1996)077<0437:TNYRP>2.0.CO;2, 1996.

Kim, D., Chin, M., Yu, H., Diehl, T., Tan, Q., Kahn, R. A., Tsigaridis, K., Bauer, S. E., Takemura, T., Pozzoli, L., Bellouin, N., Schulz, M., Peyridieu, S., Chédin, A., and Koffi, B.: Sources, sinks, and transatlantic transport of North African dust aerosol: A multimodel analysis and comparison with remote sensing data, J. Geophys. Res.-Atmos., 119, 6259–6277, https://doi.org/10.1002/2013JD021099, 2014.

Klonecki, A.: Seasonal changes in the transport of pollutants into the Arctic troposphere-model study, J. Geophys. Res., 108, 8367, https://doi.org/10.1029/2002JD002199, 2003.

Kwok, R.: Outflow of Arctic Ocean Sea Ice into the Greenland and Barents Seas: 1979–2007, J. Climate, 22, 2438–2457, https://doi.org/10.1175/2008JCLI2819.1, 2009.

Kwok, R. and Rothrock, D. A.: Decline in Arctic sea ice thickness from submarine and ICESat records: 1958–2008, Geophys. Res. Lett., 36, 1–5, https://doi.org/10.1029/2009GL039035, 2009.

Mann, H. B.: Nonparametric Tests Against Trend, Econometrica, 13, 245–259, https://doi.org/10.2307/1907187, 1945.

Maslanik, J., Stroeve, J., Fowler, C., and Emery, W.: Distribution and trends in Arctic sea ice age through spring 2011, Geophys. Res. Lett., 38, 2–7, https://doi.org/10.1029/2011GL047735, 2011.

Mori, M., Watanabe, M., Shiogama, H., Inoue, J., and Kimoto, M.: Robust Arctic sea-ice influence on the frequent Eurasian cold winters in past decades, Nat. Geosci., 7, 869–874, https://doi.org/10.1038/ngeo2277, 2014.

Mosley-Thompson, E., Readinger, C. R., Craigmile, P., Thompson, L. G., and Calder, C. A.: Regional sensitivity of Greenland precipitation to NAO variability, Geophys. Res. Lett., 32, 3–6, https://doi.org/10.1029/2005GL024776, 2005.

Moteki, N. and Kondo, Y.: Effects of Mixing State on Black Carbon Measurements by Laser-Induced Incandescence Effects of Mixing State on Black Carbon Measurements by Laser-Induced Incandescence, Aerosol Sci. Tech., 41, 398–417, https://doi.org/10.1080/02786820701199728, 2007.

Overland, J. E. and Wang, M.: Large-scale atmospheric circulation changes are associated with the recent loss of Arctic sea ice, Tellus A, 62, 1–9, https://doi.org/10.1111/j.1600-0870.2009.00421.x, 2010.

Overland, J. E., Dethloff, K., Francis, J. A., Hall, R. J., Hanna, E., Kim, S., Screen, J. A., Shepherd, T. G., and Vihma, T.: Nonlinear response of mid-latitude weather to the changing Arctic, Nat. Clim. Chang., 6, 992–999, https://doi.org/10.1038/nclimate3121, 2016.

Pan, X., Chin, M., Gautam, R., Bian, H., Kim, D., Colarco, P. R., Diehl, T. L., Takemura, T., Pozzoli, L., Tsigaridis, K., Bauer, S., and Bellouin, N.: A multi-model evaluation of aerosols over South Asia: common problems and possible causes, Atmos. Chem. Phys., 15, 5903–5928, https://doi.org/10.5194/acp-15-5903-2015, 2015.

Pausata, F. S. R., Pozzoli, L., Vignati, E., and Dentener, F. J.: North Atlantic Oscillation and tropospheric ozone variability in Europe: model analysis and measurements intercomparison, Atmos. Chem. Phys., 12, 6357–6376, https://doi.org/10.5194/acp-12-6357-2012, 2012.

Pausata, F. S. R., Pozzoli, L., Van Dingenen, R., Vignati, E., Cavalli, F., and Dentener, F. J.: Impacts of changes in North Atlantic atmospheric circulation on particulate matter and human health in Europe, Geophys. Res. Lett., 40, 4074–4080, 2013.

Petoukhov, V. and Semenov, V. A.: A link between reduced Barents-Kara sea ice and cold winter extremes over northern continents, J. Geophys. Res.-Atmos., 115, 1–11, https://doi.org/10.1029/2009JD013568, 2010.

Pozzoli, L., Bey, I., Rast, S., Schultz, M. G., Stier, P., and Feichter, J.: Trace gas and aerosol interactions in the fully coupled model of aerosol-chemistry-climate ECHAM5-HAMMOZ: 1. Model description and insights from the spring 2001 TRACE-P experiment, J. Geophys. Res.-Atmos., 113, D07308, https://doi.org/10.1029/2007JD009007, 2008a.

Pozzoli, L., Bey, I., Rast, S., Schultz, M. G., Stier, P., and Feichter, J.: Trace gas and aerosol interactions in the fully coupled model of aerosol-chemistry-climate ECHAM5-HAMMOZ: 2. Impact of heterogeneous chemistry on the global aerosol distributions, J. Geophys. Res.-Atmos., 113, 1–15, https://doi.org/10.1029/2007JD009008, 2008b.

Pozzoli, L., Janssens-Maenhout, G., Diehl, T., Bey, I., Schultz, M. G., Feichter, J., Vignati, E., and Dentener, F.: Re-analysis of tropospheric sulfate aerosol and ozone for the period 1980–2005 using the aerosol-chemistry-climate model ECHAM5-HAMMOZ, Atmos. Chem. Phys., 11, 9563–9594, https://doi.org/10.5194/acp-11-9563-2011, 2011.

Qi, L., Li, Q., Li, Y., and He, C.: Factors controlling black carbon distribution in the Arctic, Atmos. Chem. Phys., 17, 1037–1059, https://doi.org/10.5194/acp-17-1037-2017, 2017.

Rast, S., Schultz, M., Bey, I., van Noije, T., Aghedo, A., Brasseur, G., Diehl, T., Esch, M., Ganzeveld, L., Kirchner, I., Kornblueh, L., Rhodin, A., Roeckner, E., Schmidt, H., Schröder, S., Schulzweida, U., Stier, P., Thomas, K., and Walters, S.: Evaluation of the tropospheric chemistry general circulation model ECHAM5-MOZ and its application to the analysis of the chemical composition of the troposphere with an emphasis on the late RETRO period 1990–2000, Max-Planck-Institut für Meteorologie Berichte zur Erdsystemforschung, Hamburg, 114, https://doi.org/10.17617/2.2058065, 2014.

Roeckner, E., Bäuml, G., Bonaventura, L., Brokopf, R., Esch, M., Giorgetta, M., Hagemann, S., Kirchner, I., Kornblueh, L., Manzini, E., Rhodin, A., Schlese, U., Schulzweida, U., and Tompkins, A.: The atmospheric general circulation model ECHAM 5. PART I: Model description https://www.mpimet.mpg.de/fileadmin/publikationen/Reports/max_scirep_349.pdf (last access: 25 September 2017), 2003.

Rothrock, D. A., Percival, D. B., and Wensnahan, M.: The decline in arctic sea-ice thickness: Separating the spatial, annual, and interannual variability in a quarter century of submarine data, J. Geophys. Res.-Oceans, 113, 1–9, https://doi.org/10.1029/2007JC004252, 2008.

Sand, M., Berntsen, T. K., Seland, Ø., and Kristjánsson, J. E.: Arctic surface temperature change to emissions of black carbon within Arctic or midlatitudes, J. Geophys. Res.-Atmos., 118, 7788–7798, https://doi.org/10.1002/jgrd.50613, 2013a.

Sand, M., Berntsen, T. K., Kay, J. E., Lamarque, J. F., Seland, Ø., and Kirkevåg, A.: The Arctic response to remote and lo-

cal forcing of black carbon, Atmos. Chem. Phys., 13, 211–224, https://doi.org/10.5194/acp-13-211-2013, 2013b.

Sand, M., Berntsen, T. K., von Salzen, K., Flanner, M. G., Langner, J., and Victor, D. G.: Response of Arctic temperature to changes in emissions of short-lived climate forcers, Nat. Clim. Chang., 6, 1–5, https://doi.org/10.1038/nclimate2880, 2015.

Sand, M., Samset, B. H., Balkanski, Y., Bauer, S., Bellouin, N., Berntsen, T. K., Bian, H., Chin, M., Diehl, T., Easter, R., Ghan, S. J., Iversen, T., Kirkevåg, A., Lamarque, J.-F., Lin, G., Liu, X., Luo, G., Myhre, G., van Noije, T., Penner, J. E., Schulz, M., Seland, Ø., Skeie, R. B., Stier, P., Takemura, T., Tsigaridis, K., Yu, F., Zhang, K., and Zhang, H.: Aerosols at the Poles: An AeroCom Phase II multi-model evaluation, Atmos. Chem. Phys. Discuss., https://doi.org/10.5194/acp-2016-1120, in review, 2017.

Sato, K., Inoue, J., and Watanabe, M.: Influence of the Gulf Stream on the Barents Sea ice retreat and Eurasian coldness during early winter, Environ. Res. Lett., 9, 084009, https://doi.org/10.1088/1748-9326/9/8/084009, 2014.

Schultz, M. (Eds.): REanalysis of the TROpospheric chemical composition over the past 40 years: Final report, Berichte zur Erdsystemforschung, 48, https://doi.org/10.17617/2.994467, 2007.

Schultz, M. G., Heil, A., Hoelzemann, J. J., Spessa, A., Thonicke, K., Goldammer, J. G., Held, A. C., Pereira, J. M. C., and van Het Bolscher, M.: Global wildland fire emissions from 1960 to 2000, Global Biogeochem. Cy., 22, 1–17, https://doi.org/10.1029/2007GB003031, 2008.

Schwarz, J. P., Gao, R. S., Fahey, D. W., Thomson, D. S., Watts, L. A., Wilson, J. C., Reeves, J. M., Darbeheshti, M., Baumgardner, D. G., Kok, G. L., Chung, S. H., Schulz, M., Hendricks, J., Lauer, A., Ka, B., and Slowik, J. G.: Single-particle measurements of midlatitude black carbon and light-scattering aerosols from the boundary layer to the lower stratosphere, J. Geophys. Res.-Atmos., 111, D16207, https://doi.org/10.1029/2006JD007076, 2006.

Screen, J. A. and Simmonds, I.: Exploring links between Arctic amplification and mid-latitude weather, Geophys. Res. Lett., 40, 1–6, https://doi.org/10.1002/GRL.50174, 2013.

Screen, J. A., Simmonds, I., Deser, C., and Tomas, R.: The atmospheric response to three decades of observed arctic sea ice loss, J. Climate, 26, 1230–1248, https://doi.org/10.1175/JCLI-D-12-00063.1, 2013.

Semenov, V. A. and Latif, M.: Nonlinear winter atmospheric circulation response to Arctic sea ice concentration anomalies for different periods during 1966–2012, Environ. Res. Lett., 10, 054020, https://doi.org/10.1088/1748-9326/10/5/054020, 2015.

Sen, P. K.: Estimates of the Regression Coefficient Based on Kendall's Tau, J. Am. Stat. Assoc., 63, 1379–1389, https://doi.org/10.2307/2285891, 1968.

Serreze, M. C. and Barry, R. G.: Processes and impacts of Arctic amplification: A research synthesis, Global Planet. Change, 77, 85–96, https://doi.org/10.1016/j.gloplacha.2011.03.004, 2011.

Sharma, S., Andrews, E., Barrie, L. A., Ogren, J. A., and Lavoue, D.: Variations and sources of the equivalent black carbon in the high Arctic revealed by long-term observations at Alert and Barrow?: 1989–2003, J. Geophys. Res.-Atmos., 111, D14208, https://doi.org/10.1029/2005JD006581, 2006.

Sharma, S., Ishizawa, M., Chan, D., Lavoué, D., Andrews, E., Eleftheriadis, K., and Maksyutov, S.: 16-year simulation of Arctic black carbon?: Transport, source contribution, and sensitivity analysis on deposition, J. Geophys. Res.-Atmos., 118, 943–964, https://doi.org/10.1029/2012JD017774, 2013.

Shaw, G. E. and Stamnes, K.: ARCTIC HAZE: PERTURBATION OF THE POLAR RADIATION BUDGET*, Ann. N. Y. Acad. Sci., 338, 533–539, https://doi.org/10.1111/j.1749-6632.1980.tb17145.x, 1980.

Shindell, D. and Faluvegi, G.: Climate response to regional radiative forcing during the twentieth century, Nat. Geosci, 2, 294–300, https://doi.org/10.1038/ngeo473, 2009.

Shindell, D., Kuylenstierna, J. C. I., Vignati, E., van Dingenen, R., Amann, M., Klimont, Z., Anenberg, S. C., Muller, N., Janssens-Maenhout, G., Raes, F., Schwartz, J., Faluvegi, G., Pozzoli, L., Kupiainen, K., Hoglund-Isaksson, L., Emberson, L., Streets, D., Ramanathan, V., Hicks, K., Oanh, N. T. K., Milly, G., Williams, M., Demkine, V., and Fowler, D.: Simultaneously Mitigating Near-Term Climate Change and Improving Human Health and Food Security, Science, 335, 183–189, https://doi.org/10.1126/science.1210026, 2012.

Stier, P., Feichter, J., Kinne, S., Kloster, S., Vignati, E., Wilson, J., Ganzeveld, L., Tegen, I., Werner, M., Balkanski, Y., Schulz, M., Boucher, O., Minikin, A., and Petzold, A.: The aerosol-climate model ECHAM5-HAM, Atmos. Chem. Phys., 5, 1125–1156, https://doi.org/10.5194/acp-5-1125-2005, 2005.

Stohl, A.: Characteristics of atmospheric transport into the Arctic troposphere, J. Geophys. Res.-Atmos., 111, 1–17, https://doi.org/10.1029/2005JD006888, 2006.

Stohl, A., Klimont, Z., Eckhardt, S., Kupiainen, K., Shevchenko, V. P., Kopeikin, V. M., and Novigatsky, A. N.: Black carbon in the Arctic: the underestimated role of gas flaring and residential combustion emissions, Atmos. Chem. Phys., 13, 8833–8855, https://doi.org/10.5194/acp-13-8833-2013, 2013.

Tsigaridis, K., Daskalakis, N., Kanakidou, M., Adams, P. J., Artaxo, P., Bahadur, R., Balkanski, Y., Bauer, S. E., Bellouin, N., Benedetti, A., Bergman, T., Berntsen, T. K., Beukes, J. P., Bian, H., Carslaw, K. S., Chin, M., Curci, G., Diehl, T., Easter, R. C., Ghan, S. J., Gong, S. L., Hodzic, A., Hoyle, C. R., Iversen, T., Jathar, S., Jimenez, J. L., Kaiser, J. W., Kirkevåg, A., Koch, D., Kokkola, H., Lee, Y. H., Lin, G., Liu, X., Luo, G., Ma, X., Mann, G. W., Mihalopoulos, N., Morcrette, J.-J., Müller, J.-F., Myhre, G., Myriokefalitakis, S., Ng, N. L., O'Donnell, D., Penner, J. E., Pozzoli, L., Pringle, K. J., Russell, L. M., Schulz, M., Sciare, J., Seland, Ø., Shindell, D. T., Sillman, S., Skeie, R. B., Spracklen, D., Stavrakou, T., Steenrod, S. D., Takemura, T., Tiitta, P., Tilmes, S., Tost, H., van Noije, T., van Zyl, P. G., von Salzen, K., Yu, F., Wang, Z., Wang, Z., Zaveri, R. A., Zhang, H., Zhang, K., Zhang, Q., and Zhang, X.: The AeroCom evaluation and intercomparison of organic aerosol in global models, Atmos. Chem. Phys., 14, 10845–10895, https://doi.org/10.5194/acp-14-10845-2014, 2014.

Uppala, S. M., KÅllberg, P. W., Simmons, A. J., Andrae, U., Bechtold, V. D. C., Fiorino, M., Gibson, J. K., Haseler, J., Hernandez, A., Kelly, G. A., Li, X., Onogi, K., Saarinen, S., Sokka, N., Allan, R. P., Andersson, E., Arpe, K., Balmaseda, M. A., Beljaars, A. C. M., Berg, L. Van De, Bidlot, J., Bormann, N., Caires, S., Chevallier, F., Dethof, A., Dragosavac, M., Fisher, M., Fuentes, M., Hagemann, S., Hólm, E., Hoskins, B. J., Isaksen, L., Janssen, P. A. E. M., Jenne, R., Mcnally, A. P., Mahfouf, J.-F., Morcrette, J.-J., Rayner, N. A., Saunders, R. W., Simon, P., Sterl, A., Trenberth, K. E., Untch, A., Vasiljevic, D., Viterbo, P., and Woollen, J.: The ERA-40 re-analysis, Q. J. R. Meteor. Soc., 131, 2961–3012, https://doi.org/10.1256/qj.04.176, 2005.

Yasunari, T. J., Koster, R. D., Lau, K. M., Aoki, T., Sud, Y. C., Yamazaki, T., Motoyoshi, H., and Kodama, Y.: Influence of dust and black carbon on the snow albedo in the NASA Goddard Earth Observing System version 5 land surface model, J. Geophys. Res.-Atmos., 116, 1–15, https://doi.org/10.1029/2010JD014861, 2011.

# Near-real-time processing of a ceilometer network assisted with sun-photometer data: monitoring a dust outbreak over the Iberian Peninsula

Alberto Cazorla[1,2], Juan Andrés Casquero-Vera[1,2], Roberto Román[1,2], Juan Luis Guerrero-Rascado[1,2], Carlos Toledano[3], Victoria E. Cachorro[3], José Antonio G. Orza[4], María Luisa Cancillo[5,6], Antonio Serrano[5,6], Gloria Titos[7], Marco Pandolfi[7], Andres Alastuey[7], Natalie Hanrieder[8], and Lucas Alados-Arboledas[1,2]

[1] Andalusian Institute for Earth System Research, IISTA-CEAMA, University of Granada, Junta de Andalucía, Granada, Spain
[2] Department of Applied Physics, University of Granada, Granada, Spain
[3] Grupo de Óptica Atmosférica (GOA), Universidad de Valladolid, Valladolid, Spain
[4] SCOLAb, Física Aplicada, Universidad Miguel Hernández, Elche, Spain
[5] Department of Physics, University of Extremadura, Badajoz, Spain
[6] Institute of Water Research, Climate Change and Sustainability, IACYS, University of Extremadura, Badajoz, Spain
[7] Institute of Environmental Assessment and Water Research (IDAEA-CSIC), Barcelona, Spain
[8] German Aerospace Center (DLR), Institute of Solar Research, Plataforma Solar de Almería, Almería, Spain

*Correspondence to:* Alberto Cazorla (cazorla@ugr.es)

**Abstract.** The interest in the use of ceilometers for optical aerosol characterization has increased in the last few years. They operate continuously almost unattended and are also much less expensive than lidars; hence, they can be distributed in dense networks over large areas. However, due to the low signal-to-noise ratio it is not always possible to obtain particle backscatter coefficient profiles, and the vast number of data generated require an automated and unsupervised method that ensures the quality of the profiles inversions.

In this work we describe a method that uses aerosol optical depth (AOD) measurements from the AERONET network that it is applied for the calibration and automated quality assurance of inversion of ceilometer profiles. The method is compared with independent inversions obtained by co-located multiwavelength lidar measurements. A difference smaller than 15 % in backscatter is found between both instruments. This method is continuously and automatically applied to the Iberian Ceilometer Network (ICENET) and a case example during an unusually intense dust outbreak affecting the Iberian Peninsula between 20 and 24 February 2016 is shown. Results reveal that it is possible to obtain quantitative optical aerosol properties (particle backscatter coefficient) and discriminate the quality of these retrievals with ceilometers over large areas. This information has a great potential for alert systems and model assimilation and evaluation.

## 1 Introduction

Atmospheric aerosol is one of the main responsible factors of climate radiative forcing through multiple processes including aerosol–radiation and aerosol–cloud interactions (IPCC, 2014). The aerosol direct effects depend on the optical properties and spatial and vertical distribution of the aerosol in the atmosphere. In spite of the recent advances on instrumentation that has improved the ability of characterizing key aerosol properties and increase the spatial resolution, the associated uncertainties are still considered to be one of the majors in climate forcing (Boucher et al., 2013).

In this sense, the implementation of observational networks is crucial for spatial characterization of aerosol properties. Ground-level aerosol measurement networks represent key tools in the study of aerosol radiative forcing. These observational networks provide surface measurements distributed over large areas, e.g., the Global Atmospheric Watch, GAW (GAW, 2011), and ACTRIS (www.actris.eu) for Europe. In addition, one of the recognized instruments for the retrieval of column-integrated aerosol properties is the robotic sun and sky photometer that is used in the global Aerosol Robotic NETwork (AERONET; Holben et al., 1998; Dubovik et al., 2006). Lidar systems are well-known active remote sensing instruments for the vertically resolved characterization of aerosol optical and microphysical properties (Winker et al., 2003). GAW Atmospheric Lidar Observation Network (GALION) has emerged as an initiative of the GAW aerosol program (GAW, 2008). Its main objective is to provide the vertical component of the aerosol distribution through advanced laser remote sensing in a network of ground-based stations. Among other networks, GALION includes the European Aerosol Research Lidar Network (EARLINET) that provides vertical aerosol profile observations over Europe based on 27 instruments in 16 countries (Pappalardo et al., 2014), the Micro-Pulse Lidar Network, MPLNET (Welton et al., 2001), and the Latin American Lidar Network, LALINET (Guerrero-Rascado et al., 2016).

In order to obtain a larger spatial coverage than ground-based networks, in the last few years some space missions have been promoted focusing on aerosol measurements from satellites, e.g the Lidar in Space Technology Experiment, LITE (McCormick, 1997), and the Cloud-Aerosol Lidar and Infrared Pathfinder Satellite Observation, CALIPSO (Winker et al., 2003). The main disadvantage of measurements from spaceborne platforms is the low temporal resolution, since the measurements are limited to the satellite passes over a region.

The usefulness of vertically resolved aerosol characterization has been proven by monitoring dust outbreaks (e.g., Guerrero-Rascado et al., 2008, 2009; Cordoba-Jabonero et al., 2011; Bravo-Aranda et al., 2015; Preißler et al., 2011, 2013, 2017; Granados-Muñoz et al., 2016), biomass burning plumes (e.g., Alados-Arboledas et al., 2011; Ortiz-Amezcua et al., 2017) or the volcanic ash plume from the Eyjafjallajökull eruption on April 2010 (Navas-Guzmán et al., 2013; Pappalardo et al., 2013; Sicard et al., 2012). Precisely, this singular event caused aviation problems and drew the attention to the use of ceilometers for vertically resolved aerosol characterization (Flentje et al., 2010).

The complexity of lidar systems requires staff to be trained in their operation, and the analysis procedures are not fully automated in many stations. In this sense, continuous operation of lidar systems is not feasible for most stations. In addition, economic and operational costs hinder the implementation of dense lidar networks. On the other hand, ceilometers are one-wavelength (near infrared) lidars with simple technical specifications (eye-safe low pulse energy and high pulse repetition frequencies) allowing for unattended and continuous operation. Originally designed for cloud base determination, their performance has been improved in the last few years. Their capabilities have been shown for determining

planetary boundary layer (e.g., Wiegner et al., 2006; Münkel et al., 2007; Haeffelin et al., 2012; Pandolfi et al., 2013), detection and forecast of fog (Haeffelin et al., 2016), and recent efforts have been conducted to quantify the aerosol optical information that can be derived from ceilometers (Frey et al., 2010; Heese et al., 2010, Wiegner et al., 2014).

The main advantage of the use of ceilometers for aerosol characterization is, on the one hand, the automatic and much simpler operation compared to lidars and, on the other hand, the possibility of installing them distributed over large areas. Meteorological services such as those in Germany, France, the Netherlands or the United Kingdom are deploying ceilometers networks to cover their national territories with the objective of reaching a spatial density of nearly one device every 100 km (e.g., de Haij and Klein-Baltink, 2007; Flentje et al., 2010). Due to a dense number of instruments and continuous measurements, operative networks need an automated processing and a protocol that ensures the quality of the data.

In this sense, two programs in Europe are dealing with the use of automated lidars and ceilometers for aerosol and cloud properties characterization. The COST Action ES1303 TOPROF (TOwards operational ground based PROFiling with ceilometers, doppler lidars and microwave radiometers for improving weather forecasts) aims in one of its working groups at better characterizing the parameters that can be derived from ceilometer measurements and related uncertainties. At the same time, E-PROFILE, a program of EUMETNET (EUropean METeorological services NETwork), focuses on the harmonization of ceilometer measurements and data provision across Europe.

In this study we present the implementation of procedures to manage a regional ceilometer network for aerosol characterization over the Iberian Peninsula, the Iberian Ceilometer Network (ICENET). An automatic calibration procedure is applied to the ceilometers and this calibration is used to validate the elastic inversion automatically applied to the profiles. This method uses additional aerosol optical depth (AOD) information during the calibration for the quality assurance of the data.

All processes can be performed unattended and in near-real time with the objective of obtaining reliable vertically resolved aerosol optical properties. This information is especially useful for strong events, such as mineral dust outbreaks, volcanic plumes, severe biomass burning episodes or contamination episodes. Thus, the aerosol information obtained can be potentially used as an alert system for aviation or weather services or to feed models for assimilation and validation in near-real time.

The capabilities of this distributed network are explored by characterizing an unusually intense dust outbreak affecting the Iberian Peninsula on 20–24 February 2016 and a multi-wavelength (MW) Raman lidar is used to validate the retrievals from ceilometers.

The next section describes the Iberian Ceilometer Network and related instrumentation used in this study. Section 3 presents the methodology, including the calibration of ceilometers (Sect. 3.1) and the use of the calibration for inversion validation, as well as a validation with an independent lidar system (Sect. 3.2). The results are presented in Sect. 4, with a description of the dust event (Sect. 4.1 and 4.2) and retrievals from ceilometers (Sect. 4.3). Finally, conclusions are presented in Sect. 5.

## 2 Instrumentation: the Iberian Ceilometer Network

An initiative of the Atmospheric Physics Group of the University of Granada has been the coordination of a network of ceilometers (ICENET) combined with sun photometers for the characterization of atmospheric aerosol with the objective of obtaining reliable vertically resolved aerosol optical properties in near-real time. The first goal is obtaining the total attenuated backscatter for all ceilometers in the network, i.e., to obtain calibrated output from ceilometers, and the second one is applying an inversion algorithm to the ceilometer profiles in order to obtain the particle backscatter coefficient. All sites of this new network have a co-located AERONET CE318 sun–sky photometer (Cimel Electronique) that is used to constrain the ceilometers calibration and inversion retrievals. In addition, the high-performance lidar system MULHACEN, located at the EARLINET Granada station, is used as an independent validation of the inversions. This nested approach combining high-performance systems like those operated in EARLINET and the distributed ceilometer plus sun photometer is an example of synergy among active and passive remote sensing observations in the ACTRIS research infrastructure (www.actris.eu).

Figure 1 shows a map of the ceilometer distribution over the Iberian Peninsula, and Table 1 presents the characteristics of each site.

All sites operate a Jenoptik (now Lufft) CHM15k-Nimbus ceilometer and have a co-located AERONET sun photometer, except Montsec station (MSA), which has the photometer 770 m above the ceilometer and at a horizontal distance of 2 km approximately (Titos et al., 2017). The ceilometer at Murcia (UMH) was not operative during the outbreak studied in this work.

The CHM15k is a ceilometer that operates with a pulsed Nd : YAG laser emitting at 1064 nm. The energy per pulse is 8.4 μJ with a repetition frequency in the range of 5–7 kHz. The laser beam divergence is less than 0.3 mrad and the laser backscattered signal is collected on a telescope with a field of view of 0.45 mrad. The signal is detected by an avalanche photodiode in photon-counting mode. Complete overlap of the telescope and the laser beam is found about 1500 m above the instrument (Heese et al., 2010). According to the overlap function provided by the manufacturer, the overlap is 90 % complete between 555 and 885 m a.g.l. The vertical

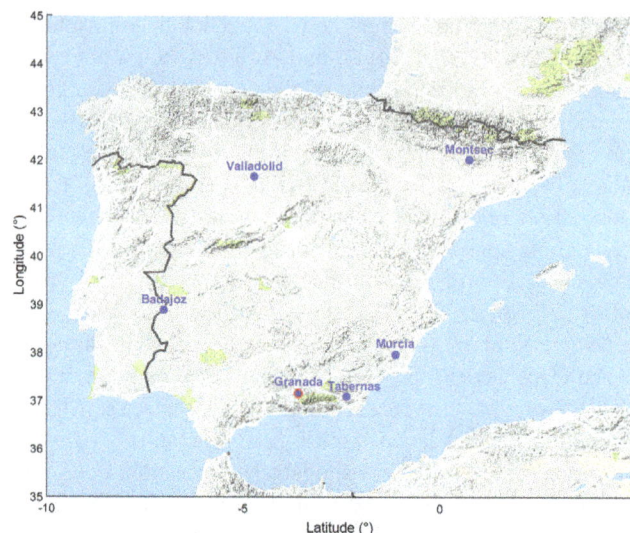

**Figure 1.** Map of the Iberian Peninsula showing the location of the ceilometers. At Granada station (circled in red) a co-located multi-wavelength Raman lidar is also available.

resolution used is 15 m and the maximum height recorded is 15360 m a.g.l. Ceilometers at Granada (UGR), Tabernas (PSA) and Valladolid (UVA) operate at a temporal resolution of 15 s, while ceilometers at Montsec (MSA) and Badajoz (UEX) operate at a temporal resolution of 1 min.

The process of calibration for ceilometers described on the next section is assisted with AOD data from co-located AERONET stations. All sun photometers near the ceilometers belong to the Iberian network for aerosol measurements (RIMA), a regional network associated with AERONET. This means that all instruments are routinely calibrated following the same protocol and the data are quality-controlled. The sun photometer provides solar extinction measurements at 340, 380, 440, 675, 870, 936 and 1020 nm, allowing for computing the AOD at these wavelengths (except 936 nm). The AOD uncertainty ranges from ±0.01 in the infrared–visible to ±0.02 in the ultraviolet channels (Holben et al., 1998). For comparison with the ceilometers the AOD is extrapolated to 1064 nm by the Ångström law (Ångström, 1964) using the AOD measurements at 870 and 1020 nm. Level 1.5 AERONET data, which are automatically cloud-screened and delivered in near-real time, are used in this analysis.

At UGR station a multi-wavelength Raman lidar system (MULHACEN) is used for validation of the ceilometer inversions. The upgraded LR331-D400 (Raymetrics Inc.) operated at IISTA-CEAMA (Andalusian Institute for Earth System Research) has been part of EARLINET since April 2005. This lidar system is a ground-based, six-wavelength system with a pulsed Nd : YAG laser. The emitted wavelengths are 355, 532 and 1064 nm with output energies per pulse of 60, 65 and 110 mJ, respectively. It has elastic backscatter chan-

**Table 1.** Description of the Iberian ceilometer network sites.

| Site (code) | Managed by | Location (°lat, °long) | Height (m a.s.l.) | Additional instruments |
|---|---|---|---|---|
| Granada (UGR) | Atmospheric Physics Group, University of Granada | 37.16° N, 3.58° W | 680 | CIMEL CE 318 Multi-wavelength lidar |
| Plataforma Solar de Almería-Tabernas (PSA) | Institute of Solar Research, German Aerospace Center | 37.09° N, 2.36° W | 500 | CIMEL CE 318 |
| Badajoz (UEX) | AIRE Group, University of Extremadura | 38.88° N, 7.01° W | 199 | CIMEL CE 318 |
| Valladolid (UVA) | Atmospheric Optics Group, University of Valladolid | 41.66° N, 4.71° W | 705 | CIMEL CE 318 |
| Montsec (MSA) | Institute of Environmental Assessment and Water Research, Spanish Research Council. | 42.02° N, 0.74° E | 800 | CIMEL CE 318 (42.05° N, 0.73° E; 1570 m a.s.l.) |
| Murcia (UMH) | Statistical and Computational Physics Lab, Miguel Hernández University | 39.98° N, 1.13° W | 69 | CIMEL CE 318 |

nels at 355, 532 and 1064 nm and Raman channels at 387 (from $N_2$), 408 (from $H_2O$) and 607 nm (from $N_2$). Full overlap is reached around 1220 m a.g.l., although the overlap is complete at 90 % between 520 and 820 m a.g.l. (Navas-Guzmán et al., 2011). Appropriate overlap corrections are derived following the procedure of Wandinger et al. (2002).

## 3 Methodology

The principle of measurement for elastic lidars and ceilometers is the same, and retrieval of optical properties in both systems follows the lidar equation (the dependency with the wavelength has been omitted for simplicity since it is always the same in ceilometers):

$$P(z) = C_L \cdot \frac{O(z)}{z^2} \beta(z) \cdot T^2(z), \tag{1}$$

where $P(z)$ is the backscattered power received in the telescope from a distance $z$, $C_L$ is a parameter that depends on the geometry and characteristics of the instrument and universal constants, and the term $z^2$ accounts for the acceptance solid angle of the receiver optics with the distance to the laser. The backscattered signal collected by the telescope depends on the overlap between the laser beam and the telescope field of view, and the degree of overlap is quantified by $O(z)$, ranging from 0, if there is no overlap, to 1, if overlap is complete. $\beta(z)$ is the atmospheric backscatter coefficient and $T(z)$ estimates the atmospheric transmittance of the laser signal (squared due to travel back and forth). Also, both properties can be split into contributions of particles and molecules ($\beta(z) = \beta_m(z) + \beta_p(z)$; $T(z) = T_m(z) \cdot T_p(z)$) (Fernald, 1984).

In Eq. (1) the only properties depending on the medium are $\beta(z)$ and $T(z)$. Thus, the atmospheric attenuated backscatter is defined as

$$\beta_{att}(z) = \beta(z) \cdot T^2(z). \tag{2}$$

### 3.1 Ceilometer calibration

The ceilometers used in this study provide the range-corrected signal ($RCS(z) = P(z) \cdot z^2$) as output, using an overlap function determined by the manufacturer and corrected for the number of laser shots. Therefore, the only parameter that needs to be addressed is $C_L$.

Wiegner et al. (2014) describe a method to find the $C_L$ parameter in ceilometers, commonly referred as ceilometer calibration. This method compares the RCS from the ceilometer in a particle-free region with the molecular attenuated backscatter that can be calculated using Rayleigh theory. The Rayleigh fit compares the gradient with altitude (the slope) of both profiles and looks for a region in the ceilometer profile that has the same trend as the expected molecular profile. In this study, we select regions of 990 m with a difference in gradients below 1 %. Thus, in that region or reference height ($z_{ref}$), $C_L$ can be calculated:

$$C_L(z_{ref}) = \frac{RCS(z_{ref})}{\beta_m(z_{ref}) \cdot T_m^2(z_{ref}) \cdot T_p^2(z_{ref})}. \tag{3}$$

At this reference height, the backscattering is only due to molecules. The transmittance due to molecules ($T_m$) can be easily determined from Rayleigh theory but the transmittance due to particles ($T_p$) is unknown. However, if a co-located sun photometer is available, $T_p$ can be calculated, using the

AOD at 1064 nm:

$$T_p^2(z_{ref}) = e^{-2 \times AOD}.\qquad(4)$$

When trying to automate the calibration process, the main problem is that $z_{ref}$ must be a particle-free region and, due to the low signal-to-noise ratio, finding $z_{ref}$ is not always possible. In some cases, the region might be a non-particle-free region that, on average, follows the molecular trend (they have a similar gradient). Also, we might find several regions that meet the criteria, but it is complicated to discriminate automatically which one is the most appropriate. A way to ensure that the $z_{ref}$ selected is a molecular region is by applying the Klett–Fernald inversion algorithm (Klett, 1981, 1985; Fernald et al., 1972; Fernald, 1984) as follows.

First, we need to determine the $z_{ref}$. Thus, after finding $z_{ref}$, the Klett–Fernald inversion is applied. Heese et al. (2010) and Wiegner et al. (2012, 2014) showed the capabilities of ceilometers by applying this inversion algorithm to study a few cases. Using the AOD measurements, the lidar ratio (Lr) of the inversion can be adjusted. This can be done by matching the integral of the particle extinction coefficient profile (i.e., particle backscatter coefficient profile multiplied by the Lr) with the AOD. Wiegner et al. (2012) applied this procedure to a Jenoptik CHM15kx ceilometer, obtaining reasonable values for the Lr.

In summary, the calibration process is carried out as follows:

1. First, temporal averaging of the profiles is performed (hourly averages are used for the calibration). The first 300 m of the profile are assigned to the value at 300 m to avoid large overlap correction.

2. Second, for each profile a set of potential $z_{ref}$ is obtained by comparing the profiles of the RCS and $\beta_m$, which is obtained from a standard atmosphere profile scaled to ground temperature and pressure. The slopes are calculated over a 990 m window. All regions with slope differences below 1 % are selected.

3. For each $z_{ref}$, and Lr from 20 to 80 sr, Klett–Fernald inversion is applied and the resulting profile for the backscattering coefficient is integrated and multiplied by the Lr and compared to the AOD. The pair $z_{ref}$ and Lr that minimizes the difference between the integral of the particle extinction coefficient profile (i.e., particle backscatter coefficient profile multiplied by the Lr) with the AOD is selected.

4. Finally, $C_L$ is calculated using Eq. (3) if the minimum difference calculated in step 3 is below 10 %.

$C_L$ calculated with this method uses Eq. (4) to calculate $T_p^2$. In the case of MSA the sun photometer is 770 m above the ceilometer and the $T_p^2$ calculated would not be representative of the entire column. However, MSA is a remote mountain site and the effect of accounting for $T_p^2$ values obtained

**Table 2.** Mean calibration factors for ceilometers in ICENET for the period 1 May 2014 to 1 May 2016.

| Site code | $C_L$ (m³ sr) |
|---|---|
| UGR | $(1.6 \pm 0.4) \times 10^{11}$ |
| PSA | $(3.3 \pm 0.7) \times 10^{11}$ |
| UEX | $(3.7 \pm 0.9) \times 10^{11}$ |
| UVA | $(3.8 \pm 0.5) \times 10^{11}$ |
| MSA | $(3.1 \pm 0.7) \times 10^{11}$ |
| UMH | $(4.4 \pm 0.9) \times 10^{11}$ |

using the AOD measured from 770 m above the ceilometer can be considered negligible.

Thus, the total attenuated backscatter can be calculated by applying the following equation:

$$\beta_{att}(z) = \frac{RCS(z)}{C_L(z_{ref})}.\qquad(5)$$

Calibration values can be used individually or averaged over a period of time. Table 2 shows a mean calibration factor ($\pm$ standard deviation) calculated using this method for all sites in ICENET for the period 1 May 2014 to 1 May 2016.

### 3.2 Ceilometer inversion

Total attenuated backscatter obtained by applying the calibration factor allows the comparison between ceilometers since the signal is corrected for instrument characteristics. Also, a long time series of the calibration allows determining possible problems or degradation of the systems. However, the total attenuated backscatter is influenced by transmission, so in order to be able to monitor and compare singular events at multiple sites, the backscattering coefficient is more appropriate. Section 3.1 showed that it is possible to apply the Klett–Fernald inversion to ceilometer data, but the challenge is to determine automatically, without human supervision, whether the inversion is correct or not.

A common step between the calibration proposed in Sect. 3.1 and Klett–Fernald inversion is finding a reference height ($Z_{ref}$). At the $Z_{ref}$ selected for the inversion with Rayleigh fit, applying Eq. (3), we can obtain a value that has to be close to the $C_L$ calculated for the instrument on a longer period of time. If a simultaneous AOD measurement is available, the calibration process itself provides the inverted backscattering coefficient profile (steps 3 and 4 of the calibration process) and the inversion can be marked as valid or invalid based on the value of the $C_L$ compared to a long-term $C_L$ value. If no simultaneous AOD measurement is available (e.g., during nighttime or partially cloudy skies), an approximation of the AOD needs to be used in order to apply Eq. (4). In this case, an interpolated value or an averaged value can be used.

The next section quantifies the differences between these backscattering coefficient inversions and the inversions calculated independently with a multi-wavelength Raman lidar.

**Lidar – ceilometer comparison**

During the dust outbreak affecting the Iberian Peninsula between 20 and 24 February 2016, the multi-wavelength lidar operated on 22 February between 07:30 and 14:00 UTC and on 23 February between 08:00 and 13:30 UTC. Elastic inversions using the Klett–Fernald method were applied to 30 min average profiles at 1064 nm using a fixed lidar ratio of 50 sr. Thus, a total of 24 particle backscatter coefficient profiles were obtained. Coherence of the inversion at 1064 nm was checked against the Klett–Fernald and Raman methods at 355 and 532 nm. The resolution of the multi-wavelength lidar (7.5 m) has been downscaled to 15 m for the comparison with the ceilometer.

The ceilometer elastic inversion, using the Klett–Fernald method, was also applied to 30 min average profiles for the same period; a total of 15 profiles were successfully inverted (a reference height was found automatically). The calibration factor at the reference height was calculated using the average AOD for the entire dust event. If negative $C_L$ values are discarded, a total of 11 profiles are comparable with lidar inversions.

Each one of the derived $C_L$ values at the reference height selected for the inversion is compared with long-term $C_L$ calculated for the Granada station ceilometer over a long period of time (see Table 2), classifying the situation according to statistical parameters measuring the agreement between the ceilometer and lidar retrievals. The normalized mean bias (NMB) in particle backscatter of the ceilometer and lidar profile is calculated following Eq. (6). The center of mass of the profiles is calculated with Eq. (7), as is the relative difference between ceilometer and lidar center of mass. Finally, the coefficient of correlation ($R$) of the profiles is determined.

$$\text{NMB} = \frac{\overline{\beta}_{ceil} - \overline{\beta}_{lidar}}{\overline{\beta}_{lidar}} \tag{6}$$

$$C_{mass} = \frac{\int_{z_{min}}^{z_{max}} z \cdot \beta(z)\mathrm{d}z}{\int_{z_{min}}^{z_{max}} \beta(z)\mathrm{d}z} \tag{7}$$

In Eq. (6), $\overline{\beta}_{ceil}$ and $\overline{\beta}_{lidar}$ are the mean particle backscatter coefficient from ceilometer and lidar, respectively, for the entire retrieved profile, and $\beta$ in Eq. (7) may refer to ceilometer or lidar particle backscatter coefficient depending on the case.

Figure 2 shows for the 11 comparable profiles, the retrieved calibration factors at the reference height on the ceilometer profile versus the NMB (panel a), center of mass relative differences (panel b) and $R$ (panel c). It is evident that ceilometer profiles with a calibration factor closer to the mean calibration factor have inversions closer to the lidar inversions. Figure 2 also shows that, for the statistics NMB

and $R$, the difference between the calibration factor and the mean calibration factor is related, and the farther the profile calibration factor is from the mean value, the worse the mentioned statistics. Thus, it seems feasible to determine the quality of the profiles by selecting an appropriate threshold for this difference. Considering a maximum discrepancy between the particular calibration factor and the long-term calibration factor equal to 1 standard deviation of the mean value of the calibration factor (dotted lines in Fig. 2), we obtain four profiles that have a NMB smaller than 15 %; the center of mass of the profiles is practically the same, with a relative difference smaller than 2 %, and finally $R$ of the profiles is above 0.92.

A sequence of ceilometer and lidar particle backscatter profiles from 23 February 2016 is shown in Fig. 3. The first ceilometer profile (marked in blue) has a calibration factor of $2.57 \times 10^{11}$ (m³ sr) and hence is rejected according to the threshold described above. For this case, the NMB of the ceilometer and lidar profiles is $-0.31$, the center of mass relative difference is $-0.06$ and the $R$ is 0.84. The other four ceilometer profiles (marked in red) have calibration factors within the standard deviation of the mean calibration factor. The profiles on 23 February at 09:00 and 09:30 UTC correspond with a decoupled dust layer. Those profiles have a NMB of $-0.08$ and 0.1, respectively, the center of mass relative difference is $-0.01$ and $-0.03$, respectively, and $R$ is 0.95 and 0.97, respectively. The profiles on 23 February at 12:00 and 12:30 UTC show that the previous dust layer is mixed with the boundary layer. In these cases, profiles have a NMB of 0.14 and $-0.12$, respectively, the center of mass relative difference is 0.006 and $-0.01$, respectively, and $R$ is 0.99 and 0.93, respectively.

## 4 Results

The capabilities of the ceilometer network for aerosol optical properties characterization and the near-real-time processing have been tested with the analysis of the African dust outbreak that affected the Iberian Peninsula on 20 February 2016 and persisted until 24 February 2016.

Sorribas et al. (2017) studied the same event and compared it with meteorological parameters, aerosol properties and ozone from historical data sets on a site in southern Spain. They concluded that the event was exceptional because of its unusual intensity, its impact on surface measurements and the month of occurrence. In addition, Titos et al. (2017) also analyzed this event using 250 air quality monitoring stations over Spain to investigate the impact and temporal evolution of the event on surface $PM_{10}$ levels. They also investigated aerosol optical properties, including attenuated backscatter from ceilometer during the event at Montsec station (one of the station included in ICENET). They concluded that the impact on surface $PM_{10}$ was exceptional and highlighted the complexity of the event.

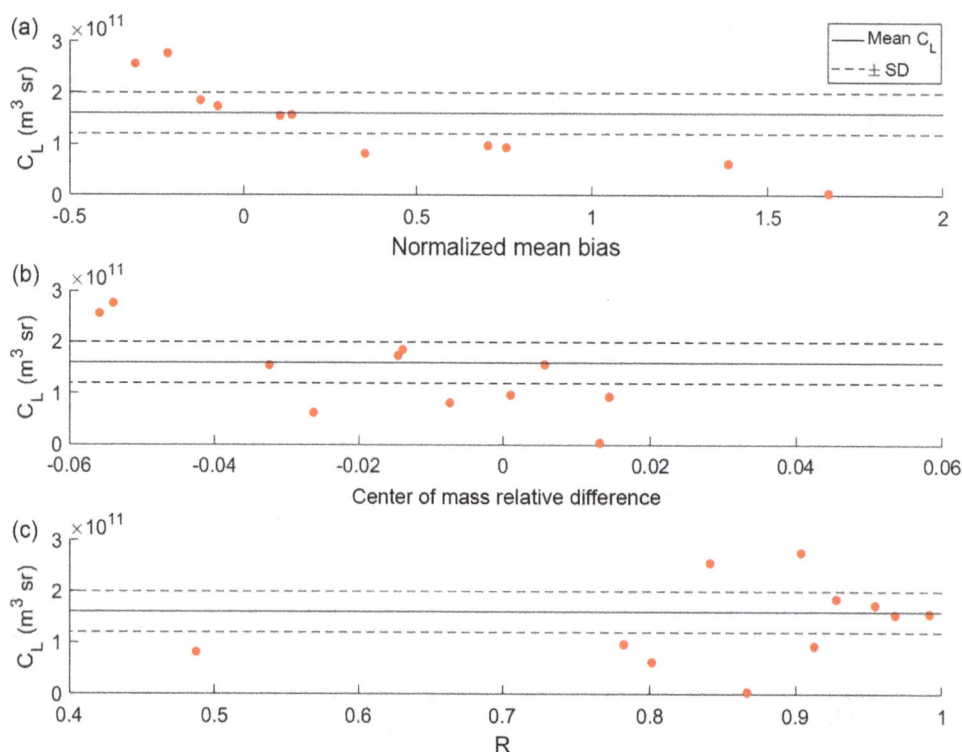

**Figure 2.** Ceilometer calibration factor ($C_L$) vs. normalized mean bias (NMB) **(a)**, relative difference in center of mass **(b)** and coefficient of correlation ($R$) **(c)**. The solid horizontal line indicates the mean $C_L$ for the dust event period and dashed lines indicate the 33 % around this mean value.

Next section provides a detailed description of the dust mobilization and arrival of the plumes to the Iberian Peninsula.

## 4.1 Description of the dust episode

The evolution of the dust outbreak is illustrated in Fig. 4, where a sequence of the false-color RGB dust images from MSG-SEVIRI is shown. This product makes use of three thermal satellite channels to contrast the brightness temperature signal between surface, cloud and dust (Lensky and Rosenfeld, 2008), in a color scheme in which dust appears in magenta. The presence of dust plumes over the High Plateau located between the Sahara and Tell Atlas in Algeria at 12:00 UTC on 20 February 2016 is shown in Fig. 4a. Dust migrated northwestward and passed over the Alboran Sea from the Algerian–Moroccan border around 14:00 UTC, reaching the southern Iberian Peninsula at 18:00 UTC (Fig. 4b) and continued moving northwestward (Fig. 4c). A second dust plume migrated northwards on 21 February 2016 at 16:00 UTC (Fig. 4d).

SYNOP meteorological observations and aerodrome routine (METAR) and special (SPECI) reports in northern Africa recorded a strong reduction in horizontal visibility (MOR, meteorological optical range), down to 2 km, between 07:00 and 08:00 UTC (20 February) at distant loca-

tions situated at the edges of the Great Western Erg in Algeria. At the eastern part of the Western Erg and in the Great Eastern Erg, visibility lowered to less than 5 km between 09:00 and 11:30 UTC. In good accordance with the satellite images, at the Saharan Atlas and the High Plateau area, with heights over 1000 m, visibility less than 2 km was recorded at 10:00 UTC at Mecheria, while at the other stations in that area values went down to 2–3 km at 12:00 UTC. High relative humidity and clouds were found in the westernmost sites, also in agreement with the satellite images. It is remarkable that no significant visibility reduction was reported in the north-facing downslope areas of the Tell Atlas and in the Rif Mountains, close or at the coast of Algeria and Morocco. This indicates that dust was uplifted before passing over the northern slope of the Atlas Mountains and the North African coast. Correspondingly, no station in the southern Iberian Peninsula reported a reduction in visibility when the dust plume reached Spain.

The entrance of dust-laden air masses above the ground level in the Iberian Peninsula is confirmed by a back-trajectory analysis performed with HYSPLIT (Hybrid Single-Particle Lagrangian Integrated Trajectory) model (Draxler and Hess, 1998; Stein et al, 2015) using ERA-Interim data of 0.5° resolution. The trajectory analysis provides an estimate of the range of heights at which the dust-laden air masses passed over the study sites. This is illus-

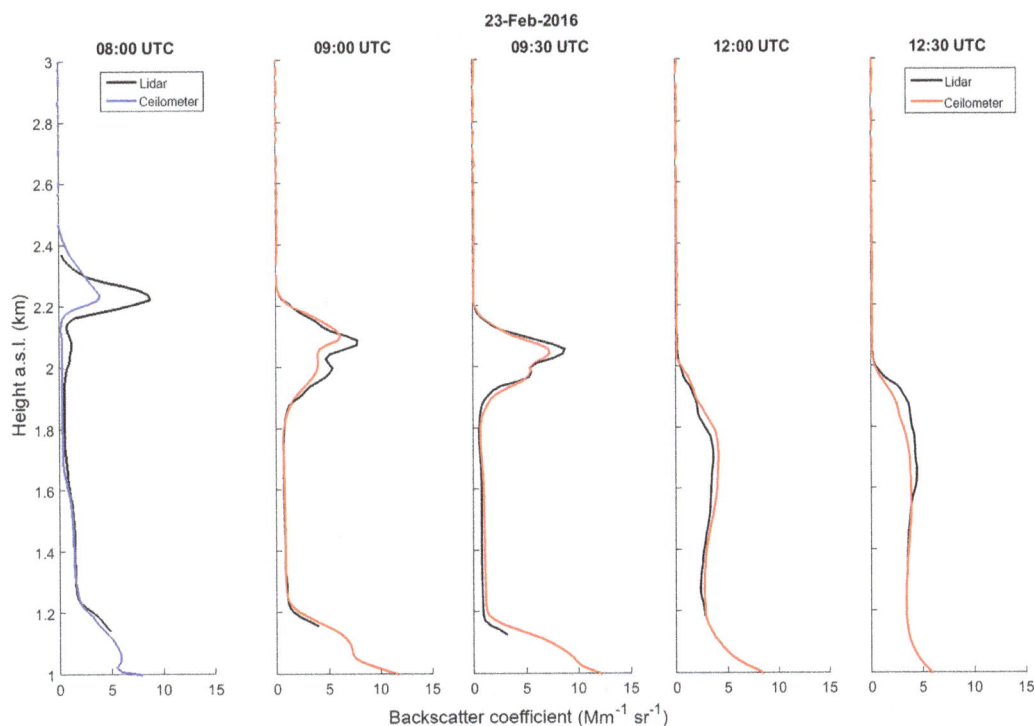

**Figure 3.** Lidar and ceilometer particle backscatter profiles for five cases on 23 February 2016. The first case (marked in blue) is a rejected ceilometer profile and the other four cases (marked in red) are cases with a ceilometer calibration factor within the 33 % of the median calibration factor. Errors on lidar profiles have been estimated with a Monte Carlo technique (Pappalardo et al., 2004; Mattis et al., 2016) and are on the order of $10^{-9}$ ($m^{-1}$ $sr^{-1}$); therefore, they are not shown.

trated in Fig. 5 for two stations: in the south where dust reached first the Iberian Peninsula (Granada) and in the westernmost station, on the cyclonic track of the dust (Badajoz). The height vs. latitude plots show that the dust plumes reaching Granada on 20 February, 18:00 UTC and Badajoz on 21 February, 00:00 UTC, arrived at mid-levels in the lower troposphere after being uplifted in the southern slope of the Saharan Atlas from heights between 500 and 1200 m above the ground in that area (black trajectory in Fig. 5). Those trajectories end at around 3250 and 2500 m a.s.l. in Granada and Badajoz, respectively. Figure 5 also shows that the trajectories reaching Granada at about the height of 2250 m a.s.l. passed over Africa before the dust mobilization took place, while the ones below had no African origin. Finally, the trajectory reaching Granada at 4250 m a.s.l. followed the mid–upper-level tropospheric circulations. In Badajoz, results are analogous: at 1000 m a.s.l. and below, trajectories arrived from the north; between 1250 and 2000 m a.s.l. they arrived from Africa, but the air parcels were located north of the area where dust was observed on the morning of 20 February. The air parcels reaching Badajoz between 2250 and 3250 m a.s.l. were previously located in the area where dust was being observed, while at upper levels trajectories followed the mid–upper circulation pattern.

In terms of aerosol load, Fig. 6 shows the time series of AOD at 1064 nm (a) and Ångström exponent between 440 and 880 nm (b) for the entire period of the event for all sites in ICENET. We observe that the increase in AOD and decrease in Ångström exponent correspond with the arrival of the plumes at each site. The strongest part of the event occurs on 21–22 February and the second plume is observed clearly for Granada station on 22 February. For dust events following a similar pattern that the one described here during the period 2005–2010, Valenzuela et al. (2012) reported a maximum AOD of 0.98 at Granada, which is significantly lower than the maximum measured at Granada station during this event.

## 4.2 Synoptic scenario and context

Several atmospheric features at mid–upper levels were relevant for this episode as they promoted instability near the surface and induced dust transport in the lower free troposphere. The first feature is the amplification and break-up of a Rossby wave in the eastern Atlantic resulted in a trough that became isolated as a cut-off low over the Atlas Mountains on 19 February. A shallow cyclone then originated leeward of the Atlas Mountains. From early 21 February, the cut-off low displaced off the Moroccan coast and centered southwest of Cape St. Vincent. On 22 February it decayed bringing

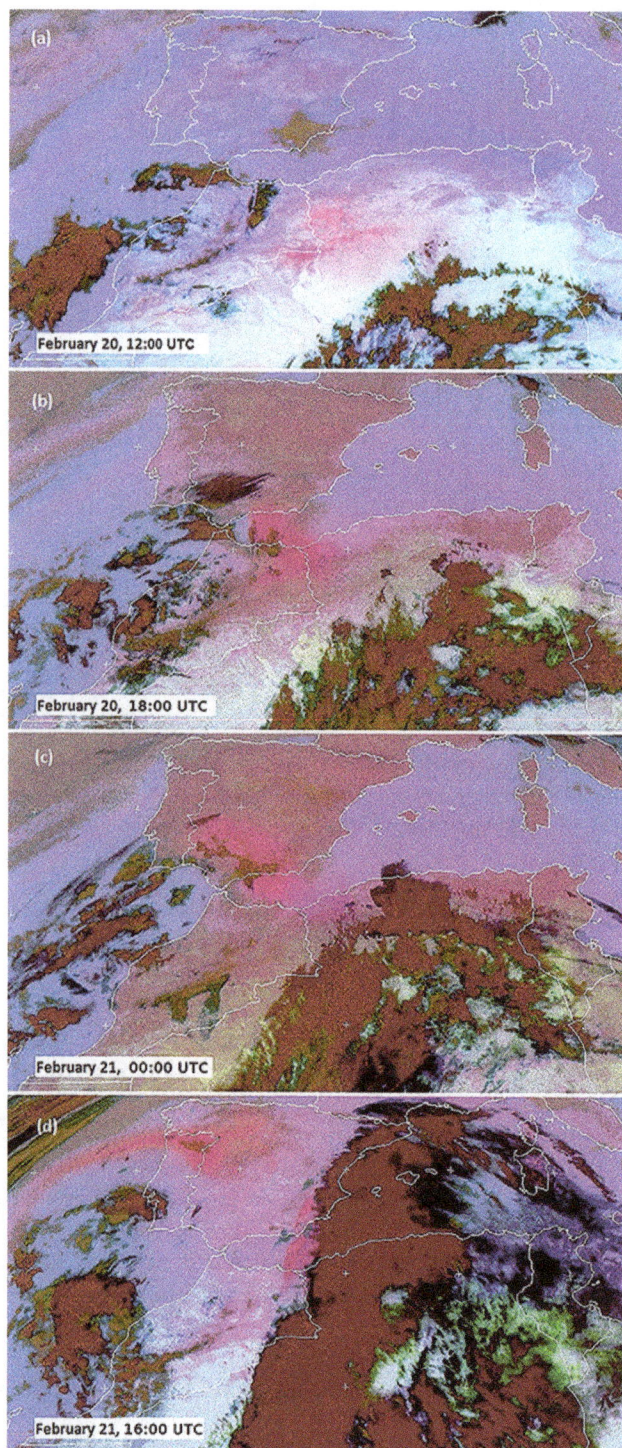

**Figure 4.** False-color RGB dust image from MSG-SEVIRI showing different stages of the dust outbreak (dust appears magenta).

fied and extended northwards to the western Mediterranean. This high pressure influenced circulation at mid–upper levels in combination with the cut-off low. The third is moisture flux at mid-tropospheric levels, which entered from the central Atlantic into the African continent below 20° N and was transported to northern Africa (at 400–550 hPa according to the radiosoundings in the area) between the upper-level trough and the high-pressure system. The tropical air masses are well visible in the satellite imagery as an elongated cloud band moving north and eastward, and so are the convective clouds formed ahead of the band. The tropical–extratropical interaction between the advected tropical moisture and the upper-level trough located over the Atlas Mountains is linked to convective precipitation in northwestern Africa; see Knippertz (2003) and references therein. Divergence at upper levels (250 hPa) and low-level (850 hPa) convergence are found over the area where the gust front mobilized the dust on 20 February. The interaction with the Ahaggar Mountains in southern Algeria possibly enhanced convection and low-level instability. Convective precipitation was registered at several locations of eastern Spain when the cloud band passed over the area in the second half of the episode. From 22 February onwards the cloud band and local convective situations were gradually displaced to the Mediterranean, as zonal flows began to dominate.

At low levels, the low pressure that formed in the lee of the Atlas Mountains moved to the SW of Cape St. Vincent on 21 February following the upper-level instability. The low was then intensified and influenced northern Africa and most of the Iberian Peninsula. In addition, high pressures over the western Mediterranean were formed when the Rossby wave train progressed to the east and retreated poleward. Then, the North African high, which was previously located over Libya at 850 hPa, extended to Tunisia and Algeria and was gradually intensified in connection with the northward extension of the high pressures at upper levels, which arrived (along with the cloud band on its western flank) at the western Mediterranean Basin.

The advection of dust-laden air masses to the Iberian Peninsula was driven by both the low located to the SW of the Iberian Peninsula and the North African high. The presence of these two synoptic systems corresponds to one of the typical synoptic situations leading to dust transport over the Iberian Peninsula (Rodríguez et al., 2003; Escudero et al., 2005). During the episode, however, two distinct strong plumes were transported from northern Africa to the Iberian Peninsula in consecutive days and showed a different evolution. Dust mobilized by the gust front on 20 February south of the Saharan Atlas and north of the Ahaggar Mountains migrated west and northward to the Iberian Peninsula, as shown in the satellite images, forming a curve-shaped plume over Iberia due to the cyclonic shear imposed by the low. The second strong dust plume was mobilized and transported northwards on 21 February on the western side of the North African high, driven by the intensification of this high-

the Iberian Peninsula under the influence of the Azores and North African subtropical highs, with dominant zonal flows. The second is an upper-level anticyclone over a wide area centered over Niger–Chad, which during the episode intensi-

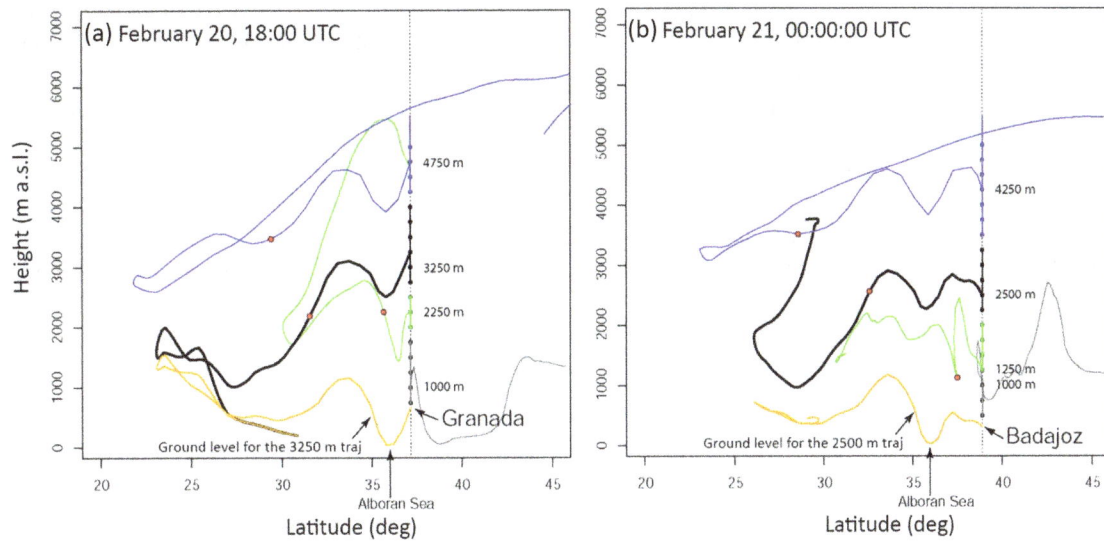

**Figure 5.** Evolution of the air parcels reaching Granada on 20 February at 18:00 UTC (**a**) and Badajoz on 21 February at 00:00 UTC (**b**) at different heights. Back-trajectories were calculated from the ground level to 5000 m a.s.l., at every 250 m. Lines in grey indicate trajectories arriving at the lowest levels, with no African history; in green are trajectories that passed over the southern slope of the Saharan Atlas before the observed dust mobilization; in black are the trajectories followed by the parcels residing at the times and area where dust was observed; in blue are the trajectories residing at higher levels. One representative trajectory is shown for each evolution and the altitude interval is shown in the same color as the representative trajectory. The brown line corresponds to the ground level for the trajectories more associated to the dust advection (thick black lines). The location of the air parcels around the time of observation (12:00 UTC) of the dust plumes is shown as a red circle.

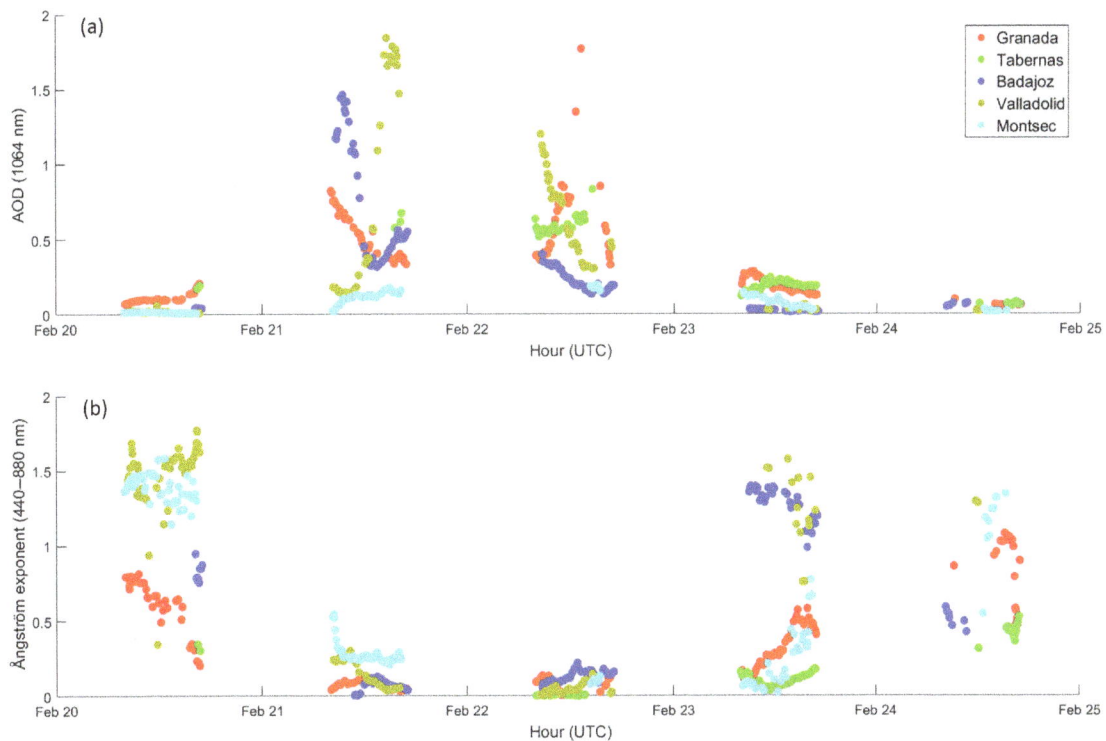

**Figure 6.** Sun-photometer time series representing the AOD at 1064 nm (**a**) and Ångström exponent between 440 and 800 nm (**b**) for all sites in ICENET during the dust event.

pressure system, which was the dominating feature in the second half of the episode. In this second case, dust was advected mostly below the cloud band and affected the eastern part of the Iberian Peninsula as well as most of the western Mediterranean Basin.

The low-pressure system weakened on early 22 February and the region was increasingly dominated by the Azores and the North African highs. As a consequence, zonal flow swept the first dust plume along northern Spain from west to east in subsequent days. The second dust plume, which was moving northward along eastern Iberia, was also displaced to the Mediterranean. The study region was then under high pressures and the event ended.

## 4.3   Ceilometer data analysis

The vertical structure of the dust event described above has been monitored with ICENET: firstly, by obtaining the total attenuated backscatter profiles using the calibration factors from Table (2) and, secondly, by applying the inversion and obtaining particle backscatter profiles. In addition, the integral of the backscatter profiles multiplied by the lidar ratio is used to estimate the AOD during the event and the center of mass of the backscatter profiles is considered as an indicator of the presence of a decoupled aerosol layer (a dust plume in this case) or the entrainment of the aerosol layer into the boundary layer. All these products were calculated in near-real time and serve as an example of the promising capabilities for real-time characterization of singular events with a network of distributed ceilometers.

Figure 7 shows time series of total attenuated backscatter profiles, i.e., calibrated profiles, for the five ceilometers used in this study. From top to bottom the series correspond to Granada, Tabernas, Badajoz, Valladolid and Montsec stations, respectively. Tabernas station is covered by clouds during most of the event and Montsec station is also affected by clouds during part of the event.

Dust arrives first at the stations in Granada and Tabernas (on 20 February at 18:00 UTC). As the dust plume moves northwestward we observe the dust plume in Badajoz (on 21 February at 00:00 UTC) and Valladolid (on 21 February at 06:00 UTC). At Montsec, the dust plume is detected on 21 February at 12:00 UTC. The second plume brings the cloud band and this is visible at Tabernas station around 12:00 UTC on 21 February and a bit later at 21:00 UTC on 21 February at Montsec station. Finally, the displacement of the dust from west to east at the end of the event, when the cut-off low weakens, appears as a dust plume at Valladolid on 22 February at 15:00 UTC, at Badajoz station on 22 February at 21:00 UTC, and at Granada station on 23 February at 06:00 UTC. Tabernas and Montsec are influenced by the second dust plume and the cloud band, and this is not as clearly visible as at the other stations. Another feature that is observed in Fig. 7 is that the dust plumes, especially the first one, are entrained into the boundary layer rapidly.

After applying the inversion, a quantitative comparison of stations is possible, as shown in Fig. 8, for different stages of the dust outbreak (the specific times are shown on the vertical lines in Fig. 7). The beginning of the outbreak, when the first plume arrives at the different stations, is shown in Fig. 8a. The center of mass of the dust plumes is about 3000 m a.s.l. for all stations. Additionally, for Granada and Badajoz, we observe that the height of the peak in particle backscatter coefficient is in accordance with the backward trajectory analysis shown in Sect. 4.1. The arrival of the second plume is shown in Fig. 8b for all sites on 22 February at 06:00 UTC. At this stage, we observe that Granada and Tabernas stations (which are only 100 km apart) show very different behavior in particle backscatter and also in the height of the dust plume. Finally, Fig. 8c shows the final part of the outbreak when dust is mobilized eastwards to the Mediterranean Sea. In this case, dust is below 2000 m a.s.l. for Granada and Tabernas, whereas for the rest of the stations it is still observed above 3000 m a.s.l. In general, the particle backscatter coefficient profiles indicate a stronger intensity of the event at this stage of the event, after the second plume arrives, especially for Granada and Tabernas.

For the entire dust outbreak period and all stations the integral of the backscatter profiles is shown in Fig. 9a. This parameter allows identifying the beginning of the dust event for each station. Thus, an increase is observed in the integral of the backscatter in Granada around 20 February at 19:30 UTC, at Badajoz it is detected around 21 February at 05:30 UTC, and at Valladolid and Montsec it is observed at 16:30 UTC and 17:00 UTC, respectively. Due to clouds, this increase in the integral of the backscatter is not observed in Tabernas. The influence of the dust load after the first plume masks the arrival of the second plume, but the dust mobilization towards the Mediterranean sea is observed again at Badajoz (around 22 February at 20:00 UTC) and in Montsec at 23:00 UTC. The change in the integral of backscatter to larger values is coincident with the starting time observed in the total attenuated backscatter temporal series, and it is in accordance with the satellite observations and backtrajectory analysis. Additionally, the center of mass of the particle backscatter coefficient profiles is used to monitor the evolution of the profile region with more predominance of aerosol particles. Thus, in Fig. 9b for Granada before the event, the center of mass is about 1500 m a.s.l., and when the dust arrives the center of mass is elevated to 2500 m. After 9 h the center of mass is about the same as before the event, indicating that, possibly, the dust plume is no longer decoupled, and it is entrained into the boundary layer. A similar behavior is observed for Badajoz, Valladolid and Montsec stations. Again, the second plume is not observed in changes in the center of mass, but the mobilization of dust towards the Mediterranean Sea is observed as an increase in the center of mass of the profiles for Badajoz, Valladolid and Tabernas.

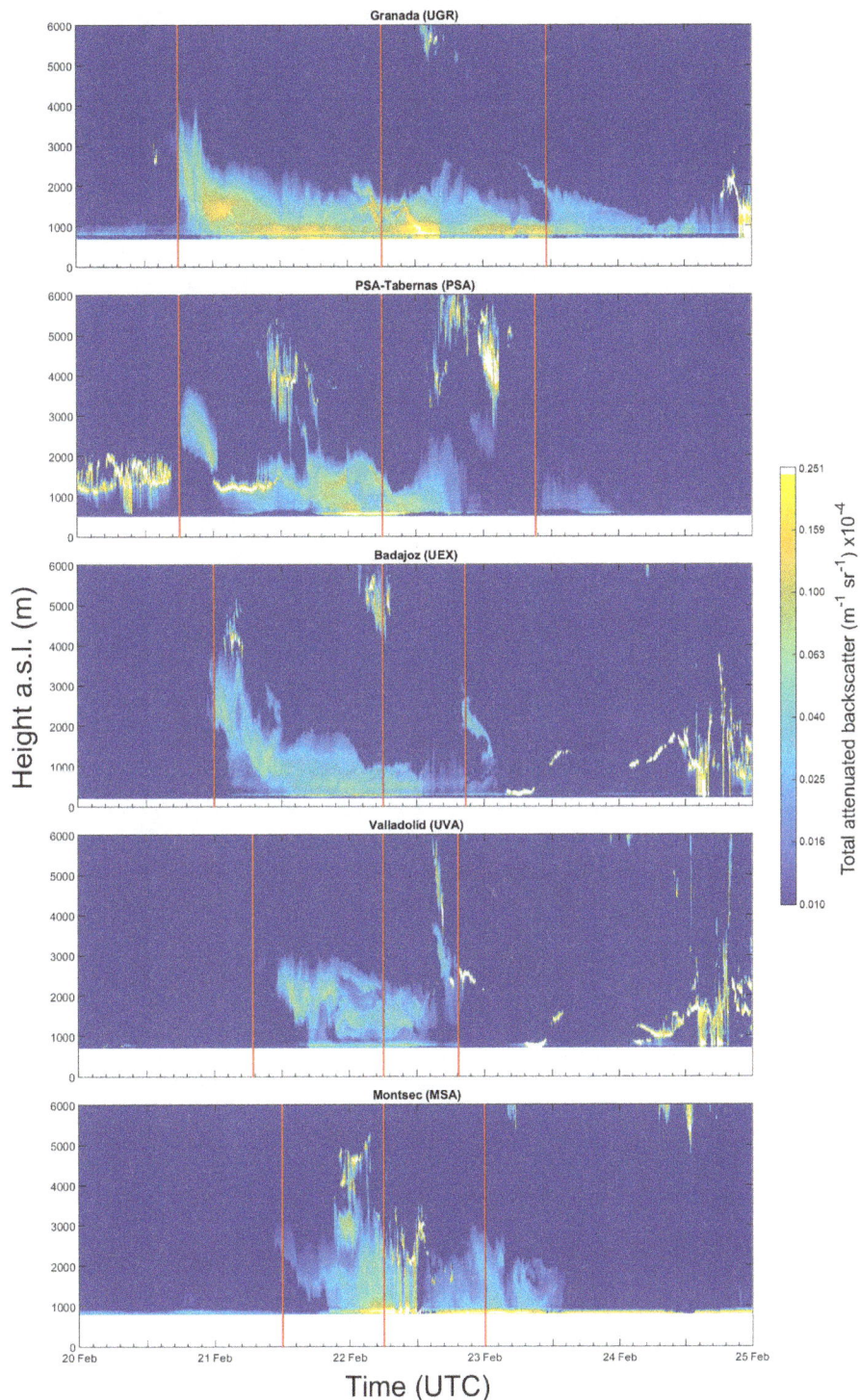

**Figure 7.** Ceilometer time series of total attenuated backscatter representing the evolution of the dust outbreak between 20 and 24 February 2016 (the color scale is logarithmic). Red vertical lines indicate the time of the profiles in Fig. 8: the first line of each site indicates the times for Fig. 8a, second line those for Fig. 8b, and third line those for Fig. 8c.

## 5   Conclusions

The use of ceilometers for the characterization of optical aerosol properties is possible but, due to the weak signal, it is important to screen out profiles in order to ensure the quality of the inversion. In addition, due to the vast number of data, it is important to perform all these operations in an automated, unsupervised way and, preferably, in near-real time.

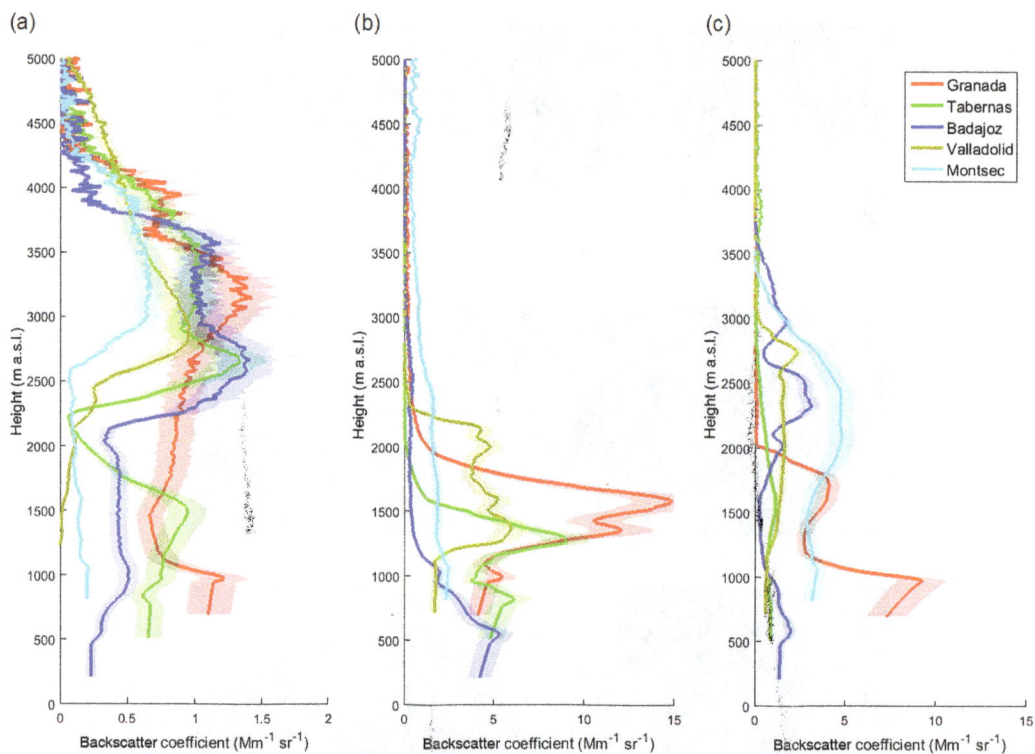

**Figure 8.** Particle backscatter coefficient profiles for all stations at the beginning (**a**), middle (**b**) and final stage (**c**) of the outbreak (note that the *x* axis has a different scale and the profiles start at ground level). The shaded areas represent the 15 % uncertainty.

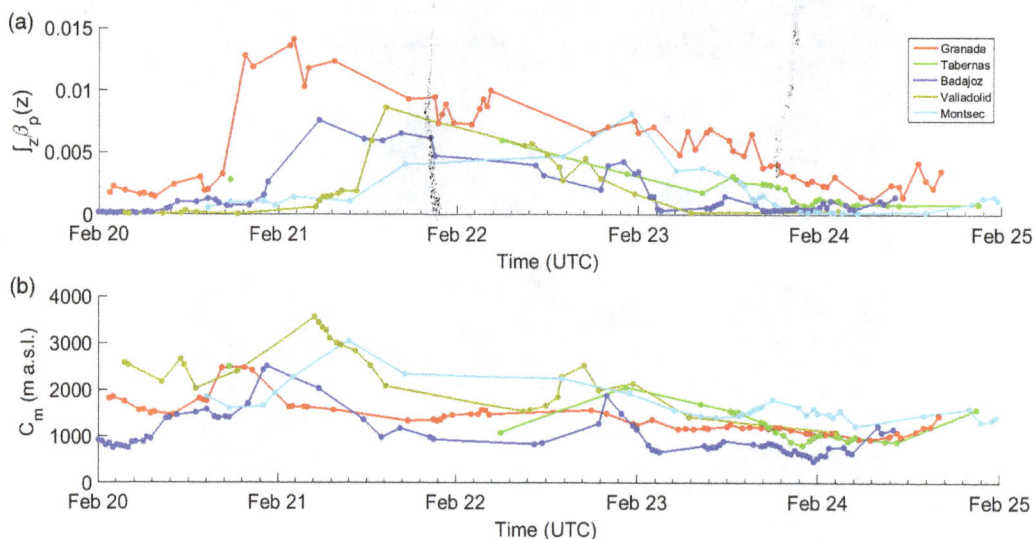

**Figure 9.** Time series of the integral of the particle backscatter coefficient for all stations (**a**) and time series of the center of mass of the backscatter profiles for all stations (**b**).

The methodology proposed uses ancillary data from sun photometers in order to constrain the calibration of the ceilometers. The time series of this calibration is used to determine the quality of the inversions, selecting those that present, at the reference height, a ratio of the backscattering signal to molecular attenuated backscatter within the mean calibration factor ± the standard deviation. A comparison with independent lidar measurements indicates that this method allows the automatic discrimination of the quality of the inversions with ceilometers. During this comparison a difference smaller than 15 % in backscatter coefficient is observed.

Thus, it is feasible to routinely provide particle backscatter coefficient profiles with ceilometers.

The inverted profiles obtained with ceilometers could be used for elevated aerosol layer alert by setting a threshold on the particle backscatter coefficient values of the profile and are potentially useful for model assimilation and evaluation since all the processing is automated and in near-real time.

This method has been applied to a group of ceilometers (ICENET) and tested during a dust outbreak reaching Spain on 20 February 2016 and lasting until 24 February 2016. This dust event affected all ICENET stations, with two distinct plumes reaching the Iberian Peninsula following different paths and a final stage where zonal flows swept the dust towards the Mediterranean Sea. This scheme of dust mobilization is unusual for this season of the year, and the intensity, spatial coverage and duration of the event make it perfect as a test for monitoring purposes with the ceilometer network. The calibration of the ceilometers allows a qualitative monitoring of the event, while the inversions provide quantitative information. Thus, ceilometers can complement lidar stations that, in principle, would operate intermittently and with less spatial density. It is worth noting that differences have been observed on profiles 100 km apart. This reinforces the need for providing vertical profiles of aerosol optical properties with a dense spatial resolution.

Parameters extracted from the particle backscatter coefficient profiles such as the integral or the center of mass can also give a quantitative idea of the presence of an elevated aerosol layer. These parameters are expected to increase with an elevated aerosol layer, and the second one can be used as a rough indicator for the deposition velocity of an elevated aerosol layer by comparing a time series of these values.

*Competing interests.* The authors declare that they have no conflict of interest.

*Acknowledgements.* This work was supported by the Spanish Ministry of Economy and Competitiveness through projects CGL2012-39623-C02-01, CGL2013-45410-R, CGL2014-56255-C2-1-R, CMT2015-66742-R, CGL2015-70741-R, CGL2015-73250-JIN and CGL2016-81092-R; by the regional government of Andalusia through project P12-RNM-2409; by the regional government of Castilla y Leon through project VA100U14; by the Junta de Extremadura (Ayuda a Grupos de Investigación GR15137); and by the European Union's Horizon 2020 research and innovation program through project ACTRIS-2 (grant agreement no. 654109). The authors thankfully acknowledge the FEDER program for the instrumentation used in this work. This work was also partially funded by the University of Granada through the contract "Plan Propio. Programa 9. Convocatoria 2013". Marco Pandolfi is funded by a Ramón y Cajal Fellowship (RYC-2013-14036) awarded by the Spanish Ministry of Economy and Competitiveness. The authors would like to acknowledge the valuable contribution through the discussions in the working group meetings organized by COST Action ES1303 (TOPROF).

Edited by: Bernhard Vogel

## References

Alados-Arboledas, L., Müller, D., Guerrero-Rascado, J. L., Navas-Guzmán, F., Pérez-Ramírez, D., and Olmo, F. J.: Optical and microphysical properties of fresh biomass burning aerosol retrieved by Raman lidar, and star-and sun-photometry, Geophys. Res. Lett., 38, L01807, https://doi.org/10.1029/2010GL045999, 2011.

Ångström, A.: The parameters of atmospheric turbidity, Tellus, 16, 64–75, 1964.

Boucher, O., Randall, D., Artaxo, P., Bretherton, C., Feingold, G., Forster, P., Kerminen, V.-M., Kondo, Y., Liao, H., Lohmann, U., Rasch, P., Satheesh, S. K., Sherwood, S., Stevens B., and Zhang, X. Y.: Clouds and Aerosols, in: Climate Change 2013: The Physical Science Basis. Contribution of Working Group I to the Fifth Assessment Report of the Intergovernmental Panel on Climate Change, edited by: Stocker, T. F., Qin, D., Plattner, G.-K., Tignor, M., Allen, S. K., Boschung, J., Nauels, A., Xia, Y., Bex, V., and Midgley, P. M., Cambridge University Press, Cambridge, UK and New York, NY, USA, 2013.

Bravo-Aranda, J., Titos, G., Granados-Muñoz, M., Guerrero-Rascado, J., Navas-Guzmán, F., Valenzuela, A., Lyamani, H., Olmo, F., Andrey, J., and Alados-Arboledas, L.: Study of mineral dust entrainment in the planetary boundary layer by lidar depolarization technique, Tellus B, 67, 1–13, https://doi.org/10.3402/tellusb.v67.26180, 2015.

Córdoba-Jabonero, C., Sorribas, M., Guerrero-Rascado, J. L., Adame, J. A., Hernández, Y., Lyamani, H., Cachorro, V., Gil, M., Alados-Arboledas, L., Cuevas, E., and de la Morena, B.: Synergetic monitoring of Saharan dust plumes and potential impact on surface: a case study of dust transport from Canary Islands to Iberian Peninsula, Atmos. Chem. Phys., 11, 3067–3091, https://doi.org/10.5194/acp-11-3067-2011, 2011.

de Haij, M., Wauben, W., and Klein-Baltink, H.: Continuous mixing layer height determination using the LD-40 ceilometer: a feasibility study. KNMI Scientific Report WR 2007-01, 98 pp., 2007.

Draxler, R. R. and Hess, G. D.: An Overview of the HYSPLIT_4 Modelling System for Trajectories, Dispersion, and Deposition, Aust. Meteorol. Mag., 47, 295–308, 1998.

Dubovik, O., Sinyuk, A., Lapyonok, T., Holben, B.N., Mishchenko, M., Yang, P., Eck, T.F., Volten, H., Muñoz, O., Veihelmann, B., van der Zande, W.J., Leon, J.-F., Sorokin, M., and Slutsker, I.: Application of spheroid models to account for aerosol particle nonsphericity in remote sensing of desert dust, J. Geophys. Res., 111, D11208, https://doi.org/10.1029/2005JD006619, 2006.

Escudero, M., Castillo, S., Querol, X., Avila, A., Alarcón, M., Viana, M. M., Alastuey, A., Cuevas, E., and Rodríguez, S.: Wet and dry African dust episodes over eastern Spain, J. Geophys. Res.-Atmos., 110, D18S08, https://doi.org/10.1029/2004JD004731, 2005.

Fernald, F. G.: Analysis of atmospheric lidar observations: some comments, Appl. Optics, 23, 652–653, 1984.

Fernald, F. G., Herman, B. M., and Reagan, J. A.: Determination of Aerosol Height Distributions by Lidar, J. Appl. Meteorol., 11, 482–489, 1972.

Flentje, H., Heese, B., Reichardt, J., and Thomas, W.: Aerosol profiling using the ceilometer network of the German

Meteorological Service, Atmos. Meas. Tech. Discuss., https://doi.org/10.5194/amtd-3-3643-2010, 2010.

Frey, S., Poenitz, K., Teschke, G., and Wille, H.: Detection of aerosol layers with ceilometers and the recognition of the mixed layer depth, ISARS, 3646/3647, 1–4, 2010.

GAW report 178: Plan for the implementation of the GAW Aerosol Lidar Observation Network GALION (Hamburg, Germany, 27–29 March 2007) (WMO TD No. 1443), 52 pp., 2008.

GAW report 197: World Meteorological Organization Global Atmospheric Watch, Addendum for the period 2012–2015 to the WMO Global Atmospheric Watch (GAW) Strategic Plan 2008–2015, 63 pp., 2011.

Granados-Muñoz, M. J., Bravo-Aranda, J. A., Baumgardner, D., Guerrero-Rascado, J. L., Pérez-Ramírez, D., Navas-Guzmán, F., Veselovskii, I., Lyamani, H., Valenzuela, A., Olmo, F. J., Titos, G., Andrey, J., Chaikovsky, A., Dubovik, O., Gil-Ojeda, M., and Alados-Arboledas, L.: A comparative study of aerosol microphysical properties retrieved from ground-based remote sensing and aircraft in situ measurements during a Saharan dust event, Atmos. Meas. Tech., 9, 1113–1133, https://doi.org/10.5194/amt-9-1113-2016, 2016.

Guerrero-Rascado, J. L., Ruiz, B., and Alados-Arboledas, L.: Multispectral Lidar characterization of the vertical structure of Saharan dust aerosol over southern Spain, Atmos. Environ., 42, 2668–2681, 2008.

Guerrero-Rascado, J. L., Olmo, F. J., Avilés-Rodríguez, I., Navas-Guzmán, F., Pérez-Ramírez, D., Lyamani, H., and Alados Arboledas, L.: Extreme Saharan dust event over the southern Iberian Peninsula in september 2007: active and passive remote sensing from surface and satellite, Atmos. Chem. Phys., 9, 8453–8469, https://doi.org/10.5194/acp-9-8453-2009, 2009.

Guerrero-Rascado, J. L., Landulfo, E., Antuña, J. C., Barbosa, H. M. J., Barja, B., Bastidas, A. E., Bedoya, A. E., da Costa, R., Estevan, R., Forno, R. N., Gouveia, D. A., Jimenez, C., Larroza, E. G., Lopes, F. J. S., Montilla-Rosero, E., Moreira, G. A., Nakaema, W. M., Nisperuza, D., Alegria, D., Manera, M., Otero, L., Papandrea, S., Pallota, J. V., Pawelko, E., Quel, E. J., Ristori, P., Rodrigues, P. F., Salvador, J., Sanchez, M., and Silva, A.: Latin American Lidar Network (LALINET): diagnosis on network instrumentation, J. Atmos. Sol.-Terr. Phys., 138/139, 112–120, https://doi.org/10.1016/j.jastp.2016.01.001, 2016.

Haeffelin, M., Angelini, F., Morille, Y., Martucci, G., Frey, S., Gobbi, G. P., Lolli, S., O'Dowd, C. D., Sauvage, L., Xueref-Rémy, I., Wastine, B., and Feist, D. G.: Evaluation of Mixing-Height Retrievals from Automatic Profiling Lidars and Ceilometers in View of Future Integrated Networks in Europe, Bound.-Lay. Meteorol., 143, 49, 49–75, https://doi.org/10.1007/s10546-011-9643-z, 2012.

Haeffelin, M., Laffineur, Q., Bravo-Aranda, J.-A., Drouin, M.-A., Casquero-Vera, J.-A., Dupont, J.-C., and De Backer, H.: Radiation fog formation alerts using attenuated backscatter power from automatic lidars and ceilometers, Atmos. Meas. Tech., 9, 5347–5365, https://doi.org/10.5194/amt-9-5347-2016, 2016.

Heese, B., Flentje, H., Althausen, D., Ansmann, A., and Frey, S.: Ceilometer lidar comparison: backscatter coefficient retrieval and signal-to-noise ratio determination, Atmos. Meas. Tech., 3, 1763–1770, https://doi.org/10.5194/amt-3-1763-2010, 2010.

Holben, B. N., Eck, T. F., Slutsker, I., Tanre, D., Buis, J. P., Setzer, A., Vermote, E., Reagan, J. A., Kaufman, Y. J., Nakajima, T., Lavenu, F., Jankowiak, I., and Smirnov, A.: AERONET – A federated instrument network and data archive for aerosol characterization, Remote Sens. Env., 66, 1–16, 1998.

IPCC, 2014: Climate Change 2014: Synthesis Report, Contribution of Working Groups I, II and III to the Fifth Assessment Report of the Intergovernmental Panel on Climate Change, edited by: Core Writing Team, Pachauri, R. K., and Meyer, L. A., IPCC, Geneva, Switzerland, 151 pp. 2014.

Klett, J. D.: Stable analytic inversion solution for processing lidar returns, Appl. Optics, 20, 211–220, 1981.

Klett, J. D.: Lidar inversion with variable backscatter/extinction ratios, Appl. Optics, 24, 1638–1643, 1985.

Knippertz, P.: Tropical-extratropical interactions causing precipitation in Northwest Africa: Statistical analysis and seasonal variations, Mon. Weather Rev., 131, 3069–3076, https://doi.org/10.1175/1520-0493(2003)131<3069:TICPIN>2.0.CO;2, 2003.

Lensky, I. M. and Rosenfeld, D.: Clouds-Aerosols-Precipitation Satellite Analysis Tool (CAPSAT), Atmos. Chem. Phys., 8, 6739–6753, https://doi.org/10.5194/acp-8-6739-2008, 2008.

Mattis, I., D'Amico, G., Baars, H., Amodeo, A., Madonna, F., and Iarlori, M.: EARLINET Single Calculus Chain – technical – Part 2: Calculation of optical products, Atmos. Meas. Tech., 9, 3009–3029, https://doi.org/10.5194/amt-9-3009-2016, 2016.

McCormick, M. P.: The Flight of the Lidar In-Space Technology Experiment (LITE), in: Advances in Atmospheric Remote Sensing with Lidaredited by: Ansmann, A., Neuber, R., Rairoux, P., and Wandinger, U., Springer, Berlin, Heidelberg, 1997.

Münkel, C., Eresmaa, N., Räsänen, J., and Karppinene, A.: Retrieval of mixing height and dust concentration with lidar ceilometer, Bound.-Lay. Meteorol., 124, 117–128, 2007.

Navas-Guzmán, F., Guerrero-Rascado, J. L., and Alados-Arboledas, L.: Retrieval of the lidar overlap function using Raman signals, Óptica Pura y Aplicada, 44, 71–75, 2011.

Navas-Guzmán, F., Müller, D., Bravo-Aranda, J. A., Guerrero-Rascado, J. L., Granados-Muñoz, M. J., Pérez-Ramírez, D., Olmo, F. J. and Alados-Arboledas, L.: Eruption of the Eyjafjallajökull Volcano in spring 2010: Multiwavelength Raman lidar measurements of sulphate particles in the lower troposphere, J. Geophys. Res.-Atmos., 118, 1804–1813, https://doi.org/10.1002/jgrd.50116, 2013.

Ortiz-Amezcua, P., Guerrero-Rascado, J. L., Granados-Muñoz, M. J., Benavent-Oltra, J. A., Böckmann, C., Samaras, S., Stachlewska, I. S., Janicka, L., Baars, H., Bohlmann, S., and Alados-Arboledas, L.: Microphysical characterization of long-range transported biomass burning particles from North America at three EARLINET stations, Atmos. Chem. Phys., 17, 5931–5946, https://doi.org/10.5194/acp-17-5931-2017, 2017.

Pandolfi, M., Martucci, G., Querol, X., Alastuey, A., Wilsenack, F., Frey, S., O'Dowd, C. D., and Dall'Osto, M.: Continuous atmospheric boundary layer observations in the coastal urban area of Barcelona during SAPUSS, Atmos. Chem. Phys., 13, 4983–4996, https://doi.org/10.5194/acp-13-4983-2013, 2013

Pappalardo, G., Amodeo, A., Pandolfi, M., Wandinger, U., Ansmann, A., Bösenberg, J., Matthias, V., Amiridis, V., De Tomasi, F., Frioud, M., Larlori, M., Komguem, L., Papayannis, A., Rocadenbosch, F., and Wang X.: Aerosol lidar intercomparison in the framework of the EARLINET project, 3. Raman lidar algorithm for aerosol extinction, backscatter, and lidar ratio, Appl. Optics, 43, 5370–5385, https://doi.org/10.1364/AO.43.005370, 2004.

Pappalardo, G., Mona, L., D'Amico, G., Wandinger, U., Adam, M., Amodeo, A., Ansmann, A., Apituley, A., Alados Arboledas, L., Balis, D., Boselli, A., Bravo-Aranda, J. A., Chaikovsky, A., Comeron, A., Cuesta, J., De Tomasi, F., Freudenthaler, V., Gausa, M., Giannakaki, E., Giehl, H., Giunta, A., Grigorov, I., Groß, S., Haeffelin, M., Hiebsch, A., Iarlori, M., Lange, D., Linné, H., Madonna, F., Mattis, I., Mamouri, R.-E., McAuliffe, M. A. P., Mitev, V., Molero, F., Navas-Guzman, F., Nicolae, D., Papayannis, A., Perrone, M. R., Pietras, C., Pietruczuk, A., Pisani, G., Preißler, J., Pujadas, M., Rizi, V., Ruth, A. A., Schmidt, J., Schnell, F., Seifert, P., Serikov, I., Sicard, M., Simeonov, V., Spinelli, N., Stebel, K., Tesche, M., Trickl, T., Wang, X., Wagner, F., Wiegner, M., and Wilson, K. M.: Four-dimensional distribution of the 2010 Eyjafjallajökull volcanic cloud over Europe observed by EARLINET, Atmos. Chem. Phys., 13, 4429–4450, https://doi.org/10.5194/acp-13-4429-2013, 2013.

Pappalardo, G., Amodeo, A., Apituley, A., Comeron, A., Freudenthaler, V., Linné, H., Ansmann, A., Bösenberg, J., D'Amico, G., Mattis, I., Mona, L., Wandinger, U., Amiridis, V., Alados-Arboledas, L., Nicolae, D., and Wiegner, M.: EARLINET: towards an advanced sustainable European aerosol lidar network, Atmos. Meas. Tech., 7, 2389–2409, https://doi.org/10.5194/amt-7-2389-2014, 2014.

Preißler, J., Wagner, F., Pereira, S. N., and Guerrero-Rascado, J. L.: Multi-instrumental observation of an exceptionally strong Saharan dust outbreak over Portugal, J. Geophys. Res., 116, D24204, https://doi.org/10.1029/2011JD016527, 2011.

Preißler, J., Wagner, F., Guerrero-Rascado, J. L., and Silva, A. M.: Two years of free-tropospheric aerosol layers observed over Portugal by lidar, J. Geophys. Res.-Atmos., 118, 3676–3686, https://doi.org/10.1002/jgrd.50350, 2013.

Rodríguez, S., Querol, X., Alastuey, A. Viana, M. M., and Mantilla, E.: Events Affecting Levels and Seasonal Evolution of Airborne Particulate Matter Concentrations in the Western Mediterranean, Environ. Sci. Technol., 37, 216–222, https://doi.org/10.1021/es020106p, 2003.

Sicard, M., Guerrero-Rascado, J. L., Navas-Guzmán, F., Preißler, J., Molero, F., Tomás, S., Bravo-Aranda, J. A., Comerón, A., Rocadenbosch, F., Wagner, F., Pujadas, M., and Alados-Arboledas, L.: Monitoring of the Eyjafjallajökull volcanic aerosol plume over the Iberian Peninsula by means of four EARLINET lidar stations, Atmos. Chem. Phys., 12, 3115–3130, https://doi.org/10.5194/acp-12-3115-2012, 2012.

Sorribas, M., Adame, J. A., Andrews, E., and Yela, Y.: An anomalous African dust event and its impact on aerosol radiative forcing on the Southwest Atlantic coast of Europe in February 2016, Sci. Total Environ., 583, 269–279, https://doi.org/10.1016/j.scitotenv.2017.01.064, 2017.

Stein, A. F., Draxler, R. R., Rolph, G. D., Stunder, B. J. B., Cohen, M. D., and Ngan, F.: NOAA's HYSPLIT Atmospheric Transport and Dispersion Modeling System, Bull. Am. Meteor. Soc., 96, 2059–2077, https://doi.org/10.1175/BAMS-D-14-00110.1, 2015.

Titos, G., Ealo, M., Pandolfi, M., Pérez, N., Sola, Y., Sicard, M., Comerón, A., Querol, X., and Alastuey, A.: Spatiotemporal evolution of a severe winter dust event in the western Mediterranean: Aerosol optical and physical properties, J. Geophys. Res.-Atmos., 122, 4052–4069, https://doi.org/10.1002/2016JD026252, 2017.

Valenzuela, A., Olmo, F.J., Lyamani, H., Antón, M., Quirantes, A., and Alados-Arboledas, L.: Classification of aerosol radiative properties during African desert dust intrusions over southeastern Spain by sector origins and cluster analysis, J. Geophys. Res., 117, D06214, https://doi.org/10.1029/2011JD016885, 2012.

Valenzuela, A., Costa, M. J., Guerrero-Rascado, J. L., Bortoli, D., and Olmo, F. J.: Solar and thermal radiative effects during the 2011 extreme desert dust episode over Portugal, Atmos. Environ., 148, 16–29, 2017.

Wandinger, U., Müller, D., Böckmann, C., Althausen, D., Matthias, V., Bösenberg, J., Weiß, V., Fiebig, M., Wendisch, M., Stohl, A., and Ansmann, A.: Optical and microphysical characterization of biomass-burning and industrial-pollution aerosols from multiwavelength lidar and aircraft measurements, J. Geophys. Res., 107, 1–20, https://doi.org/10.1029/2000JD000202,2002

Welton, E. J., Campbell, J. R., Spinhirne, J. D., and Scott, V. S.: Global monitoring of clouds and aerosols using a network of micro-pulse lidar systems, in: Lidar Remote Sensing for Industry and Environmental Monitoring, edited by: Singh, U. N., Itabe, T., and Sugimoto, N., Proc. SPIE, 4153, 151–158, 2001.

Wiegner, M. and Geiß, A.: Aerosol profiling with the Jenoptik ceilometer CHM15kx, Atmos. Meas. Tech., 5, 1953–1964, https://doi.org/10.5194/amt-5-1953-2012, 2012.

Wiegner, M., Emeis, S., Freudenthaler, V., Heese, B., Junkermann, W., Munkel, C., Schafer, K., Seefeldner, M., and Vogt, S.: Mixing layer height over Munich, Germany: Variability and comparisons of different methodologies, J. Geophys. Res., 111, D13201, https://doi.org/10.1029/2005JD006593, 2006.

Wiegner, M., Madonna, F., Binietoglou, I., Forkel, R., Gasteiger, J., Geiß, A., Pappalardo, G., Schäfer, K., and Thomas, W.: What is the benefit of ceilometers for aerosol remote sensing? An answer from EARLINET, Atmos. Meas. Tech., 7, 1979–1997, https://doi.org/10.5194/amt-7-1979-2014, 2014.

Winker, D. M., Pelon, J. R., and McCormick, M. P.: CALIPSO mission: spaceborne lidar for observation of aerosols and clouds, Proc. SPIE 4893, Lidar Remote Sensing for Industry and Environment Monitoring III, (21 March 2003), 4893, 1–11, https://doi.org/10.1117/12.466539, 2003.

# Impacts of stratospheric sulfate geoengineering on tropospheric ozone

**Lili Xia[1], Peer J. Nowack[2,a], Simone Tilmes[3], and Alan Robock[1]**

[1]Department of Environmental Sciences, Rutgers University, New Brunswick, New Jersey, USA
[2]Department of Chemistry, Centre for Atmospheric Science, University of Cambridge, Cambridge, UK
[3]Atmospheric Chemistry Observations and Modeling Laboratory, National Center for Atmospheric Research, Boulder, Colorado, USA
[a]now at: Grantham Institute and Department of Physics, Faculty of Natural Sciences, Imperial College London, London, UK

*Correspondence to:* Lili Xia (lxia@envsci.rutgers.edu)

**Abstract.** A range of solar radiation management (SRM) techniques has been proposed to counter anthropogenic climate change. Here, we examine the potential effects of stratospheric sulfate aerosols and solar insolation reduction on tropospheric ozone and ozone at Earth's surface. Ozone is a key air pollutant, which can produce respiratory diseases and crop damage. Using a version of the Community Earth System Model from the National Center for Atmospheric Research that includes comprehensive tropospheric and stratospheric chemistry, we model both stratospheric sulfur injection and solar irradiance reduction schemes, with the aim of achieving equal levels of surface cooling relative to the Representative Concentration Pathway 6.0 scenario. This allows us to compare the impacts of sulfate aerosols and solar dimming on atmospheric ozone concentrations. Despite nearly identical global mean surface temperatures for the two SRM approaches, solar insolation reduction increases global average surface ozone concentrations, while sulfate injection decreases it. A fundamental difference between the two geoengineering schemes is the importance of heterogeneous reactions in the photochemical ozone balance with larger stratospheric sulfate abundance, resulting in increased ozone depletion in mid- and high latitudes. This reduces the net transport of stratospheric ozone into the troposphere and thus is a key driver of the overall decrease in surface ozone. At the same time, the change in stratospheric ozone alters the tropospheric photochemical environment due to enhanced ultraviolet radiation. A shared factor among both SRM scenarios is decreased chemical ozone loss due to reduced tropospheric humidity. Under insolation reduction, this is the dominant factor giving rise to the global surface ozone increase. Regionally, both surface ozone increases and decreases are found for both scenarios; that is, SRM would affect regions of the world differently in terms of air pollution. In conclusion, surface ozone and tropospheric chemistry would likely be affected by SRM, but the overall effect is strongly dependent on the SRM scheme. Due to the health and economic impacts of surface ozone, all these impacts should be taken into account in evaluations of possible consequences of SRM.

## 1 Introduction

### 1.1 Atmospheric ozone

It is well known that sulfate aerosols in the stratosphere enhance heterogeneous chemical reactions that lead to enhanced ozone depletion after larger volcanic eruptions (Solomon, 1999). With present-day anthropogenic halogen loading, the aerosols provide additional surface area for heterogeneous reactions that activate halogens and hence increase catalytic ozone destruction, especially in high latitudes (Tie and Brasseur, 1995). This has been modeled and observed following the large 1982 El Chichón and 1991 Pinatubo volcanic eruptions (Tie and Brasseur, 1995; Portman et al., 1996).

However, volcanic eruptions do not only affect stratospheric ozone but also impact tropospheric composition, often due to stratosphere–troposphere coupled effects. The 1991 Pinatubo eruption, for example, has been linked to changes in the stratosphere–troposphere exchange (STE) of ozone (Aquila et al., 2012, 2013; Pitari et al., 2016). In addition, the stratospheric ozone decrease led to an invigorated photochemical environment in the troposphere due to enhanced downward chemically active ultraviolet (UV) radiation (Tang et al., 2013).

This study focuses on tropospheric ozone, in particular surface ozone concentration changes. Surface ozone is of central importance to Earth's environment, and as an air pollutant it adversely impacts human health (e.g., Kampa and Castanas, 2008) and the ecosystem (e.g., Mauzeral and Wang, 2001; Ashmore, 2005; Ainsworth et al., 2012). There have been numerous studies of the observed surface ozone trend (e.g., Cooper et al., 2014), identifying ozone sources and sinks (e.g., Wild, 2007), predicting future changes (e.g., Young et al., 2013), and understanding the impacts of such changes (e.g., Silva et al., 2013). Global surface ozone concentrations are estimated to have doubled since the preindustrial period (Vingarzan, 2004), mainly due to increased emissions of ozone precursors associated with industrialization (e.g., Forster et al., 2007). Differences in future tropospheric ozone concentrations will be strongly dependent on the emission pathway followed (Stevenson et al., 2006), which will determine both in situ tropospheric chemical production of ozone and transport from the ozone-rich stratosphere (Collins et al., 2003; Wild et al., 2012; Neu et al., 2014).

## 1.2 Differences between sulfate and solar geoengineering

The progression of global warming, slow mitigation efforts, and our relatively limited adaptive capacity force consideration of solar radiation management (SRM) geoengineering as one possible strategy to avoid many of the most undesirable consequences of global warming (Crutzen, 2006; Wigley, 2006; Tilmes et al., 2016a). The above discussed factors controlling tropospheric ozone concentrations could be affected by SRM schemes (Nowack et al., 2016). Here we compare a proposed geoengineering scheme, stratospheric sulfur injection, to solar irradiance reduction. Both schemes would cool Earth's surface by reducing sunlight reaching the surface, either by aerosols reflecting sunlight or by artificially reducing the solar constant in a climate model, but sulfate geoengineering would strongly heat the stratosphere and provide aerosol surfaces for chemical reactions. Previous studies have shown that injected sulfur chemically forms sulfate aerosols within a couple of weeks. The aerosol layer absorbs near infrared solar radiation as well as outgoing longwave radiation and results in stratospheric warming (e.g., Tilmes et al., 2009; Ammann et al., 2010; Jones et al., 2011). Additionally, changes in ozone and advection impact the warming in the stratosphere (Richter et al., 2017). Under solar

reduction, the stratosphere would be cooler due to reduced shortwave heating (Govindasamy and Caldeira, 2000), although simultaneous stratospheric ozone changes (if considered) may buffer this effect (Nowack et al., 2016).

One of the most important differences between the two scenarios is that if a permanently enhanced stratospheric aerosol layer is artificially created in an attempt to reduce anthropogenic global warming, the resulting strong ozone depletion, in particular in mid- and high latitudes, would have serious impacts on the biosphere, similar to the effects observed after large volcanic eruptions described above (Crutzen, 2006; Rasch et al., 2008a, b; Tilmes et al., 2008, 2009, 2012). This effect would have to be expected as long as there is anthropogenic halogen in the stratosphere. In the remote future, the decreasing burden of anthropogenic halogen will eventually result in the recovery of the ozone layer. Under such conditions additional stratospheric aerosols could actually have the opposite effect by deactivating ozone-depleting nitrogen oxides, thus leading to an increase in ozone in the stratosphere (Tie and Brasseur, 1995; Pitari et al., 2014). Overall, such changes to the stratosphere would also have important implications for tropospheric composition. Decreasing stratospheric ozone leads to more UV propagating through, with increasing ozone having the opposite effect, which would thus alter the photochemical environment of the troposphere in different ways (Tilmes et al., 2012; Nowack et al., 2016).

In the following sections, we describe the experimental set-up of the two geoengineering schemes and discuss some general climate impacts, followed by a detailed discussion of tropospheric and surface ozone changes. We also show that sulfate and solar geoengineering would impact the stratosphere differently, which implies further key differences in their potential influences on tropospheric composition. In this study, we examine the impacts of stratospheric sulfate geoengineering on tropospheric ozone for the first time.

## 2 Model and experiment design

We simulated both types of SRM schemes using the full tropospheric and stratospheric chemistry version of the Community Earth System Model–Community Atmospheric Model 4 (CESM CAM4-chem) with a horizontal resolution of $0.9° \times 1.25°$ (lat $\times$ long) and 26 levels from the surface to about 40 km (3.5 mb). The model has been shown to give a good representation of present-day atmospheric composition in the troposphere (Tilmes et al., 2016b) and stratosphere at 2° resolution (Fernandez et al., 2017). Similar to the 2° model version, the 1° horizontal resolution version of the model also produces reasonable stratosphere and troposphere ozone chemistry (Figs. S1–S2 in the Supplement). CAM4-chem is fully coupled to the Community Land Model version 4.0 with prescribed satellite phenology (CLM4SP), the Parallel Ocean Program version 2 (POP2) ocean model, and

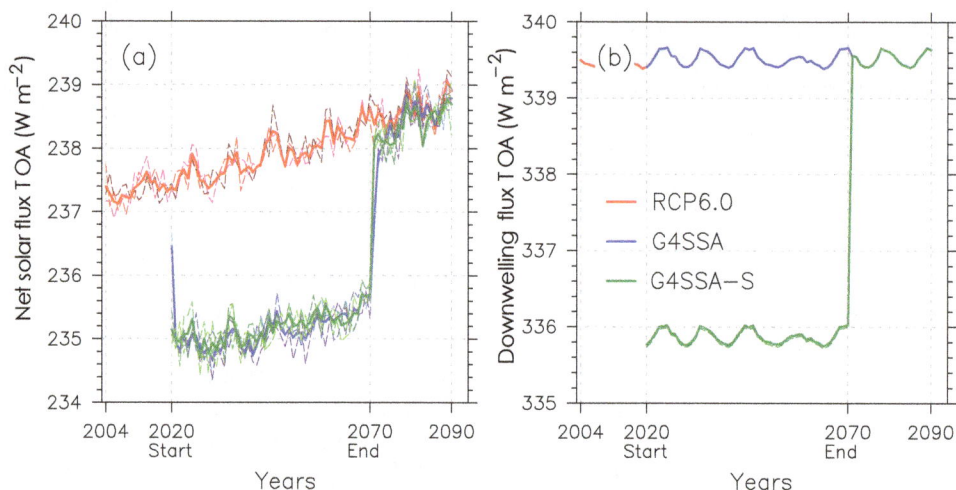

**Figure 1. (a)** Global averaged annual net solar flux at the top of the atmosphere $(W\,m^{-2})$ and **(b)** downwelling solar flux at the top of the atmosphere $(W\,m^{-2})$. Dashed lines are ensemble members, and solid lines are the average of three ensemble members. Geoengineering starts on 1 January 2020 and ends on 1 January 2070. The 11-year periodicity is imposed as a prediction of the sunspot cycle. In **(b)** the G4SSA curve exactly covers the RCP6.0 curve.

the Los Alamos sea ice model (CICE version 4). The tropospheric chemical mechanism in CAM4-chem is based on the Model for Ozone and Related chemical Tracers (MOZART) version 4 (Emmons et al., 2010). The stratospheric chemical mechanism is described in Kinnison et al. (2007), Lamarque et al. (2012), and Tilmes et al. (2015), and the complete chemical reactions included (photolysis, gas-phase chemistry, and heterogeneous chemistry) are listed in Tilmes et al. (2016b, Table A2). Reaction rates are updated following Jet Propulsion Laboratory 2010 recommendations (Sander et al., 2011). The model uses a nudged quasi-biennial oscillation (QBO), which means the QBO will not be modified by the radiative interaction of the aerosols. Interaction between aerosol burden and photolysis rates is not included in the model. Changes in photolysis rates in the troposphere depend on the stratospheric ozone column change (Kinnison et al., 2007). Increased ozone depletion as the result of geoengineering would therefore lead to an increase in UV in mid- and high latitudes. Since our model does not include the aerosol scattering effect on UV, expected UV reductions from the increased sulfate aerosol layer are not taken into account, which might result in an overestimation of the tropospheric photolysis. Volatile organic compound (VOC) emissions are simulated by the Model of Emission of Gases and Aerosols from Nature (MEGAN v2.1) (Guenther et al., 2012). The dynamical ocean model does not include any biogeochemical feedbacks and only the atmospheric and land models are coupled to the atmospheric chemistry component. The model configuration used here, but at 2° resolution, is participating in the current phase of the Chemistry-Climate Model Initiative (Tilmes et al., 2016b; Morgenstern et al., 2017).

We compare three ensemble members each of the two geoengineering scenarios to a three-ensemble reference run with

Representative Concentration Pathway 6.0 (RCP6.0; Meinshausen et al., 2011) anthropogenic forcing from 2020 to 2089. Both geoengineering scenarios include RCP6.0 forcings. Our sulfate aerosol implementation is the G4 Specified Stratospheric Aerosol (G4SSA) experiment (Tilmes et al., 2015), whereas solar reduction geoengineering is the solar analog (hereafter G4SSA-S) by imposing a solar irradiance reduction with the same negative radiative forcing at the top of the atmosphere (TOA) as in G4SSA. G4SSA uses a prescribed stratospheric aerosol surface area distribution to mimic the effects of continuous emission into the tropical stratosphere at 60 mb of $8\,Tg\,SO_2\,yr^{-1}$ from 2020 to 2069. More details of this prescribed stratospheric aerosol distribution are given in Tilmes et al. (2015) and Xia et al. (2016). The G4SSA scenario then continues from 2070 to 2089 without imposed aerosols to study the termination effect of geoengineering. During the sulfate injection period, the net solar flux at the TOA was decreased by $2.5\,W\,m^{-2}$ compared to RCP6.0 (Fig. 1a). This number was obtained by a double radiation call in the model in calculating the direct forcing of the prescribed aerosol layer. To attain the same TOA solar flux reduction in G4SSA-S, we reduced the total solar insolation by $14.7\,W\,m^{-2}$ during 2020–2069 assuming a global average planetary albedo of 0.32 ($14.7\,W\,m^{-2} = \frac{2.5\,W\,m^{-2} \times 4}{1.0 - 0.32}$) (Fig. 1b). From 2070 on, we accordingly reset the total solar insolation back to the reference level to simulate the abrupt termination of geoengineering.

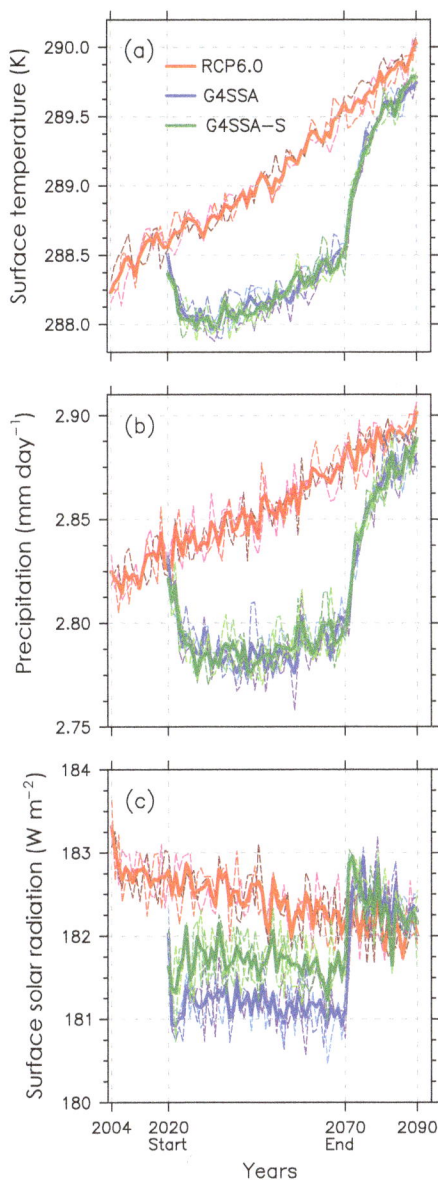

**Figure 2.** **(a)** Global averaged annual surface air temperature (K), **(b)** global averaged annual precipitation (mm day$^{-1}$), and **(c)** downwelling surface solar radiation (W m$^{-2}$). Dashed lines are ensemble members, and solid lines are the average of the three ensemble members. Geoengineering starts on 1 January 2020 and ends on 1 January 2070.

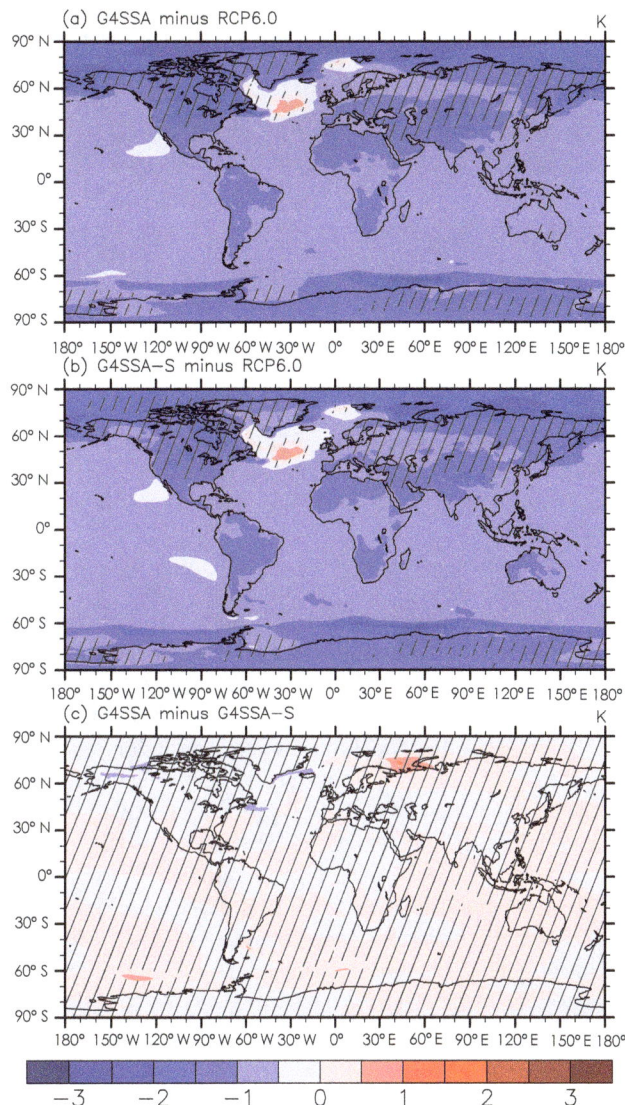

**Figure 3.** Global maps of surface temperature differences (K) between **(a)** G4SSA and RCP6.0, **(b)** G4SSA-S and RCP6.0, and **(c)** G4SSA and G4SSA-S over the period 2030–2069. Hatched regions are areas with $p > 0.05$ (where changes are not statistically significant based on a paired $t$ test).

## 3    Results and discussion

### 3.1    Climatology in G4SSA and G4SSA-S

As a consequence of the same net all-sky TOA solar flux reduction in G4SSA and G4SSA-S (Fig. 1a), the two scenarios show approximately the same global mean surface temperature reduction of 0.8 K compared with RCP6.0 (Fig. 2a) (all values below are the average of the years 2030–2069, the last 40 years of geoengineering). After the termination of

geoengineering on 1 January 2070, the global mean surface temperature rapidly increases. Figure 3 shows the surface temperature differences between G4SSA, G4SSA-S, and RCP6.0 in the years 2030–2069. Consistent with the global average temperature change, the two geoengineering scenarios have similar temperature reduction patterns (Fig. 3a and b), and the differences between them are not significant in most regions (Fig. 3c). The similar warming in the North Atlantic under G4SSA and G4SSA-S relative to RCP6.0 (Fig. 3a and b) is due to the regional cooling under RCP6.0 as a result of slowing down of the Gulf Stream (Hartmann et al., 2013). The temperature difference between G4SSA and

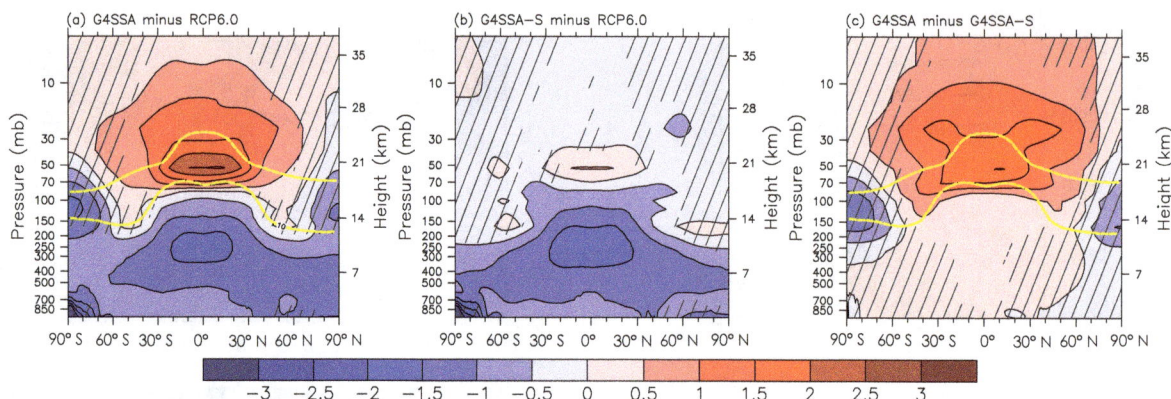

**Figure 4.** Zonal mean temperature differences (K) in the geoengineering experiments (a) G4SSA minus RCP6.0, (b) G4SSA-S minus RCP6.0, and (c) G4SSA minus G4SSA-S. These are averaged for the three ensemble members for the years 2030–2069. Hatched regions are insignificant, with $p > 0.05$. The yellow dashed lines in (a) and (c) are upper and lower limits of the prescribed sulfate aerosol (surface area density: $10\,\mu m^2\,cm^{-3}$).

G4SSA-S (Fig. 3c) is larger in the Northern Hemisphere winter (Fig. S3). The warming over northern Europe and Asia in G4SSA relative to G4SSA-S is the characteristic "winter warming" from volcanic stratospheric aerosols (Robock, 2000). However, the zonal mean stratospheric temperatures in G4SSA and G4SSA-S differ substantially (Fig. 4). Sulfate aerosols in the stratosphere result in strong warming by 3 K in the tropics (Fig. 4a), while in G4SSA-S there is slight cooling (Fig. 4b), consistent with previous studies (Tilmes et al., 2009; Ammann et al., 2010; Jones et al., 2011). The slight warming in the lower stratosphere under G4SSA-S (Fig. 4b) might be a result of ozone changes and dynamical heating (discussion in Sect. 3.3.2). In both cases, the troposphere shows strong temperature reduction with similar patterns and ranges.

Global averaged precipitation and evaporation have similar reductions of $0.07\,mm\,day^{-1}$ in the two scenarios (Figs. 2b and S4), with no statistically significant difference between them. Most of the evaporation terms show a larger reduction in G4SSA than in G4SSA-S, except for plant transpiration, which has the opposite pattern (Fig. S4). As shown by Xia et al. (2016), enhanced diffuse radiation in G4SSA increases photosynthesis, which produces stronger transpiration. Therefore, transpiration in G4SSA reduces less than in G4SSA-S.

The similar evaporation reduction in G4SSA and G4SSA-S can also be explained by the surface energy budget (Fig. 5b). Although we keep the net shortwave radiation at the TOA the same in the two schemes (Figs. 1a and 5a), surface net solar radiation reduces more in G4SSA than in G4SSA-S (Figs. 2c and 5b) due to the absorption by sulfate aerosols in the near-infrared. This stronger surface solar forcing in G4SSA-S is mainly balanced by larger net longwave radiation to the atmosphere (Fig. 5). As a result, latent heat changes in the two scenarios are similar.

Here, precipitation and evaporation changes are very similar under sulfate and solar geoengineering. This is different from previous studies by Niemeier et al. (2013) and Ferraro et al. (2014), who found that the effect on the hydrological cycle is larger for sulfate geoengineering. These differences are related to the experimental design. Niemeier et al. (2013) bias-corrected all geoengineering scenarios to keep the net total flux at the TOA the same as that in 2020, while we keep the same net solar flux at the TOA in G4SSA and G4SSA-S (Fig. 1a). However, we found that the net total fluxes at the top of the model in G4SSA and G4SSA-S are similar as well (Figs. 5a and S5). Therefore, differences in the TOA boundary conditions might not be the main reason for the different hydrological cycle responses. In their studies, with the same magnitude of surface cooling, the sulfate injection scenario led to a greater reduction in globally averaged evaporation and precipitation as compared with the solar case. Ferraro et al. (2014) attributed the enhanced hydrological cycle response to sulfate geoengineering to extra downwelling longwave radiation because of stratospheric heating from the injected aerosols. Sulfate geoengineering thus led to a relative stabilization of the troposphere (by heating the upper troposphere more than the mid–lower troposphere) compared with the solar reduction case (which we do not find; Fig. 4c). A more stratified troposphere, in turn, results in a stronger reduction in latent heat fluxes and precipitation (similar to theoretical considerations by Bala et al., 2008). We find two possible reasons for the different response in our experiments. (1) The column ozone change could play an important role. In Niemeier et al. (2013) and Ferraro et al. (2014), the same prescribed ozone was used in all scenarios, while we used a fully coupled atmosphere–chemistry model. As shown in Sect. 3.2, total column ozone in G4SSA reduces by about 5 Dobson Units (DU) (mainly in the lower stratosphere) compared with RCP6.0 and G4SSA-S (Fig. 6). Less ozone in G4SSA will change its radiative forcing, surface radiative

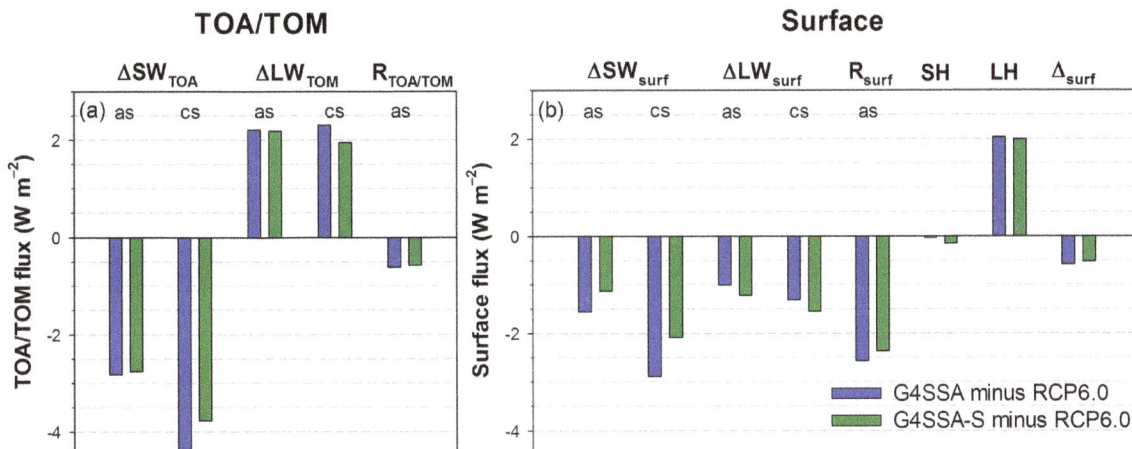

**Figure 5.** Energy flux **(a)** the top of the atmosphere (TOA)/the top of the model (TOM) **(b)** and at the surface (surf), shown as G4SSA minus RCP6.0 and G4SSA-S minus RCP6.0 for 2030–2069. For all fluxes, downwelling is positive. $\Delta$SW is the net shortwave flux, $\Delta$LW is the net longwave flux, R is the sum of $\Delta$SW and $\Delta$LW, SH is sensible heat, LH is latent heat, and $\Delta_{surf}$ is the sum of $\Delta SW_{surf}$, $\Delta LW_{surf}$, SH, and LH; as is all sky and cs is clear sky.

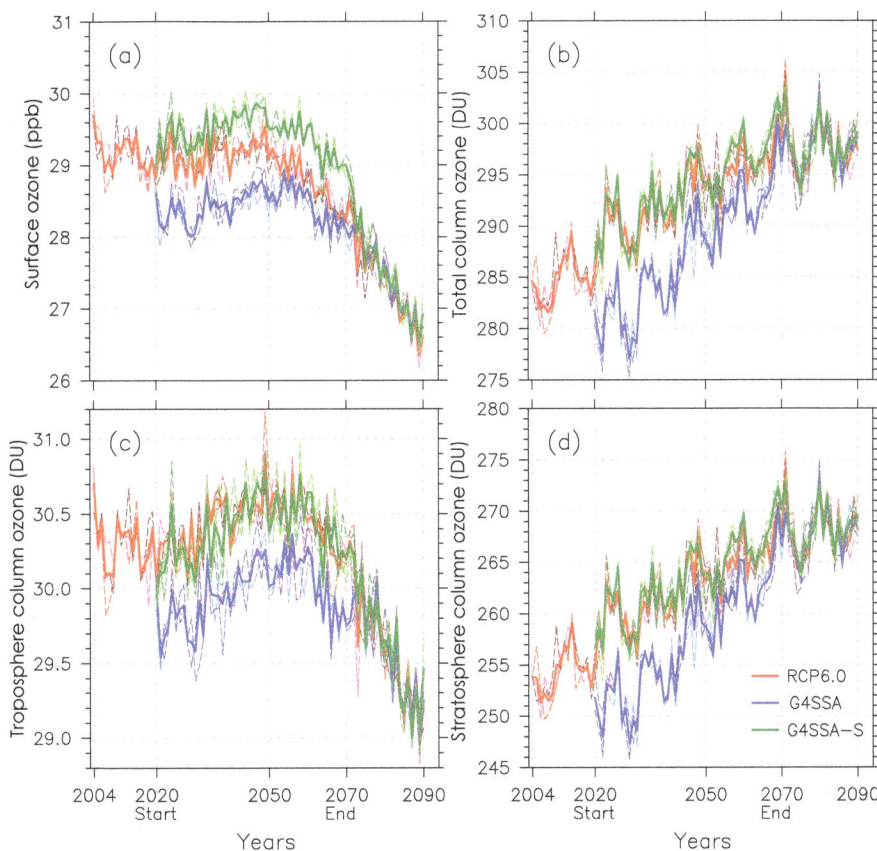

**Figure 6.** **(a)** Global averaged annual surface ozone concentrations (ppb), **(b)** total column ozone (DU), **(c)** tropospheric column ozone (DU), and **(d)** stratospheric column ozone (DU). An ozone concentration of 150 ppb is used as the boundary of tropospheric ozone and stratospheric ozone. Dashed lines are ensemble members, and solid lines are the average of the three ensemble members. Geoengineering starts on 1 January 2020 and ends on 1 January 2070.

fluxes and atmospheric lapse rate (Chiodo and Polvani, 2015; MacIntosh et al., 2016; Nowack et al., 2015, 2017) and thus contribute to the differences between the two studies. (2) Enhanced transpiration in G4SSA due to enhanced diffuse radiation reduces the evaporation difference in the two SRM schemes as discussed above.

## 3.2 Surface and tropospheric ozone response

The ozone response is remarkably different in G4SSA and G4SSA-S. Global mean surface ozone increases under G4SSA-S and decreases under G4SSA, relative to RCP6.0 (Fig. 6a). The total ozone column is dominated by stratospheric column ozone and shows strong reduction under G4SSA compared to RCP6.0, while the increase under G4SSA-S is small (Fig. 6b and d). The underlying upward trends of total column ozone as well as stratospheric ozone in all three scenarios are in line with the gradually declining stratospheric halogen content over time (Fig. 6b and d). The halogen loading in the three scenarios is the same, and more information can be found in Morgenstern et al. (2017). As there is less halogen in the stratosphere toward the end of the geoengineering, stratospheric ozone is recovering (Fig. 6d) and there is less reduction in the total ozone column in G4SSA relative to RCP6.0 (Fig. 6b). The agreement across the simulations concerning interannual and decadal variations is because of the imposed QBO and 11-year solar cycles in all the runs. The decreasing tropospheric ozone column and surface ozone after the year 2060 in all scenarios results from decreases in global ozone precursor emissions following the RCP6.0 scenario (Young et al., 2013).

The surface ozone concentration distributions in the three scenarios are similar, with the highest concentration over the continents in the Northern Hemisphere (NH) (Fig. S6), while the concentration differences as well as the percentage difference between scenarios are spatially variable (Figs. 7 and S7). This highlights that the key driver behind the absolute surface ozone abundances is the underlying ozone precursor emissions following the RCP6.0 scenario. SRM is only a modulating factor, but depending on the SRM scheme even the sign of its impact can differ; global mean surface ozone concentrations in G4SSA are lower, relative to RCP6.0, whereas there are moderate surface ozone increases over the tropics (Fig. 7a). The strongest surface ozone reductions compared with RCP6.0 occur in NH midlatitudes across all seasons (Fig. S8a–d) and Southern Hemisphere (SH) mid- to high latitudes in MAM and JJA (Fig. S8b, c). In G4SSA-S, surface ozone also increases in the tropics relative to RCP6.0 (Fig. 7b), and this regional change is greater than in G4SSA (Fig. 7c). Surface ozone decreases, however, are found at NH midlatitudes over the continents during all seasons (Fig. S8e–h). Comparing the two types of geoengineering experiments directly, surface ozone concentrations are generally lower in G4SSA than in G4SSA-S (Fig. 7c), with peak differences in terms of absolute changes (ppb) at

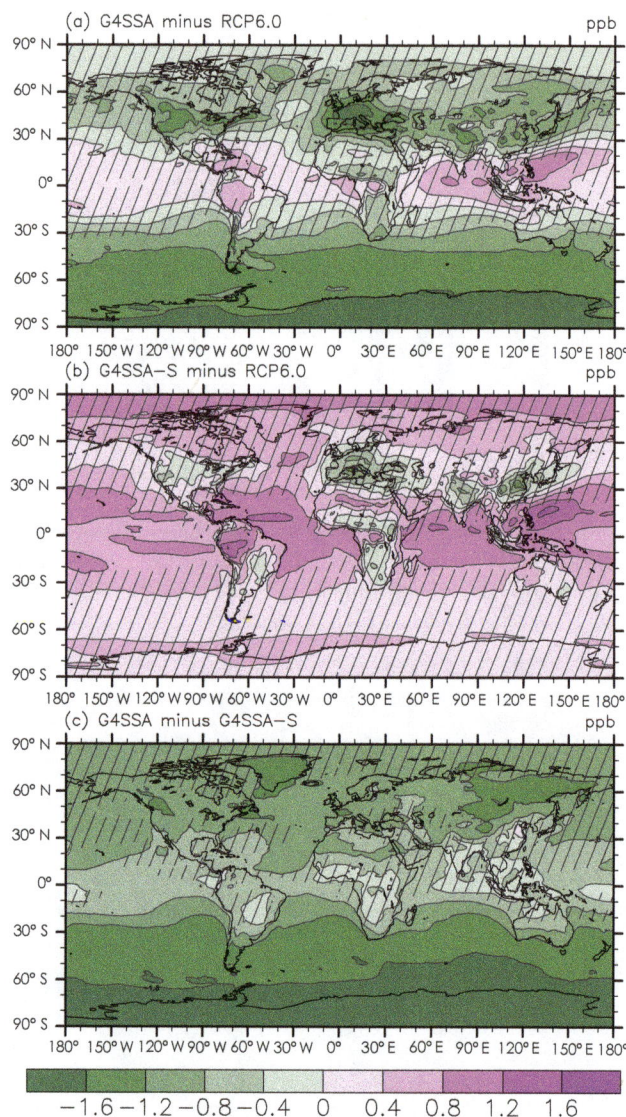

**Figure 7.** Global maps of surface ozone concentration differences (ppb) between **(a)** G4SSA and RCP6.0, **(b)** G4SSA-S and RCP6.0, and **(c)** G4SSA and G4SSA-S for 2030–2069. Hatched regions are insignificant, with $p > 0.05$.

SH mid- to high latitudes in MAM and JJA (Fig. S8i and j) and at NH mid- to high latitudes in DJF (Fig. S8l).

## 3.3 Mechanisms of surface ozone change

Surface ozone concentrations are determined by chemical production and loss controlled by emissions of ozone precursors and the composition of the atmosphere, loss due to surface deposition of ozone, and the transport of ozone from other regions of the atmosphere (Monks et al., 2015). Since all scenarios considered here are based on the same RCP6.0 emission scenario of ozone precursors, such as nitrogen oxide ($NO_x$) and methane ($CH_4$), the differences in surface ozone must necessarily be driven by changes in cli-

**Table 1.** Tropospheric ozone production and loss rates $(\mathrm{Tg\,yr^{-1}})$ over the period of the years 2030–2069 (average of three ensemble members). For chemical ozone production and ozone loss the net impacts of only the most important reaction pathways are listed.

| | RCP6.0 | G4SSA | G4SSA-S |
|---|---|---|---|
| $O_3$ net chemical change | 346.1 | 472.7 | 384.8 |
| $O_3$ tropospheric dry deposition | 901.5 | 891.5 | 909.4 |
| $O_3$ STE* | 555.4 | 418.8 | 524.6 |
| $O_3$ production | 4895.8 | 4764.0 | 4671.8 |
| $r$-NO-HO$_2$ | 3087.3 | 3031.0 | 2964.8 |
| $r$-CH$_3$O$_2$-NO | 1132.3 | 1105.2 | 1083.1 |
| $r$-PO$_2$-NO | 21.8 | 20.1 | 19.9 |
| $r$-CH$_3$CO$_3$-NO | 183.1 | 172.2 | 171.2 |
| $r$-C$_2$H$_5$O2-NO | 6.6 | 6.7 | 6.7 |
| 0.92*$r$-ISOPO$_2$-NO | 149.8 | 135.3 | 134.0 |
| $r$-MACRO$_2$-NOa | 76.1 | 69.8 | 69.5 |
| $r$-MCO$_3$-NO | 34.5 | 30.5 | 30.3 |
| $r$-RO$_2$-NO | 12.2 | 11.5 | 11.5 |
| $r$-XO$_2$-NO | 66.5 | 60.8 | 60.5 |
| 0.9*$r$-TOLO$_2$-NO | 4.1 | 4.1 | 4.1 |
| $r$-TERPO$_2$-NO | 18.1 | 16.9 | 16.8 |
| 0.9*$r$-ALKO$_2$-NO | 22.9 | 23.0 | 22.9 |
| $r$-ENEO$_2$-NO | 12.5 | 11.6 | 11.7 |
| $r$-EO$_2$-NO | 36.8 | 34.6 | 34.5 |
| $r$-MEKO$_2$-NO | 17.7 | 17.9 | 17.8 |
| 0.4*$r$-ONITR-OH | 7.5 | 6.8 | 6.8 |
| $r$-jonitr | 1.4 | 1.2 | 1.2 |
| $O_3$ loss | 4421.1 | 4158.6 | 4151.6 |
| $r$-O(1D)-H$_2$O | 2430.4 | 2286.5 | 2263.5 |
| $r$-OH-O$_3$ | 548.2 | 528.3 | 527.0 |
| $r$-HO$_2$-O$_3$ | 1288.9 | 1216.7 | 1232.9 |
| $r$-C$_3$H$_6$-O$_3$ | 13.8 | 11.5 | 11.5 |
| 0.9*$r$-ISOP-O$_3$ | 71.4 | 58.0 | 57.6 |
| $r$-C$_2$H$_4$-O$_3$ | 9.3 | 7.8 | 8.0 |
| 0.8*$r$-MVK-O$_3$ | 18.6 | 15.5 | 15.7 |
| 0.8*$r$-MACR-O$_3$ | 3.5 | 2.9 | 2.9 |
| $r$-C$_{10}$H$_{16}$-O$_3$ | 37.0 | 31.5 | 31.6 |

* $O_3$ STE is ozone transported through the stratosphere–troposphere exchange. We calculated this value using the following equation:

$O_3\,\text{STE} + O_3\,\text{net tropospheric chemical change} + O_3\,\text{dry tropospheric deposition} = 0$.

Tropospheric ozone is defined by the 150 ppb isopleth.

mate in response to the geoengineering interventions, which include changes in temperature, humidity, atmospheric dynamics, and the photochemical environment. To understand the differences mechanistically, it is helpful to consider the impact of geoengineering on the tropospheric ozone budget.

The upper part of Table 1 shows the sources (production and net transport from the stratosphere (STE)) and sinks (loss rates and dry deposition) of tropospheric ozone. Both G4SSA and G4SSA-S show a positive net chemical change in tropospheric ozone (chemical production minus loss) and negative change in the STE of ozone relative to RCP6.0. However, the magnitude of these changes is significantly different. Compared with RCP6.0, tropospheric ozone net chemical change increases by $\sim 125$ and $\sim 40\,\mathrm{Tg\,yr^{-1}}$ in G4SSA and G4SSA-S, respectively, whereas the STE of ozone decreases

by $\sim 140\,\mathrm{Tg\,yr^{-1}}$ ($\sim 25\%$) and $\sim 30\,\mathrm{Tg\,yr^{-1}}$ ($\sim 5\%$) in G4SSA and G4SSA-S, respectively. The positive net chemical changes are the result of reductions in both chemical ozone production and loss under G4SSA and G4SSA-S relative to RCP6.0, with larger reductions in ozone loss reactions (Table 1). Specifically, G4SSA-S shows a $\sim 90\,\mathrm{Tg\,yr^{-1}}$ larger decrease in ozone chemical production, whereas ozone loss budgets are reduced by similar magnitudes for the two SRM schemes (262.5 and $269.5\,\mathrm{Tg\,yr^{-1}}$). Combining the chemical and transport changes, the tropospheric ozone budget decreases under G4SSA and increases under G4SSA-S relative to RCP6.0, which is consistent with the overall surface ozone changes.

The reasons for these specific changes are discussed in detail in the following two sections. Then, the impacts of the factors are combined to explain regional surface ozone differences, as shown in Fig. 7.

### 3.3.1 Chemical ozone production and loss in the troposphere

Changes in tropospheric water vapor concentrations and the tropospheric photolysis environment under G4SSA and G4SSA-S are key to understand the differences in tropospheric ozone production and loss. This result is consistent with the results of a previous study for the case of solar geoengineering under a more idealized forcing scenario (Nowack et al., 2016).

To explain this, we briefly reiterate that tropospheric ozone ($O_3$) production is driven by the photolysis of nitrogen dioxide ($NO_2$) and the subsequent formation of ozone via a three-body reaction with resulting ground-state atomic oxygen $O(^3P)$ (Monks, 2005):

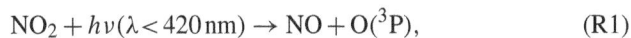

$$NO_2 + h\nu(\lambda < 420\,\mathrm{nm}) \rightarrow NO + O(^3P), \quad (R1)$$

$$O(^3P) + O_2 + M \rightarrow O_3 + M, \quad (R2)$$

where $M$ is an inert collision partner (mostly molecular nitrogen). $NO_2$ formation in turn is crucially dependent on the oxidation of NO by reaction with peroxides present in the troposphere, for example,

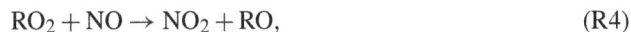

$$HO_2 + NO \rightarrow NO_2 + OH, \quad (R3)$$

$$RO_2 + NO \rightarrow NO_2 + RO, \quad (R4)$$

where $R$ represents general organic residues such as $CH_3$ (row 6 in Table 1). $RO_2$ in turn is produced by oxidation reactions between VOCs and the hydroxyl radical OH. Tropospheric OH is formed primarily by ozone photolysis and the subsequent reaction of excited atomic oxygen $O(^1D)$ with water vapor:

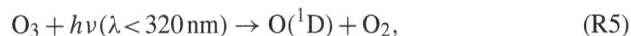

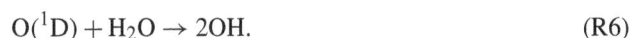

$$O_3 + h\nu(\lambda < 320\,\mathrm{nm}) \rightarrow O(^1D) + O_2, \quad (R5)$$

$$O(^1D) + H_2O \rightarrow 2OH. \quad (R6)$$

Reaction (R6) competes with several other reactions due to the high reactivity of $O(^1D)$. However, most importantly, the majority of $O(^1D)$ is quenched by collision with inert molecules such as molecular nitrogen to ground-state atomic oxygen $O(^3P)$, which subsequently undergoes reformation to $O_3$ via a three-body Reaction (R2). Therefore, tropospheric ozone production and loss are strongly linked to concentrations of water vapor and the photochemical environment (availability of UV) in the troposphere.

In the case of clean (low $NO_x$) environments, lower water vapor concentrations (mainly in the tropical region 30° N–30° S) (Fig. S9) lead to less ozone loss via Reactions (R5) and (R6) and additional reactions with the formed $HO_x$ species ($r$-O(1D)-$H_2O$, $r$-OH-$O_3$, and $r$-HO2-$O_3$ in Table 1). This happens at the expense of more quenching of $O(^1D)$ and subsequent recycling of ozone, thus resulting in ozone increases. In contrast, in polluted (high $NO_x$) environments, less OH formation under lower atmospheric water vapor concentrations leads to reduced formation of $RO_2$ and $HO_2$. Therefore, less $NO_2$ is produced via Reactions (R3) and (R4), resulting in less catalytic ozone production via Reactions (R1) and (R2) ($r$-NO-HO2 and, e.g., $r$-CH3O2-NO in Table 1). Consequently, ozone production is reduced in $NO_x$-polluted environments under lower atmospheric water vapor concentrations.

With these fundamentals in mind, it is possible to understand the sign of the tropospheric ozone loss and production rate changes summarized in Table 1. Under both G4SSA and G4SSA-S, the key objective is to achieve surface temperature decreases. Tropospheric water vapor concentrations (or specific humidity) are strongly coupled to surface temperatures because relative humidity does not change much with climate change (Soden and Held, 2006; Dessler and Sherwood, 2009), so that the surface cooling simultaneously reduces tropospheric specific humidity by 5–20 % depending on region and altitude. As a result, less water vapor in both G4SSA and G4SSA-S reduces ozone chemical loss by $\sim 150\,\mathrm{Tg\,yr^{-1}}$ through Reactions (R5) and (R6). The resulting decrease in $HO_x$ leads to further reductions in ozone loss, i.e., via reaction with OH ($\sim 20\,\mathrm{Tg\,yr^{-1}}$) and HO2 ($\sim 60$–$70\,\mathrm{Tg\,yr^{-1}}$). Overall, these water vapor/$HO_x$-related reactions explain $\sim 90\,\%$ of the overall reduction in ozone loss under SRM compared to a future RCP6.0 simulation.

The reduction in atmospheric humidity also affects ozone production but to a smaller degree. Here, ozone production via reaction between NO and $HO_2$ is the key factor in driving these changes, with reductions of $\sim 55$ and $120\,\mathrm{Tg\,yr^{-1}}$ for G4SSA and G4SSA-S, respectively. The signal of reduced OH production propagates through all other $NO_x$-catalyzed ozone production pathways involving $RO_2$ via Reactions (R4) and subsequently (R1) and (R2). NO oxidation via the $CH_3O_2$-NO pathway decreases by $\sim 27$ and $49\,\mathrm{Tg\,yr^{-1}}$ in G4SSA and G4SSA-S. Changes in natural $NO_x$ emissions by lightning play a minor role in comparison. In both SRM schemes, the reduction in lightning-induced $NO_x$ is not significant in most regions, and there is no significant difference between the two SRM schemes (Fig. S10).

The changes in chemical ozone production rates tend to be smaller in the sulfate G4SSA experiment than in the case of a solar constant reduction in G4SSA-S. There are three possible factors that contribute to this:

1. The entire reaction cycle depends on the availability of sunlight to photolyze $O_3$ and $NO_2$. Since SRM schemes modulate the intensity of sunlight (here by 1 %) reaching the troposphere in order to mitigate tropospheric warming, this will necessarily also play a role in all changes to ozone production and loss reactions in our SRM simulations. More importantly, however, the sulfate injection geoengineering alters stratospheric ozone concentrations, which ultimately impacts the photochemical environment of the troposphere by changing radiative fluxes into the troposphere (DeMore et al., 1997; Nowack et al., 2016). For example, a reduced stratospheric column will help to stimulate the tropospheric photochemistry by allowing the more radiation-relevant Reactions (R1) and (R5) to propagate into the troposphere.

2. Diffuse radiation under G4SSA promotes the photosynthesis rate and increases canopy transpiration (Fig. S4). Therefore, we expect that water vapor concentration over the continents with plants would be slightly higher in G4SSA relative to G4SSA-S (Fig. S11). Those regions with higher water vapor (East Asia, South Asia, North America, South Africa) often overlap with regions of high $NO_x$ pollution (Fig. S12). $HO_x$–$NO_x$ coupling in these regions will thus contribute to the smaller reduction in ozone production in G4SSA than in G4SSA-S.

3. Different biogenic VOC emissions occur under G4SSA and G4SSA-S, which, due to their central role in forming $NO_2$, are highly important for ozone production. In both scenarios, lower temperatures reduce the heat stress on the emitting plants and therefore reduce their VOC emissions (Tingey et al., 1980; Sharkey and Yeh, 2001; Lathière et al., 2005; Bornman et al., 2015) (e.g., bio-emitted isoprene; Fig. S13). However, at the same time enhanced diffuse radiation under G4SSA increases biogenic VOC emissions compared with G4SSA-S (Wilton et al., 2011) (Fig. S13i, j, k and l). In Table 1, biogenic VOC-related ozone chemical production is generally very similar between G4SSA with G4SSA-S (e.g., $r$-ISOPO2-NO, $r$-MACRO2-NOa, $r$-MCO3-NO, and $r$-TERPO2-NO) and contributes less than 2 % to the overall difference between G4SSA and G4SSA-S.

### 3.3.2 Changes in the stratosphere–troposphere exchange

Stratospheric chemical and dynamical changes can impact tropospheric ozone not only by changing the tropospheric

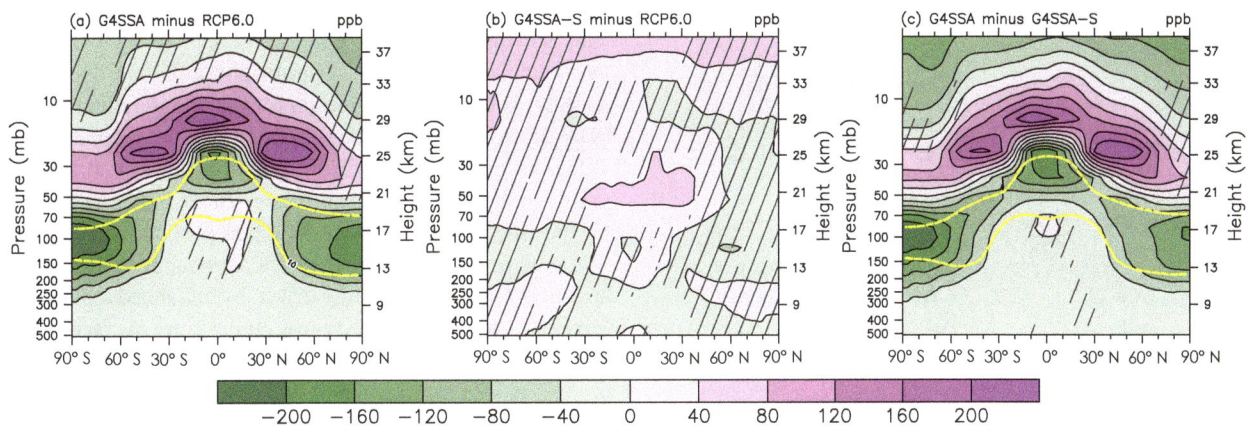

**Figure 8.** Zonal mean ozone concentration differences (ppb) in the geoengineering experiments, averaged for the three ensemble members for 2030–2069. Hatched regions are insignificant, with $p > 0.05$. The yellow dashed lines in **(a)** and **(c)** are the upper and lower limits of the prescribed sulfate aerosol (surface area density: $10\,\mu m^2\,cm^{-3}$).

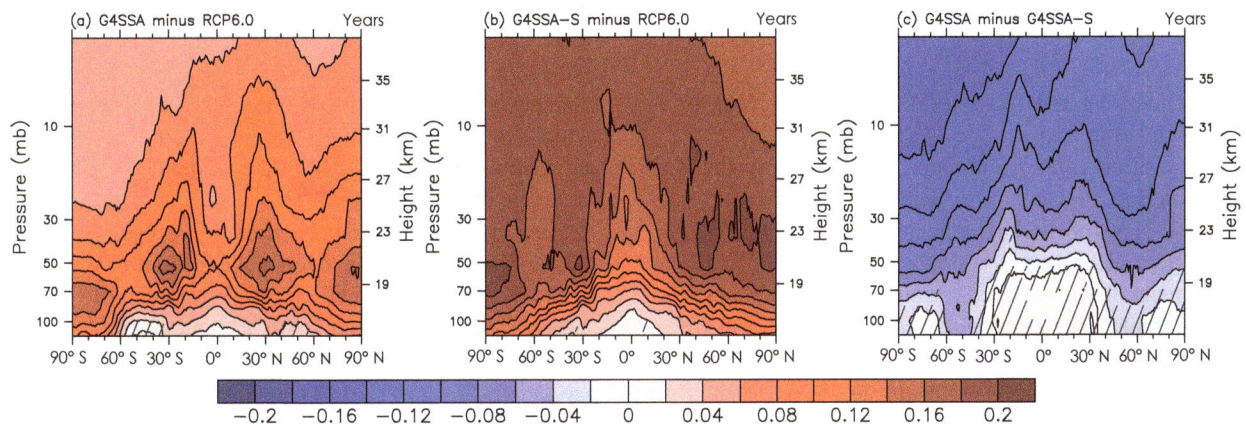

**Figure 9.** Zonal mean age of air differences (years) between **(a)** G4SSA and RCP6.0, **(b)** G4SSA-S and RCP6.0, and **(c)** G4SSA and G4SSA-S. They are averaged for the three ensemble members for 2030–2069. Hatched regions are insignificant, with $p > 0.05$.

photochemical environment but also by changing the actual transport of ozone from the stratosphere into the troposphere (Hegglin and Shepherd, 2009; Neu et al., 2014). This can be either caused by changes in ozone concentrations in the stratosphere or by changes in the rate of the exchange of air masses between the stratosphere and the troposphere (i.e., the strength of the Brewer–Dobson (B-D) circulation and tropopause folds).

Figure 8 shows seasonal latitude–height cross sections of differences in ozone volume mixing ratios between G4SSA and RCP6.0 as well as G4SSA-S and RCP6.0 for altitudes above the 500 mb pressure level. Under G4SSA, heterogeneous reactions on the aerosol surfaces lead to increased halogen activation and with that an enhancement of ozone depletion in mid- to high latitudes (60–90° N/S) in the lower stratosphere (70–150 mb) (Tilmes et al., 2008, 2009, 2012; Heckendorn et al., 2009). On the other hand, heterogenous reactions reduce the $NO_x$ to $NO_y$ ratio, which results in an increase in ozone mixing ratios, mainly in the middle strato-

sphere (10–30 mb) (Tie and Brasseur, 1995) (Fig. 8a). In addition, changes in stratospheric temperature (warming in G4SSA and cooling in G4SSA-S) also change the photochemistry of ozone. Altogether, this results in year-round lower stratospheric ozone loss worldwide that peaks during the return of sunlight at high SH latitudes (Fig. S14). In comparison, the solar reduction in G4SSA-S does not enhance stratospheric heterogeneous reactions. The much smaller change (increase) in ozone (Fig. 8b) is driven by the change in homogeneous chemistry due to slight temperature reduction (Fig. 4b). However, in Fig. 4b, there is a slight warming around 50 mb in the tropics, where ozone concentration also shows a stronger increase (Fig. 8b). As tropospheric cooling results in a slowdown of the B-D circulation (Fig. 9b) (Lin and Fu, 2013; Nowack et al., 2015; Shepherd and McLandress, 2011), there is an increase in ozone in the tropical upwelling region, which leads to increasing temperatures there as ozone is a strong shortwave and longwave absorber. The net result is small ozone increases in the tropical

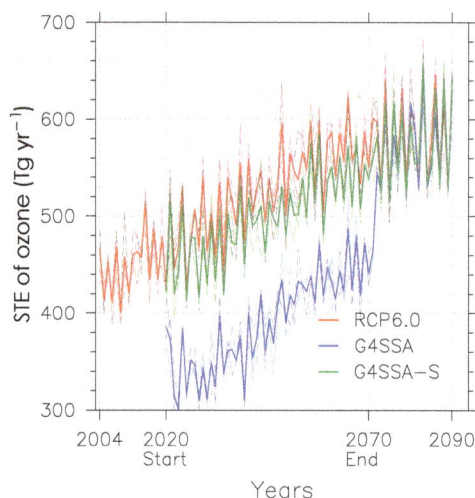

**Figure 10.** Global annual averaged ozone transported from the stratosphere to the troposphere (STE of ozone) in $\mathrm{Tg\,yr^{-1}}$. Geoengineering starts on 1 January 2020 and ends on 1 January 2070.

lower stratosphere and decreases in both extratropical lower stratospheres (Fig. 8b).

The age of air is used to indicate the strength of the B-D circulation (Fig. 9). Here, it is calculated relative to the zonal mean of $1°$ N at 158.1 mb (Garcia and Randel, 2008; Waugh, 2002). Older air indicates a slowdown of the B-D circulation. Compared with RCP6.0, both G4SSA and G4SSA-S show older air in the stratosphere indicating a slowdown of the circulation. The cooling effect in two SRM scenarios correlates with a weakening of tropical upwelling. However, in G4SSA, the heating of the tropical stratosphere results in enhanced lifting, which counteracts the weakening of the B-D circulation (Fig. 9a and c). Previous studies show controversial results on how the B-D circulation changes due to extra aerosols in the atmosphere. Aquila et al. (2012) modeled stronger tropic upwelling after the eruption of Mt Pinatubo, and other studies also found enhanced simulated B-D circulation after this volcano eruption (Aquila et al., 2013; Pitari et al., 2016). The differences between previous studies and our result may be because some previous studies used fixed ozone, with different stratospheric heating rates. In addition, in previous studies, the QBO was interactively simulated and the models had a higher model top. However, with extra black carbon in the stratosphere, the tropical upwelling weakens due to the simultaneous effect of tropospheric cooling (Shepherd and McLandress, 2011; Mills et al., 2014). We hope that future studies will address the potential model dependency of this result.

The sum of both effects, stratospheric chemical changes and the impact of B-D circulation change on the STE of ozone, is shown in Fig. 10. In G4SSA, ozone transported from the stratosphere to the troposphere is significantly decreased by $\sim 25\%$ relative to RCP6.0. In G4SSA-S the reduction is small. Since the air mass transported from the

stratosphere to the troposphere is reduced in both scenarios, and is even more strongly reduced under G4SSA-S (Fig. 9), we find that the enhanced stratospheric ozone depletion in G4SSA is the dominant reason for the strong reduction in the STE of ozone. This is also supported by a stratospheric ozone tracer from the model, $O_3^{\mathrm{Strat}}$, which is set to ozone mixing ratios in the stratosphere and experiences only chemical loss in the troposphere without chemical production (Fig. S15). We thus conclude that the significant changes in the STE of ozone in G4SSA are mainly driven by enhanced stratospheric ozone depletion catalyzed through the aerosols (see also Table 1).

### 3.4 Balance of the different mechanisms and uncertainties

In summary, there are two main factors that determine the tropospheric ozone responses in our SRM and RCP6.0 simulations: (a) changes in tropospheric ozone chemical production/loss due to water vapor changes and impacts on the photochemical environment of the troposphere as a result of both changes in stratospheric ozone and (to a smaller degree) the actual dimming of sunlight depending on the geoengineering scheme and (b) changes in the stratosphere–troposphere exchange of ozone.

These factors can also be used to explain the big picture behind the surface ozone changes shown in Fig. 7. In G4SSA-S the reduced tropospheric humidity leads to stronger reductions in ozone loss than the decreases in ozone production, leading to global increases in surface ozone but particularly in clean-air oceanic environments in the tropics. This net increase in ozone chemical change is not canceled out by the slight reduction in ozone transport from the stratosphere (Fig. 10). In G4SSA, the reduction in ozone transport from the stratosphere is the controlling factor, which overwhelms the increase in net ozone production. The effect is particularly pronounced at mid- to high latitudes (Fig. S14a), thus giving rise to surface ozone decreases there (Fig. 7). In contrast, the effect of reduced tropospheric humidity is relatively more important in the tropics than in other regions, which results in a local increase in surface ozone under G4SSA. Regionally, $HO_x$-$NO_x$-induced reductions in ozone production (Table 1) can become important to explain surface ozone decreases in $NO_x$-polluted land areas in the NH for both scenarios (Figs. 7 and S6). Further minor contributions to the differences in surface ozone between G4SSA and G4SSA-S are a consequence of changes in water vapor due to regional canopy transpiration effects and biogenic VOC emissions (e.g., isoprene; Table 1 and Fig. S13).

This study may be biased by the following factors. (1) Using prescribed stratospheric aerosols does not allow the simulation of the full interactions between chemistry, aerosol microphysics, and dynamics. A fully interactive model including those interactions would be important. (2) The vertical resolution is not sufficient to produce an interactive

QBO in the model used, which may also affect transport processes. (3) The model does not include the scattering effect of aerosols on tropospheric photolysis rates, which might lead to an overestimate of the UV enhancement in the troposphere.

## 4 Conclusions

Tropospheric ozone changes are to be expected in a geoengineered climate with consequent impacts on air pollution and crop yields. However, for the scenarios considered here, solar and sulfate geoengineering could have entirely different impacts, even in terms of the sign of the response, a rare discrepancy for a surface signal between these two types of geoengineering. There have been many studies using solar irradiance reductions to illustrate SRM. However, it turns out that different SRM strategies would have different impacts on hydrology, atmospheric dynamics, the terrestrial carbon sink, and, as investigated in this paper, tropospheric chemistry. These results also depend on the scenario of future ozone precursor and halogen emissions.

We have identified and explained the mechanisms by which stratospheric sulfate geoengineering would change surface ozone concentrations. We find that geoengineering might have the potential to significantly reduce some climate impacts, but it cannot fix the problem of air pollution. To reduce air pollution effectively, changes in surface emissions are key, with changes in climate (including geoengineering) being only a modulating factor (Monks et al., 2015; Stevenson et al., 2013; Young et al., 2013). More importantly, the surface ozone reduction between 2030 and 2070 in G4SSA is primarily the result of the decreased STE of ozone following ozone depletion in the stratosphere. The rather mild pollution benefit under the RCP6.0 background would thus be bought at the expense of a delay in stratospheric ozone recovery, which would result in enhanced UV penetration to Earth's surface and with that serious impacts on human health (e.g., skin cancer) and the ecosystem. In the future, potential increases in stratospheric ozone as a result of geoengineering may result in an increase in surface ozone, causing more ozone pollution. However, further analysis on air pollutants other than ozone is needed.

As shown by Pitari et al. (2014), impacts on ozone from stratospheric geoengineering can be highly model dependent. Therefore, we consider the results here to be a Geoengineering Model Intercomparison Project (GeoMIP) test bed experiment and encourage others to compare our results to those from other climate models to evaluate the robustness of the results presented here.

*Competing interests.* The authors declare that they have no conflict of interest.

*Special issue statement.* This article is part of the special issue "The Geoengineering Model Intercomparison Project (GeoMIP): Simulations of solar radiation reduction methods (ACP/GMD interjournal SI)". It is not associated with a conference.

*Acknowledgements.* This work is supported by US National Science Foundation (NSF) grants AGS-1157525, GEO-1240507, AGS-1430051, and AGS-1617844. Computer simulations were conducted on the National Center for Atmospheric Research (NCAR) Yellowstone supercomputer. NCAR is funded by NSF. The CESM project is supported by NSF and the Office of Science (BER) of the US Department of Energy. Peer Nowack is supported by the European Research Council through the ACCI project, project number 267760 and is now supported through an Imperial College Research Fellowship. We thank Doug Kinnison for helpful comments and Jean-Francois Lamarque, Daniel Marsh, Andrew Conley, Louisa K. Emmons, Rolando R. Garcia, Anne K. Smith, and Douglas E. Kinnison for the CAM4-Chem development.

Edited by: Lynn M. Russell

## References

Ainsworth, E. A., Yendrek, C. R., Sitch, S., Collins, W. J., and Emberson, L. D.: The effects of tropospheric ozone on net primary productivity and implications for climate change, Ann. Rev. Plant Biol., 63, 637–661, https://doi.org/10.1146/annurev-arplant-042110-103829, 2012.

Ammann, C. M., Washington, W. M., Meehl, G. A., Buja, L., and Teng, H.: Climate engineering through artificial enhancement of natural forcings: Magnitudes and implied consequences, J. Geophys. Res., 115, D22109, https://doi.org/10.1029/2009JD012878, 2010.

Aquila, V., Oman, L. D., Stolarski, R. S., Colarco, P. R., and Newman, P. A.: Dispersion of the volcanic sulfate cloud from a Mount Pinatubo-like eruption, J. Geophys. Res., 117, D06216, https://doi.org/10.1029/2011JD016968, 2012.

Aquila, V., Oman, L. D., Stolarski, R. S., Douglass, A. R., and Newman, P. A.: The response of ozone and nitrogen dioxide to the eruption of Mt. Pinatubo at southern and northern midlatitudes, J. Atmos. Sci., 70, 894–900, https://doi.org/10.1175/JAS-D-12-0143.1, 2013.

Ashmore, M. R.: Assessing the future global impacts of ozone on vegetation, Plant, Cell Environ., 28, 949–964, https://doi.org/10.1111/j.1365-3040.2005.01341.x, 2005.

Bala, G., Duffy, P. B., and Taylor, K. E.: Impact of geoengineering schemes on the global hydrological cycle, PNAS, 105, 7664–7669, https://doi.org/10.1073/pnas.0711648105, 2008.

Bornman, J. F., Barnes, P. W., Robinson, S. A., Ballaré, C. L., Flint, S. D., and Caldwell, M. M.: Solar ultraviolet radiation and ozone depletion-driven climate change: effects on terrestrial ecosystems, Photochem. Photobiol. Sci., 14, 88–107, https://doi.org/10.1039/C4PP90034K, 2015.

Chiodo, G. and Polvani, L. M.: Reduction of climate sensitivity of solar forcing due to stratospheric ozone feedback, J. Clim., 29, 4651–4663, https://doi.org/10.1175/JCLI-D-15-0721.1, 2015.

Collins, W. J., Derwent, R. G., Garnier, B., Johnson, C. E., and Sanderson, M. G.: Effect of stratosphere-troposphere exchange on the future tropospheric ozone trend, J. Geophys. Res., 108, 8528, https://doi.org/10.1029/2002JD002617, 2003.

Cooper, O. R., Parrish, D. D., Ziemke, J., Balashov, N. V., Cupeiro, M., Galbally, I. E., Gilge, S., Horowitz, L., Jensen, N. R., Lamarque, J.-F., Naik, V., Oltmans, S. J., Schwab, J., Shindell, D. T., Thompson, A. M., Thouret, V., Wang, Y., and Zbinden, R. M.: Global distribution and trends of tropospheric ozone: An observation-based review, Elem. Sci. Anth., 2–29, https://doi.org/10.12952/journal.elementa.000029, 2014.

Crutzen, P.: Albedo enhancement by stratospheric sulfur injections: A contribution to resolve a policy dilemma?, Climatic Change, 77, 211–220, https://doi.org/10.1007/s10584-006-9101-y, 2006.

DeMore, W. B., Sander, S. P., Golden, D. M., Hampson, R. F., Kurylo, M. J., Howard, C. J., Ravishankara, A. R., Kolb, C. E., and Molina, M. J.: Chemical Kinetics and Photochemical Data for Use in Stratospheric Modeling, Evaluation No. 12, JPL Publication 97-4, Jet Propulsion Laboratory, Pasadena, https://jpldataeval.jpl.nasa.gov, 1997.

Dessler, A. E. and Sherwood, S. C.: A matter of humidity, Science, 323, 1020–1021, https://doi.org/10.1126/science.1171264, 2009.

Emmons, L. K., Walters, S., Hess, P. G., Lamarque, J.-F., Pfister, G. G., Fillmore, D., Granier, C., Guenther, A., Kinnison, D., Laepple, T., Orlando, J., Tie, X., Tyndall, G., Wiedinmyer, C., Baughcum, S. L., and Kloster, S.: Description and evaluation of the Model for Ozone and Related chemical Tracers, version 4 (MOZART-4), Geosci. Model Dev., 3, 43–67, https://doi.org/10.5194/gmd-3-43-2010, 2010.

Fernandez, R. P., Kinnison, D. E., Lamarque, J. F., Tilmes, S., and Saiz-Lopes, A.: Impact of biogenic very short-lived bromine on the Antarctic ozone hole during the 21st century, Atmos. Chem. Phys., 17, 1673–1688, https://doi.org/10.5194/acp-17-1673-2017, 2017.

Ferraro, A. J., Highwood, E. J., and Andrew J Charlton-Perez, A. J.: Weakened tropical circulation and reduced precipitation in response to geoengineering, Env. Res. Lett., 9, 014001, https://doi.org/10.1088/1748-9326/9/1/014001, 2014.

Forster, P., Ramaswamy, V., Artaxo, P., Berntsen, T., Betts, R., Fahey, D. W., Haywood, J., Lean, J., Lowe, D. C., Myhre, G., Nganga, J., Prinn, R., Raga, G., Schulz, M., and Van Dorland, R.: Changes in Atmospheric Constituents and in Radiative Forcing in: Climate Change 2007: The Physical Science Basis, Contribution of Working Group I to the Fourth Assessment Report of the Intergovernmental Panel on Climate Change, Cambridge University Press, Cambridge, UK and New York, NY, USA, https://www.ipcc.ch/pdf/assessment-report/ar4/wg1/ar4-wg1-chapter2.pdf, 2007.

Garcia, R. R. and Randel, W. J.: Acceleration of the Brewer-Dobson Circulation due to increases in greenhouse gases, J. Atmos. Sci., 65, 2731–2739, https://doi.org/10.1175/2008JAS2712.1, 2008.

Govindasamy, B. and Calderia, K.: Geoengineering Earth's radiation balance to mitigate $CO_2$-induced climate change, Geophys. Res. Lett., 27, 2141–2144, https://doi.org/10.1029/1999GL006086, 2000.

Guenther, A. B., Jiang, X., Heald, C. L., Sakulyanontvittaya, T., Duhl, T., Emmons, L. K., and Wang, X.: The Model of Emissions of Gases and Aerosols from Nature version 2.1

(MEGAN2.1): an extended and updated framework for modeling biogenic emissions. Geosci. Model Dev., 5, 1471–1492, https://doi.org/10.5194/gmd-5-1471-2012, 2012.

Hartmann, D. L., Klein Tank, A. M. G., Rusticucci, M., Alexander, L. V., Brönnimann, S., Charabi, Y., Dentener, F. J., Dlugokencky, E. J., Easterling, D. R., Kaplan, A., Soden, B. J., Thorne, P. W., Wild, M., and Zhai, P. M.: Observations: Atmosphere and surface, in: Climate Change 2013: The Physical Science Basis, Contribution to Working Group 1 to the Fifth Assessment Report of the Intergovernmental Panel on Climate Change, edited by: Stocker, T. F., Qin. D., Plattner, G.-K., Tignor, M., Allen, S. K., Boschung, J., Nauels, A., Xia, Y., Bex, Y., and Midgley, P. M., Cambridge University Press, Cambridge, UK and New York, NY, USA, http://www.ipcc.ch/report/ar5/wg1, 2013.

Heckendorn, P., Weisenstein, D., Fueglistaler, S., Luo, B. P., Rozanov, E., Schraner, M., Thomason, L. W., and Peter, T.: The impact of geoengineering aerosols on stratospheric temperature and ozone, Environ. Res. Lett., 4, 045108, https://doi.org/10.1088/1748-9326/4/4/045108, 2009.

Hegglin, M. I. and Shepherd, T. G.: Large climate-induced changes in ultraviolet index and stratosphere-to-troposphere ozone flux, Nat. Geosci., 2, 687–691, https://doi.org/10.1038/ngeo604, 2009.

Jones, A., Haywood, J., and Boucher, O.: A comparison of the climate impacts of geoengineering by stratospheric $SO_2$ injection and by brightening the marine stratocumulus cloud, Atmos. Sci. Let., 12, 176–183, https://doi.org/10.1002/asl.291, 2011.

Kampa, M. and Castanas, E.: Human health effects of air pollution, Environ. Pollut., 151, 362–367, https://doi.org/10.1016/j.envpol.2007.06.012, 2008.

Kinnison, D. E., Brasseur, G. P., Walters, S., Garcia, R. R., Marsh, D. R., Sassi, F., Harvey, V. L., Randall, C. E., Emmons, L., Lamarque, J.-F., Hess, P., Orlando, J. J., Tie, X. X., Randel, W., Pan, L. L., Gettelman, A., Granier, C., Diehl, T., Niemeier, U., and Simmons, A. J.: Sensitivity of chemical tracers to meteorological parameters in the MOZART-3 chemical transport model, J. Geophys. Res., 112, D20302, https://doi.org/10.1029/2006JD007879, 2007.

Lamarque, J.-F., Bond, T. C., Eyring, V., Granier, C., Heil, A., Klimont, Z., Lee, D., Liousse, C., Mieville, A., Owen, B., Schultz, M. G., Shindell, D., Smith, S. J., Stehfest, E., Van Aardenne, J., Cooper, O. R., Kainuma, M., Mahowald, N., Mc- Connell, J. R., Naik, V., Riahi, K., and van Vuuren, D. P.: Historical (1850–2000) gridded anthropogenic and biomass burning emissions of reactive gases and aerosols: methodology and application, Atmos. Chem. Phys., 10, 7017–7039, https://doi.org/10.5194/acp-10-7017-2010, 2010.

Lamarque, J.-F., Emmons, L. K., Hess, P. G., Kinnison, D. E., Tilmes, S., Vitt, F., Heald, C. L., Holland, E. A., Lauritzen, P. H., Neu, J., Orlando, J. J., Rasch, P. J., and Tyndall, G. K.: CAM-Chem: Description and evaluation of interactive atmospheric chemistry in the Community Earth System Model, Geosci. Model Dev., 5, 369–411, https://doi.org/10.5194/gmd-5-369-2012, 2012.

Lathière, J., Hauglustaine, D. A., De Noblet-Ducoudré, N., Krinner, G., and Folberth, G. A.: Past and future changes in biogenic volatile organic compound emissions simulated with a global dynamic vegetation model, Geophys. Res. Lett., 32, L20818, https://doi.org/10.1029/2005GL024164, 2005.

Lin, P. and Fu, Q.: Changes in various branches of the Brewer-Dobson circulation from an ensemble of chemistry climate models, J. Geophys. Res.-Atmos., 118, 73–84, https://doi.org/10.1029/2012JD018813, 2013.

MacIntosh, C. R., Allan, R. P., Baker, L. H., Bellouin, N., Collins, W., Mousavi, Z., and Shine, K. P.: Contrasting fast precipitation responses to tropospheric and stratospheric ozone forcing, Geophys. Res. Lett., 43, 1263–1271, https://doi.org/10.1002/2015GL067231, 2016.

Mauzerall, D. L. and Wang, X. P.: Protecting agricultural crops from the effects of tropospheric ozone exposure: Reconciling science and standard setting in the United States, Europe, and Asia, Annu. Rev. Energ. Env., 26, 237–268, https://doi.org/10.1146/annurev.energy.26.1.237, 2001.

Meinshausen, M., Smith, S. J., Calvin, K., Daniel, J. S., Kainuma, M. L., T., Lamarque, J.-F., Matsumoto, K., Montzka, S. A., Raper, S. C. B., Riahi, K., Thomson, A., Velders, G. J. M., and van Vuuren, D. P. P.: The RCP greenhouse gas concentrations and their extension from 1765 to 2300, Climatic Change, 109, 213–241, https://doi.org/10.1007/s10584-011-0156-z, 2011.

Mills, M., Toon, O. B., Lee-Taylor, J., and Robock, A.: Multidecadal global cooling and unprecedented ozone loss following a regional nuclear conflict, Earth's Future, 2, 161–176, https://doi.org/10.1002/2013EF000205, 2014.

Monks, P. S.: Gas-phase radical chemistry in the troposphere, Chem. Soc. Rev., 34, 376–395, https://doi.org/10.1039/b307982c, 2005.

Monks, P. S., Archibald, A. T., Colette, A., Cooper, O., Coyle, M., Derwent, R., Fowler, D., Granier, C., Law, K. S., Mills, G. E., Stevenson, D. S., Tarasova, O., Thouret, V., Von Schneidemesser, E., Sommariva, R., Wild, O., and Williams, M. L.: Tropospheric ozone and its precursors from the urban to the global scale from air quality to short-lived climate forcer, Atmos. Chem. Phys., 15, 8889–8973, https://doi.org/10.5194/acp-15-8889-2015, 2015.

Morgenstern, O., Hegglin, M. I., Rozanov, E., O'Connor, F. M., Abraham, N. L., Akiyoshi, H., Archibald, A. T., Bekki, S., Butchart, N., Chipperfield, M. P., Deushi, M., Dhomse, S. S., Garcia, R. R., Hardiman, S. C., Horowitz, L. W., Jöckel, P., Josse, B., Kinnison, D., Lin, M., Mancini, E., Manyin, M. E., Marchand, M., Marécal, V., Michou, M., Oman, L. D., Pitari, G., Plummer, D. A., Revell, L. E., Saint-Martin, D., Schofield, R., Stenke, A., Stone, K., Sudo, K., Tanaka, T. Y., Tilmes, S., Yamashita, Y., Yoshida, K., and Zeng, G.: Review of the global models used within phase 1 of the Chemistry-Climate Model Initiative (CCMI), Geosci. Model Dev., 10, 639–671, https://doi.org/10.5194/gmd-10-639-2017, 2017.

Neu, J. L., Flury, T., Manney, G. L., Santee, M. L., Livesey, N. J., and Worden, J.: Tropospheric ozone variations governed by changes in stratospheric circulation, Nat. Geosci., 7, 340–344, https://doi.org/10.1038/NGEO2138, 2014.

Niemeier, U., Schmidt, H., Alterskjær, K., and Kristjánsson, J. E.: Solar irradiance reduction via climate engineering: Impact of different techniques on the energy balance and the hydrological cycle, J. Geophys. Res.-Atmos., 118, 11905–11917, https://doi.org/10.1002/2013JD020445, 2013.

Nowack, P. J., Abraham, N. L., Maycock, A. C., Braesicke, P., Gregory, J. M., Joshi, M. M., Osprey, A., and Pyle, J. A.: A large ozone-circulation feedback and its implications for global warming assessments, Nature Climate Change, 5, 41–45, https://doi.org/10.1038/nclimate2451, 2015.

Nowack, P. J., Abraham, N. L., Braesicke, P., and Pyle, J. A.: Stratospheric ozone changes under solar geoengineering: implications for UV exposure and air quality, Atmos. Chem. Phys., 16, 4191–4203, https://doi.org/10.5194/acpd-15-31973-2015, 2016.

Nowack, P. J., Braesicke, P., Abraham, N. L., and Pyle, J. A.: On the role of ozone feedback in the ENSO amplitude response under global warming, Geophys. Res. Lett., 44, 3858–3866, https://doi.org/10.1002/2016GL072418, 2017.

Pitari, G., Aquila, V., Kravitz, B., Robock, A., Watanabe, S., Cionni, I., De Luca, N., Genova, G. I., Mancini, E., and Tilmes, S.: Stratospheric ozone response to sulfate geoengineering: Results from the Geoengineering Model Intercomparison Project (GeoMIP), J. Geophys. Res.-Atmos., 119, 2629–2653, https://doi.org/10.1002/2013JD020566, 2014.

Pitari, G., Cionni, I., Genova, G. D., Visioni, D., Gandolfi, I., and Mancini, E.: Impact of stratospheric volcanic aerosols on age-of-air and transport of long lived species, Atmosphere, 7, 149 pp., https://doi.org/10.3390/atmos7110149, 2016.

Portmann, R. W., Solomon, S., Garcia, R. R., Thomason, L. W., Poole, L. R., and McCormick, M. P.: Role of aerosol variations in anthropogenic ozone depletion in the polar regions, J. Geophys. Res., 101, 22991–23006, https://doi.org/10.1029/96JD02608, 1996.

Rasch, P. J., Crutzen, P. J., and Coleman, D. B.: Exploring the geoengineering of climate using stratospheric sulfate aerosols: The role of particle size, J. Geophys. Res., 35, L02809, https://doi.org/10.1029/2007GL032179, 2008a.

Rasch, P. J., Tilmes, S., Turco, R. P., Robock, A., Oman, L., Chen, C-C., Stenchikov, G. L., and Garcia, R. R.: An overview of geoengineering of climate using stratospheric sulphate aerosols, Phil. Trans. R. Soc. A, 366, 4007–4037, https://doi.org/10.1098/rsta.2008.0131, 2008b.

Richter, J. H., Tilmes, S., Mills, M. J., Kravitz, B., MacMartin, D. G., Vitt, F., and Lamarque, J.-F.: Stratospheric dynamical response and ozone feedbacks in the presence of $SO_2$ injection, J. Geophys. Res.-Atmos., in press, 2017.

Robock, A.: Volcanic eruption and climate, Rev. Geophys., 38, 191–219, https://doi.org/10.1029/1998RG000054, 2000.

Sander, S. P., Abbatt, J., Barker, J. R., Burkholder, J. B., Friedl, R. R., Golden, D. M., Huie, R. E., Kolb, C. E., Kurylo, M. J., Moortgat, G. K., Orkin, V. L., and Wine, P. H.: Chemical Kinetics and Photochemical Data for Use in Atmospheric Studies Evaluation No. 17, JPL Publication 10-6, Jet Propulsion Laboratory, Pasadena, https://jpldataeval.jpl.nasa.gov/pdf/JPL10-6Final15June2011.pdf, 2011.

Sharkey, T. D. and Yeh, S. S.: Isoprene emission from plants. Ann. Rev. Plant Phys. Plant Mol. Biol., 52, 407–436, https://doi.org/10.1146/annurev.arplant.52.1.407, 2001.

Shepherd, T. G. and McLandress, C.: A Robust Mechanism for Strengthening of the Brewer–Dobson Circulation in Response to Climate Change: Critical-Layer Control of Subtropical Wave Breaking, J. Atmos. Sci., 68, 784–797, https://doi.org/10.1175/2010JAS3608.1, 2011.

Silva, R. A., West, J. J., Zhang, Y., Anenberg, S. C., Lamarque, J.-F., Shindell, D. T., Collins, W. J., Dalsoren, S., Faluvegi, G., Folberth, G., Horowitz, L. W., Nagashima, T., Naik, V., Rumbold, S., Skeie, R., Sudo, K., Takemura, T., Bergmann, D., Cameron-

Smith, P., Cionni, I., Doherty, R. M., Eyring, V., Josse, B., MacKenzie, I. A., Plummer, D., Righi, M., Stevenson, D. S., Strode, S., Szopa, S. and Zeng, G.: Global premature mortality due to anthropogenic outdoor air pollution and the contribution of past climate change, Environ. Res. Lett., 8, 34005, https://doi.org/10.1088/1748-9326/8/3/034005, 2013.

Soden, B. and Held, I.: An Assessment of Climate Feedbacks in Coupled Ocean – Atmosphere Models, J. Clim., 19, 3354–3360, https://doi.org/10.1175/JCLI9028.1, 2006.

Solomon, S.: Stratospheric ozone depletion: A review of concepts and history, Rev. Geophys., 37, 275–316, https://doi.org/10.1029/1999RG900008, 1999.

Stevenson, D. S., Dentener, F. J., Schultz, M. G., Ellingsen, K., van Noije, T. P. C., Wild, O., Zeng, G., Amann, M., Atherton, C. S., Bell, N., Bergmann, D. J., Bey, I., Butler, T., Cofala, J., Collins, W. J., Derwent, R. G., Doherty, R. M., Drevet, J., Eskes, H. J., Fiore, A. M., Gauss, M., Hauglustaine, D. A., Horowitz, L. W., Isaksen, I. S. A., Krol, M. C., Lamarque, J.-F., Lawrence, M. G., Montanaro, V., Müller, J.-F., Pitari, G., Prather, M. J., Pyle, J. A., Rast, S., Rodriguez, J. M., Sanderson, M. G., Savage, N. H., Shindell, D. T., Strahan, S. E., Sudo, K., and Szopa, S.: Multimodel ensemble simulations of present-day and near-future tropospheric ozone, J. Geophys. Res., 111, D08301, https://doi.org/10.1029/2005JD006338, 2006.

Stevenson, D. S., Young, P. J., Naik, V., Lamarque, J. F., Shindell, D. T., Voulgarakis, A., Skeie, R. B., Dalsoren, S. B., Myhre, G., Berntsen, T. K., Folberth, G. A., Rumbold, S. T., Collins, W. J., MacKenzie, I. A., Doherty, R. M., Zeng, G., Van Noije, T. P. C., Strunk, a., Bergmann, D., Cameron-Smith, P., Plummer, D. A., Strode, S. A., Horowitz, L., Lee, Y. H., Szopa, S., Sudo, K., Nagashima, T., Josse, B., Cionni, I., Righi, M., Eyring, V., Conley, A., Bowman, K. W., Wild, O., and Archibald, A.: Tropospheric ozone changes, radiative forcing and attribution to emissions in the Atmospheric Chemistry and Climate Model Intercomparison Project (ACCMIP), Atmos. Chem. Phys., 13, 3063–3085, https://doi.org/10.5194/acp-13-3063-2013, 2013.

Tang, Q., Hess, P. G., Brown-steiner, B., and Kinnison, D. E.: Tropospheric ozone decrease due to the Mount Pinatubo eruption?: Reduced stratospheric influx, Geophys. Res. Lett., 40, 5553–5558, https://doi.org/10.1002/2013GL056563, 2013.

Tie, X. and Brasseur, G.: The response of stratospheric ozone to volcanic eruptions: Sensitivity to atmospheric chlorine loading, Geophys. Res. Lett., 22, 3035–3038, https://doi.org/10.1029/95GL03057, 1995.

Tilmes, S., Müller, R., and Salawitch, R.: The sensitivity of polar ozone depletion to proposed geoengineering schemes, Science, 320, 1201–1204, https://doi.org/10.1126/science.1153966, 2008.

Tilmes, S., Garcia, R. R., Kinnison, D. E., Gettelman, A., and Rasch, P. J.: Impact of geoengineered aerosols on the troposphere and stratosphere, J. Geophys. Res., 114, D12305, https://doi.org/10.1029/2008JD011420, 2009.

Tilmes, S., Kinnison, D. E., Garcia, R. R., Salawitch, R., Canty, T., Lee-Taylor, J., Madronich, S., and Chance, K.: Impact of very short-lived halogens on stratospheric ozone abundance and UV radiation in a geo-engineered atmosphere, Atmos. Chem. Phys., 12, 10945–10955, https://doi.org/10.5194/acp-12-10945-2012, 2012.

Tilmes, S., Mills, M. J., Niemeier, U., Schmidt, H., Robock, A.,

Kravitz, B., Lamarque, J.-F., Pitari, G., and English, J. M.: A new Geoengineering Model Intercomparison Project (GeoMIP) experiment designed for climate and chemistry models, Geosci. Model Dev., 8, 43–49, https://doi.org/10.5194/gmd-8-43-2015, 2015.

Tilmes, S., Sanderson, B. M., and O'Neill, B. C.: Climate impacts of geoengineering in a delayed mitigation scenario, Geophys. Res. Lett., 43, 8222–8229, https://doi.org/10.1002/2016GL070122, 2016a.

Tilmes, S., Lamarque, J.-F., Emmons, L. K., Kinnison, D. E., Marsh, D., Garcia, R. R., Smith, A. K., Neely, R. R., Conley, A., Vitt, F., Val Martin, M., Tanimoto, H., Simpson, I., Blake, D. R., and Blake, N.: Representation of the Community Earth System Model (CESM1) CAM4-chem within the Chemistry-Climate Model Initiative (CCMI), Geosci. Model Dev., 9, 1853–1890, https://doi.org/10.5194/gmd-9-1853-2016, 2016b.

Tingey, D. T., Manning, M., Grothaus, L. C., and Burns, W. F.: Influence of light and temperature on monoterpene emission rates from Slash Pine, Plant Physiol. 65, 797–801, https://doi.org/10.1104/pp.65.5.797, 1980.

Vingarzan, R.: A review of surface ozone background levels and trends, Atmos. Env., 38, 3431–3442, https://doi.org/10.1016/j.atmosenv.2004.03.030, 2004.

Waugh, D.: Age of stratospheric air: Theory, observations, and models, Rev. Geophys., 40, 1–26, https://doi.org/10.1029/2000RG000101, 2002.

Wigley, T. M. L.: A combined mitigation/geoengineering approach to climate stabilization, Science, 314, 452–454, https://doi.org/10.1126/science.1131728, 2006.

Wild, O.: Modelling the global tropospheric ozone budget: exploring the variability in current models, Atmos. Chem. Phys., 7, 2643–2660, https://doi.org/10.5194/acp-7-2643-2007, 2007.

Wild, O., Fiore, A. M., Shindell, D. T., Doherty, R. M., Collins, W. J., Dentener, F. J., Schultz, M. G., Gong, S., MacKenzie, I. A., Zeng, G., Hess, P., Duncan, B. N., Bergmann, D. J., Szopa, S., Jonson, J. E., Keating, T. J., and Zuber, A.: Modelling future changes in surface ozone: a parameterized approach, Atmos. Chem. Phys., 12, 2037–2054, https://doi.org/10.5194/acp-12-2037-2012, 2012.

Wilton, D. J., Hewitt, C. N., and Beerling, D. J.: Simulating effect of changes in direct and diffuse radiation on canopy scale isoprene emissions from vegetation following volcanic eruptions, Atmos. Chem. Phys., 11, 11723–11731, https://doi.org/10.5194/acp-11-11723-2011, 2011.

Xia, L., Robock, A., Tilmes, S., and Neely III, R. R.: Stratospheric sulfate geoengineering could enhance the terrestrial photosynthesis rate, Atmos. Chem. Phys., 16, 1479–1489, https://doi.org/10.5194/acp-16-1479-2016, 2016.

Young, P. J., Archibald, A. T., Bowman, K. W., Lamarque, J.-F., Naik, V., Stevenson, D. S., Tilmes, S., Voulgarakis, A., Wild, O., Bergmann, D., Cameron-Smith, P., Cionni, I., Collins, W. J., Dalsøren, S. B., Doherty, R. M., Eyring, V., Faluvegi, G., Horowitz, L. W., Josse, B., Lee, Y. H., MacKenzie, I. A., Nagashima, T., Plummer, D. A., Righi, M., Rumbold, S. T., Skeie, R. B., Shindell, D. T., Strode, S. A., Sudo, K., Szopa, S., and Zeng, G.: Pre-industrial to end 21st century projections of tropospheric ozone from the Atmospheric Chemistry and Climate Model Intercomparison Project (ACCMIP), Atmos. Chem. Phys., 13, 2063–2090, https://doi.org/10.5194/acp-13-2063-2013, 2013.

# Projected global ground-level ozone impacts on vegetation under different emission and climate scenarios

**Pierre Sicard[1], Alessandro Anav[2], Alessandra De Marco[3], and Elena Paoletti[2]**

[1]ACRI-HE, Sophia Antipolis, France
[2]Institute of Sustainable Plant Protection, National Research Council, Sesto Fiorentino, Italy
[3]Italian National Agency for New Technologies, Energy and the Environment, C.R. Casaccia, Italy

*Correspondence to:* Pierre Sicard (pierre.sicard@acri-he.fr)

**Abstract.** The impact of ground-level ozone ($O_3$) on vegetation is largely under-investigated at the global scale despite large areas worldwide that are exposed to high surface $O_3$ levels. To explore future potential impacts of $O_3$ on vegetation, we compared historical and projected surface $O_3$ concentrations simulated by six global atmospheric chemistry transport models on the basis of three representative concentration pathways emission scenarios (i.e. RCP2.6, 4.5, 8.5). To assess changes in the potential surface $O_3$ threat to vegetation at the global scale, we used the AOT40 metric. Results point out a significant exceedance of AOT40 in comparison with the recommendations of UNECE for the protection of vegetation. In fact, many areas of the Northern Hemisphere show that AOT40-based critical levels will be exceeded by a factor of at least 10 under RCP8.5. Changes in surface $O_3$ by 2100 worldwide range from about +4–5 ppb in the RCP8.5 scenario to reductions of about 2–10 ppb in the most optimistic scenario, RCP2.6. The risk of $O_3$ injury for vegetation, through the potential $O_3$ impact on photosynthetic assimilation, decreased by 61 and 47 % under RCP2.6 and RCP4.5, respectively, and increased by 70 % under RCP8.5. Key biodiversity areas in southern and northern Asia, central Africa and North America were identified as being at risk from high $O_3$ concentrations.

## 1  Introduction

Tropospheric ozone ($O_3$) is a secondary air pollutant; that is, $O_3$ is not emitted as such in the air but it is formed by reactions among precursors (e.g. $CH_4$, VOCs, $NO_x$). Ozone is an important greenhouse gas resulting in a direct radiative forcing of $0.35$–$0.37\,W\,m^{-2}$ on climate (Shindell et al., 2009; Ainsworth et al., 2012). Despite significant control efforts and legislation to reduce $O_3$ precursor emissions, tropospheric $O_3$ pollution is still a major air quality issue over large regions of the globe (Lefohn et al., 2010; Langner et al., 2012; Young et al., 2013; Cooper et al., 2014; EEA, 2015; Sicard et al., 2016a, b; Ochoa-Hueso et al., 2017). Long-range transport of $O_3$ and precursors of $O_3$ can elevate the local and regional $O_3$ background concentrations (Ellingsen et al., 2008; Sicard et al., 2009; Wilson et al., 2012; Paoletti et al., 2014; Derwent et al., 2015; Xing et al., 2015; Sicard et al., 2016a). Therefore, remote areas such as the Arctic region can be affected (Langner et al., 2012). The current surface $O_3$ levels (35–50 ppb in the Northern Hemisphere, NH;) are high enough to damage both forests and crops by reducing growth rates and productivity (Wittig et al., 2009; Anav et al., 2011; Mills et al., 2011; Sicard et al., 2011; Ashworth et al., 2013; Proietti et al., 2016).

Increasing atmospheric $CO_2$, nitrogen deposition and temperatures enhance plant growth and increase primary production and greening of plants (Nemani et al., 2003; Zhu et al., 2016). At the global scale, a widespread increase of greening and net primary production (NPP) is observed over 25–50 % of the vegetated area, while a decrease is observed over only 7 % of the globe (Nemani et al., 2003; Zhu et al., 2016). In contrast, a previous modelling study over Europe shows how surface $O_3$ reduces the mean annual gross primary production (GPP) by about 22 % and the leaf area index (LAI) by 15–20 % (Anav et al., 2011). Similarly, Proietti et al. (2016), using different in situ measurements collected over 37 European forest sites, found a GPP decrease (up to 30 %) caused by $O_3$ during the time period of 2000–2010. At the global scale, over the time period of 1901–2100, GPP is projected to decrease by 14–23 % (Sitch et al., 2007). As a consequence of reduced photosynthetic assimilation, the total biomass of trees is estimated to be decreased by 7 % under the current ground-level $O_3$ mean concentrations (40 ppb on average) and by 17 % at mean $O_3$ concentrations expected in 2100 (97 ppb based on a meta-analysis) compared to preindustrial $O_3$ levels in NH (about 10 ppb, Wittig et al., 2009). From experiments, Wittig et al. (2009) also reported that the total tree biomass of angiosperms was reduced by 23 % at $O_3$ mean concentrations of 74 ppb and by 7 % at 92 ppb for gymnosperms. High surface $O_3$ levels, exceeding 40 ppb, do occur in many regions of the globe with associated economic costs of several billion dollars per year (Wang and Mauzerall, 2004; Ashmore, 2005). Ashworth et al. (2013) reported an annual loss of 3.5 % for wheat (very $O_3$ sensitive) and 1.0 % for maize (more $O_3$ tolerant) for Europe in 2010 relative to 2000, while Holland et al. (2006) estimated a EUR 4.5 billion loss in the production of 23 common crop species, due to surface $O_3$ exposure by 2020 relative to 2000.

The international Tropospheric Ozone Assessment Report (TOAR) establishes a state-of-the-art of global $O_3$ metrics for climate change, human health and crop/ecosystem research (Lefohn et al., 2017). To assess the potential $O_3$ risk and protect vegetation from $O_3$, different metrics are used: the European and US standard (AOT40 and W126, respectively) are based on exposure-based metrics, while flux-based metrics have been introduced only recently (UNECE, 2010; Klingberg et al., 2014; EEA, 2015). Unlike the exposure-based metrics, which only rely on the surface $O_3$ concentration, the flux-based metrics were developed to quantify the accumulation of damaging $O_3$ taken up by vegetation through the stomata over a species-specific phenological time window. These metrics also provide an information-rich tool in assessing the relative effectiveness of air pollution control strategies in lowering surface $O_3$ levels worldwide (Monks et al., 2015). By reducing plant photosynthesis and growth, high surface $O_3$ levels will result in reduction in carbon storage by vegetation and, finally, an indirect radiative forcing as a consequence of the $CO_2$ rising in the atmosphere (Sitch et al., 2007; Ainsworth et al., 2012). This rising $CO_2$ reduces stomatal conductance, which decreases $O_3$ flux into plants, leading to increased $O_3$ levels in the air of 3–4 ppb during the growing season over the NH by doubling of $CO_2$ concentration (Fiscus et al., 2005; Sanderson et al., 2007).

Projected changes in ground-level $O_3$ vary considerably among models (Stevenson et al., 2006; Wild, 2007) and emission scenarios. In earlier studies, the emissions of $O_3$ precursors were based on a high population growth, leading to very high projected surface $O_3$ concentrations by 2100 (Stevenson et al., 2000; Zeng and Pyle, 2003; Shindell et al., 2006). The last emission scenarios, i.e. the Representative Concentration Pathways (RCPs), were developed as part of the Fifth

Assessment Report of the Intergovernmental Panel on Climate Change (Meinshausen et al., 2011; van Vuuren et al., 2011; Cubasch et al., 2013; Myhre et al., 2013). These scenarios include different assumptions on climate, energy access policies, and land cover and land use changes (Arneth et al., 2008; Kawase et al., 2011; Kirtman et al., 2013). Until now, studies on $O_3$ pollution impacts on terrestrial ecosystems are either limited to a single model or to particular regions (e.g. Clifton et al., 2014; Rieder et al., 2015) and only a few applications of global or regional models under the new RCPs scenarios were carried out (Kelly et al., 2012). In the framework of the Atmospheric Chemistry and Climate Model Intercomparison Project (ACCMIP), different simulations were performed by Lamarque et al. (2013) and Young et al. (2013) from 16 global chemistry models.

A few issues about surface $O_3$, such as a better understanding of spatial changes and a better assessment of $O_3$ impacts worldwide, are still challenging. To overcome these issues, the aim of this study is to quantify, for the first time, the spatial and temporal changes in the projected potential $O_3$ impacts on photosynthetic carbon assimilation of vegetation at the global scale by comparing the $O_3$ potential injury at present with that expected at the end of the 21st century from different global chemistry models. The purpose of this study is not to provide a quantitative estimation of the ecosystem injury due to $O_3$ but to highlight the world areas at higher risk and changes by 2100.

## 2  Materials and methods

### 2.1  ACCMIP models and RCP scenarios

The global chemistry models used in this work were developed under ACCMIP. A detailed description of the selected models and of the emission scenarios (i.e. RCPs) is included in the Supplement. ACCMIP models were widely validated and used to evaluate projected changes in atmospheric chemistry and air quality under different emission and climate assumptions (e.g. Lamarque et al., 2010; Prather et al., 2012; Bowman et al., 2013; Lee et al., 2013; Voulgarakis et al., 2013). Lamarque et al. (2013) and Young et al. (2013) provided the main characteristics of 16 models and details for the ACCMIP simulations. Although within ACCMIP 16 models are available, due to the lack of hourly $O_3$ concentration here we only focus on six global chemistry models with different configurations as presented in Table 1.

The length of historical and RCP simulations varies between models, but for all models the historical runs cover a period centred around 2000, while the time slice of RCPs is centred around 2100 (Table 1). As for each model we compare the relative mean change between the historical and RCP simulations, a different length in the number of years used in the analysis, the uncertainty is limited.

**Table 1.** Characteristics of the models, including simulation time slice, spatial resolution, simulated gas species and associated bibliographic references (from Lamarque et al., 2013 and Young et al., 2013). BC is black carbon, OC is organic carbon, SOA is secondary organic aerosols, DMS is dimethyl sulfide, CCM is chemistry climate model, CTM is chemistry transport model and CGCM is chemistry general circulation model.

| Models | Type | Simulation length | Resolution (lat/long) | Number of vertical pressure levels and top level | Species simulated | References |
|---|---|---|---|---|---|---|
| CESM-CAM | CCM | 2000–2009 and 2100–2109 | 1.875/2.5 | 26 levels 3.5 hPa | 16 gas species; constant present-day isoprene, soil $NO_x$, DMS and volcanic sulfur, oceanic CO. | Lamarque et al. (2012) |
| GFDL-AM3 | CCM | 2001–2010 and 2101–2110 | 2.0/2.5 | 48 levels 0.017 hPa | 81 gas species; $SO_x$, BC, OC, SOA, $NH_3$, $NO_3$; constant pre-industrial soil $NO_x$; constant present-day soil and oceanic CO, and biogenic VOC; climate-sensitive dust, sea salt, and DMS. | Donner et al. (2011) Naik et al. (2012) |
| GISS-E2-R | CCM | 2000–2004 and 2101–2105 | 2.0/2.5 | 40 levels 0.14 hPa | 51 gas species; interactive sulfate, BC, OC, sea salt, dust, $NO_3$, SOA, alkenes; constant present-day soil $NO_x$; climate-sensitive dust, sea salt, and DMS; climate-sensitive isoprene based on present-day vegetation. | Lee and Adams (2011) Shindell et al. (2012) |
| MIROC-CHEM | CCM | 2000–2010 and 2100–2104 | 2.8/2.8 | 80 levels 0.003 hPa | 58 gas species; $SO_4$, BC, OC; constant present-day VOCs, soil-$NO_x$, oceanic-CO; climate-sensitive dust, sea salt and DMS. | Watanabe et al. (2011) |
| MOCAGE | CTM | 2000–2003 and 2100–2103 | 2.0/2.0 | 47 levels 6.9 hPa | 110 gas species; constant present-day isoprene, other VOCs, oceanic CO and soil $NO_x$. | Josse et al. (2004) Krinner et al. (2005) Teyssèdre et al. (2007) |
| UM-CAM | CGCM | 2000–2005 and 2094–2099 | 2.50/3.75 | 19 levels 4.6 hPa | 60 gas species; constant present-day biogenic isoprene, soil $NO_x$, biogenic and oceanic CO. | Zeng et al. (2008, 2010) |

## 2.2 Potential ozone injury on vegetation

The $O_3$ exposure-based index, i.e. AOT40 (ppb h), is a metric used to assess the potential $O_3$ risk to vegetation from local to global scales (Emberson et al., 2014). In literature, AOT40 is computed as sum of the hourly exceedances above 40 ppb, for hours between 08:00 and 20:00 or for hours with a solar radiation exceeding $50\,W\,m^{-2}$ over species-specific growing seasons (UNECE, 2010). Conventionally, two major growing season time windows are used, namely 6 months (April to September) for temperate climates, for example in Europe, and all-year round for Mediterranean, subtropical and tropical-type climates where vegetation is physiologically active all along the year (Paoletti et al., 2007).

UNECE (2010) supports the use of a growing season, but a fixed time window does not allow incorporating the changes in the growing season due to climate change and would thus not be well suited for investigating changes over time. A recent study over Europe showed how computing AOT40 only over the growing season (i.e. April–September) would lead to an underestimation of AOT40 up to 50 % for conifer trees, while in the case of deciduous trees the underestimation is much smaller (< 5 %, Anav et al., 2016). Also, it should be noted that in Anav et al. (2016) the AOT40 is computed year-round. We computed the AOT40 for a model grid for hours between 08:00 and 20:00 (local time) for all days of the year. Therefore, we computed AOT40 as follows:

$$AOT40 = \int_{01jan}^{31dec} \int_{08:00\,a.m.}^{08:00\,p.m.} \max\left(([O_3] - 40), 0\right) \cdot dt, \quad (1)$$

where [$O_3$] is hourly $O_3$ concentration (ppb) simulated by the models at the lower model layer and d$t$ is time step (1 h). The function "maximum" ensures that only values exceeding 40 ppb are taken into account. For the protection of forests, a critical level of 5 ppm h calculated over the growing season is recommended by UNECE (2010). Within the 2008/50/CE Directive, the critical level for agricultural crops (3 ppm h) is adopted as the long-term objective value for the protection of vegetation by 2020.

The current chemistry models cannot predict changes in phenology over time (Anav et al., 2017), and thus the growing season length is the same between the historical period and different RCPs. The use of a common fixed time window (08:00–20:00) all year-round at global level allows skipping the use of a latitude-dependent model, which would increase the level of complexity. Because the growing season is highly variable across the latitude, rather than introducing further uncertainties by using a single model to simulate the growing season at all latitudes, we applied a simplified approach here with a year-long growing season which should be considered as a worst-case scenario. This approach is valuable and can be easily applied at the global scale to compare the historical and projected potential risk to vegetation.

The $O_3$ concentration to be used in AOT40 calculation should be at the top of the canopy; however, most of models used here provide $O_3$ concentrations at 90–120 m. Nevertheless, even if the $O_3$ concentration is simulated at different elevations above the sea level, because for each model we compare the variation between present and future, the change is consistent because the elevation is the same. In the case of risk assessment, by calculating AOT40 year-round, an overestimation can be observed over polluted region of NH. Since the aim of this study is to compare how $O_3$ stress to vegetation changes between historical period and future, even if the AOT40 is mis-estimated at a given model grid point, the relative mean change is consistent because we compared the changes in AOT40 at the same model grid point.

From the AOT40, a factor of risk for forests and crops can be computed (Anav et al., 2011; Proietti et al., 2016). Thus, the potential $O_3$ impact on photosynthetic carbon assimilation (I$O_3$), in the worst-case scenario, is expressed through a dimensionless value as following:

$$IO_3 = \alpha \times AOT40, \tag{2}$$

where $\alpha$ is an empirically derived $O_3$ response coefficient representing the proportional change in net photosynthesis per unit of AOT40 (Anav et al., 2011). From the Global Land Cover Facility (GLCF) data at 1° of spatial resolution, we grouped the vegetation in three categories: conifers, crops (including grassland) and deciduous (including tropical forests and shrubs) trees. Even dynamic global vegetation models make use of plant functional types rather than complex and specific vegetation to simulate shifts in potential vegetation as a response to shifts in climate (Sitch et al.,

2007). The relationships between cumulative ozone exposure and reductions in net photosynthesis vary among and even within species (Reich, 1987; Ollinger et al., 1997). Differences in response per unit uptake tend to be greater in magnitude between functional groups (e.g. hardwoods vs. conifers) where leaf structure and plant growth strategy differ most widely (Reich, 1987). The dimensionless coefficient for coniferous trees ($0.7 \times 10^{-6}$) and crops ($3.9 \times 10^{-6}$) are based on the regressions of the photosynthesis response to $O_3$ (Reich, 1987), while the coefficient for deciduous trees ($2.6 \times 10^{-6}$) is based on Ollinger et al. (1997). From simulated changes in the risk factor, we can highlight potential risk areas for vegetation.

## 3 Results and discussion

We show the simulated global $O_3$ spatial pattern of mean annual $O_3$ concentration at the lower model layer in Fig. 1 explaining AOT40 patterns. Then, in Fig. 2 we show and discuss the AOT40 spatial and temporal distribution from the ACCMIP models for the historical and RCPs simulations, and finally in Fig. 3 we show the percentage of variation of I$O_3$, i.e. the change in the potential impact of $O_3$ on photosynthetic carbon assimilation for the ACCMIP models computed comparing the RCPs simulations with historical runs. A detailed description of each figure, model by model, is included in the Supplement. Table 2 show the annual total emissions and changes of CO, NMVOCs, $NO_x$, total lightning $NO_x$ emissions ($LNO_x$) and global atmospheric methane ($CH_4$) burden for the historical simulations in each model. The averaged values (simulated percentage) of global, NH and Southern Hemisphere (SH) mean surface $O_3$, AOT40 and I$O_3$ are derived from averaging values over the global/NH/SH land areas only are presented in Table 3.

### 3.1 Spatial pattern of historical ozone concentration and AOT40

The highest surface $O_3$ concentrations (Fig. 1) and potential $O_3$ impacts (Fig. 2) are found in the NH, highlighting a hemispheric asymmetry. AOT40 was used widely during the last 2 decades, not only in Europe but also in South America (Moura et al., 2014) and Asia (Hoshika et al., 2011), when environmental factors are not limiting, e.g. water availability, air temperature, solar radiation affecting stomata opening (Anav et al., 2016; De Marco et al., 2016).

The multi-model $O_3$ mean concentration, averaged over the land points of the domain, is $37.9 \pm 4.3$ ppb in NH and $22.9 \pm 3.8$ ppb in SH (Table 3a). Over land surfaces, the NH extratropics (i.e. mid-latitudes beyond the tropics) have 65 % more $O_3$ than the SH extratropics (data not shown). Similarly, the highest AOT40 values are found in the NH, with an averaged AOT40 of $24.8 \pm 10.1$ ppm h in NH and $2.5 \pm 1.7$ ppm h in SH (Table 3a).

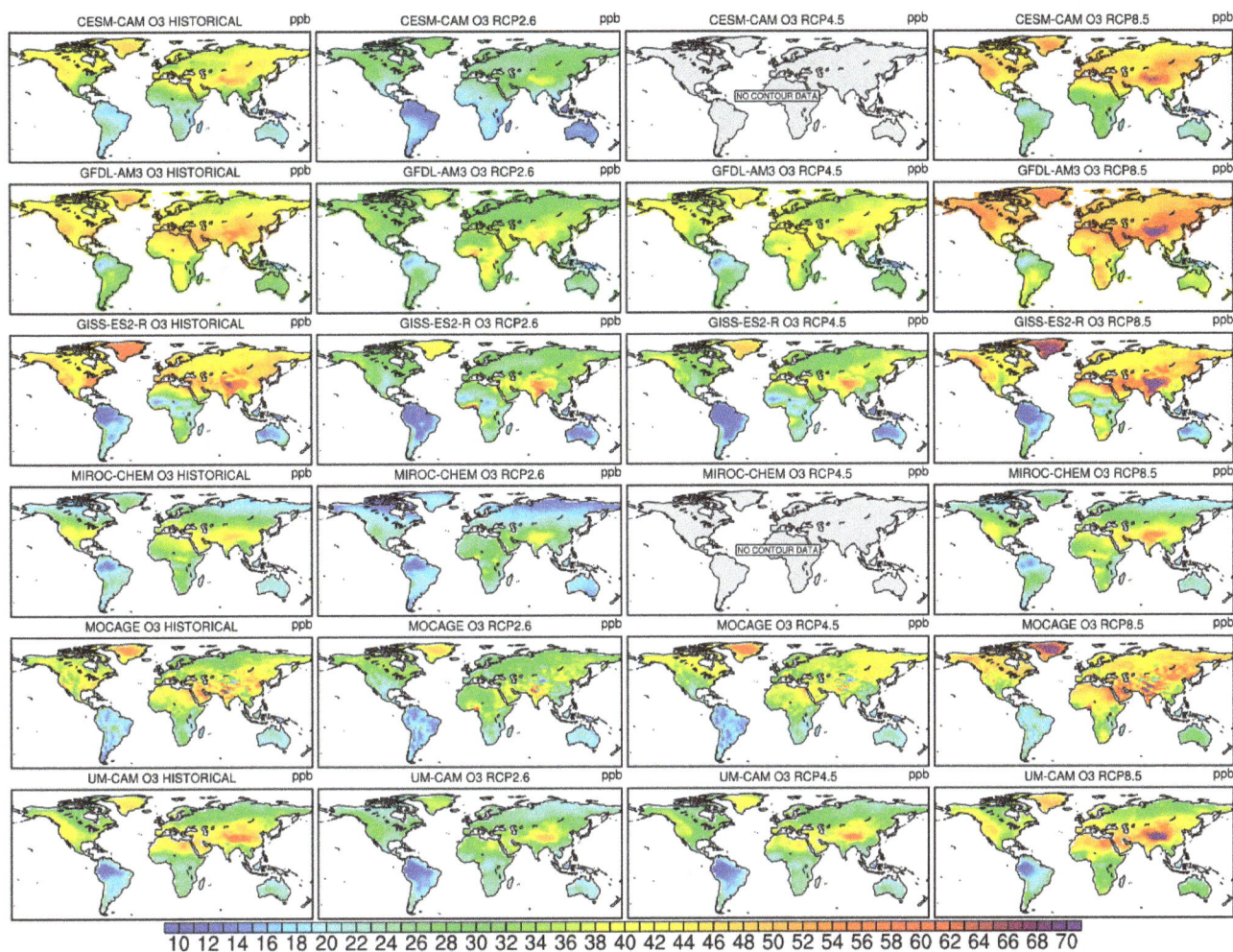

**Figure 1.** Surface ozone average concentrations (in ppb) at the lower model layer for each ACCMIP model over the historical period and for RCP2.6, RCP4.5 and RCP8.5 simulations by 2100. The data are missing for two models under RCP4.5 ("no contour data").

According to previous studies, the annual mean background $O_3$ concentrations at NH mid-latitude range between 35 and 50 ppb during the end of the 20th century (e.g. Cooper et al., 2012; IPCC, 2014; Lefohn et al., 2014). Similarly, we found historical surface $O_3$ mean concentrations ranging between 35 and 50 ppb and between 35 and 50 ppm h for AOT40 in the NH, with the highest values occurring over Greenland and in the latitude band 15–45° N, particularly around the Mediterranean basin, Near East, North America and over the Tibetan Plateau ($> 50$ ppb and 70 ppm h), while the lowest $O_3$ burden (15–30 ppb, $< 20$ ppm h) was recorded in SH, particularly over Amazonian, African and Indonesian rainforests, where the $O_3$ dry deposition rate is maximum, up to 1.80 cm s$^{-1}$ for mixed wood forests (Wesely and Hicks, 2000). Tropospheric $O_3$ has a significant source from stratospheric $O_3$ (Parrish et al., 2012) and it can be transported by the large-scale Brewer–Dobson overturning circulation, i.e. an upward motion from the tropics and downward at higher latitudes, resulting in higher $O_3$ concentrations in the extrat-

ropics (Hudson et al., 2006; Seidel et al., 2008; Parrish et al., 2012). The six models are able to reproduce the spatial pattern of $O_3$ concentration and thus AOT40 worldwide.

The highest historical $O_3$ mean concentrations are observed in GFDL-AM3 and the lowest are found in MIROC-CHEM. In the early 2000s, the maximum global $O_3$ mean concentration (39 ppb) in GFDL-AM3 is associated to the lowest annual total $NO_x$ emissions (46.2 Tg; Table 2a) and low $LNO_x$ (4.4 Tg) while the minimum global $O_3$ mean concentration (28 ppb) in MIROC-CHEM is related to the highest emissions of total $NO_x$ per year (57.3 Tg) and erroneously high $LNO_x$ (9.7 Tg per year; Lamarque et al., 2013). MIROC-CHEM simulates 58 gaseous species in the chemical scheme with constant present-day biogenic VOCs emissions while GFDL-AM3 simulates 81 species (Stevenson et al., 2013; Lamarque et al., 2013). In GISS-E2-R, the hemispheric asymmetry in $O_3$ is more important with e.g. a mean concentration of 22 ppb in SH and 42 ppb in NH. A stronger global AOT40 mean (26 ppm h) is observed in GISS-E2-R

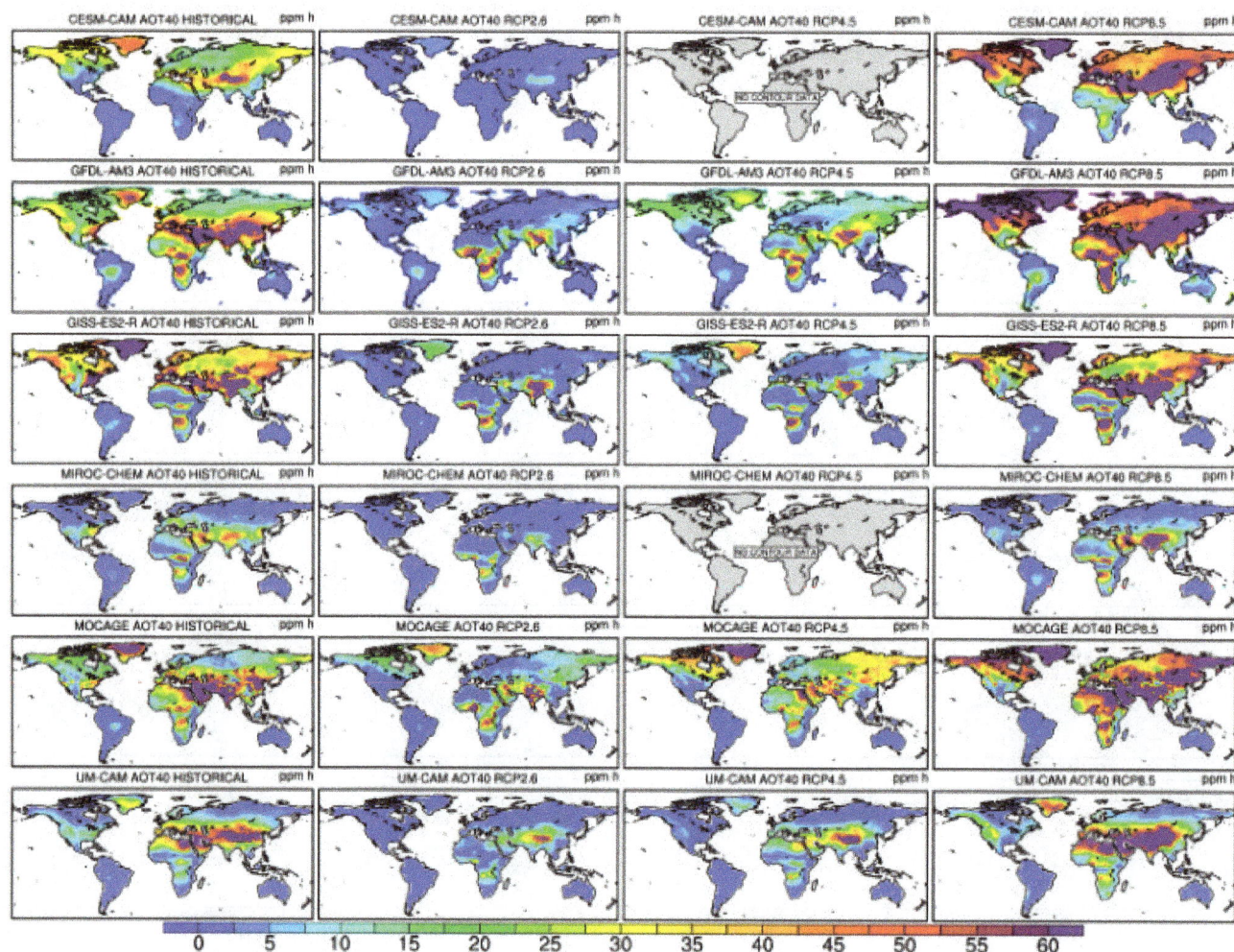

**Figure 2.** Surface mean AOT40 (in ppm h) at the lower model layer for each ACCMIP model over the historical period and for RCP2.6, RCP4.5 and RCP8.5 simulations by 2100. The data are missing for two models under RCP4.5 ("no contour data").

and the lowest (7 ppm h) in MIROC-CHEM for historical simulations. Model-to-model differences are observed due to different natural emissions of $O_3$ precursors (e.g. lightning $NO_x$) and the different chemical schemes used.

Higher $O_3$ burdens (mean concentration > 50 ppb, AOT40 > 70 ppm h) are simulated at high-elevation areas, e.g. at Rocky and Appalachian mountains and over the Tibetan Plateau (Figs. 1 and 2). At high elevation, solar radiation, biogenic VOC emission, exchange between free troposphere and boundary layer and stratospheric $O_3$ intrusion within the troposphere are more important that at the surface layer (Steinbacher et al., 2004; Kulkarni et al., 2011; Lefohn et al., 2012). Altitude reduces the $O_3$ destruction by deposition and NO (Chevalier et al., 2007). In addition, due to the high elevation, ambient air remains colder and dryer in summer, leading to lower summertime $O_3$ losses from photolysis (Helmig et al., 2007). The high-elevation areas, characterized by higher $O_3$ burdens, are well simulated in GISS-E2-R and MOCAGE models.

The Tibetan Plateau, a so-called "ozone valley", is the highest plateau in the world, with a mean height of 4000 m a.s.l. (Tian et al., 2008) with strong thermal and dynamic influences on regional and global climate (Chen et al., 2011). High surface $O_3$ mean concentrations (40–60 ppb) were reported in previous studies (e.g. Zhang et al., 2004; Bian et al., 2011; Guo et al., 2015; Wang et al., 2015). Although this region is remote, road traffic, biofuel energy source, coal mines and trash burning are prevalent. These pollution sources contribute to significant amount of $NO_x$, CO and VOCs (Wang et al., 2015). The high $O_3$ levels are attributed to the combined effects of high-elevation surface, thermal and dynamical forcing of the Tibetan Plateau and in situ photochemical production in the air trapped in the plateau by surrounding mountains (Guo et al., 2015; Wang et al., 2015). The dynamic effect, associated with the large-scale circulation, is more important than the chemical effect (Tian et al., 2008; Liu et al., 2010) and responsible for the high $O_3$ levels over the Tibetan Plateau. The six models are

**Table 2. (a)** Annual total emissions of CO ($TgCO\,year^{-1}$), NMVOCs ($TgC\,year^{-1}$), $NO_x$ ($TgN\,year^{-1}$, including lightning and soil $NO_x$), total lightning $NO_x$ emissions ($LNO_x$) and global atmospheric methane ($CH_4$) burden (Tg) for the historical simulations in each model (from Young et al., 2013, and * from Voulgarakis et al., 2013). **(b)** Simulated percentage (%) changes in total emissions of CO, NMVOCs, $NO_x$ (including lightning and soil $NO_x$), total lightning $NO_x$ emissions ($LNO_x$) and global atmospheric $CH_4$ burden for each model between 2100 and historical simulation for RCPs (from Young et al., 2013, and * Voulgarakis et al., 2013). The last row shows means and SDs. Missing or not available data are identified (n.a.).

**(a)**

| Models | Historical | | | | |
|---|---|---|---|---|---|
| | CO | *CH$_4$ | NMVOCs | NO$_x$ | *LNO$_x$ |
| CESM-CAM | 1248 | 4902 | 429 | 50.0 | 4.2 |
| GFDL-AM3 | 1246 | 4809 | 830 | 46.2 | 4.4 |
| GISS-E2-R | 1070 | 4793 | 830 | 48.6 | 7.7 |
| MIROC-CHEM | 1064 | 4805 | 833 | 57.3 | 9.7 |
| MOCAGE | 1168 | 4678 | 1059 | 47.9 | 5.2 |
| UM-CAM | 1148 | 4879 | 535 | 49.2 | 5.1 |

**(b)**

| Models | RCP2.6 scenario | | | | | RCP4.5 scenario | | | | | RCP8.5 scenario | | | | |
|---|---|---|---|---|---|---|---|---|---|---|---|---|---|---|---|
| | CO | VOCs | NO$_x$ | *LNO$_x$ | *CH$_4$ | CO | VOCs | NO$_x$ | *LNO$_x$ | *CH$_4$ | CO | VOCs | NO$_x$ | *LNO$_x$ | *CH$_4$ |
| CESM-CAM | −36.7 | 0 | −52.8 | +7.1 | −27.1 | n.a. | n.a. | n.a. | n.a. | n.a. | −30.1 | 0 | −33.0 | +29.7 | +112.1 |
| GFDL-AM3 | −36.9 | −5.0 | −47.0 | +12.6 | −27.9 | −47.4 | −3.6 | −41.5 | +23.5 | −9.3 | −30.3 | −1.9 | −22.4 | +38.2 | +116.1 |
| GISS-E2-R | −42.8 | +0.5 | −44.2 | +3.8 | −21.0 | −54.9 | +6.9 | −39.2 | +12.2 | +4.6 | −35.1 | +19.8 | −20.0 | +26.2 | +152.7 |
| MIROC-CHEM | −43.1 | −7.1 | −36.0 | +7.5 | −28.2 | n.a. | n.a. | n.a. | n.a. | n.a. | −35.4 | −3.4 | −6.9 | +38.0 | +116.0 |
| MOCAGE | −39.4 | −6.5 | −45.7 | +5.2 | −28.8 | n.a. | n.a. | n.a. | n.a. | n.a. | −32.3 | −2.8 | −22.9 | +19.9 | +113.4 |
| UM-CAM | −39.0 | −11.3 | −40.6 | +8.1 | −27.9 | −50.4 | −9.2 | −36.0 | +17.5 | −8.7 | −32.0 | −4.2 | −17.2 | +43.6 | +112.1 |
| Mean ± SD | −39.7 | −4.9 | −44.4 | +7.4 | −26.8 | −50.9 | −2.0 | −38.9 | +17.7 | −4.5 | −32.5 | +1.3 | −20.4 | +32.6 | +120.4 |
| | ±2.2 | ±4.9 | ±4.3 | ±2.0 | ±3.7 | ±3.2 | ±11.4 | ±2.3 | ±3.7 | ±9.4 | ±1.8 | ±11.6 | ±7.0 | ±10.8 | ±19.5 |

able to reproduce the high surface $O_3$ mean concentrations ($> 50\,ppb$) over the Tibetan Plateau.

Higher $O_3$ mean concentrations ($> 60\,ppb$) are also observed in southwestern USA, at the stations inland close to Los Angeles, in northeastern USA and eastern Asia (e.g. Beijing) in Fig. 1. The American southwest is an $O_3$ precursor hotspot where the industrial sources emit $CH_4$ and VOCs into the air (Jeričević et al., 2013) and the eastern and northern desert areas have higher ambient $O_3$ concentrations than urban areas of southern California due to four factors: on-shore winds, gasoline reformulation, eastward population expansion and nighttime air chemistry (Arbaugh and Bytnerowicz, 2003). The surface concentrations show higher $O_3$ levels in areas downwind of $O_3$ precursor sources, i.e. urban and well-industrialized areas, at distances of hundreds or even thousands of kilometres due to transport of $O_3$ and precursors, including "reservoir" species such as peroxyacetyl nitrate (PAN), lower $O_3$ titration by NO and higher biogenic VOC emission (Wilson et al., 2012; Paoletti et al., 2014; Monks et al., 2015; Sicard et al., 2016a). The higher $O_3$ levels in areas downwind of $O_3$ precursor sources are well simulated in GISS-E2-R and MOCAGE models.

Over Greenland, mean $O_3$ concentrations during the historical runs ranged from 40 to 55 ppb (Fig. 1) except in MIROC-CHEM (20–25 ppb). Similarly, Helmig et al. (2007) reported annual mean of surface $O_3$ concentrations of 47 ppb over Greenland between 2000 and 2005, particularly at the high-elevation Summit Station (3200 m a.s.l.). Several inves-

tigations of snow photochemical and oxidation processes over Greenland concluded that photochemical $O_3$ production can be attributed to high levels of reactive compounds (e.g. oxidized nitrogen species) present in the surface layer during the sunlit periods due to local sources, e.g. $NO_x$ enhancement from snowpack emissions, PAN decomposition, boreal forest fires or ship emissions (Granier et al., 2006; Stohl et al., 2007; Legrand et al., 2009; Walker et al., 2012). The PAN to $NO_x$ ratio increases with increasing altitude and latitude (Singh et al., 1992). The PAN reservoir for $NO_x$ may be responsible for the increase in surface $O_3$ concentrations at high latitudes (Singh et al., 1992). Local $O_3$ production does not appear to have an important contribution to the ambient high $O_3$ levels (Helmig et al., 2007), but the long-range $O_3$ transport can elevate the background concentrations measured at remote sites, e.g. Greenland (Ellingsen et al., 2008; Derwent et al., 2010). Low dry deposition rates for $O_3$, from 0.01 to $0.05\,cm\,s^{-1}$ over oceans and snow, the downward transport of stratospheric $O_3$, the photochemical local production and the large-scale transport (Zhang et al., 2003; Legrand et al., 2009; Walker et al., 2012; Hess and Zbinden, 2013) are known factors to explain higher $O_3$ pollution over Greenland.

The surface $O_3$ concentrations ($> 40\,ppb$) and AOT40 ($> 60\,ppm\,h$) are higher over deserts, downwind of $O_3$ precursor sources (e.g. Near East, Sierra Nevada, Colorado Desert), due to lower $O_3$ dry deposition fluxes (Wesely and Hicks, 2000), $O_3$ precursors long-range transport from urbanized ar-

**Table 3. (a)** Global and hemispheric (averaged over the land points of the domain) mean annual-average surface ozone concentrations (in ppb) and mean AOT40 (in ppm h) for the historical simulations in each model (Northern and Southern hemispheres, i.e. NH and SH). The last row shows means and SDs. **(b)** Simulated percentage (%) changes in global and hemispheric mean annual-average surface ozone concentrations (over the land points of the domain) and in global mean stratospheric ozone column (* from Voulgarakis et al., 2013) for each model between 2100 and historical simulation for RCPs (NH and SH). The last row shows means and SDs. Missing or not available data are identified (n.a.). **(c)** Simulated percentage (%) changes in global and hemispheric mean AOT40 (over the land points of the domain) for each model between 2100 and historical simulation for RCPs (NH and SH). Missing or not available data are identified (n.a.). **(d)** Simulated percentage (%) changes in potential $O_3$ impact on vegetation ($IO_3$, over the land points of the domain) for each model between 2100 and historical simulation for RCPs (NH and SH). Missing or not available data are identified (n.a.).

**(a)**

| Models | Ozone conc. global | Ozone conc. SH | Ozone conc. NH | AOT40 global | AOT40 SH | AOT40 NH |
|---|---|---|---|---|---|---|
| CESM-CAM | 31.3 | 20.9 | 36.4 | 12.8 | 0.2 | 18.9 |
| GFDL-AM3 | 38.6 | 30.6 | 42.9 | 21.8 | 4.7 | 30.8 |
| GISS-E2-R | 35.8 | 22.3 | 42.3 | 26.0 | 3.6 | 36.8 |
| MIROC-CHEM | 27.9 | 20.4 | 31.4 | 7.3 | 1.9 | 9.8 |
| MOCAGE | 32.9 | 21.5 | 38.3 | 22.9 | 3.5 | 31.8 |
| UM-CAM | 31.3 | 21.4 | 36.0 | 14.4 | 1.3 | 20.6 |
| Mean $\pm$ SD | 33.0 $\pm$ 3.8 | 22.9 $\pm$ 3.8 | 37.9 $\pm$ 4.3 | 17.5 $\pm$ 7.2 | 2.5 $\pm$ 1.7 | 24.8 $\pm$ 10.1 |

**(b)**

| Models | Surface ozone mean concentrations | | | | | | | | | * Stratospheric ozone | | |
|---|---|---|---|---|---|---|---|---|---|---|---|---|
| | RCP2.6 global | RCP2.6 SH | RCP2.6 NH | RCP4.5 global | RCP4.5 SH | RCP4.5 NH | RCP8.5 global | RCP8.5 SH | RCP8.5 NH | RCP2.6 global | RCP4.5 global | RCP8.5 global |
| CESM-CAM | −29.1 | −20.6 | −31.3 | n.a. | n.a. | n.a. | +21.9 | +22.5 | +20.5 | n.a. | n.a. | +5.3 |
| GFDL-AM3 | −20.5 | −10.8 | −24.5 | −11.7 | −6.9 | −13.5 | +15.5 | +18.6 | +14.5 | +3.3 | +3.9 | +8.4 |
| GISS-E2-R | −23.5 | −5.8 | −27.9 | −20.4 | −6.3 | −23.9 | +7.0 | +19.3 | +3.8 | +8.0 | +8.8 | +15.1 |
| MIROC-CHEM | −23.3 | −12.3 | −26.8 | n.a. | n.a. | n.a. | +3.9 | +10.3 | +2.2 | +2.6 | n.a. | +4.2 |
| MOCAGE | −12.8 | +7.4 | −18.5 | −1.8 | +17.7 | −7.0 | +20.1 | +40.4 | +16.7 | +19.9 | n.a. | +23.6 |
| UM-CAM | −17.3 | −4.7 | −21.1 | −8.3 | +0.9 | −10.8 | +14.4 | +24.3 | +11.4 | +6.7 | +6.9 | +7.4 |
| Mean $\pm$ SD | −21.1 | −7.8 | −25.0 | −10.5 | +1.4 | −13.8 | +13.8 | +22.6 | +11.5 | +8.1 | +6.5 | +10.7 |
| | $\pm$5.6 | $\pm$9.4 | $\pm$4.7 | $\pm$7.7 | $\pm$11.5 | $\pm$7.2 | $\pm$7.1 | $\pm$10.0 | $\pm$7.3 | $\pm$7.0 | $\pm$2.5 | $\pm$7.4 |

**(c)**

| Models | AOT40 | | | | | | | | |
|---|---|---|---|---|---|---|---|---|---|
| | RCP2.6 global | RCP2.6 SH | RCP2.6 NH | RCP4.5 global | RCP4.5 SH | RCP4.5 NH | RCP8.5 global | RCP8.5 SH | RCP8.5 NH |
| CESM-CAM | −96.9 | −99.9 | −96.8 | n.a. | n.a. | n.a. | +138.3 | +150.0 | +134.9 |
| GFDL-AM3 | −75.2 | −25.5 | −78.9 | −53.2 | −36.2 | −54.5 | +96.3 | +242.5 | +85.1 |
| GISS-E2-R | −78.1 | −13.9 | −81.2 | −75.0 | −27.8 | −77.2 | +22.3 | +83.3 | +19.5 |
| MIROC-CHEM | −74.0 | −10.5 | −80.6 | n.a. | n.a. | n.a. | +20.5 | +78.9 | +16.3 |
| MOCAGE | −53.7 | +68.6 | −59.7 | −17.5 | +202.9 | −28.3 | +85.1 | +448.6 | +67.0 |
| UM-CAM | −73.6 | +92.3 | −76.7 | −52.8 | +7.7 | −54.8 | +49.3 | +176.9 | +45.1 |
| Mean $\pm$ SD | −75.2 | +1.9 | −79.0 | −49.6 | +36.6 | −53.7 | +68.6 | +196.7 | +61.3 |
| | $\pm$13.7 | $\pm$69.5 | $\pm$11.8 | $\pm$23.8 | $\pm$112.4 | $\pm$20.0 | $\pm$46.3 | $\pm$137.7 | $\pm$44.8 |

**(d)**

| Models | Risk factor $IO_3$ | | | | | | | | |
|---|---|---|---|---|---|---|---|---|---|
| | RCP2.6 global | RCP2.6 SH | RCP2.6 NH | RCP4.5 global | RCP4.5 SH | RCP4.5 NH | RCP8.5 global | RCP8.5 SH | RCP8.5 NH |
| CESM-CAM | −97.2 | −91.8 | −97.5 | n.a. | n.a. | n.a. | +129.6 | +146.8 | +127.5 |
| GFDL-AM3 | −69.4 | −49.1 | −74.8 | −50.1 | −61.1 | −47.2 | +91.9 | +95.5 | +90.4 |
| GISS-E2-R | −66.1 | −20.7 | −74.3 | −71.7 | −53.3 | −74.6 | +21.5 | +56.6 | +14.2 |
| MIROC-CHEM | −41.4 | −18.9 | −51.9 | n.a. | n.a. | n.a. | +41.0 | +103.8 | +25.5 |
| MOCAGE | −46.6 | −22.8 | −51.4 | −7.0 | −38.0 | −1.0 | +77.7 | +68.2 | +80.0 |
| UM-CAM | −45.8 | −9.2 | −71.3 | −59.5 | +2.0 | −69.0 | +61.3 | +84.2 | +56.0 |
| Mean $\pm$ SD | −61.1 | −35.5 | −70.2 | −47.1 | −37.6 | −47.9 | +70.5 | +92.5 | +65.6 |
| | $\pm$21.1 | $\pm$30.7 | $\pm$17.2 | $\pm$28.1 | $\pm$28.1 | $\pm$33.4 | $\pm$38.4 | $\pm$31.7 | $\pm$42.4 |

eas and high insolation. Around the Mediterranean basin, elevated AOT40 values (> 60 ppm h) are recorded, mainly due to the industrial development, road traffic increment, high insolation, sea–land breeze recirculation and long-range trans- port of $O_3$ precursors and $O_3$ (Sicard et al., 2013). All models, except MIROC-CHEM, reproduce well the high surface $O_3$ mean concentrations over Greenland and over deserts.

## 3.2    Projected changes in ozone concentration and AOT40

Recent studies display a mean global increase in background $O_3$ concentration from a current level of 35–50 ppb (e.g. IPCC, 2014; Lefohn et al., 2014) to 55–65 ppb (e.g. Wittig et al., 2007) and up to 85 ppb at NH mid-latitudes by 2100 (IPCC, 2014). During the latter half of the 20th century surface $O_3$ concentrations have increased markedly at NH mid-latitudes (e.g. Oltmans et al., 2006; Parrish et al., 2012; Paoletti et al., 2014), mainly related to increasing anthropogenic precursor emissions related to economic growth of industrialized countries (e.g. Lamarque et al., 2005). Our results indicate that the future projections of the mean surface $O_3$ concentrations and AOT40 vary considerably with the different scenarios and models (Figs. 1 and 2). The six models simulate a decrease of $O_3$ concentration by 2100 under the RCP2.6 and RCP4.5 scenarios and an increase under the RCP8.5 scenario (Lamarque et al., 2011). In our study, the averaged relative changes in surface $O_3$ concentration means (and AOT40) for the different RCPs are −21 % (−75 %) for RCP2.6, −10 % (−50 %) for RCP4.5 and +14 % (+69 %) for RCP8.5 with a strong disparity between both hemispheres, e.g. −8 % in SH and −25 % in NH for RCP2.6 (Table 3b and c). RCP8.5 is the only scenario to show an increase in global background $O_3$ levels by 2100 (+23 % in SH and +11 % in NH).

Under the RCP2.6 scenario, all models predict that surface $O_3$ will strongly decrease worldwide, except in equatorial Africa, where higher $O_3$ levels are observed in GFDL-AM3, GISS-E2-R and MOCAGE. In CESM-CAM, GFDL-AM3 and MIROC-CHEM, a homogeneous decrease in $O_3$ burden is simulated worldwide while in GISS-E2-R, MOCAGE and UM-CAM, the strongest decrease in surface $O_3$ mean concentrations are found where high historical $O_3$ concentrations were reported. Under RCP4.5 scenario, the surface $O_3$ mean concentrations and AOT40 values are lower than historical runs worldwide for all models except in MOCAGE, where deterioration is observed over Canada, Greenland and eastern Asia. For all models, the surface $O_3$ levels and AOT40 are higher for RCP8.5 as compared to historical runs and the highest increase is found in northwestern USA, Greenland, Mediterranean basin, Near East and eastern Asia. The AOT40 values, exceeding 70 ppm h, are found over the Tibetan Plateau, in the Near East and over Greenland. For RCP8.5, GFDL-AM3 is the most pessimistic model and MIROC-CHEM the most optimistic. By the end of the 21st century, similar patterns are evident for RCP4.5 compared to RCP2.6 and RCP4.5 simulation is intermediate between RCP2.6 and RCP8.5 scenarios.

For all models and RCPs, the $O_3$ hotspots (mean concentrations > 50 ppb and AOT40 > 70 ppm h) are over Greenland and southern Asia, in particular over the Tibetan Plateau. The highest increases are observed in NH, in particular in northwestern USA, Greenland, Near East and south-

ern Asia (> 65 ppb). For the three RCPs, no significant change in ground-level $O_3$ is observed in SH and the SH extratropics makes a small contribution to the overall change.

A recent global study showed the geographical patterns of surface air temperature differences for late 21st century relative to the historical run (1986–2005) in all RCP scenarios (Nazarenko et al., 2015). The global warming in the RCP2.6 scenario is 2–3 times weaker than RCP4.5 scenario and 4–5 times weaker than RCP8.5 scenario (Nazarenko et al., 2015). For the three RCPs, the greatest change is observed over the Arctic, above latitude 60° N and in the latitude band 15–45° N (IPCC, 2014; Nazarenko et al., 2015). The weaker warming is simulated over the large area of the Southern Ocean. For RCP8.5 scenario, the global pattern of surface $O_3$ levels and AOT40 (Figs. 1 and 2) is similar to surface air temperature increase distribution. For RCP8.5, significant increases in air temperature are simulated over latitude 60° N and over the Tibetan Plateau (more than 5 °C). An increase of 4–5 °C over the Near East, eastern and southern Asia, northern and southern Africa and Canada are simulated as well as +1–3 °C for the rest of the world (Nazarenko et al., 2015). The tropospheric warming is stronger in the latitude band 15–45° N (Seidel et al., 2008) and Hudson et al. (2006) have demonstrated that $O_3$ trends over a 24-year period in the NH are due to trends observed in tropics and mid-latitudes areas and polar regions. The models are able to reproduce the global pattern of air temperature changes distribution in agreement with surface $O_3$ concentrations changes.

The spread in precursor emissions (e.g. VOCs, $NO_x$ and CO) is due to the range of representation of biogenic emissions ($NO_x$ from soils and lightning, CO from oceans and vegetation) as well as the complexity of chemical schemes in particular for NMVOCs simulations (e.g. isoprene) from explicitly specified to fully interactive with climate. RCP2.6 scenario has the lowest $O_3$ precursor concentrations, and RCP8.5 has relatively low $NO_x$, CO and VOCs emissions but very high $CH_4$ (Table 2b). The global emissions of $NO_x$ (−44 %), VOCs (−5 %) CO (−40 %) and $CH_4$ burden (−27 %) decline, while $LNO_x$ increases by e.g. 7 % under RCP2.6 (Table 2b). The CO (−32 %) and $NO_x$ (−20 %) emissions have decreased while LNOX (+33 %), VOCs (+1 %) and $CH_4$ burden have increased (+120 %) under RCP8.5 scenario (Table 2b). The GISS-E2-R model shows a greater degree of variation than other models, with a stronger increase in $CH_4$ burden (+153 %) and in VOCs emissions (+20 %) for RCP8.5 (Table 2b).

Excluding $CH_4$ burden and VOCs emissions, all the RCP scenarios include reductions and redistributions of $O_3$ precursor emissions throughout the 21st century due to the air pollution control strategies worldwide. The changes in $CH_4$ burden are due to the different climate policies in model assumptions. In RCP2.6, $CH_4$ emissions decrease steadily throughout the century; in RCP4.5 they remain steady until 2050 and then decrease (Voulgarakis et al., 2013) and in RCP8.5 (no climate policy) it rapidly increases compared to 2000. Methane burdens are fixed in the models with no sources, except for the GISS-E2-R simulations in which surface $CH_4$ emissions are used rather than $CH_4$ concentrations (Shindell et al., 2013). The model chemical schemes are greatly different due mainly to the NMVOCs simulations (Young et al., 2013). Isoprene dominates the total NMVOCs emissions (Guenther et al., 1995). In contrast to other models with constant present-day isoprene emissions, the GISS-ES2-R simulations incorporate climate-driven isoprene emissions, with greater BVOC emissions by 2100 and a positive change in total VOCs emissions across RCPs, related to the positive correlation between air temperature and isoprene emission (e.g. Guenther et al., 2006; Arneth et al., 2011; Young et al., 2013).

For RCP2.6 and RCP4.5 scenarios, there is a widespread decrease in $O_3$ in NH by 2100. The overall decrease in $O_3$ concentration and AOT40 means for RCP4.5 are about half of that between RCP2.6 and the historical simulation. For both scenarios, the changes are dominated by the decrease in $O_3$ precursor emissions in the NH extratropics compared to historical simulations (Table 2b). In $NO_x$-saturated areas, annual mean $O_3$ will slightly increase as a result of a less efficient titration by NO, but the overall $O_3$ burden will decrease substantially at hemispheric scale over time (Gao et al., 2013; Querol et al., 2014; Sicard et al., 2016a). In RCP4.5, Gao et al. (2013) showed that the largest decrease in $O_3$ (4–10 ppb) occurs in summer at mid-latitudes in the lower troposphere while the $O_3$ concentrations undergo an increase in winter. During the warm period, the photochemistry plays a major role in the $O_3$ production, suggesting that the reduction in surface $O_3$ concentrations is in agreement with the large reduction in anthropogenic $O_3$ precursor emissions (Sicard et al., 2016a) reducing the extent of regional photochemical $O_3$ formation (e.g. Derwent et al., 2013; Simpson et al., 2014). Titration effect was also reported by Collette et al. (2012) over Europe as analysed from six chemistry transport models.

The $O_3$ increase can be also driven by the net impacts of climate change, i.e. increase in stratospheric $O_3$ intrusion, changing $LNO_x$ and impacting reaction rates, through sea surface temperatures and relative humidity changes (Lau et al., 2006; Voulgarakis et al., 2013; Young et al., 2013).

Under the RCP8.5 scenario, the increase in surface $O_3$ concentrations, by 14 % on average, can be attributed to the higher $CH_4$ emissions coupled with a strong global warming, exceeding 2 °C, and a weakened NO titration by reducing $NO_x$ emissions (Stevenson et al., 2013; Young et al., 2013). The global $CH_4$ burden is 27 and 5 % lower than 2000, for the RCP2.6 and RCP4.5 scenarios, respectively, while for RCP8.5 the total $CH_4$ burden has more than doubled compared to early 2000s and $LNO_x$ emissions increased by 33 % (Table 2b). In addition, stronger increases are found over the high-elevation Himalayan Plateau reflecting increased exchange with the free troposphere or stratosphere (Lefohn et al., 2012; Schnell et al., 2016). Several studies reported

an increase in the stratospheric $O_3$ influx and higher stratospheric $O_3$ levels in response to a warming climate (e.g. Hegglin and Shepherd, 2009; Zeng et al., 2010). The downwards $O_3$ transport from the stratosphere is an important source of tropospheric $O_3$ (Hsu and Prather, 2009; Tang et al., 2011); therefore, stratospheric $O_3$ recovery also plays a partial role (e.g. +11 % for RCP8.5) in surface $O_3$ burden pattern. As an example, in MOCAGE, a smaller reduction in global $O_3$ mean concentrations (−13 %) and higher increase in stratospheric $O_3$ inputs (+20 %) are observed for RCP2.6 (Table 3b). Similarly, for RCP8.5, the highest increases in $O_3$ mean concentrations (+23 %) and stratospheric $O_3$ (+24 %) are recorded in MOCAGE. In addition, $LNO_x$ emissions show significant upward trend from 2000 to 2100, in particular for the strongest warming scenario (RPC8.5) with greater convective and lightning activity (e.g. Williams, 2009; Lamarque et al., 2013). For RCP8.5, a reduction in surface $O_3$ concentrations is also simulated over the equatorial region, where the increased relative humidity, in a warmer climate, increases the $O_3$ loss rate (e.g. Johnson et al., 1999; Zeng and Pyle, 2003).

For RCP2.6 and RCP4.5, absolute decreases are observed for the Mediterranean basin and the western USA due to less precursor emissions in the NH extratropics (e.g. reduction of 5–7 ppb over Europe). Smaller reduction in surface $O_3$ levels in southern and eastern Asia highlight the smaller changes in $O_3$ precursor emissions due to the recent emission growth in this region (e.g. Zhang et al., 2009; Xing et al., 2015). For RCP8.5, the high $O_3$ increase (up to 10 ppb) in southern Asia can be attributed to substantial increase in $CH_4$ emissions coupled with a strong global warming, exceeding 2 °C, and a weakened NO titration and a greater stratospheric $O_3$ influx (Kawase et al., 2011; Wild et al., 2012; Young et al., 2013).

### 3.3  Risk areas for vegetation under RCP scenarios

Figure 3 shows the changes in the potential $O_3$ impact on photosynthetic carbon assimilation between present and future. It should be noted that a zero percentage of change (i.e. no change) for $IO_3$ is simulated in sparsely vegetated regions (e.g. Gobi, Sahara, Near East, Western Plateau and Greenland), while the change can be higher than 100 % when the historical $O_3$ concentrations are lower than 40 ppb (i.e. AOT40 = 0 and $IO_3$ = 0) and the $O_3$ concentrations exceed 40 ppb under RCPs (i.e. AOT40 > 0, $IO_3$ > 0). If the AOT40 during the historical period is 0 then the percentage of change is undefined and we have considered and set these grid points as missing values.

The potential $O_3$ impact for vegetation strongly decreases in NH for RCP2.6 except in MOCAGE, where a slight increase in the risk factor (+15 %) is simulated at high latitudes and in southern Asia. Conversely, the areas where the risk for vegetation increases (> 60 %) occur over Africa (+15 to +60 %) for all models except in CESM-CAM, where no change is observed across Africa. Under RCP4.5 scenario,

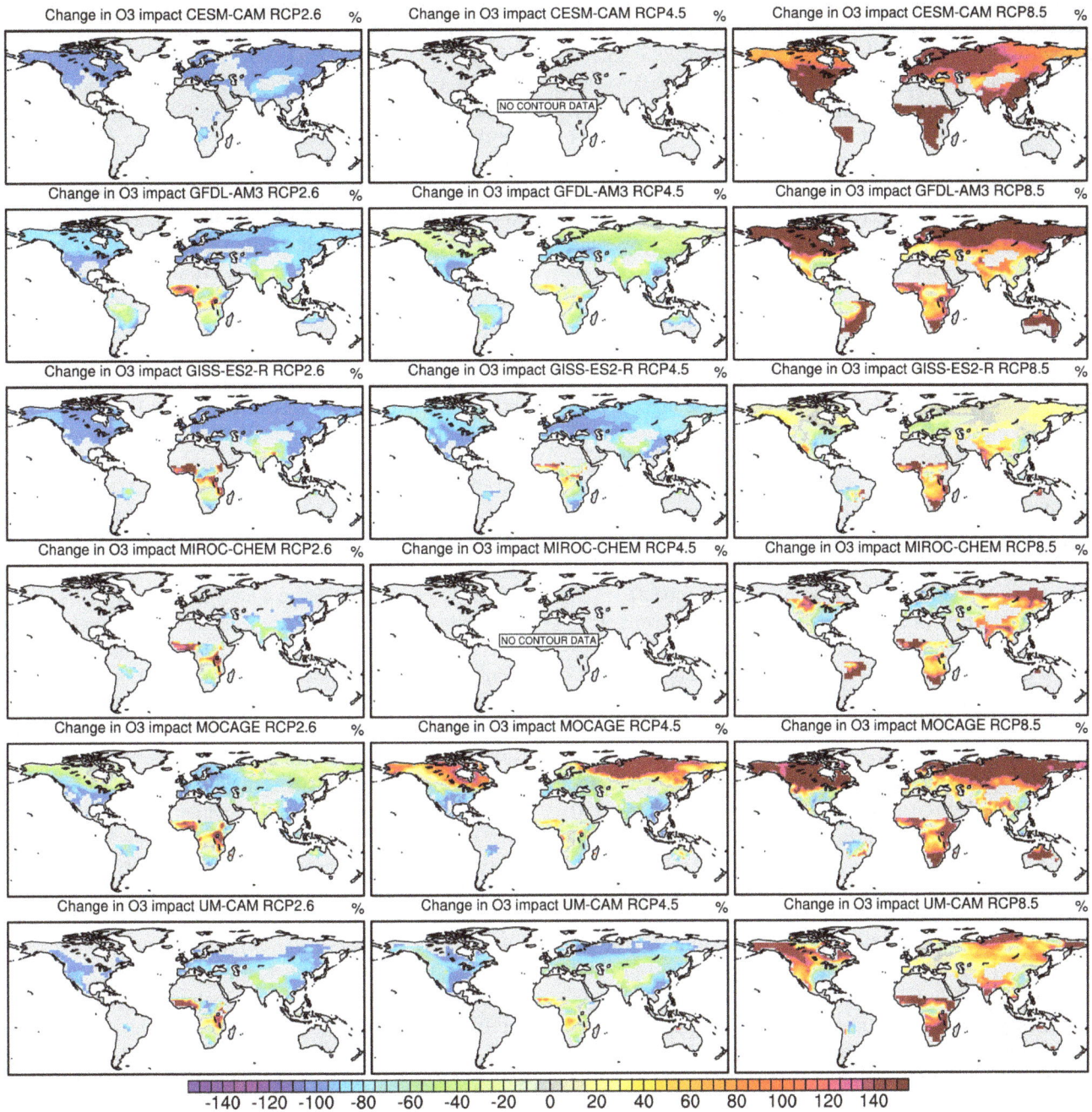

**Figure 3.** Simulated percentage changes (%) in the potential ozone impact on photosynthetic carbon assimilation ($IO_3$) for each ACCMIP model between RCP2.6, RCP4.5 and RCP8.5 simulations and the historical run. The data are missing for two models under RCP4.5 ("no contour data").

the strongest increase in potential risk for vegetation (> + 60 %) is simulated by MOCAGE, markedly different from the other models, above the latitude 50° N. For all models, the potential $O_3$ impact for vegetation increases across Africa, from −15 to +60 %, while slight decreases or no change occur over other parts of the world. Under RCP8.5 scenario, an increase of average $O_3$ over a significant part of the domain is simulated; therefore the exposure to $O_3$ pol-

lution and impacts on vegetation will increase worldwide by 2100. An increase of the $O_3$ impacts on vegetation is simulated in northern USA, South America, Asia and Africa, while a reduction in particular over eastern USA and south-eastern China and a slight increase (+15 %) or decrease (−15 %) over Europe, depending on the model, are simulated.

In summary, compared to the historical simulations, the averaged relative changes in the $O_3$ risk factor for the different RCPs are $-61\%$ for RCP2.6, $-47\%$ for RCP4.5 and $+70\%$ for RCP8.5 (Table 3d). We thus find a significant reduction in risk for vegetation for both RCP2.6 and RCP4.5 scenarios, except in South Africa and at high latitudes in MOCAGE simulations, and a strong increase in global risk under RCP8.5. Under RCP2.6 and RCP4.5 scenarios, $IO_3$ slightly increases in Africa and over North America and Asia (> latitude 60° N) in MOCAGE. The risk increases over the few areas where the $O_3$ concentrations increased between the historical period and 2100. Under both scenarios, the strongest reductions in risk are observed over Amazon, central Africa and southern Asia, i.e. where the $O_3$ concentrations have strongly declined between historical period and 2100. Under the RCP8.5, the areas where the highest projected $O_3$ mean concentrations are simulated (e.g. Greenland and deserts) are not associated with an increase in $IO_3$ due to the absence of vegetation. Under RCP8.5, $IO_3$ increases worldwide while a reduction is simulated over southeastern North America, northern Amazon, central Africa and Southeast Asia, and a slighter reduction or a slight increase is simulated over western Europe (depending on the model).

The spatial pattern of $IO_3$ is consistent with previous analyses of global environmental change (climate, land cover, nitrogen deposition and $CO_2$ fertilization) impacts on vegetation (Nemani et al., 2003; Zhu et al., 2016), i.e. the highest reduction in risk for vegetation, in particular under RCP8.5, occurs over areas where a strong increase in greening, LAI and NPP is observed due to global changes and where a reduction in surface $O_3$ mean concentrations is found by 2100 (Fig. 1). The regions with the largest greening trends are in southeastern North America, northern Amazon, Europe, central Africa and Southeast Asia with an average increase of the observed LAI exceeding $0.25\,m^2\,m^{-2}$ per year (Zhu et al., 2016). The $CO_2$ fertilization effects (70 %), nitrogen deposition (9 %) and climate change (8 %) explain the observed greening trend (Zhu et al., 2016). The changing climate alone produces persistent NPP increases and the regions with the highest increase in NPP, ranging from 1.0 to 1.5 % per year, are in southeastern North America, northern Amazon, western Europe, central Africa and southern Asia (Nemani et al., 2003). From 1982 to 1999, the largest increase is observed in tropical regions, with more than 1.5 % per year over the Amazon rainforest, which accounts for 42 % of the global NPP increase (Nemani et al., 2003). The Amazon rainforest is one those regions where the effects are statistically significant. This is particularly important owing to the role of the Amazon rainforests in the global carbon cycle (Zhu et al., 2016). In these areas, we observed a strong increase in NPP and LAI due to warming climate while a reduction in GPP (from $-10$ to $-20\%$) due to $O_3$ is observed (Sitch et al., 2007). Inversely, the risk for vegetation $IO_3$ increases in particular in Africa, e.g. western Africa along the Gulf of Guinea, in southern Brazil and over high-latitude regions (> 60° N) in

North America and Asia, where a reduction or a slight increase in LAI (from $-0.05$ to $+0.03\,m^2\,m^{-2}$ per year) and strong decreases in NPP (1.0–1.5 % per year) are simulated (Nemani et al., 2003; Zhu et al., 2016).

Sitch et al. (2007) reported a high GPP reduction due to $O_3$ effects, between 1901 and 2100 under the A2 emissions scenario of the Special Report on Emissions Scenarios, exceeding 30 % in summer over western Europe, eastern North America, Amazon, central Africa and southern Asia. Previous studies have reported that the reductions in GPP simulated by Sitch et al. (2007) are overestimated up to 6 times due to the lack of empirical data about the response of different species to $O_3$. Indeed, Sitch et al. (2007) focused on broad-leaved tree, needle-leaved tree, C3 crops, C4 crops and shrubs. The fact that a few experiments have shown no response, e.g. grasslands (Bassin et al., 2013), and the non-inclusion of the nitrogen limitation of growth are additional reasons of this overestimation (Zak et al., 2011; Kvalevg and Myhre, 2013). In addition, the simulated $O_3$ concentrations over Amazon forest exceed 90 ppb in summer in Sitch et al. (2007), while the annual $O_3$ mean is around 15–20 ppb by 2100 in our study.

The projected land covers widely vary under RCPs (Betts et al., 2015). In the RCP2.6 scenario, the ground surface covered by croplands increases as a result of bioenergy production, with a more-or-less constant use of grassland. The RCP4.5 scenario focuses on global reforestation programs as part of global climate policy, as a result, the use of cropland and grassland decreases. Under RCP8.5, an increase in croplands and grasslands is applied mostly driven by an increasing global population (van Vuuren et al., 2011). About 50 % of forests, grasslands and croplands might be exposed to high $O_3$ levels by the end of the 21st century (Sitch et al., 2007; Wittig et al., 2009).

Generally, deciduous broadleaf are highly $O_3$-sensitive risk areas and needleleaf forests are moderately $O_3$-sensitive risk areas. Crops and grasslands are more sensitive to $O_3$ exposure than trees and deciduous trees are more sensitive than coniferous trees with lower stomatal conductance (Felzer et al., 2004; Ren et al., 2007; Wittig et al., 2009; Anav et al., 2011). Based on a comparison between Fig. 2 and the Global Land Cover Facility maps, we can observe that generally the AOT40, i.e. the potential $O_3$ risk to vegetation, is high over shrublands (e.g. high-latitude region), broadleaf forests (e.g. central Africa), needleleaf forests (e.g. North America) and crops (e.g. southern Asia). Under RCP2.6 and RCP4.5, the risk decreases over areas covered by shrublands, savannas and slightly decreases over areas with needleleaf forests in North America and northern Asia. The risk strongly increases over broadleaf forest in Africa and the risk slightly decreases or slightly increases over grasslands (central Asia and central Africa and USA). Under RCP8.5, the largest decreases in risks occur in eastern USA, Europe and southeastern China, where the ground is mainly dominated by croplands, in all models except CESM-CAM.

# 4 Conclusions

From six global atmospheric chemistry transport models, we illustrate the changes, i.e. differences for late 21st century relative to the historical run, in ground-level $O_3$ concentrations and vegetation impact metric (AOT40). Finally, the potential $O_3$ impacts on photosynthetic carbon assimilation worldwide are investigated to define potential risk areas for vegetation at global scale by 2100. A major advantage of this study is a comparison between models and scenarios to explore future potential $O_3$ impacts.

The six models reproduce well the spatial pattern of historical $O_3$ concentration and AOT40 at global scale; in particular GISS-E2-R and MOCAGE are able to simulate the higher $O_3$ levels in areas downwind of precursor sources and at the high-elevation areas. The model outputs emphasize the strong asymmetry in the tropospheric $O_3$ distribution between NH and SH. The natural emissions of $O_3$ precursors (e.g. lightning $NO_x$, CO from oceans, isoprene) as well as the complexity of chemical schemes are significant sources of model-to-model differences.

Compared to early 2000s, the results suggest changes in surface $O_3$ of $-9.5 \pm 2.0$ ppb (NH) and $-1.8 \pm 2.1$ ppb (SH) in the cleaner RCP2.6 scenario and of $+4.4 \pm 2.8$ ppb (NH) and $+5.1 \pm 2.1$ ppb (SH) in the RCP8.5 scenario. For RCP2.6 and RCP4.5, absolute decreases are observed for the Mediterranean basin and the western USA due to less precursor emissions in the NH extratropics. For RCP8.5, all models show climate-driven increases in ground-level $O_3$ in particular over the western USA, Greenland, southern Asia and northeastern China and the changes ranged from $+1$ to $+5$ ppb over North America and Europe. This $O_3$ increase can be mainly attributed to substantial increase in $CH_4$ emissions coupled with a global warming and a weakened NO titration.

Most important results from the study are the spatial patterns and projected changes in global AOT40 and risk areas for vegetation under the RCP scenarios. Even if AOT40 was computed year-round, the global models suggest that, despite an improvement under RCP2.6 and RCP4.5, the AOT40-based critical levels for the protection of forests and crops will be exceeded over many areas of the NH and they may be much more exceeded under RCP8.5 up to a factor exceeding 10 by 2100.

Ozone may be a major threat to biodiversity over large regions of the world; however, the size of these areas remains uncertain. The potential $O_3$ impact on carbon assimilation, $IO_3$, provides a clear indicator of the potential risk to vegetation. By 2100, the potential $O_3$ impact on photosynthetic carbon assimilation decreases by 61 and 47 % under RCP2.6 and RCP4.5, respectively, and increases by 70 % under RCP8.5, compared to early 2000s over the whole domain. The strongest increase of the $O_3$ impacts on vegetation is simulated in North America, northern Asia and central Africa. The highest reduction in risk for vegetation (i.e.

southeastern North America, the northern Amazon, central Africa and Southeast Asia) occurs over areas where a strong increase in greening, LAI and NPP is observed and where a reduction in $O_3$ mean concentrations is found by 2100.

Many ecosystems worldwide are unprotected from $O_3$ due to the lack of international efforts (Emberson et al., 2014). An efficient reduction in overall $O_3$ levels is expected over North America and Europe in all RCP scenarios and worldwide if $CH_4$ emissions are reduced (e.g. Kirtman et al., 2013; Pfister et al., 2014; Schnell et al., 2016). To efficiently protect vegetation against $O_3$ pollution, suitable standards are urgently needed and the mitigation actions must be as part of international emission reduction programmes. The flux-based metric is introduced as new standard for vegetation protection against effects of $O_3$ pollution, taking into account the detoxification processes and the modifying effects of multiple climatic and phenological factors on $O_3$ uptake (Paoletti and Manning, 2007; Sicard et al., 2016b, c). Plant phenology plays a pivotal role in the climate system as it regulates the gas exchange between the biosphere and the atmosphere. Currently, in many $O_3$ risk assessment studies, the phenology function is based on a simple latitude and topography model and the chemistry models do not take into account the shifts in plant phenology and in start and end date of the growing season; however, a first attempt to study the role of phenology on stomatal ozone uptake is shown by Anav et al. (2017).

*Competing interests.* The authors declare that they have no conflict of interest.

*Acknowledgements.* This work was carried out with the contribution of the LIFE financial instrument of the European Union (LIFE15 ENV/IT/183) in the framework of the MOTTLES project "Monitoring ozone injury for setting new critical levels" and published within the International Union of Forest Research Organizations (IUFRO) Task Force on Climate Change and Forest Health and IUFRO RG 7.01.09 "Ground-level ozone".

Edited by: Jayanarayanan Kuttippurath

# References

Ainsworth, E. A., Yendrek, C. R., Sitch, S., Collins, W. J., and Emberson, L. D.: The effect of Tropospheric Ozone on net primary productivity and implications for climate change, Annu. Rev. Plant Biol., 63, 637–661, 2012.

Anav, A., Menut, L., Khvorostyanov, D., and Viovy, N.: Impact of tropospheric ozone on the Euro-Mediterranean vegetation, Glob. Change Biol., 17, 2342–2359, 2011.

Anav, A., De Marco, A., Proietti, C., Alessandri, A., Dell'Aquila, A., Cionni, I., Friedlingstein, P., Khvorostyanov, D., Menut, L., Paoletti, E., Sicard, P., Sitch, S., and Vitale, M.: Comparing concentration-based (AOT40) and stomatal uptake (PODY) metrics for ozone risk assessment to European forests, Global Change Biol., 22, 1608–1627, https://doi.org/10.1111/gcb.13138, 2016.

Anav, A., Liu, Q., De Marco, A., Proietti, C., Savi, F., Paoletti, E., and Piao, S.: The role of plant phenology in stomatal ozone flux modelling, Glob, Change Biol., https://doi.org/10.1111/gcb.13823, in press, 2017.

Arbaugh, M. J. and Bytnerowicz, A.: Ambient ozone patterns and effects over the Sierra Nevada: synthesis and implications for future research, in: Ozone Air Pollution in the Sierra Nevada: Distribution and Effects on Forests, Developments in Environmental Science, vol. 2, edited by: Bytnerowicz, A., Arbaugh, M., Alonso, R., Elsevier, Amsterdam, 249–261, 2003.

Arneth, A., Schurgers, G., Hickler, T., and Miller, P. A.: Effects of species composition, land surface cover, $CO_2$ concentration and climate on isoprene emissions from European forests, Plant Biol., 10, 150–162, 2008.

Arneth, A., Schurgers, G., Lathiere, J., Duhl, T., Beerling, D. J., Hewitt, C. N., Martin, M., and Guenther, A.: Global terrestrial isoprene emission models: sensitivity to variability in climate and vegetation, Atmos. Chem. Phys., 11, 8037–8052, https://doi.org/10.5194/acp-11-8037-2011, 2011.

Ashmore, M. R.: Assessing the future global impacts of ozone on vegetation, Plant Cell Environ., 28, 949–964, 2005.

Ashworth, K., Wild, O., and Hewitt, C. N.: Impacts of biofuel cultivation on mortality and crop yields, Nat. Clim. Change, 3, 492–496, 2013.

Bassin, S., Volk, M., and Fuhrer, J.: Species composition of subalpine grassland is sensitive to nitrogen deposition, but not ozone, after seven years of treatment, Ecosystems, 16, 1105–1117, 2013.

Betts, R. A., Golding, N., Gonzalez, P., Gornall, J., Kahana, R., Kay, G., Mitchell, L., and Wiltshire, A.: Climate and land use change impacts on global terrestrial ecosystems and river flows in the HadGEM2-ES Earth system model using the representative concentration pathways, Biogeosciences, 12, 1317–1338, https://doi.org/10.5194/bg-12-1317-2015, 2015.

Bian, J., Yan, R., Chen, H., Lü, D., and Massie, S. T.: Formation of the summertime ozone valley over the Tibetan Plateau: the Asian summer monsoon and air column variations, Adv. Atmos. Sci., 28, 1318–1325, 2011.

Bowman, K. W., Shindell, D. T., Worden, H. M., Lamarque, J. F., Young, P. J., Stevenson, D. S., Qu, Z., de la Torre, M., Bergmann, D., Cameron-Smith, P. J., Collins, W. J., Doherty, R., Dalsøren, S. B., Faluvegi, G., Folberth, G., Horowitz, L. W., Josse, B. M., Lee, Y. H., MacKenzie, I. A., Myhre, G., Nagashima, T., Naik, V., Plummer, D. A., Rumbold, S. T., Skeie, R. B., Strode, S. A., Sudo, K., Szopa, S., Voulgarakis, A., Zeng, G., Kulawik, S. S., Aghedo, A. M., and Worden, J. R.: Evaluation of ACCMIP outgoing longwave radiation from tropospheric ozone using TES satellite observations, Atmos. Chem. Phys., 13, 4057–4072, https://doi.org/10.5194/acp-13-4057-2013, 2013.

Clifton, O. E., Fiore, A. M., Correa, G., Horowitz, L. W., and Naik, V.: Twenty-first century reversal of the surface ozone seasonal cycle over the northeastern United States, Geophys. Res. Lett., 41, 7343–7350, 2014.

Chen, X. L., Ma, Y. M., Kelder, H., Su, Z., and Yang, K.: On the behaviour of the tropopause folding events over the Tibetan Plateau, Atmos. Chem. Phys., 11, 5113–5122,

https://doi.org/10.5194/acp-11-5113-2011, 2011.

Chevalier, A., Gheusi, F., Delmas, R., Ordóñez, C., Sarrat, C., Zbinden, R., Thouret, V., Athier, G., and Cousin, J.-M.: Influence of altitude on ozone levels and variability in the lower troposphere: a ground-based study for western Europe over the period 2001–2004, Atmos. Chem. Phys., 7, 4311–4326, https://doi.org/10.5194/acp-7-4311-2007, 2007.

Colette, A., Granier, C., Hodnebrog, Ø., Jakobs, H., Maurizi, A., Nyiri, A., Rao, S., Amann, M., Bessagnet, B., D'Angiola, A., Gauss, M., Heyes, C., Klimont, Z., Meleux, F., Memmesheimer, M., Mieville, A., Rouïl, L., Russo, F., Schucht, S., Simpson, D., Stordal, F., Tampieri, F., and Vrac, M.: Future air quality in Europe: a multi-model assessment of projected exposure to ozone, Atmos. Chem. Phys., 12, 10613–10630, https://doi.org/10.5194/acp-12-10613-2012, 2012.

Cooper, O. R., Sweeney, C., Gao, R. S., Tarasick, D., and Leblanc, T.: Long-term ozone trends at rural ozone monitoring sites across the United States, 1990–2010, J. Geophys. Res.-Atmos., 117, D22307, https://doi.org/10.1029/2012JD018261, 2012.

Cooper, O. R., Parrish, D. D., Ziemke, J., Balashov, N. V., and Cupeiro, M.: Global distribution and trends of tropospheric ozone: an observation-based review, Elementa: Sci. Anthropocene, 2, 29, https://doi.org/10.12952/journal.elementa.000029, 2014.

Cubasch, U., Wuebbles, D., Chen, D., Facchini, M. C., Frame, D., Mahowald, N., and Winther, J. G.: Introduction, in Climate Change 2013: The Physical Science Basis. Contribution of Working Group I to the Fifth Assessment Report of the Intergovernmental Panel on Climate Change, edited by: Stocker, T. F., Qin, D., Plattner, G.-K., Tignor, M., Allen, S. K., Boschung, J., Nauels, A., Xia, Y., Bex, V., and Midgley, P. M., Cambridge Univ. Press, Cambridge, UK, and New York, 2013.

De Marco, A., Sicard, P., Vitale, M., Carriero, G., Renou, C., and Paoletti, E.: Metrics of ozone risk assessment for Southern European forests: canopy moisture content as a potential plant response indicator, Atmos. Environ., 120, 182–190, 2015.

Derwent, R. G., Witham, C. S., Utembe, S. R., Jenkin, M. E., and Passant, N. R.: Ozone in Central England: the impact of 20 years of precursor emission controls in Europe, Environ. Sci. Policy, 13, 195–204, 2010.

Derwent, R. G., Manning, A. J., Simmonds, P. G., Spain, T. G., and O'Doherty, S.: Analysis and interpretation of 25 years of ozone observations at the Mace Head Atmospheric Research Station on the Atlantic Ocean coast of Ireland from 1987 to 2012, Atmos. Environ., 80, 361–368, 2013.

Derwent, R. G., Utembe, S. R., Jenkin, M. E., and Shallcross, D. E.: Tropospheric ozone production regions and the intercontinental origins of surface ozone over Europe, Atmos. Environ., 112, 216–224, 2015.

Donner, L. J., Wyman, B. L., Hemler, R. S., Horowitz, L. W., Ming, Y., Zhao, M., Golaz, J. C., Ginoux, P., Lin, S. J., Schwarzkopf, M. D., Austin, J., Alaka, G., Cooke, W. F., Delworth, T. L., Freidenreich, S. M., Gordon, C. T., Griffies, S. M., Held, I. M., Hurlin, W. J., Klein, S. A., Knutson, T. R., Langenhorst, A. R., Lee, H. C., Lin, Y., Magi, B. I., Malyshev, S. L., Milly, P. C. D., Naik, V., Nath, M. J., Pincus, R., Ploshay, J. J., Ramaswamy, V., Seman, C. J., Shevliakova, E., Sirutis, J. J., Stern, W. F., Stouffer, R. J., Wilson, R. J., Winton, M., Wittenberg, A. T., and Zeng, F.: The dynamical core, physical parameterizations, and basic simulation characteristics of the atmospheric component AM3 of the GFDL Global Coupled Model CM3, J. Climate, 24, 3484–3519, 2011.

European Environment Agency: Air quality in Europe – 2015 report, Report No. 5/2015, Copenhagen, Denmark, https://doi.org/10.2800/62459, 2015.

Ellingsen, K., Gauss, M., Van Dingenen, R., Dentener, F. J., Emberson, L., Fiore, A. M., Schultz, M. G., Stevenson, D. S., Ashmore, M. R., Atherton, C. S., Bergmann, D. J., Bey, I., Butler, T., Drevet, J., Eskes, H., Hauglustaine, D. A., Isaksen, I. S. A., Horowitz, L. W., Krol, M., Lamarque, J. F., Lawrence, M. G., van Noije, T., Pyle, J., Rast, S., Rodriguez, J., Savage, N., Strahan, S., Sudo, K., Szopa, S., and Wild, O.: Global ozone and air quality: a multi-model assessment of risks to human health and crops, Atmos. Chem. Phys. Discuss., 8, 2163–2223, https://doi.org/10.5194/acpd-8-2163-2008, 2008.

Emberson, L. D., Fuhrer, J., Ainsworth, L., and Ashmore, M. R.: Biodiversity and Ground-level Ozone, Report UNEP/CBD/SBSTTA/18/INF/17, Convention on Biological Diversity, 18th Meeting, Montreal, 23–28 June 2014.

Felzer, B. S. F., Kicklighter, D. W., Melillo, J. M., Wang, C., Zhuan, Q. and Prinn, R. G.: Ozone effects on net primary production and carbon sequestration in the conterminous United States using a biogeochemistry model, Tellus B, 56, 230–248, 2004.

Fiscus, E. L., Booker, F. L., and Burkey, K. O.: Crop responses to ozone: uptake, modes of action, carbon assimilation and partitioning, Plant Cell Environ., 28, 997–1011, 2005.

Gao, Y., Fu, J. S., Drake, J. B., Lamarque, J.-F., and Liu, Y.: The impact of emission and climate change on ozone in the United States under representative concentration pathways (RCPs), Atmos. Chem. Phys., 13, 9607–9621, https://doi.org/10.5194/acp-13-9607-2013, 2013.

Granier, C., Niemeier, U., Jungclaus, J. H., Emmons, L., Hess, P., Lamarque, J. F., Walters, S., and Brasseur, G. P.: Ozone pollution from future ship traffic in the Arctic northern passages, Geophys. Res. Lett., 33, L13807, https://doi.org/10.1029/2006GL026180, 2006.

Guenther, A. B., Hewitt, C. N., Erickson, D., Fall, R., Geron, C., Graedel, T., Harley, P., Klinger, L., Lerdau, M., Mckay, W. A., Pierce, T., Scholes, B., Steinbrecher, R., Tallamraju, R., Taylor, J., and Zimmerman, P.: A global model of natural volatile organic compound emissions, J. Geophys. Res.-Atmos., 100, 8873–8892, 1995.

Guenther, A., Karl, T., Harley, P., Wiedinmyer, C., Palmer, P. I., and Geron, C.: Estimates of global terrestrial isoprene emissions using MEGAN (Model of Emissions of Gases and Aerosols from Nature), Atmos. Chem. Phys., 6, 3181–3210, https://doi.org/10.5194/acp-6-3181-2006, 2006.

Guo, D., Su, Y., Shi, C., Xunn, J., and Powell Jr., A. M.: Double core of ozone valley over the Tibetan Plateau and its possible mechanisms, J. Atmos. Sol.-Terr. Phy., 130, 127–131, 2015.

Hegglin, M. I. and Shepherd, T. G.: Large climate-induced changes in ultraviolet index and stratosphere-to-troposphere ozone flux, Nat. Geosci., 2, 687–691, https://doi.org/10.1038/ngeo604, 2009.

Helmig, D., Oltmans, S. J., Morse, T. O., and Dibb, J. E.: What is causing high ozone at Summit, Greenland?, Atmos. Environ., 41, 5031–5043, 2007.

Hess, P. G. and Zbinden, R.: Stratospheric impact on tropospheric ozone variability and trends: 1990–2009, Atmos. Chem. Phys., 13, 649–674, https://doi.org/10.5194/acp-13-649-2013, 2013.

Holland, M., Kinghorn, S., Emberson, L., Cinderby, S., Ashmore, M., Mills, G., and Harmens, H.: Development of a framework for probabilistic assessment of the economic losses caused by ozone damage to crops in Europe, UNECE International Cooperative Programme on Vegetation, Project Report Number C02309, NERC/Centre for Ecology and Hydrology, 50 pp., 2006.

Hoshika, Y., Shimizu, Y., and Omasa, K.: A comparison between stomatal ozone uptake and AOT40 of deciduous trees in Japan, iForest, 4, 128–135, https://doi.org/10.3832/ifor0573-004, 2011.

Hsu, J. and Prather, M. J.: Stratospheric variability and tropospheric ozone, J. Geophys. Res.-Atmos., 114, D06102, https://doi.org/10.1029/2008JD010942, 2009.

Hudson, R. D., Andrade, M. F., Follette, M. B., and Frolov, A. D.: The total ozone field separated into meteorological regimes – Part II: Northern Hemisphere mid-latitude total ozone trends, Atmos. Chem. Phys., 6, 5183–5191, https://doi.org/10.5194/acp-6-5183-2006, 2006.

IPCC, Intergovernmental Panel on Climate Change: Summary for Policymakers, in: "Climate Change 2014: Impacts, Adaptation and Vulnerability". Contribution of Working Group II to the Fifth Assessment Report of the Intergovernmental Panel on Climate Change, Cambridge University Press, Cambridge, UK, 2014.

Jeričević, A., Koračin, D., Jiang, J., Chow, J., Watson, J., Fujita, E., and Minoura, H: Air Quality Study of High Ozone Levels in South California, in: Part of the series NATO Science for Peace and Security Series C: Environmental Security, Air Pollution Modeling and its Application XXII, edited by: Steyn, D. G., Builtjes, P. J. H., and Timmermans, R. M. A., 629–633, 2013.

Johnson, C. E., Collins, W. J., Stevenson, D. S., and Derwent, R. G.: Relative roles of climate and emissions changes on future tropospheric oxidant concentrations, J. Geophys. Res., 104, 18631–18645, 1999.

Josse, B., Simon, P., and Peuch, V. H.: Radon global simulations with the multiscale chemistry and transport model MOCAGE, Tellus B, 56, 339–356, 2004.

Kawase, H., Nagashima, T., Sudo, K., and Nozawa, T.: Future changes in tropospheric ozone under Representative Concentration Pathways (RCPs), Geophys. Res. Lett. 38, L05801, https://doi.org/10.1029/2010GL046402, 2011.

Kelly, J., Makar, P. A., and Plummer, D. A.: Projections of mid-century summer air-quality for North America: effects of changes in climate and precursor emissions, Atmos. Chem. Phys., 12, 5367–5390, https://doi.org/10.5194/acp-12-5367-2012, 2012.

Kirtman, B., Power, S. B., Adedoyin, J. A., Boer, G. J., Bojariu, R., Camilloni, I., Doblas-Reyes, F., Fiore, A. M., Kimoto, M., Meehl, G., Prather, M., Sarr, A., Schär, C., Sutton, R., van Oldenborgh, G. J., Vecchi, G., and Wang, H. J.: Near–term climate change: projections and predictability, in Climate Change 2013: The Physical Science Basis. Contribution of Working Group I to the Fifth Assessment Report of the Intergovernmental Panel on Climate Change, edited by: Stocker, T. F. et al., Cambridge Univ. Press, Cambridge, UK, and New York, 2013.

Klingberg, J., Engardt, M., Karlsson, P. E., Langner, J., and Pleijel, H.: Declining ozone exposure of European vegetation under climate change and reduced precursor emissions, Biogeosciences, 11, 5269–5283, https://doi.org/10.5194/bg-11-5269-2014, 2014.

Krinner, G., Viovy, N., de Noblet–Ducoudré, N., Ogée, J., Polcher, J., Friedlingstein, P., Ciais, P., Sitch, S., and Prentice, I. C.: A dynamic global vegetation model for studies of the coupled atmosphere-biosphere system, Global Biogeochem. Cy., 19, GB1015, https://doi.org/10.1029/2003GB002199, 2005.

Kulkarni, P. S., Bortoli, D., Salgado, R., Anton, M., Costa, M. J., and Silva, A. M.: Tropospheric ozone variability over the Iberian Peninsula, Atmos. Environ., 45, 174–182, 2011.

Kvalevag, M. M. and Myrhe, G.: The effect of carbon-nitrogen coupling on the reduced land carbon sink caused by ozone, Geophys. Res. Lett., 40, 3227–3231, 2013.

Lamarque, J. F., Hess, P. G., Emmons, L. K., Buja, L. E., Washington, W. M., and Granier, C.: Tropospheric ozone evolution between 1890 and 1990, J. Geophys. Res.-Atmos., 110, D08304, https://doi.org/10.1029/2004JD00553, 2005.

Lamarque, J.-F., Bond, T. C., Eyring, V., Granier, C., Heil, A., Klimont, Z., Lee, D., Liousse, C., Mieville, A., Owen, B., Schultz, M. G., Shindell, D., Smith, S. J., Stehfest, E., Van Aardenne, J., Cooper, O. R., Kainuma, M., Mahowald, N., McConnell, J. R., Naik, V., Riahi, K., and van Vuuren, D. P.: Historical (1850–2000) gridded anthropogenic and biomass burning emissions of reactive gases and aerosols: methodology and application, Atmos. Chem. Phys., 10, 7017–7039, https://doi.org/10.5194/acp-10-7017-2010, 2010.

Lamarque, J. F., Kyle, G. P., Meinshausen, M., Riahi, K., Smith, S. J., van Vuuren, D. P., Conley, A. J., and Vitt, F.: Global and regional evolution of short-lived radiatively-active gases and aerosols in the Representative Concentration Pathways, Climatic Change, 109, 191–212, https://doi.org/10.1007/s10584-011-0155-0, 2011.

Lamarque, J.-F., Emmons, L. K., Hess, P. G., Kinnison, D. E., Tilmes, S., Vitt, F., Heald, C. L., Holland, E. A., Lauritzen, P. H., Neu, J., Orlando, J. J., Rasch, P. J., and Tyndall, G. K.: CAM-chem: description and evaluation of interactive atmospheric chemistry in the Community Earth System Model, Geosci. Model Dev., 5, 369–411, https://doi.org/10.5194/gmd-5-369-2012, 2012.

Lamarque, J.-F., Shindell, D. T., Josse, B., Young, P. J., Cionni, I., Eyring, V., Bergmann, D., Cameron-Smith, P., Collins, W. J., Doherty, R., Dalsoren, S., Faluvegi, G., Folberth, G., Ghan, S. J., Horowitz, L. W., Lee, Y. H., MacKenzie, I. A., Nagashima, T., Naik, V., Plummer, D., Righi, M., Rumbold, S. T., Schulz, M., Skeie, R. B., Stevenson, D. S., Strode, S., Sudo, K., Szopa, S., Voulgarakis, A., and Zeng, G.: The Atmospheric Chemistry and Climate Model Intercomparison Project (ACCMIP): overview and description of models, simulations and climate diagnostics, Geosci. Model Dev., 6, 179–206, https://doi.org/10.5194/gmd-6-179-2013, 2013.

Langner, J., Engardt, M., Baklanov, A., Christensen, J. H., Gauss, M., Geels, C., Hedegaard, G. B., Nuterman, R., Simpson, D., Soares, J., Sofiev, M., Wind, P., and Zakey, A.: A multi-model study of impacts of climate change on surface ozone in Europe, Atmos. Chem. Phys., 12, 10423–10440, https://doi.org/10.5194/acp-12-10423-2012, 2012.

Lau, N. C., Leetmaa, A., and Nath, M. J.: Attribution of atmospheric variations in the 1997–2003 period to SST anomalies in the Pacific and Indian Ocean basins, J. Climate, 19, 3607–3628, 2006.

Lee, Y. H. and Adams, P. J.: A fast and efficient version of the two-moment aerosol sectional (TOMAS) global aerosol microphysics model, Aerosol Sci. Tech., 46, 678–689, 2011.

Lee, Y. H., Lamarque, J. F., Flanner, M. G., Jiao, C., Shindell, D. T., Berntsen, T., Bisiaux, M. M., Cao, J., Collins, W. J., Curran, M., Edwards, R., Faluvegi, G., Ghan, S., Horowitz, L. W., McConnell, J. R., Ming, J., Myhre, G., Nagashima, T., Naik, V., Rumbold, S. T., Skeie, R. B., Sudo, K., Takemura, T., Thevenon, F., Xu, B., and Yoon, J. H.: Evaluation of preindustrial to present-day black carbon and its albedo forcing from Atmospheric Chemistry and Climate Model Intercomparison Project (ACCMIP), Atmos. Chem. Phys., 13, 2607–2634, https://doi.org/10.5194/acp-13-2607-2013, 2013.

Lefohn, A. S., Shadwick, D., and Oltmans, S. J.: Characterizing changes in surface ozone levels in metropolitan and rural areas in the United States for 1980–2008 and 1994–2008, Atmos. Environ., 44, 5199–5210, 2010.

Lefohn, A. S., Wernli, H., Shadwick, D., Oltmans, S. J., and Shapiro, M.: Quantifying the frequency of stratospheric-tropospheric transport affecting enhanced surface ozone concentrations at high- and low-elevation monitoring sites in the United States, Atmos. Environ., 62, 646–656, 2012.

Lefohn, A. S., Emery, C., Shadwick, D., Wernli, H., Jung, J., and Oltmans, S. J.: Estimates of background surface ozone concentrations in the United States based on model-derived source apportionment, Atmos. Environ., 84, 275–288, 2014.

Lefohn, A. S., Malley, C. S., Simon, H., Wells, B., Xu, X., Zhang, L., and Wang, T.: Responses of human health and vegetation exposure metrics to changes in ozone concentration distributions in the European Union, United States, and China, Atmos. Environ., 152, 123–145, 2017.

Legrand, M., Preunkert, S., Jourdain, B., Gallée, H., Goutail, F., Weller, R., and Savarino, J.: Year-round record of surface ozone at coastal (Dumont d'Urville) and inland (Concordia) sites in East Antarctica, J. Geophys. Res.-Atmos., 114, D20306, https://doi.org/10.1029/2008JD011667, 2009.

Liu, C., Liu, Y., Cai, Z., Gao, S., Bian, J., Liu, X., and Chance, K.: Dynamic formation of extreme ozone minimum events over the Tibetan Plateau during northern winters 1987–2001, J. Geophys. Res.-Atmos., 115, D18311, https://doi.org/doi:10.1029/2009JD013130, 2010.

Meinshausen, M., Wigley, T. M. L., and Raper, S. C. B.: Emulating atmosphere-ocean and carbon cycle models with a simpler model, MAGICC6 – Part 2: Applications, Atmos. Chem. Phys., 11, 1457–1471, https://doi.org/10.5194/acp-11-1457-2011, 2011.

Mills, G., Hayes, F., Simpson, D., Emberson, L., Norris, D., Harmens, H., and Buker, P.: Evidence of widespread effects of ozone on crops and (semi-)natural vegetation in Europe (1990–2006) in relation to AOT40 and flux-based risk maps, Glob. Change Biol., 17, 592–613, 2011.

Monks, P. S., Archibald, A. T., Colette, A., Cooper, O., Coyle, M., Derwent, R., Fowler, D., Granier, C., Law, K. S., Mills, G. E., Stevenson, D. S., Tarasova, O., Thouret, V., von Schneidemesser, E., Sommariva, R., Wild, O., and Williams, M. L.: Tropospheric ozone and its precursors from the urban to the global scale from air quality to short-lived climate forcer, Atmos. Chem. Phys., 15, 8889–8973, https://doi.org/10.5194/acp-15-8889-2015, 2015.

Moura, B. B., Alves, E. S., de Souza, S. R., Domingos, M., and Vollenweider, P.: Ozone phytotoxic potential with regard to fragments of the Atlantic Semi-deciduous Forest downwind of Sao Paulo, Brazil, Environ. Pollut., 192, 65–73, 2014.

Myhre, G., Shindell, D., Bréon, F. M., Collins, W., Fuglestvedt, J., Huang, J., Koch, D., Lamarque, J. F., Lee, D., Mendoza, B., Nakajima, T., Robock, A., Stephens, G., Takemura, T., and Zhang, H.: Anthropogenic and natural radiative forcing, in: Climate Change 2013: The Physical Science Basis, Contribution of Working Group I to the Fifth Assessment Report of the Intergovernmental Panel on Climate Change, Cambridge University Press, Cambridge, UK and New York, USA, 2013.

Naik, V., Voulgarakis, A., Fiore, A. M., Horowitz, L. W., Lamarque, J.-F., Lin, M., Prather, M. J., Young, P. J., Bergmann, D., Cameron-Smith, P. J., Cionni, I., Collins, W. J., Dalsøren, S. B., Doherty, R., Eyring, V., Faluvegi, G., Folberth, G. A., Josse, B., Lee, Y. H., MacKenzie, I. A., Nagashima, T., van Noije, T. P. C., Plummer, D. A., Righi, M., Rumbold, S. T., Skeie, R., Shindell, D. T., Stevenson, D. S., Strode, S., Sudo, K., Szopa, S., and Zeng, G.: Preindustrial to present-day changes in tropospheric hydroxyl radical and methane lifetime from the Atmospheric Chemistry and Climate Model Intercomparison Project (ACCMIP), Atmos. Chem. Phys., 13, 5277–5298, https://doi.org/10.5194/acp-13-5277-2013, 2013.

Nazarenko, L., Schmidt, G. A., Miller, R. L., Tausnev, N., Kelley, M., Ruedy, R., Russell, G. L., Aleinov, I., Bauer, M., Bauer, S., Bleck, R., Canuto, V., Cheng, Y., Clune, T. L., Del Genio, A. D., Faluvegi, G., Hansen, J. E., Healy, R. J., Kiang, N. Y., Koch, D., Lacis, A. A., Le Grande, A. N., Lerner, J., Lo, K. K., Menon, S., Oinas, V., Perlwitz, J., Puma, M. J., Rind, D., Romanou, A., Sato, M., Shindell, D. T., Sun, S., Tsigaridis, K., Unger, N., Voulgarakis, A., Yao, M. S., and Zhang, J.: Future climate change under RCP emission scenarios with GISS ModelE2, J. Adv. Model. Earth Sy., 7, 244–267, 2015.

Nemani, R. R., Keeling, C. D., Hashimoto, H., Jolly, W. M., and Piper, S. C.: Climate-driven increases in global terrestrial net primary production from 1982 to 1999, Science, 300, 1560–1563, 2003.

Ochoa-Hueso, R., Munzi, S., Alonso, R., Arroniz-Crespo, M., Avila, A., Bermejo, V., Bobbink, R., Branquinho, C., Concostrina-Zubiri, L., Cruz, C., Cruz de Carvalho, R., De Marco, A., Dias, T., Elustondo, D., Elvira, S., Estebanez, B., Fusaro, L., Gerosa, G., Izquieta-Rojano, S., Lo Cascio, M., Marzuoli, R., Matos, P., Mereu, S., Merino, J., Morillas, L., Nunes, A., Paoletti, E., Paoli, L., Pinho, P., Rogers, I. B., Santos, A., Sicard, P., Stevens, C. J., and Theobald, M. R.: Ecological impacts of atmospheric pollution and interactions with climate change in terrestrial ecosystems of the Mediterranean Basin: current research and future directions, Environ. Pollut., 227, 194–206, https://doi.org/10.1016/j.envpol.2017.04.062, 2017.

Ollinger, S. V., Aber, J. D., and Reich, P. B.: Simulating ozone effects on forest productivity: interactions among leaf, canopy, and stand-level processes, Ecol. Appl., 7, 1237–1251, 1997.

Oltmans, S. J., Lefohn, A. S., Harris, J. M., Galbally, I., Scheel, H. E., Bodeker, G., Brunke, E., Claude, H., Tarasick, D., Johnson, B. J., Simmonds, P., Shadwick, D., Anlauf, K., Hayden, K., Schmidlin, F., Fujimoto, T., Akagi, K., Meyer, C., Nichol, S., Davies, J., Redondas, A., and Cuevas, E.: Long-term changes in tropospheric ozone, Atmos. Environ., 40, 3156–3173, 2006.

Paoletti, E. and Manning, W. J.: Toward a biologically significant and usable standard for ozone that will also protect plants, Environ. Pollut., 150, 85–95, 2007.

Paoletti, E., De Marco, A., Beddows, D. C. S., Harrison, R. M., and Manning, W. J.: Ozone levels in European and USA cities are increasing more than at rural sites, while peak values are decreasing, Environ. Pollut., 192, 295–299, 2014.

Parrish, D. D., Law, K. S., Staehelin, J., Derwent, R., Cooper, O. R., Tanimoto, H., Volz-Thomas, A., Gilge, S., Scheel, H.-E., Steinbacher, M., and Chan, E.: Long-term changes in lower tropospheric baseline ozone concentrations at northern mid-latitudes, Atmos. Chem. Phys., 12, 11485–11504, https://doi.org/10.5194/acp-12-11485-2012, 2012.

Pfister, G. G., Walters, S., Lamarque, J. F., Fast, J., Barth, M. C., Wong, J., Done, J., Holland, G., and Bruyère, C. L.: Projections of future summertime ozone over the US, J. Geophys. Res.-Atmos., 119, 5559–5582, 2014.

Prather, M., Bergmann, D., Cameron-Smith, P. J., Cionni, I., Collins, W. J., Dalsøren, S., Eyring, V., Folberth, G. A., Ginoux, P., Horowitz, L. W., Josse, B., Lamarque, J. F., MacKenzie, I. A., Nagashima, T., O'Connor, F. M., Righi, M., Rumbold, S. T., Shindell, D. T., Skeie, R. B., Sudo, K., Szopa, S., Takemura, T., and Zeng, G.: Global air quality and climate, Chem. Soc. Rev., 41, 6663–6683, https://doi.org/10.1039/C2CS35095E, 2012.

Proietti, C., Anav, A., De Marco, A., Sicard, P., and Vitale, M.: A multi-sites analysis on the ozone effects on Gross Primary Production of European forests, Sci. Total Environ., 556, 1–11, 2016.

Querol, X., Alastuey, A., Pandolfi, M., Reche, C., Pérez, N., Minguillón, M. C., Moreno, T., Viana, M., Escudero, M., Orio, A., Pallarés, M., and Reina, F.: 2001–2012 trends on air quality in Spain, Sci. Total Environ., 490, 957–969, 2014.

Reich, P. B.: Quantifying plant response to ozone: a unifying theory, Tree Physiol., 3, 63–91, 1987.

Ren, W., Tian, H., Liu, M., Zhang, C., Chen, G., Pan, S., Felzer, B., and Xu, X.: Effects of tropospheric ozone pollution on net primary productivity and carbon storage in terrestrial ecosystems of China, J. Geophys. Res.-Atmos., 112, 1–17, 2007.

Rieder, H. E., Fiore, A. M., Horowitz, L. W., and Naik, V.: Projecting policy-relevant metrics for high summertime ozone pollution events over the eastern United States due to climate and emission changes during the 21st century, J. Geophys. Res.-Atmos., 120, 784–800, 2015.

Sanderson, M. G., Collins, W. J., Hemming, D. L., and Betts, R. A.: Stomatal conductance changes due to increasing carbon dioxide levels: projected impact on surface ozone levels, Tellus B, 59, 404–411, 2007.

Schnell, J. L., Prather, M. J., Josse, B., Naik, V., Horowitz, L. W., Zeng, G., Shindell, D. T., and Falugevi, G.: Effect of climate change on surface ozone over North America, Europe, and East Asia, Geophys. Res. Lett., 43, 3509–3518, https://doi.org/10.1002/2016GL068060, 2016.

Seidel, D. J., Fu, Q., Randel, W. J., and Reichler, T. J.: Widening of the tropical belt in a changing climate, Nat. Geosci., 1, 21–4, 2008.

Shindell, D. T., Faluvegi, G., Stevenson, D. S., Krol, M. C., Emmons, L. K., Lamarque, J. F., Pétron, G., Dentener, F. J., Ellingsen, K., Schultz, M. G., Wild, O., Amann, M., Atherton, C. S., Bergmann, D. J., Bey, I., Butler, T., Cofala, J., Collins, W.

J., Derwent, R. G., Doherty, R. M., Drevet, J., Eskes, H. J., Fiore, A. M., Gauss, M., Hauglustaine, D. A., Horowitz, L. W., Isaksen, I. S. A., Lawrence, M. G., Montanaro, V., Müller, J. F., Pitari, G., Prather, M. J., Pyle, J. A., Rast, S., Rodriguez, J. M., Sanderson, M. G., Savage, N. H., Strahan, S. E., Sudo, K., Szopa, S., Unger, N., van Noije, T. P. C., and Zeng, G.: Multi-model simulations of carbon monoxide: comparison with observations and projected near-future changes, J. Geophys. Res., 111, D19306, https://doi.org/10.1029/2006JD007100, 2006.

Shindell, D. T., Faluvegi, G., Koch, D. M., Schmidt, G. A., Unger, N., and Bauer, S. E.: Improved attribution of climate forcing to emissions, Science, 326, 716–718, https://doi.org/10.1126/science.1174760, 2009.

Shindell, D. T., Lamarque, J.-F., Schulz, M., Flanner, M., Jiao, C., Chin, M., Young, P. J., Lee, Y. H., Rotstayn, L., Mahowald, N., Milly, G., Faluvegi, G., Balkanski, Y., Collins, W. J., Conley, A. J., Dalsoren, S., Easter, R., Ghan, S., Horowitz, L., Liu, X., Myhre, G., Nagashima, T., Naik, V., Rumbold, S. T., Skeie, R., Sudo, K., Szopa, S., Takemura, T., Voulgarakis, A., Yoon, J.-H., and Lo, F.: Radiative forcing in the ACCMIP historical and future climate simulations, Atmos. Chem. Phys., 13, 2939–2974, https://doi.org/10.5194/acp-13-2939-2013, 2013.

Sicard, P., Coddeville, P., and Galloo, J. C.: Near-surface ozone levels and trends at rural stations in France over the 1995–2003 period, Environ. Monit. Assess., 156, 141–157, 2009.

Sicard, P., Vas, N., and Dalstein-Richier, L.: Annual and seasonal trends for ambient ozone concentration and its impact on forest vegetation in Mercantour National Park (South-eastern France) over the 2000–2008 period, Environ. Pollut., 159, 351–362, 2011.

Sicard, P., De Marco, A., Troussier, F., Renou, C., Vas, N., and Paoletti, E.: Decrease in surface ozone concentrations at Mediterranean remote sites and increase in the cities, Atmos. Environ., 79, 705–715, 2013.

Sicard, P., Serra, R., and Rossello, P.: Spatiotemporal trends of surface ozone concentrations and metrics in France, Environ. Res., 149, 122–144, 2016a.

Sicard, P, Augustaitis, A., Belyazid, S., Calfapietra, C., and De Marco, A.: Global topics and novel approaches in the study of air pollution, climate change and forest ecosystems, Environ. Pollut., 213, 977–987, 2016b.

Sicard, P., De Marco, A., Dalstein-Richier, L., Tagliaferro, F., and Paoletti, E.: An epidemiological assessment of stomatal ozone flux-based critical levels for visible ozone injury in Southern European forests, Sci. Total Environ., 541, 729–741, 2016c.

Simpson, D., Arneth, A., Mills, G., Solberg, S., and Uddling, J.: Ozone – the persistent menace: interactions with the N cycle and climate change, Curr. Opin. Env. Sust., 9–10, 9–19, 2014.

Singh, H. B., Herlth, D., O'Hara, D., Zahnle, K., Bradshaw, J. D., Sandholm, S. T., Talbot, R., Crutzen, P. J., and Kanakidou, M.: Relationship of Peroxyacetyl nitrate to active and total odd nitrogen at northern high latitudes: influence of reservoir species on $NO_x$ and $O_3$, J. Geophys. Res.-Atmos., 97, 16523–16530, 1992.

Sitch, S., Cox, P. M., Collins, W. J., and Huntingford, C.: Indirect radiative forcing of climate change through ozone effects on the land-carbon sink, Nature, 448, 791–794, 2007.

Steinbacher, M., Henne, S., Dommen, J., Wiesen, P., and Prevot, A. S. H.: Nocturnal trans-alpine transport of ozone and its

effects on air quality on the Swiss Plateau, Atmos. Environ., 38, 4539–4550, 2004.

Stevenson, D. S., Johnson, C. E., Collins, W. J., Derwent, R. G., and Edwards, J. M.: Future estimates of tropospheric ozone radiative forcing and methane turnover – the impact of climate change, Geophys. Res. Lett., 27, 2073–2076, 2000.

Stevenson, D. S., Dentener, F. J., Schultz, M. G., Ellingsen, K., van Noije, T. P. C., Wild, O., Zeng, G., Amann, M., Atherton, C. S., Bell, N., Bergmann D. J., Bey, I., Butler, T., Cofala, J., Collins, W. J., Derwent, R. G., Doherty, R. M., Drevet, J., Eskes, H. J., Fiore, A. M., Gauss, M., Hauglustaine, D. A., Horowitz, L. W., Isaksen, I. S. A., Krol, M. C., Lamarque, J. F., Lawrence, M. G., Montanaro, V., Müller, J. F., Pitari, G., Prather, M. J., Pyle, J. A., Rast, S., Rodriguez, J. M., Sanderson, M. G., Savage, N. H., Shindell, D. T., Strahan, S. E., Sudo, K., and Szopa, S.: Multi-model ensemble simulations of present-day and near-future tropospheric ozone, J. Geophys. Res.-Atmos., 111, D08301, https://doi.org/10.1029/2005JD006338, 2006.

Stevenson, D. S., Young, P. J., Naik, V., Lamarque, J.-F., Shindell, D. T., Voulgarakis, A., Skeie, R. B., Dalsoren, S. B., Myhre, G., Berntsen, T. K., Folberth, G. A., Rumbold, S. T., Collins, W. J., MacKenzie, I. A., Doherty, R. M., Zeng, G., van Noije, T. P. C., Strunk, A., Bergmann, D., Cameron-Smith, P., Plummer, D. A., Strode, S. A., Horowitz, L., Lee, Y. H., Szopa, S., Sudo, K., Nagashima, T., Josse, B., Cionni, I., Righi, M., Eyring, V., Conley, A., Bowman, K. W., Wild, O., and Archibald, A.: Tropospheric ozone changes, radiative forcing and attribution to emissions in the Atmospheric Chemistry and Climate Model Intercomparison Project (ACCMIP), Atmos. Chem. Phys., 13, 3063–3085, https://doi.org/10.5194/acp-13-3063-2013, 2013.

Stohl, A., Berg, T., Burkhart, J. F., Fjǽraa, A. M., Forster, C., Herber, A., Hov, Ø., Lunder, C., McMillan, W. W., Oltmans, S., Shiobara, M., Simpson, D., Solberg, S., Stebel, K., Ström, J., Tørseth, K., Treffeisen, R., Virkkunen, K., and Yttri, K. E.: Arctic smoke – record high air pollution levels in the European Arctic due to agricultural fires in Eastern Europe in spring 2006, Atmos. Chem. Phys., 7, 511–534, https://doi.org/10.5194/acp-7-511-2007, 2007.

Tang, Q., Prather, M. J., and Hsu, J.: Stratosphere-troposphere exchange ozone flux related to deep convection, Geophys. Res. Lett., 38, L03806, https://doi.org/10.1029/2010GL046039, 2011.

Teyssèdre, H., Michou, M., Clark, H. L., Josse, B., Karcher, F., Olivié, D., Peuch, V.-H., Saint-Martin, D., Cariolle, D., Attié, J.-L., Nédélec, P., Ricaud, P., Thouret, V., van der A, R. J., Volz-Thomas, A., and Chéroux, F.: A new tropospheric and stratospheric Chemistry and Transport Model MOCAGE-Climat for multi-year studies: evaluation of the present-day climatology and sensitivity to surface processes, Atmos. Chem. Phys., 7, 5815–5860, https://doi.org/10.5194/acp-7-5815-2007, 2007.

Tian, W., Chipperfield, M., and Huang, Q.: Effects of the Tibetan Plateau on total column ozone distribution, Tellus B, 60, 622–635, 2008.

UNECE: United Nations Economic Commission for Europe. Convention on Long-Range Trans-boundary Air Pollution: Mapping Critical Levels for Vegetation. International Cooperative Programme on Effects of Air Pollution on Natural Vegetation and Crops, Bangor, UK, 2010.

van Vuuren, D., Edmonds, J., Kainuma, M., Riahi, K., Thomson, A., Hibbard, K., Hurtt, G. C., Kram, T., Krey, V.,

Lamarque, J. F., Masui, T., Meinshausen, M., Nakicenovic, N., Smith, S. J., and Rose, S. K.: The representative concentration pathways: an overview, Climatic Change, 109, 5–31, https://doi.org/10.1007/s10584-011-0148-z, 2011.

Voulgarakis, A., Naik, V., Lamarque, J.-F., Shindell, D. T., Young, P. J., Prather, M. J., Wild, O., Field, R. D., Bergmann, D., Cameron-Smith, P., Cionni, I., Collins, W. J., Dalsøren, S. B., Doherty, R. M., Eyring, V., Faluvegi, G., Folberth, G. A., Horowitz, L. W., Josse, B., MacKenzie, I. A., Nagashima, T., Plummer, D. A., Righi, M., Rumbold, S. T., Stevenson, D. S., Strode, S. A., Sudo, K., Szopa, S., and Zeng, G.: Analysis of present day and future OH and methane lifetime in the ACCMIP simulations, Atmos. Chem. Phys., 13, 2563–2587, https://doi.org/10.5194/acp-13-2563-2013, 2013.

Walker, T. W., Jones, D. B. A., Parrington, M., Henze, D. K., Murray, L. T., Bottenheim, J. W., Anlauf, K., Worden, J. R., Bowman, K. W., Shim, C., Singh, K., Kopacz, M., Tarasick, D. W., Davies, J., von der Gathen, P., Thompson, A. M., and Carouge, C. C.: Impacts of mid-latitude precursor emissions and local photochemistry on ozone abundances in the Arctic, J. Geophys. Res.-Atmos., 117, D01305, https://doi.org/10.1029/2011JD016370, 2012.

Wang, Q. Y., Gao, R. S., Cao, J. J., Schwarz, J. P., Fahey, D. W., Shen, Z. X., Hu, T. F., Wang, P., Xu, X. B., and Huang, R. J.: Observations of high level of ozone at Qinghai Lake basin in the northeastern Qinghai–Tibetan Plateau, western China, J. Atmos. Chem., 72, 19–26, 2015.

Wang, X. and Mauzerall, D. L.: Characterizing distributions of surface ozone and its impact on grain production in China, Japan and South Korea: 1900 and 2020, Atmos. Environ., 38, 4383–4402, 2004.

Watanabe, S., Hajima, T., Sudo, K., Nagashima, T., Takemura, T., Okajima, H., Nozawa, T., Kawase, H., Abe, M., Yokohata, T., Ise, T., Sato, H., Kato, E., Takata, K., Emori, S., and Kawamiya, M.: MIROC-ESM 2010: model description and basic results of CMIP5-20c3m experiments, Geosci. Model Dev., 4, 845–872, https://doi.org/10.5194/gmd-4-845-2011, 2011.

Wesely, M. L. and Hicks, B. B.: A review of the current status of knowledge in dry deposition, Atmos. Environ., 34, 2261–2282, 2000.

Wild, O.: Modelling the global tropospheric ozone budget: exploring the variability in current models, Atmos. Chem. Phys., 7, 2643–2660, https://doi.org/10.5194/acp-7-2643-2007, 2007.

Wild, O., Fiore, A. M., Shindell, D. T., Doherty, R. M., Collins, W. J., Dentener, F. J., Schultz, M. G., Gong, S., MacKenzie, I. A., Zeng, G., Hess, P., Duncan, B. N., Bergmann, D. J., Szopa, S., Jonson, J. E., Keating, T. J., and Zuber, A.: Modelling future changes in surface ozone: a parameterized approach, Atmos. Chem. Phys., 12, 2037–2054, https://doi.org/10.5194/acp-12-2037-2012, 2012.

Williams, E. R.: The global electrical circuit: a review, Atmos. Res., 91, 140–152, 2009.

Wilson, R. C., Fleming, Z. L., Monks, P. S., Clain, G., Henne, S., Konovalov, I. B., Szopa, S., and Menut, L.: Have primary emission reduction measures reduced ozone across Europe? An analysis of European rural background ozone trends 1996–2005, Atmos. Chem. Phys., 12, 437–454, https://doi.org/10.5194/acp-12-437-2012, 2012.

Wittig, V. E., Ainsworth, E. A., and Long, S. P.: To what extent do current and projected increases in surface ozone affect pho-

tosynthesis and stomatal conductance of trees? A meta-analytic review of the last 3 decades of experiments, Plant Cell Environ., 30, 1150–1162, 2007.

Wittig, V. E., Ainsworth, E. A., Naidu, S. L., Karnosky, D. F., and Long, S. P.: Quantifying the impact of current and future tropospheric ozone on tree biomass, growth, physiology and biochemistry: a quantitative meta-analysis, Glob. Change Biol., 15, 396–424, 2009.

Xing, J., Mathur, R., Pleim, J., Hogrefe, C., Gan, C.-M., Wong, D. C., Wei, C., Gilliam, R., and Pouliot, G.: Observations and modeling of air quality trends over 1990–2010 across the Northern Hemisphere: China, the United States and Europe, Atmos. Chem. Phys., 15, 2723–2747, https://doi.org/10.5194/acp-15-2723-2015, 2015.

Young, P. J., Archibald, A. T., Bowman, K. W., Lamarque, J.-F., Naik, V., Stevenson, D. S., Tilmes, S., Voulgarakis, A., Wild, O., Bergmann, D., Cameron-Smith, P., Cionni, I., Collins, W. J., Dalsøren, S. B., Doherty, R. M., Eyring, V., Faluvegi, G., Horowitz, L. W., Josse, B., Lee, Y. H., MacKenzie, I. A., Nagashima, T., Plummer, D. A., Righi, M., Rumbold, S. T., Skeie, R. B., Shindell, D. T., Strode, S. A., Sudo, K., Szopa, S., and Zeng, G.: Pre-industrial to end 21st century projections of tropospheric ozone from the Atmospheric Chemistry and Climate Model Intercomparison Project (ACCMIP), Atmos. Chem. Phys., 13, 2063–2090, https://doi.org/10.5194/acp-13-2063-2013, 2013.

Zak, D. R., Pregitzer, K. S., Kubiske, M. E., and Burton, A. J.: Forest productivity under elevated $CO_2$ and $O_3$: positive feedbacks to soil N cycling sustain decade-long net primary productivity enhancement by $CO_2$, Ecol. Lett., 14, 1220–1226, 2011.

Zeng, G. and Pyle, J. A.: Changes in tropospheric ozone between 2000 and 2100 modeled in a chemistry-climate model, Geophys. Res. Lett., 30, 1392, https://doi.org/10.1029/2002GL016708, 2003.

Zeng, G., Pyle, J. A., and Young, P. J.: Impact of climate change on tropospheric ozone and its global budgets, Atmos. Chem. Phys., 8, 369–387, https://doi.org/10.5194/acp-8-369-2008, 2008.

Zeng, G., Morgenstern, O., Braesicke, P., and Pyle, J. A.: Impact of stratospheric ozone recovery on tropospheric ozone and its budget, Geophys. Res. Lett., 37, L09805, https://doi.org/10.1029/2010GL042812, 2010.

Zhang, L., Brook, J. R., and Vet, R.: A revised parameterization for gaseous dry deposition in air-quality models, Atmos. Chem. Phys., 3, 2067–2082, https://doi.org/10.5194/acp-3-2067-2003, 2003.

Zhang, M., Xu, Y., Uno, I., and Akimoto, H.: A numerical study of tropospheric ozone in the springtime in east Asia, Adv. Atmos. Sci., 21, 163–170, 2004.

Zhang, Q., Streets, D. G., Carmichael, G. R., He, K. B., Huo, H., Kannari, A., Klimont, Z., Park, I. S., Reddy, S., Fu, J. S., Chen, D., Duan, L., Lei, Y., Wang, L. T., and Yao, Z. L.: Asian emissions in 2006 for the NASA INTEX-B mission, Atmos. Chem. Phys., 9, 5131–5153, https://doi.org/10.5194/acp-9-5131-2009, 2009.

Zhu, Z., Piao, S., Myneni, R. B., Huang, M., Zeng, Z., Canadell, J. G., Ciais, P., Sitch, S., Friedlingstein, P., Arneth, A., Cao, C., Cheng, L., Kato, E., Koven, C., Li, Y., Lian, X., Liu, Y., Liu, R., Mao, J., Pan, Y., Peng, S., Peñuelas, J., Poulter, B., Pugh, T. A. M., Stocker, B. D., Viovy, N., Wang, X., Wang, Y., Xiao, Z., Yang, H., Zaehle, S., and Zeng, N.: Greening of the Earth and its drivers, Nat. Clim. Change, 6, 791–795, 2016.

# Classification of Arctic, midlatitude and tropical clouds in the mixed-phase temperature regime

**Anja Costa**[1], **Jessica Meyer**[1,a], **Armin Afchine**[1], **Anna Luebke**[1,b], **Gebhard Günther**[1], **James R. Dorsey**[2], **Martin W. Gallagher**[2], **Andre Ehrlich**[3], **Manfred Wendisch**[3], **Darrel Baumgardner**[4], **Heike Wex**[5], and **Martina Krämer**[1]

[1]Forschungszentrum Jülich GmbH, Jülich, Germany
[2]Centre for Atmospheric Science, University of Manchester, Manchester, UK
[3]Leipziger Institut für Meteorologie, Universität Leipzig, Germany
[4]Droplet Measurement Technologies, Longmont, CO 80503, USA
[5]Leibniz Institute for Tropospheric Research, Leipzig, Germany
[a]now at: Bundesanstalt für Arbeitsschutz und Arbeitsmedizin, Dortmund, Germany
[b]now at: Atmosphere in the Earth System Department, Max Planck Institute for Meteorology, Hamburg, Germany

*Correspondence to:* Martina Krämer (m.kraemer@fz-juelich.de)

**Abstract.** The degree of glaciation of mixed-phase clouds constitutes one of the largest uncertainties in climate prediction. In order to better understand cloud glaciation, cloud spectrometer observations are presented in this paper, which were made in the mixed-phase temperature regime between 0 and $-38\,°C$ (273 to 235 K), where cloud particles can either be frozen or liquid. The extensive data set covers four airborne field campaigns providing a total of 139 000 1 Hz data points (38.6 h within clouds) over Arctic, midlatitude and tropical regions. We develop algorithms, combining the information on number concentration, size and asphericity of the observed cloud particles to classify four cloud types: liquid clouds, clouds in which liquid droplets and ice crystals coexist, fully glaciated clouds after the Wegener–Bergeron–Findeisen process and clouds where secondary ice formation occurred. We quantify the occurrence of these cloud groups depending on the geographical region and temperature and find that liquid clouds dominate our measurements during the Arctic spring, while clouds dominated by the Wegener–Bergeron–Findeisen process are most common in midlatitude spring. The coexistence of liquid water and ice crystals is found over the whole mixed-phase temperature range in tropical convective towers in the dry season. Secondary ice is found at midlatitudes at $-5$ to $-10\,°C$ (268 to 263 K) and at higher altitudes, i.e. lower temperatures in the tropics. The

distribution of the cloud types with decreasing temperature is shown to be consistent with the theory of evolution of mixed-phase clouds. With this study, we aim to contribute to a large statistical database on cloud types in the mixed-phase temperature regime.

## 1 Introduction

Clouds can be classified according to their altitude (low, mid-level, high; see e.g. Rossow and Schiffer, 1991), their temperature (warm, cold) or their cloud particle phase (liquid, mixed-phase: both liquid and ice, ice). Especially for intermediate altitudes, these classification criteria overlap: ice particles may sediment into warm cloud layers, updraughts can transport liquid water droplets into colder cloud regions and droplet formation may produce liquid water content in a cold, formerly glaciated cloud (Findeisen et al., 2015; Korolev, 2007).

To avoid ambiguities, we refer here to all clouds observed at temperatures between 0 and $-38\,°C$ (273 to 235 K) as "clouds in the mixed-phase temperature regime" (mpt clouds). In that temperature regime, purely liquid (supercooled) clouds can be found as well as mixed-phase clouds (where liquid water droplets and ice crystals coexist) and

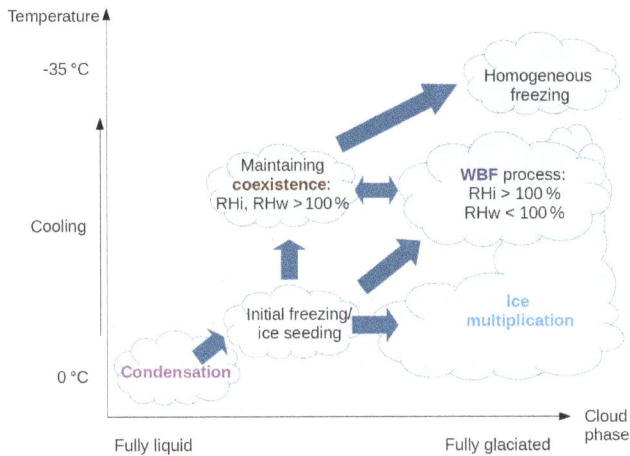

**Figure 1.** Possible paths to glaciation in the mixed-phase temperature regime.

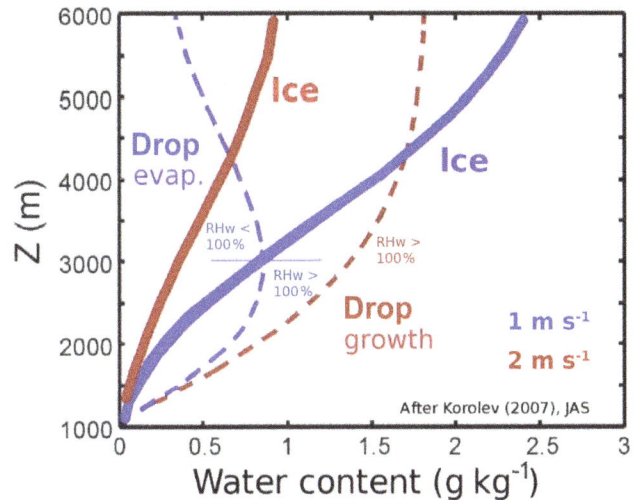

**Figure 2.** Liquid water content (dashed lines) and ice water content (solid lines) development with altitude $Z$ ($\sim$ 1/temperature) in mixed-phase clouds for different vertical velocities (adapted from Korolev, 2007, with modification). Blue lines (updraught $1\,\mathrm{m\,s^{-1}}$): the cloud glaciates when $RH_w$ falls below $100\%$ (Wegener–Bergeron–Findeisen regime); red lines (updraught $2\,\mathrm{m\,s^{-1}}$): $RH_w$ stays above $100\%$, liquid droplets and ice crystals coexist (coexistence regime).

also fully glaciated clouds (Pruppacher et al., 1998). Within this temperature range, important processes take place that transform the cloud's phase or microphysical characteristics significantly. This phase transition is not only an important part of precipitation-forming processes like the cold rain process, it also affects the cloud's radiative properties by influencing the solar albedo of mpt clouds (Curry et al., 1996; Shupe and Intrieri, 2004): with growing ice fraction, their solar albedo (cooling) effect is reduced (Ehrlich et al., 2009; Wendisch et al., 2013). Thus a correct representation of this cloud type in global climate models is of importance for an improved certainty of climate predictions (Wendisch et al., 2013, 2017).

The transformation from a fully liquid to a fully frozen cloud can follow different, sometimes non-linear paths, as illustrated in Fig. 1. After the activation of cloud condensation nuclei forms small droplets $< 50\,\mu\mathrm{m}$ (all-liquid state), initial freezing can occur in those droplets that contain or touch an ice nucleating particle (INP) that can be activated in the ambient cloud environment (resulting in a mixed-phase state: coexistence of ice and water). Different INPs can induce ice nucleation at different temperatures, depending on their nature, e.g. whether they are of biological or mineral origin, their morphology and freezing efficiency. Therefore, the number of droplets containing an INP needed to heterogeneously form ice is important for its glaciation, and the temperature of the mpt cloud is also relevant, as the freezing efficiency of different INPs varies with temperature. The INP properties that favour ice formation are a major discussion point in cloud and climate research. The conditions that favour drop freezing are, in a simplified summary, as follows: cold temperatures, high relative humidities and a "good freezing ability". For more details on the specific conditions see e.g. Kanji et al. (2017) and references therein. Biological particles are known to induce ice nucleation in the temperature range between about 0 and $-20\,^{\circ}\mathrm{C}$ (273 to 253 K), while

mineral dust particles initiate ice at temperatures below about $-20\,^{\circ}\mathrm{C}$ (Kanji et al., 2017; Augustin-Bauditz et al., 2014).

The persistence of supercooled liquid clouds in case no ice active particles are present is also reported by Korolev (2007). Moreover, the further development of the glaciation degree of a mpt cloud, in which a few ice crystals are present, is discussed in this study in relation to the environmental dynamical conditions. This is illustrated by theoretical considerations (Korolev, 2007) of the partitioning of liquid and ice water content in rising mixed-phase cloud parcels under different conditions (see Fig. 2, adapted from Korolev, 2007). The first scenario represents an intermediate vertical velocity ($1\,\mathrm{m\,s^{-1}}$; blue lines), where the Wegener–Bergeron–Findeisen process (Findeisen et al., 2015) is triggered above the altitude marked by the blue line (note that the temperature decreases with increasing altitude), which leads to full glaciation of the cloud. At that point, the relative humidity over water falls below $100\%$ ($RH_w < 100\%$), as more and more water vapour is consumed by the many small liquid cloud droplets. As a result, these droplets evaporate, decreasing the liquid water content. The RH over ice remains above $100\%$ ($RH_i > 100\%$), allowing the few ice crystals to grow to large sizes $> 50\,\mu\mathrm{m}$, thus increasing the ice water content.

In contrast, the red graphs show a scenario for higher vertical velocities ($2\,\mathrm{m\,s^{-1}}$). Here, due to the high updraught, the supersaturation is preserved over both water and ice ($RH_w$, $RH_i > 100\%$) over the complete altitude range. Subsequently, the liquid and ice water content increase in coexistence and the cloud continues to be only partly glaciated (coexistence cloud). These simulations demonstrate that vertical velocity is a major parameter controlling the occurrence of different cloud types, because the updraught is the crucial parameter for possible supersaturations. The supersaturation over water can remain at or above 100 % only in high updraughts, thus allowing coexistence clouds to survive down to about $-38\,^{\circ}\mathrm{C}$ (235 K), where the supercooled liquid cloud droplets will freeze homogeneously (Pruppacher et al., 1998; Koop et al., 2000). Also, secondary ice production can take place, producing high number concentrations of small ice particles (see overview in Field et al., 2015, 2017). Examples of known processes are the Hallett–Mossop process (also called rime splintering; see Hallett and Mossop, 1974), drop-freezing (Lawson et al., 2015) and ice–ice collisions (Yano and Phillips, 2011). When one of these processes has started, the remaining liquid fraction of a cloud can glaciate quickly via freezing initiated by ice crystals colliding with supercooled water droplets, even if the conditions for the Wegener–Bergeron–Findeisen (WBF) process are not met.

Evaporation of both numerous small liquid droplets and large ice particles occurs when the environment is subsaturated with respect to both water and ice ($\mathrm{RH_i} < 100$, $\mathrm{RH_w} < 100\,\%$), as predicted by Korolev (2007) for downdraught regions within the cloud. If this state persists for a sufficiently long time, the cloud will fully evaporate.

In summary, as illustrated in Fig. 1 and Table 1, four types of mpt cloud are expected to occur: the first type describes clouds with many small (diameter $< 50\,\mu\mathrm{m}$) liquid droplets that often appear at slightly supercooled conditions and with lower frequencies as the temperature becomes colder (Bühl et al., 2013). This cloud type may additionally contain a low concentration of large particles (large droplets from coalescence or ice particles sedimenting from above). The second cloud type is coexistence clouds with a high concentration of small cloud particles $< 50\,\mu\mathrm{m}$ that can be liquid or frozen. The coexistence cloud type appears at decreasing temperatures in higher updraughts. In case the updraughts are very strong as in tropical convective clouds, the supercooled liquid cloud droplets can reach cold temperature regions around $-38\,^{\circ}\mathrm{C}$ (235 K) and freeze homogeneously. Furthermore a third type with a high concentration of small ice particles (diameter $< 50\,\mu\mathrm{m}$) might emerge as a result of secondary ice production, e.g. due to the Hallett–Mossop process at temperatures between $-3$ and $-8\,^{\circ}\mathrm{C}$ (270 to 265 K) or ice splintering. A fourth cloud type in the case of lower updraughts are fully glaciated WBF clouds. They contain only very few or no small liquid droplets ($< 50\,\mu\mathrm{m}$), but consist mostly of large ice crystals and are expected to appear with increasing frequency when the temperature decreases.

Due to the manifold interactions between large-scale and small-scale dynamics, aerosol particle/INP availability and complex processes of formation and evolution of supercooled liquid and frozen cloud particles, mpt clouds are not well understood and therefore poorly represented in global climate models (Boucher et al., 2013). As a consequence, the uncertainties concerning the global mpt cloud cover's radiative impact are large. Of particular interest is the partitioning of ice and liquid water, i.e. the glaciation degree. An important step that improves the incomplete understanding of the phase transition processes is taking reliable observations of the different types of mpt cloud. However, cloud-particle-phase observations are limited by technical constraints: passive satellite data mostly provide information on cloud tops, and ground-based lidars cannot quantify thick layers of liquid water (Shupe et al., 2008; Storelvmo and Tan, 2015). Active remote sensing techniques have been used to derive liquid and ice water paths for the full depth of the atmosphere (reported in Boucher et al., 2013, p. 580), but are subject to large errors. In situ measurements may cover the full vertical extent (Taylor et al., 2016; Lloyd et al., 2015; Klingebiel et al., 2015), but are restricted to the flight path and have to be analysed carefully (Wendisch and Brenguier, 2013). For in situ data sets in the past, the phase identification often relied on cloud particle sizes. Small cloud particles $< 50\,\mu\mathrm{m}$ are usually regarded as liquid (see e.g. Taylor et al., 2016). With particle imaging probes like OAPs (optical array probes), more sophisticated shape recognition algorithms can be used (e.g. Korolev and Sussman, 2000), which are nevertheless limited. Usually, they require a minimum number of pixels (corresponding to cloud particles with diameters of $70\,\mu\mathrm{m}$ and more) to recognize round or aspherical particles reliably. Due to these limitations, the shape identification of small particles has not been considered in many microphysical cloud studies. In the paper presented here, we use a new detector that can measure the asphericity of small ($< 50\,\mu\mathrm{m}$) cloud particles (Baumgardner et al., 2014) together with a visual shape inspection of particles $> 50\,\mu\mathrm{m}$. We thus hope to provide new insights into the microphysical evolution of clouds in the mpt regime.

To this end, we use in situ airborne cloud measurements in the cloud particle size range from 3 to $937\,\mu\mathrm{m}$ to classify the above-described types of cloud in the mpt regime (see Fig. 1): mostly liquid clouds occur after drop formation, coexistence clouds after initial freezing, secondary ice clouds are influenced by ice multiplication and large ice clouds occur after the WBF process. This classification enables us to revisit a statistical overview published by Pruppacher et al. (1998), stating at which temperatures purely liquid or ice-containing clouds were found.

For all except the fourth cloud type, we expect high cloud particle number concentrations with a peak at cloud particle sizes $< 50\,\mu\mathrm{m}$. Thus, particle size distributions and concentrations allow a differentiation between glaciated clouds mainly formed via the WBF process and other cloud types in the mpt regime. To investigate these other types more closely, they are divided into three groups with differing aspherical cloud particle fractions, in agreement with the cloud types described above. The occurrence of the four cloud types is then quantified with regard to measurement location and tem-

**Table 1.** Characteristics of the cloud types expected in the mpt regime.

|  | Cloud particle concentration | Particles $D_p < 50\,\mu m$ | Particles $D_p > 50\,\mu m$ | Dominant mass mode |
|---|---|---|---|---|
| Mostly liquid | high | Liquid | Drizzle drops/few ice crystals possible | $D_p < 50\,\mu m$: Type 1a |
| Coexistence | high | Mostly liquid, some ice crystals | Ice crystals | $D_p < 50\,\mu m$: Type 1b |
| Secondary ice | high | Ice crystals | Ice crystals | $D_p < 50\,\mu m$: Type 1c |
| Large ice/WBF | low | Ice crystals | Ice crystals | $D_p > 50\,\mu m$: Type 2 |

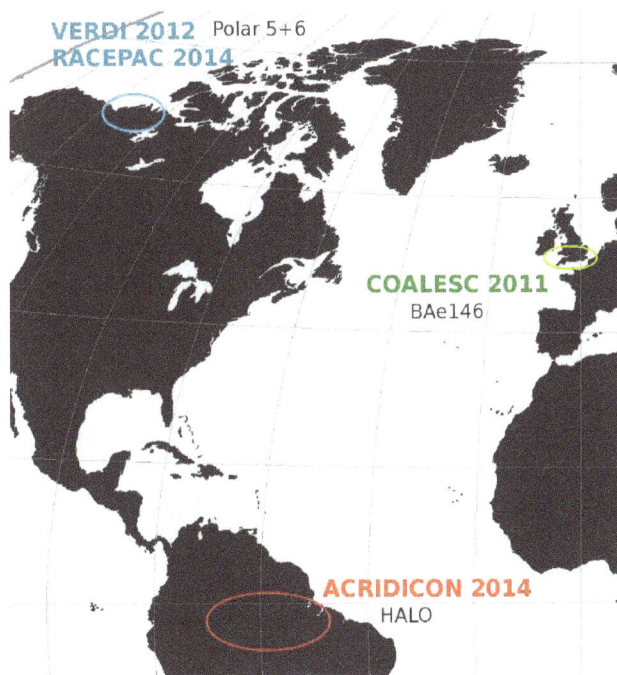

**Figure 3.** Locations of the campaigns presented in this paper.

perature by performing a statistical analysis of the data obtained by the NIXE–CAPS (New Ice eXpEriment – Cloud and Aerosol Particle Spectrometer) cloud spectrometer with 1 Hz along the flight path.

The article is structured as follows: in Sect. 2, the field campaigns are described as well as the cloud spectrometer NIXE–CAPS and its data products. In Sect. 3, the observations are evaluated with respect to the clouds' size distribution, the correlation of cloud particle concentrations to expected ice nucleating particle concentrations, the cloud particle asphericity and the associated vertical velocities. Section 4 summarises the findings of this study.

## 2  Methodology

Four airborne field campaigns were performed in Arctic, midlatitude and tropical regions (see Fig. 3 and Sect. 2.1). In total, the data set in the mixed-phase temperature regime between 0 and $-38\,°C$ (273 to 235 K) covers 38.6 h. Mpt clouds were measured using the cloud spectrometer NIXE–CAPS (see Sect. 2.2). The data analysis is described in Sect. 2.3.

### 2.1  Field campaigns

The first campaign, COALESC (Combined Observation of the Atmospheric boundary Layer to study the Evolution of StratoCumulus), was based in Exeter, UK, in February and March 2011. The NIXE–CAPS was installed as a wing probe on the BAe146 aircraft operated by the Facility for Airborne Atmospheric Measurements (FAAM), UK. All flights took place in the coastal area of south-eastern England and Wales; the main campaign targets were low stratus and stratocumulus clouds. The campaign is described in Osborne et al. (2014), Table 2 provides an overview of the flights. Out of 16 measurement flights, 14 provided observations of mpt clouds, with in total 41042 seconds (11.4 h) of data.

Measurements in Arctic clouds have been conducted during the campaigns VERDI (April and May 2012, study on the Vertical Distribution of Ice in Arctic Clouds; see also Klingebiel et al., 2015) and RACEPAC (April and May 2014, Radiation–Aerosol–Cloud Experiment in the Arctic Circle). Both campaigns took place in Inuvik, northern Canada. Research flights were performed with the Polar-5 and Polar-6 aircraft of the Alfred Wegener Institute, Germany. The 13 flights of both VERDI (see Table 3) and RACEPAC (Table 4) covered the region of the Arctic Beaufort Sea coast with its retreating sea ice in spring. VERDI yielded 59 028 s (16.4 h) of observations within mpt clouds, RACEPAC contributed 33 354 s (9.3 h). Although both campaigns took place at the same time of the year, different synoptic situations lead to different cloud characteristics: VERDI was dominated by stable anticyclonic periods with weak gradients of atmospheric parameters that allow the formation of a strong inversion in the boundary layer associated with persisting stratus, whereas during RACEPAC frontal systems frequently passed

**Table 2.** Flight table for COALESC. Dates are dd/mm/yyyy.

| Date | Probed clouds/flight objectives | Cloud $T$ in the mpt regime | Minutes in mpt clouds |
|---|---|---|---|
| 15.02.2011 | Warm clouds, mixed clouds, cirrus; test flight | −1.5 to −37.6 °C | 85.1 |
| 23.02.2011 | Warm clouds, cirrus clouds | 0 to −37.8 °C | 11.7 |
| 24.02.2011 | Warm stratocumulus | 0 to −0.1 °C | 0.1 |
| 26.02.2011 | Stratocumulus | 0 to −17.9 °C | 46.0 |
| 01.03.2011 | Stratocumulus | 0 to −6.4 °C | 124.7 |
| 02.03.2011 | Stratocumulus | 0 to −3.1 °C | 92.0 |
| 03.03.2011 | Stratocumulus | 0 to −4.4 °C | 61.9 |
| 05.03.2011 | Stratocumulus | 0 to −3.3 °C | 51.4 |
| 07.03.2011 | No clouds | − | 0 |
| 08.03.2011 | Warm stratocumulus and cirrus clouds | 0 to −38.0 °C | 47.0 |
| 11.03.2011 | Stratocumulus | 0 to −4.9 °C | 105.9 |
| 14.03.2011 | Mostly cirrus clouds | −8.9 to −37.9 °C | 10.6 |
| 15.03.2011 | Stratocumulus and cirrus | 0 to −38.0 °C | 25.8 |
| 16.03.2011 | Stratocumulus | 0 to −0.3 °C | 6.7 |
| 18.03.2011 | No clouds | − | 0 |
| 19.03.2011 | Mostly contrail cirrus | −18.1 to −38.0 °C | 11.9 |

**Table 3.** Flight table for VERDI. Dates are dd/mm/yyyy.

| Date | Probed clouds/flight objectives | Cloud $T$ in the mpt regime | Minutes in mpt clouds |
|---|---|---|---|
| 25.04.2012 | Low mostly liquid stratus; test flight | −3.7 to −9.1 °C | 47.1 |
| 27.04.2012 | Stratus (liquid and ice) over sea ice | −8.1 to −16.5 °C | 73.4 |
| 27.04.2012 | Low dissipating clouds over sea ice | −9.1 to −17.3 °C | 47.6 |
| 29.04.2012 | Stable stratus over sea ice | −8.4 to −12.5 °C | 77.9 |
| 30.04.2012 | Extensive cloud with layer structure | −6.3 to −19.1 °C | 212.8 |
| 03.05.2012 | Thin low subvisible clouds | −9.4 to −12.1 °C | 56.15 |
| 05.05.2012 | Patchy low cloud layer | −8.6 to −16.8 °C | 77.9 |
| 08.05.2012 | Mostly supercooled liquid clouds, two layers | −4.9 to −9.7 °C | 65.8 |
| 10.05.2012 | Dissolving altostratus layer | −5.5 to −11.2 °C | 45.1 |
| 14.05.2012 | Two thin stratus and cumulus | −1.4 to −5.8 °C | 41.9 |
| 15.05.2012 | Mostly liquid stratus and a cumulus | −0.7 to −14.1 °C | 73.2 |
| 16.05.2012 | Thin, mostly liquid stratus | −1.7 to −5.3 °C | 95.2 |
| 17.05.2012 | Mostly liquid stratus with large snow | 0 to −6.3 °C | 54.5 |

the area of the observations and lead to a more variable and short-lived cloud situation.

The tropical measurement campaign ACRIDICON–CHUVA (Aerosol, Cloud, Precipitation, and Radiation Interactions and Dynamics of Convective Cloud Systems – Cloud processes of the main precipitation systems in Brazil: A contribution to cloud resolving modelling and to the GPM – Global Precipitation Measurement) was carried out in September and October 2014. The instrument platform was HALO (High Altitude and Long Range Research Aircraft), a Gulfstream V aircraft operated by DLR (Deutsches Luft- und Raumfahrtszentrum/German Aerospace Centre). Based in Manaus, Brazil, ACRIDICON–CHUVA was aimed at convective clouds over tropical rainforest and deforested areas (see Table 5; for details, see Wendisch et al., 2016). The cam-

paign comprises 14 flights, 11 of which contained clouds in the mixed-phase temperature regime. Although cloud profiling at various altitudes and temperatures was a main directive of ACRIDICON–CHUVA, the total time spent within mpt clouds was only 5368 seconds (1.5 h). The relatively limited time span was caused by the high flying speed of HALO (up to $240\,\mathrm{m\,s^{-1}}$); it results in short penetration times (in the range of several seconds) of the convective towers. A second reason is the increasing danger of strong vertical winds and icing in developing cumulonimbus clouds. From certain cloud development stages on, only the cloud's anvil and outflow at cold temperatures lower than −38 °C (235 K) could be probed.

**Table 4.** Flight table for RACEPAC. Dates are dd/mm/yyyy.

| Date | Probed clouds/flight objectives | $T$ range/cloud top $T$ | Minutes in mpt clouds |
|------|------|------|------|
| 28.04.2014 | Cumulus | −12.9 to −17.8 °C | 54.1 |
| 30.04.2014 | Low-level clouds in cold sector of a low | −2.3 to −14.4 °C | 70.2 |
| 01.05.2014 | Thin fog layer | −2.0 to −9.6 °C | 5.0 |
| 03.05.2014 | Single/double layer liquid-dominated cloud | 0 to −2.4 °C | 27.2 |
| 06.05.2014 | Single/multilayer clouds | 0 to −6.3 °C | 55.6 |
| 08.05.2014 | Thick stratus | 0 to −3.8 °C | 22.5 |
| 10.05.2014 | Two stratus clouds | −3.0 to −9.1 °C | 49.0 |
| 11.05.2014 | No clouds | — — | 0 |
| 13.05.2014 | No clouds | — — | 0 |
| 14.05.2014 | Homogeneous stratus | −1.9 to −10.1 °C | 25.8 |
| 16.05.2014 | Mid-level clouds | 0 to −10.1 °C | 75.7 |
| 17.05.2014 | Liquid and ice clouds on various altitudes | 0 to −11.3 °C | 22.7 |
| 20.05.2014 | Low-level clouds | −1.5 to −9.5 °C | 54.2 |
| 22.05.2014 | Low-level clouds before front | −6.1 to −15.0 °C | 29.2 |
| 22.05.2014 | Stratus behind front | −1.5 to −11.8 °C | 29.6 |
| 23.05.2014 | Mid-level clouds | −2.3 to −15.1 °C | 14.3 |

**Table 5.** Flight table for ACRIDICON–CHUVA. Dates are dd/mm/yyyy.

| Date | Probed clouds/flight objectives | Cloud T in the mpt regime | Minutes in mpt clouds |
|------|------|------|------|
| 06.09.2014 | Convective cloud and outflow | 0 to −32.2 °C | 13.2 |
| 09.09.2014 | Convective cloud | 0 to −1.2 °C | 1.1 |
| 11.09.2014 | Convective cloud and outflow | 0 to −38.0 °C | 8.6 |
| 12.09.2014 | Cloud tops for satellite comparison | 0 to −29.6 °C | 5.5 |
| 16.09.2014 | Pyrocumulus and outflow | 0 to −38.0 °C | 18.1 |
| 18.09.2014 | Shallow convective cloud and outflow | −36.6 to −38.0 °C | 1.4 |
| 19.09.2014 | Pyrocumulus, convective outflow | −0.4 to −35.1 °C | 8.8 |
| 21.09.2014 | Albedo flight | − | 0 |
| 23.09.2014 | Convective cloud and outflow | 0 to −38.0 °C | 5.5 |
| 25.09.2014 | Convective cloud anvil/outflow | −29.4 to −38.0 °C | 13.5 |
| 27.09.2014 | Warm clouds over forested and deforested areas | − | 0 |
| 28.09.2014 | Convective cloud | 0 to −38.0 °C | 11.1 |
| 30.09.2014 | Albedo flight | − | 0 |
| 01.10.2014 | Convective cloud and outflow | 0 to −5.6 °C | 2.6 |

## 2.2 The NIXE–CAPS instrument

The observations presented here comprise particle number concentrations, size distributions and shape information obtained by NIXE–CAPS. Two instruments are incorporated in NIXE–CAPS (Baumgardner et al., 2001; Meyer, 2012; Luebke et al., 2016): the NIXE–CAS–DPOL (Cloud and Aerosol Spectrometer with detector for polarization) and the NIXE–CIPg (Cloud Imaging Probe greyscale). In combination, particles with diameters between 0.61 and 937 µm can be sized and counted. NIXE–CAPS measurements are thus split into an aerosol data set (particle diameters 0.61 to 3 µm) and cloud particle data set (i.e. hydrometeors with diameters of 3 to 937 µm). For aircraft speeds between 240 and 80 m s$^{-1}$, the instruments' sampling volumes limit the par-

ticle concentration measurements to concentrations above 0.02 to 0.05 cm$^{-3}$ (NIXE–CAS–DPOL) and about 0.0001 to 0.001 cm$^{-3}$ (NIXE-CIPg; the exact values depend on the particle size; see Knollenberg, 1970). The instrument is mounted below the aircraft wing. A detailed description of the operating principles, limitations and uncertainties can be found in Meyer (2012) and Baumgardner et al. (2017). The overall measurement uncertainties concerning particle concentrations and sizes are estimated to be approximately 20 % (Meyer, 2012).

As an improvement over former instrument versions, NIXE–CAPS was modified to minimise ice crystal shattering on the instrument housing, because those ice fragments can artificially enlarge the ice particle concentrations (Field et al.,

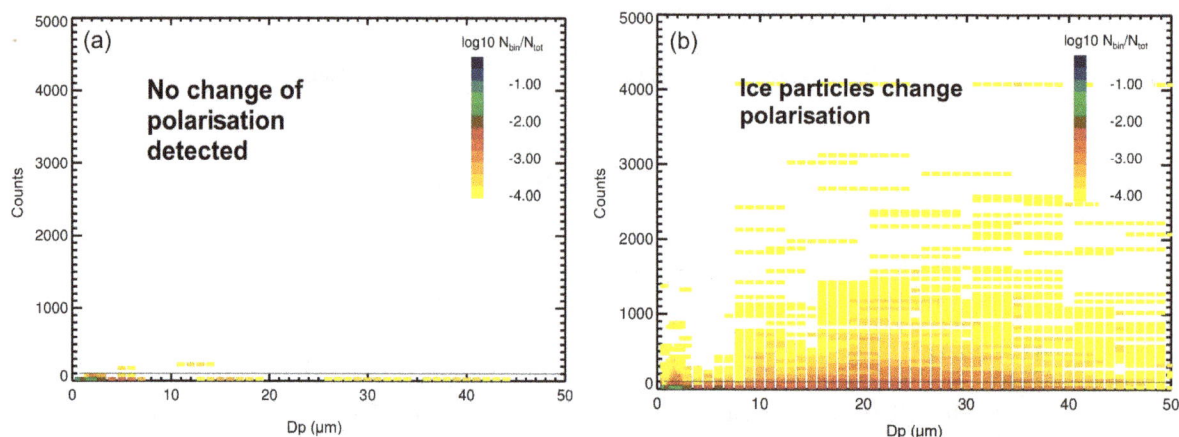

**Figure 4. (a)** Cross-polarised (S-pol) counts vs. particle size in a warm liquid cloud observed in the ACRIDICON–CHUVA campaign. The colour code denotes the relative frequency of particles in this bin ($N_{bin}$) to overall particle count ($N_{tot}$). The horizontal line in the bottom of the panels shows the signal intensity in the S-pol detector which must be exceeded for a particle to be detected as aspherical. The spherical particles cause a weak signal in the S-pol detector. Right panel shows the same, but in a cold cloud ($-60\,°\mathrm{C}$, 213 K) consisting of ice crystals. Ice crystals can cause strong signals in the S-pol detector.

2006; Korolev and Field, 2015). Therefore, the tube inlet of the NIXE–CAS–DPOL has been sharpened to a knife edge, and K-tips have been attached to the NIXE–CIPg's arms (Korolev et al., 2013; Luebke et al., 2016).

In the following, we present an overview of the two instrument components NIXE–CAS–DPOL and NIXE–CIPg as well as the data analysis.

### 2.2.1 NIXE–CAS–DPOL – particle asphericity detection

The NIXE–CAS–DPOL (hereafter referred to as the CAS) covers the small particle size range between 0.61 and 50 µm. As particles pass through the spectrometer's laser beam, the forward-scattered light intensity is used for particle sizing (Baumgardner et al., 2001). As a new feature, the CAS records the change of polarisation in the backward-scattered light, thus giving information about the particle asphericity (Baumgardner et al., 2014). Light scattered by spherical particles in the near-backward direction (168—176°) will retain the same angle of polarisation as the incident light. In contrast, depending on the amount of asphericity, light scattered by non-spherical particles will have some components that are not at the same incident light polarisation. The CAS uses a linearly polarised laser and two detectors that measure the backscattered light. One detector is configured to only detect scattered light with polarisation that is perpendicular (cross-polarised) to the incident light. This signal is referred to as S-pol. In Fig. 4, we show that the intensity of the S-pol signal generates characteristic values for both spherical and aspherical particles. The signature of spherical particles is measured in warm cloud sections ($T > 0\,°\mathrm{C}/273\,\mathrm{K}$), if possible during each measurement campaign. Figure 4 shows an example obtained during the ACRIDICON–CHUVA cam-

paign: measurements of the cross-polarised light as a function of cloud particle size are shown for both a liquid and a glaciated cloud. The liquid spherical particles cause only a very weak S-pol signal. From this measurement, we derive an asphericity threshold (see black line in Fig. 4), providing a method to distinguish between spherical and aspherical particles. This asphericity threshold is verified, if possible, during each of the airborne campaigns by analysing a flight segment in clouds warmer than 0 °C (273 K). The S-pol signal caused by ice particles is shown in Fig. 4a for a cirrus cloud (at $-60\,°\mathrm{C}/213\,\mathrm{K}$). Clearly, the ice crystals cause strong S-Pol signals above the asphericity threshold. It can also be seen that the signal strength depends on the size of the crystals. In particular, the instrument sensitivity with regard to particle asphericity decreases for particles smaller than 20 µm (note that the particles with diameters smaller than 3 µm are aerosol particles). This was found during the experiments described by Järvinen et al. (2016), who compared several asphericity detection methods, including the CAS. Järvinen et al. (2016) also show that ice crystals can be near spherical. The low signal caused in the CAS polarisation detector by this type of crystal can lead to an underestimation of the glaciation degree of a mixed-phase cloud if it is derived from aspherical cloud particle fractions (see also Nichman et al., 2016). In addition, there are variations in the S-pol signals that are caused by the orientation of the crystal with respect to the laser beam (Baumgardner et al., 2014).

Taking into account these uncertainties, we find that it is possible to use the S-pol signal for a classification of mpt clouds. Firstly, we perform the asphericity analysis only for particle sizes between 20 and 50 µm, the range with the strongest S-pol signal. For this size range, we derive aspherical fractions (AFs) as the percentage of aspherical particles per second, which means that particle bulk properties are analysed, not single particle signatures alone. Secondly, we do not interpret each aspherical fraction measurement alone, but divide the AFs into three groups: (i) AF is 0 % (zero), (ii) AF is 0–50 % (low) and (iii) AF is 50–100 % (high).

### 2.2.2 NIXE–CIPg

The NIXE–CIPg (called CIP from here on) is an optical array probe (OAP) that nominally records particles between 7.5 and 960 µm. Shadow image pixels are defined by shadow intensities of 100–65, 65–35 and 35–0 % of the incident light. Particle sizes and concentrations are derived by using the SODA2 programme (Software for OAP Data Analysis, provided by A. Bansemer, National Center for Atmospheric Research NCAR/University Corporation for Atmospheric Research UCAR, 2013). For a detailed description of SODA2, see for example Frey (2011). Pixels with shadow intensities of 35 % and higher were used for the image analysis. In the observations presented here, only the number concentrations for particles with diameters > 22 µm are taken from the CIP data set. The smaller particle fraction is covered by the CAS measurements. The shadow images can be analysed for particle asphericity using various algorithms (Korolev and Sussman, 2000); in this study, however, the occurrence of irregular (i.e. ice) particles was verified manually.

### 2.3 Data analysis

NIXE–CAPS records four individual data sets: histogram and particle by particle (PBP) data for each of the CIP and the CAS instruments. All data sets are evaluated using the NIXElib library (Meyer, 2012; Luebke et al., 2016). In the 1 Hz histogram data sets, particles are sorted into size bins according to predefined forward-scattering cross sections (CAS) or maximum shadow diameters (CIP). These histograms are created for every second.

The PBP data set recorded by the CIP consists of a time stamp and the shadow image of each individual particle. The shadow images can be analysed with regard to maximum diameter, equivalent size, area ratio and shape. The CAS PBP data are limited to 300 particles per second. For these particles, detailed information is stored: the forward-, backward P-pol- and backward S-pol-scattering intensities, a time stamp and the particle interarrival time.

Apart from the asphericity analysis, this data set also allows a diagnosis of ice crystal shattering following Field et al. (2006) and Korolev and Field (2015). Thus, an interarrival time (IAT) correction was applied (Field et al., 2006) additionally to the instrument modifications described above. This correction rejects particles if their IATs are significantly shorter than those of majority of ice crystals, as these short IATs might result from shattering. IAT histograms compiled during the data analysis showed only very few measurements

with short IATs, during which a maximum of about 5 % of the cloud particle population might result from shattering.

## 3 Results and discussion

### 3.1 Mpt cloud classification based on particle number size distributions

Four cloud types are expected in the mpt regime (see Table 1). As mentioned in the introduction, however, only two typical particle number size distributions (PSDs) are found frequently in mpt clouds. Figure 5 shows NIXE–CAPS PSDs measured during VERDI flight 08, where both types alternate: some cloud regions show very high particle concentrations of small particles with a mode diameter < 50 µm (see example of PSD in the lower-right corner). Alternatively, the clouds consist mostly of large ice crystals > 50 µm with either no small particles or concentrations below the NIXE–CAS detection limit (see example of PSD in the lower-left corner).

As a first step of the mpt cloud classification, we sort all clouds according to their particle size distribution type and address these types separately. To this end, we calculate two cloud particle number concentrations: one for particles with diameters between 3 and 50 µm ($N_{small}$) and one for all larger particles ($N_{large}$). For the classification of the first cloud type (Type 1), $N_{small}$ must exceed $1 \, cm^{-3}$, while $N_{large}$ can be zero or larger. The mode of the cloud particle mass distribution is at particle diameters < 50 µm. We assume that this type matches the young clouds after droplet condensational growth in Fig. 1. In the second cloud type (Type 2) we classify those clouds with $N_{small}$ below $1 \, cm^{-3}$ and $N_{large}$ present. The mode of the cloud particle mass distribution is here at particle diameters > 100 µm. This type matches fully glaciated clouds, e.g. as a result of the WBF process (see Fig. 1).

In Fig. 6, a histogram is provided that shows the occurrence of cloud particle concentrations throughout our data set. The spectrum of observed concentrations is continuous, but the two modes associated with the Type 1 and Type 2 clouds (as described above) are clearly visible. The area between the two modes (a total of 6 % of all observations) might result from clouds in a transition state to glaciation. In this study, these measurements were assigned to Type 1 clouds. In addition to the two modes, a small peak at very low cloud particle concentrations (about $10^{-4} \, cm^{-3}$) indicates slightly elevated concentrations around the detection limit of the CIP (a total of 5 % of all observations). We assume that these are measurements in precipitation, especially in snow that occurred frequently in the Arctic campaigns and in sedimenting aggregates of ice crystals from tropical convective clouds (see Sect. 3.3).

In the following, we discuss the cloud types described above in more detail. Type 1 cloud characteristics measured

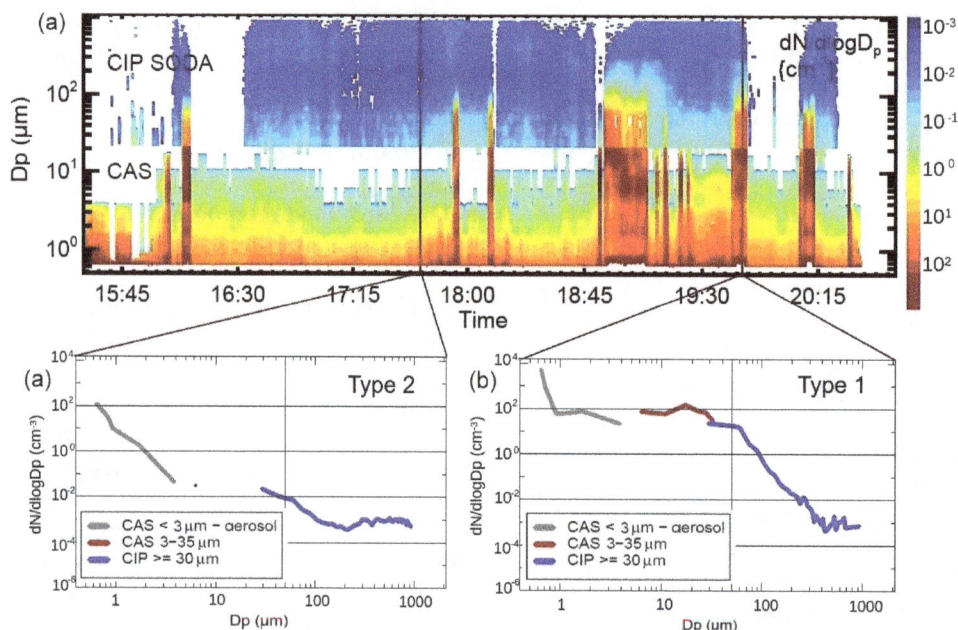

**Figure 5.** Upper panel: size distributions along time during flight 08 of the VERDI campaign (colour code: $dN / dlogD_p$). Two types of cloud can be distinguished; one is dominated by the large particle mode (Type 2, example in lower **a**), the second by small particles (Type 1, example in lower **b**). The two cloud types are also associated with strongly differing particle number concentration ranges; see Fig. 6.

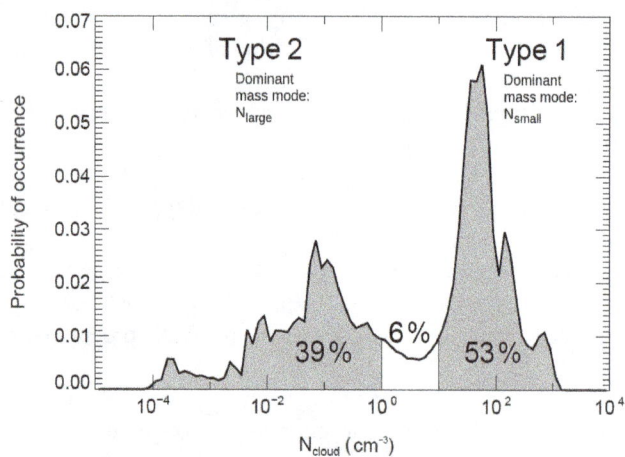

**Figure 6.** Histogram of cloud particle concentrations ($D_p$ 3 to 937 μm) of Type 1 and Type 2 clouds in the mixed-phase temperature regime between 0 and −38 °C (273 to 235 K). For cloud type definitions see Sect. 3.1. The 6 % between the two clear modes were classified as "Type 1" in this study. $N_{small}$ is particles with diameters between 3 and 50 μm. $N_{large}$ is particles with diameters > 50 μm. $N_{cloud}$ is all particles with diameters of 3 μm and larger.

during all campaigns are shown in Fig. 7. These clouds have a clear mode between 3 and 50 μm and are very dense, while cloud particle number concentrations reach average values of dozens to more than 200 cm$^{-3}$. Table 6 shows average cloud particle concentrations for the Type 1 clouds at 5 K intervals. Low number concentrations of large ice particles > 50 μm

are sometimes found, but all clouds of this type are dominated by $N_{small}$, which may consist of liquid droplets, frozen droplets or small ice from ice multiplication processes. With regard to the concentrations of $N_{small}$ in the different temperature intervals (Fig. 7 and Table 6), it can be clearly seen that they decrease with decreasing temperature. When a cloud consists of liquid droplets, they grow by condensation when lifted to higher altitudes – and thus colder temperatures – followed by an increasing coalescence of the droplets, which consequently causes a higher number of $N_{large}$ while depleting the concentration of small droplets. This is also visible in Fig. 7. Note, however, that $N_{large}$ also decreases with increasing temperature, reaches a minimum around 260 K and then rises again, possibly reflecting the increasing occurrence of sedimenting particles. Visual inspection of the CIP images indicates that in the $N_{large}$ cloud mode ice crystals can be found in addition to the drizzle drops. Three of the cloud types of the mpt regime are expected to show Type 1 cloud characteristics: liquid, coexistence and secondary ice clouds (Types 1a, 1b and 1c).

The second set of PSDs (Type 2: Fig. 8) is not strongly dominated by $N_{small}$. Here, $N_{large}$ form a distinct mode. Both mode concentration and maximum values decrease with decreasing temperatures. Clouds of this PSD type have low number concentrations of, on average, less than 0.1 cm$^{-3}$ in the size range 3 to 50 μm (see Table 6). For the sizes > 50 μm, the CIP images show ice crystals or aggregates. This is the typical appearance of a fully glaciated cloud, formed either via the WBF process during which the small

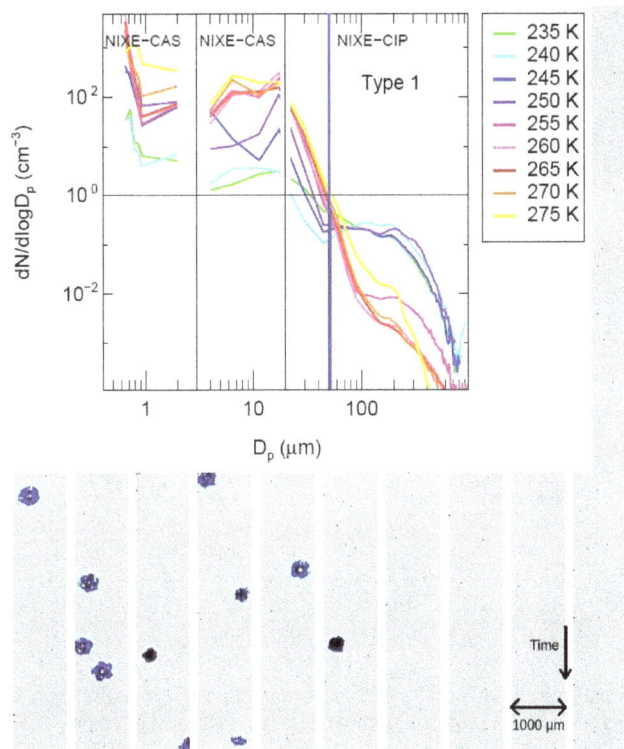

**Figure 7.** Type 1 clouds: example of CIP images (background picture). The stripes represent a series of CIP shadow images, depicting the particles that have passed subsequently through the detector. Foreground: average particle size distributions (PSDs) in 5 K intervals, all campaigns. The thin vertical line at 3 µm marks the boundary between aerosol and cloud particles. The line at 20 µm marks the transition from the NIXE–CAS–DPOL to the NIXE–CIPg instrument. The thick blue line divides the cloud particle population into particles smaller and larger than 50 µm.

**Figure 8.** Same as in Fig. 7 but for Type 2 clouds.

liquid droplets evaporate or at lower altitudes (higher temperatures), due to sedimentation, when aggregates precipitate from higher levels. Again, the two temperature groups are seen as for the Type 1 clouds (Fig. 7), with a clear accumulation of mass at larger particle sizes for temperatures below 247.5 K. An explanation could be that Type 2 clouds most probably develop from Type 1: once the environment becomes subsaturated (RH$_w$ < 100, RH$_i$ > 100 %), all liquid droplets evaporate, leaving only the ice crystals that have already formed, e.g. via immersion freezing or ice seeding. Therefore, $N_{large}$ of Type 2 is only a fraction of those of Type 1, which might reflect the number of active INPs in the respective temperature interval in the case that no ice multiplication takes place (see Sect. 3.2). Thus, the larger differences between the two temperature groups – as seen for Type 1 clouds – more or less balance out. Indeed, an increase in average ice crystal numbers can be seen (Table 6, bottom, $N_{large}$), which might be interpreted as an increasing fraction of activated INPs with decreasing temperature. Note that $N_{small}$ is still larger than $N_{large}$. Since shattering artefacts

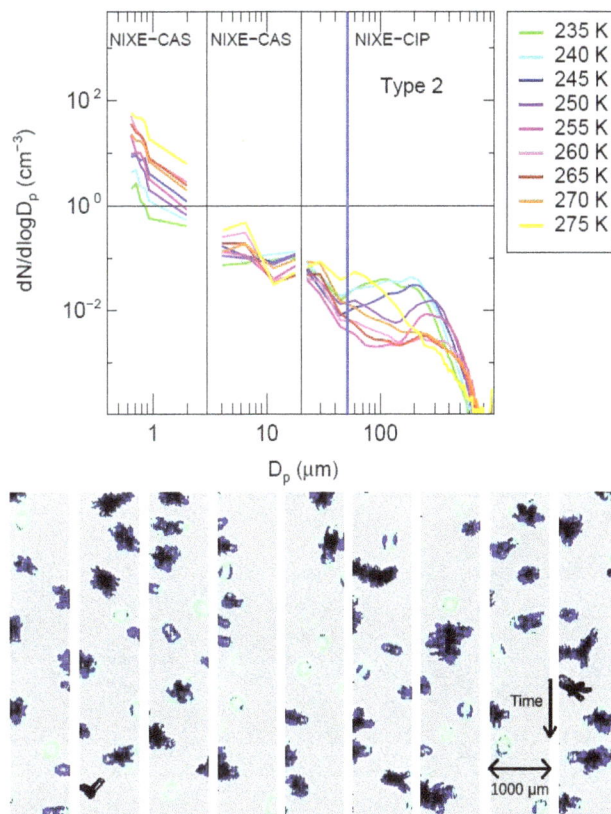

are unlikely (see Sect. 2.3), this means that in Type 2 clouds, a significant number of small particles also persist over the whole temperature range.

In addition to these two types, thin clouds with only low concentrations (less than 1 cm$^{-3}$) of small particles (< 50 µm) and no large particles are sometimes found, which are most likely evaporating clouds. They are not considered a separate cloud type, since they do not appear frequently and cannot be regarded as a distinct type, they are remnants of one of the two cloud types defined above. Further, the respective measurements stem from the CAS instrument alone and are close to its detectable concentration limit, thus suffering from an enhanced uncertainty.

## 3.2 Comparison of cloud particle with ice nucleating particle numbers

A comparison of the measured cloud particle number concentrations to INP concentrations ($N_{INP}$) can indicate whether the ice particles may result from primary ice nucleation. No direct INP measurements are available for our data set, so we estimated $N_{INP}$ using the formula provided by De-Mott et al. (2010), for which aerosol number concentrations of particles between 0.5 and 3 µm are related to INP concentrations. NIXE–CAPS records particles larger than 0.6 µm; the fraction from 0.6 to 3 µm is used as the aerosol frac-

**Table 6.** Average cloud particle concentrations for the two cloud types defined in Sect. 3.1 (see also Fig. 5), for both small ($D_p < 50\,\mu m$) and large ($D_p > 50\,\mu m$) cloud particles.

| Type 1 | $N_{small}$ (cm$^{-3}$) | $N_{large}$ (cm$^{-3}$) |
|---|---|---|
| 235 K | 2.207 | 0.162 |
| 240 K | 2.632 | 0.177 |
| 245 K | 19.894 | 0.134 |
| 250 K | 24.902 | 0.166 |
| 255 K | 109.944 | 0.035 |
| 260 K | 109.798 | 0.022 |
| 265 K | 269.979 | 0.032 |
| 270 K | 166.362 | 0.047 |
| 275 K | 67.788 | 0.098 |

| Type 2 | $N_{small}$ (cm$^{-3}$) | $N_{large}$ (cm$^{-3}$) |
|---|---|---|
| 235 K | 0.057 | 0.023 |
| 240 K | 0.08 | 0.025 |
| 245 K | 0.069 | 0.017 |
| 250 K | 0.062 | 0.01 |
| 255 K | 0.064 | 0.004 |
| 260 K | 0.14 | 0.003 |
| 265 K | 0.07 | 0.003 |
| 270 K | 0.116 | 0.005 |
| 275 K | 0.117 | 0.017 |

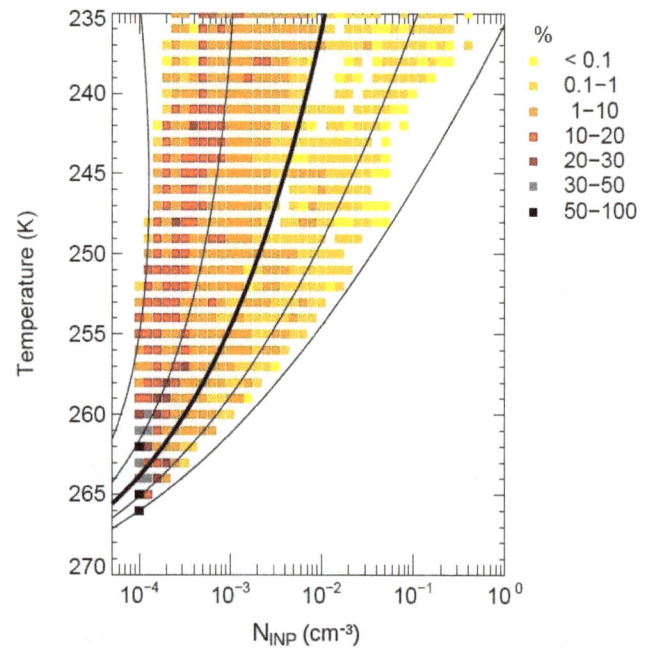

**Figure 9.** Frequencies of ice nucleating particle number concentrations ($N_{INP}$) vs. temperature for all measurement campaigns, colour coded by their frequency of occurrence. $N_{INP}$ is estimated from NIXE–CAPS measurements of aerosol concentrations ($D_p$ 0.6–3 $\mu m$) following DeMott et al. (2010). The black lines indicate INP concentrations for constant aerosol concentrations of 0.01 standard cm$^{-3}$ (leftmost line), 0.1 standard cm$^{-3}$, 1 standard cm$^{-3}$ (thick line), 10 standard cm$^{-3}$ and 100 standard cm$^{-3}$ (rightmost line).

tion. Due to the slightly smaller range of our aerosol measurements, the $N_{INP}$ might be underestimated. However, we believe that this uncertainty is small in comparison to that of the parameterisation by DeMott et al. (2010) itself, since (i) the difference at the lower sizes is only 0.1 micrometre and (ii) aerosol particles larger than 3 micrometres contribute only very little to the concentration of particles larger than 0.5 micrometre (see e.g. Lachlan-Cope et al., 2016). The purpose of using $N_{INP}$ derived in the described way is to show the differences found between the measurement campaigns and temperature ranges.

The results for $N_{INP}$ are shown in Fig. 9 as a function of temperature. Generally, $N_{INP}$ increases with decreasing temperature, as already mentioned in the last section. The most frequent $N_{INP}$ range is between the lowest calculated value of $10^{-4}$ (0.1 L$^{-1}$) and $\sim 10^{-3}$ cm$^{-3}$ (1 L$^{-1}$), while the maximum reaches up to 0.3 cm$^{-3}$ ($\sim 300$ L$^{-1}$). In comparison to a compilation of INP measurements presented recently by Kanji et al. (2017), the estimated range of $N_{INP}$ is shifted to somewhat smaller concentrations.

In Fig. 10, $N_{small}$ and $N_{large}$ for both Type 1 and Type 2 clouds are now presented in the same way as before $N_{INP}$. In Type 1 clouds, especially for $N_{small}$ (upper left panel), we find concentrations between 2 cm$^{-3}$ and more than 200 cm$^{-3}$ down to temperatures of $-20\,°C$ (253 K), well exceeding all INP estimations in this temperature range. However, also for $N_{large}$ (upper right panel), the cloud particle concentrations exceed the expected $N_{INP}$ by several orders of magnitude.

For colder temperatures, where the measured cloud particle number concentrations are lower, the estimated $N_{INP}$ are also mostly lower than the cloud particle concentrations. In general, we can exclude primary ice nucleation as a main contributor for cloud particles in the Type 1 clouds.

The $N_{large}$ of Type 2 clouds (lower-right panel) agree quite well with $N_{INP}$ for a wide range of temperatures. However, in warm areas, the cloud particle concentrations can be higher – they might represent large ice crystals sedimenting from upper layers, as mentioned in Sect. 3.1. For the colder regions, the agreement is consistent with the assumption that the Type 2 clouds we observed were formed by the WBF process (see Sect. 3.1) and that the formation of the initial ice crystals has been likely initiated by INPs immersed in the cloud droplets. $N_{small}$ is slightly increased in comparison with $N_{INP}$. Again, it is possible that this is an effect of the CAS limited detectable concentration range, as discussed in Sect. 2.2. Detailed microphysical cloud simulations might help to further investigate this concentration range.

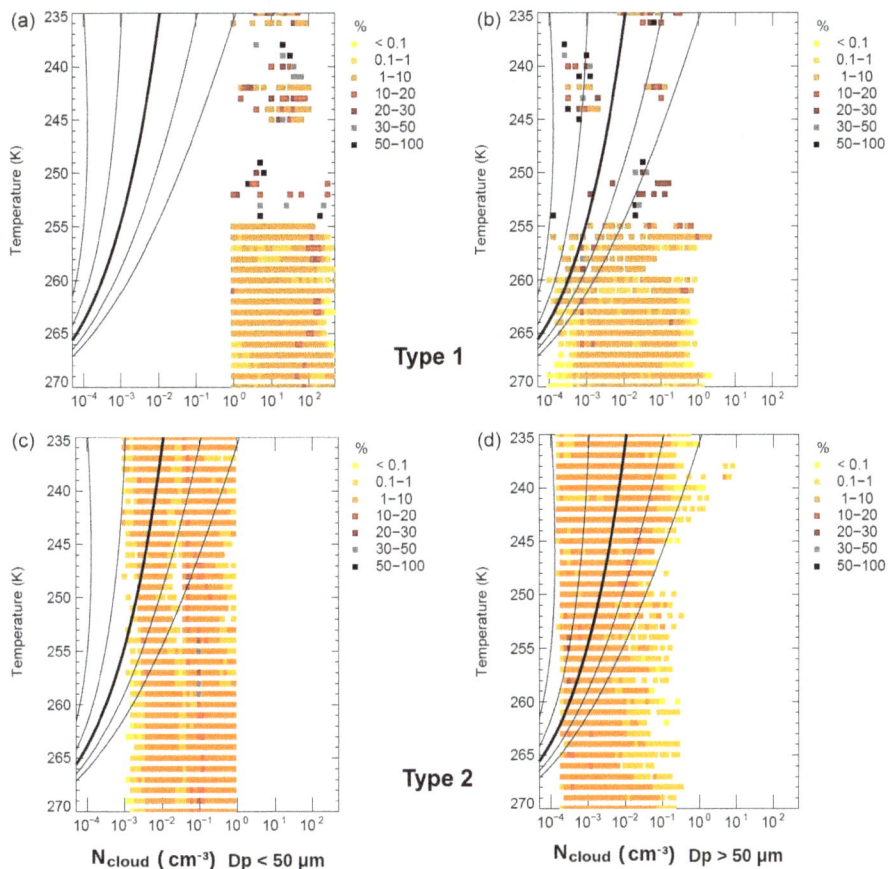

**Figure 10.** Same as Fig. 9, but frequencies of cloud particle number concentrations for $N_{small}$ (**a, c**) and $N_{large}$ (**b, d**). Top row: Type 1 clouds, bottom row: Type 2 clouds.

### 3.3   Mpt cloud classification based on particle asphericity

Size distributions, cloud particle number concentrations and comparisons with expected INP number concentrations provide little information on the cloud particle phase (see Sects. 3.1, 3.2). For further insights on the nature of the observed clouds, information on cloud particle asphericity is used.

As described in Sect. 2.2.1, for $N_{small}$ we define three groups with regard to AFs (1 Hz data of aspherical fractions) to help to classify the mpt clouds: (i) AF is 0 % (zero), (ii) AF is 0 to 50 % (low) and (iii) AF is 50 to 100 % (high). AFs found in group (i)   zero AF are classified as liquid, while AF observations in group (ii)   low AF are regarded as mixed-phase clouds (liquid and ice). Group (iii)   high AF is most likely fully frozen. Particles > 100 µm are mostly irregular (i.e. ice) in group (ii) and (iii). In group (i), large ice particles can occasionally be found. In the size range between 50 and 100 µm, the distinction between drizzle drops and ice particles is not possible, because the shadow images do not contain enough pixels to differentiate between spherical and irregular particles (see Sect. 2.2).

Figure 11 shows the aspherical fractions of Type 1 and 2 cloud particles vs. temperature; the data points are colour coded by the respective field campaigns. The horizontal lines show the 0 (liquid) and −38 °C (ice) temperature thresholds (273 and 235 K, respectively). Looking at the data points in pure ice clouds below −38 °C (235 K) it can be seen that most of the measurements are found in group (iii) high AF range. These AFs can therefore be associated with fully glaciated clouds. Note that Type 2 clouds show AFs comparable to those of cirrus clouds. The small particles found in relatively large number concentrations in this cloud type (see Sect. 3.1 and Table 6) must therefore be small ice crystals.

Due to the ambiguities of the polarisation measurement discussed in Sect. 2.2.1, AF covers a broad range, most often between 70 and 80 %. Note that even in the cirrus clouds the AF never reaches 100 %. A possible reason for this deviation can be columnar ice crystals: these are not well recognised by the CAS sensor (see Järvinen et al., 2016). Alternatively, frozen droplets might have maintained their compact, quasi-spherical shape. All aspherical fractions derived from CAS measurements must therefore be seen as minimum aspherical fractions.

### 3.4 Cloud type detection in the mpt regime

The different cloud types that can be expected in the mpt regime (Table 1) can be identified by the combination of information about $N_{small}$, $N_{large}$ and the respective aspherical fractions (AFs) in each size range. Following this line, we developed algorithms to sort the mpt clouds – second by second – into the four cloud types, using the following criteria:

1. Mostly liquid Type 1a clouds are classified where $N_{small}$ is $> 1\,cm^{-3}$ and AF is zero (liquid).

2. Coexistence Type 1b clouds are classified where $N_{small}$ is $> 1\,cm^{-3}$ and AF is low ($< 50\%$, liquid and ice) and large ice crystals $N_{large}$ are present.

3. Secondary ice Type 1c clouds are classified where $N_{small}$ is $> 1\,cm^{-3}$, AF is high (ice) and large ice crystals $N_{large}$ are present.

4. WBF and large ice Type 2 clouds are classified where $N_{small}$ is $< 1\,cm^{-3}$, AF is high ($> 50\%$, ice) and large ice crystals $N_{large}$ are present.

### 3.5 Mpt cloud classification: results

The mpt clouds observed in this study were probed under a wide range of meteorological conditions (see Sect. 2.1). We can therefore assume that these clouds have formed and evolved in different environments with regard to INP properties and updraughts, which are shown in the previous section to be the major parameters influencing the mpt cloud glaciation process.

For a comprehensive interpretation of the observed clouds, we divided the clouds into Arctic, midlatitude and tropical clouds, analysed the vertical velocities from the aircraft's meteorological data for the different cloud types (Fig. 12), estimated INP numbers (Fig. 13) and finally established distributions of the four mpt cloud categories (see Sect. 3.4) as a function of temperature (note that the temperatures are related to different altitudes depending on the geographical region, Fig. 14). The results are presented in Fig. 15.

### 3.5.1 Arctic clouds

The cloud types found during the field campaigns VERDI and RACEPAC are shown in Fig. 15a. For the probed temperature ranges (253 to 273 K – note that the temperature values in the figure indicate midbins), 50 to 80 % of the mpt clouds belong to the Type 1a/mostly liquid (pink) category. Further, we find a low number of Type 1b/coexistence clouds (brown) and a small percentage of glaciated Type 2/WBF clouds (dark blue). As the estimated INP concentration based on aerosol measurements do not show clear conditions in the Arctic (see Fig. 13), a possible explanation for the large number of Type 1a/mostly liquid clouds could be a lack of biological INP

at the time and location of our Arctic measurements as predicted in a model study by Wilson et al. (2015), so those clouds might not freeze at low temperatures (Shupe et al., 2008; Augustin-Bauditz et al., 2014). This might explain the lack of ice crystals, even though – possibly due to the low altitude of those warm layers (see Fig. 14) – the overall aerosol concentration is comparable to the midlatitudes.

The INP estimations for the Arctic (see Fig. 13a) have to be used with caution, because the "out of cloud" probed altitude range only covers warm temperatures, for which the INP estimation is not very sensitive to the measured aerosol concentrations.

However, an inspection of the vertical velocities measured during the Arctic campaigns in Fig. 12a indicates that 60 % of the Type 1a/mostly liquid (pink) clouds are found in areas with very low updraughts, of $0.1\,m\,s^{-1}$ and lower while mostly fluctuating around zero, and 40 % are found in weak updraughts/downdraughts. Comparably weak updraughts are also frequently found in the Type 2/WBF (dark blue) clouds. This is to be expected, because the WBF regime develops in weak updraughts, implying that the trigger to transform a cloud from liquid to ice is the available INP concentration. The coexistence (brown) and secondary ice clouds were observed with low frequency ($< 1\%$) in the Arctic and show a slightly wider spread in updraught velocities. In particular, higher updraughts occurred more often ($\sim 30\%$) in these clouds, which is consistent with the theoretical considerations shown in Fig. 2 for the coexistence regime. Note that, due to the uncertainties in the vertical velocity measurements, the statistical differences found between the cloud types should be regarded as an incentive for future investigations. While single data points might thus contain measurement errors, the distribution of observed vertical velocities is smooth and centred near zero, which is expected for the meteorological situations discussed in Sect. 2.1. Due to this, and because our data set consists of a large number of observations, we would like to point out the systematic differences found between cloud types and campaigns

### 3.5.2 Midlatitude clouds

At midlatitudes (COALESC field campaign), the largest cloud fractions are the fully glaciated Type 2/WBF clouds (dark blue in Fig. 15, middle panel). This is consistent with the assumption that at midlatitudes, the WBF process is the dominant process for cloud evolution (Boucher et al., 2013). More INP seem to be available that are ice active below $-10\,°C$ (263 K). At temperatures warmer than $-20\,°C$ (253 K), the fraction of this cloud type is slowly reduced, while more and more Type 1a/mostly liquid clouds (pink in Fig. 15) and Type 1b/coexistence clouds (brown in Fig. 15) are found for higher temperatures. The WBF process depends on the presence of INPs (or seed ice from higher cloud layers), which are likely available in higher quantities at midlatitudes than in the Arctic (compare Sect. 3.5.1

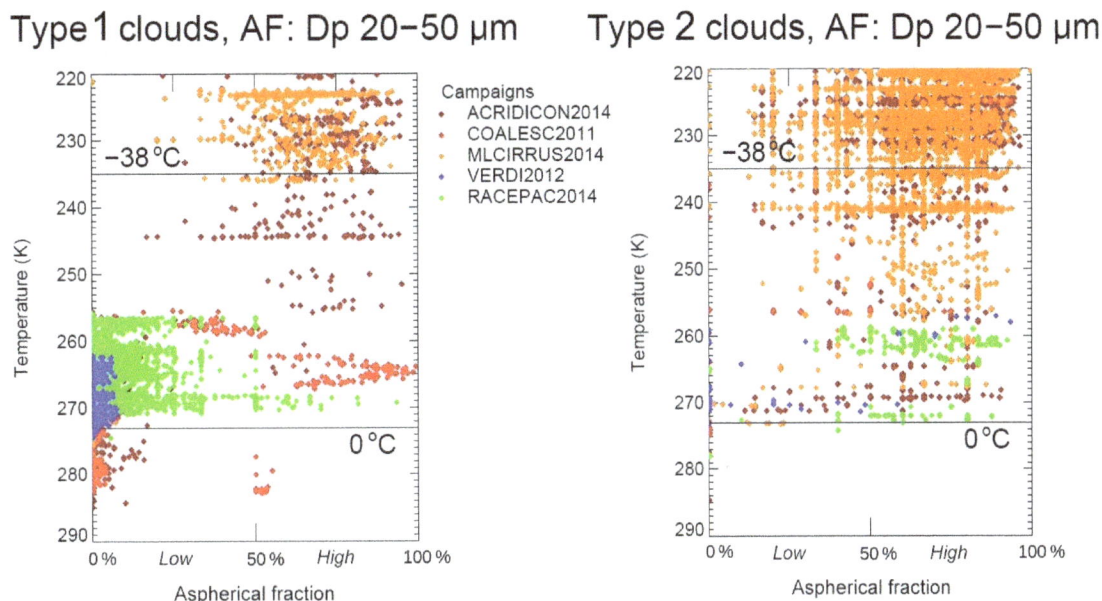

**Figure 11.** Aspherical fractions (AFs) for $D_p = 20$ to $50\,\mu\mathrm{m}$. Type 1 clouds show a variety of AF. Type 2 shows AFs comparable to cirrus clouds – which is illustrated by observations from the ML-Cirrus campaign – throughout the temperature range.

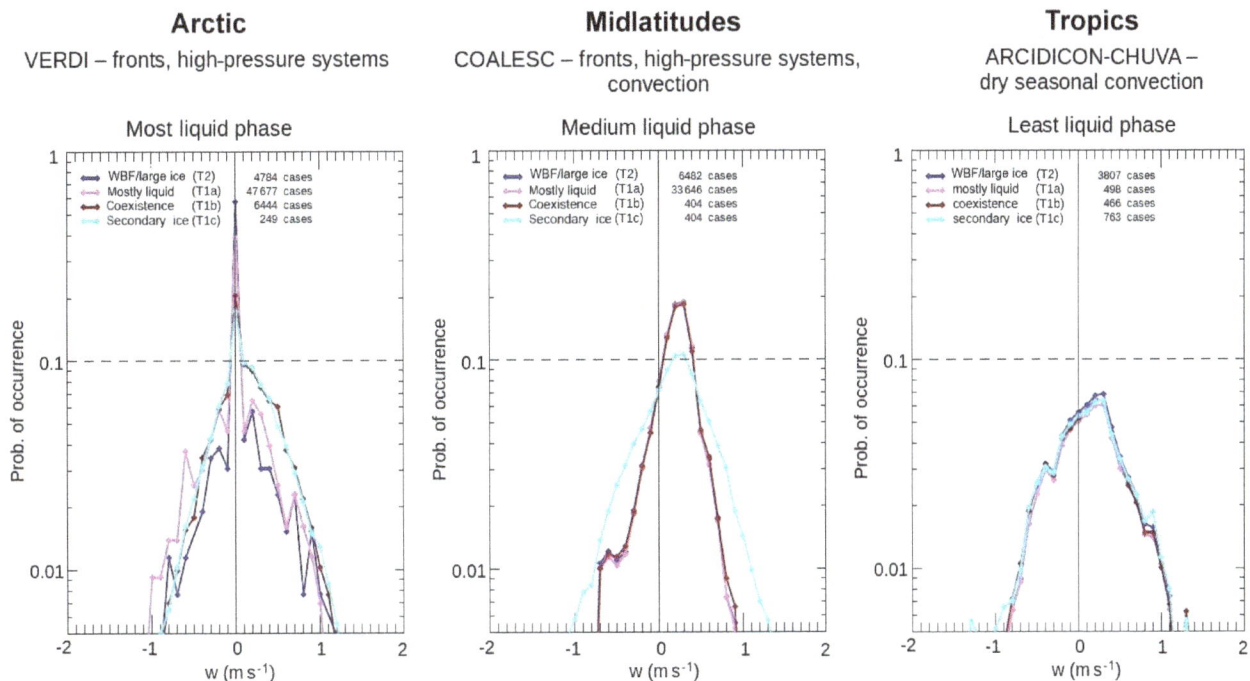

**Figure 12.** Frequency of occurrence for vertical velocities ($w$) within mpt clouds during the campaigns VERDI (Arctic), COALESC (midlatitudes) and ACRIDICON–CHUVA (tropics) for the different cloud types ($T$).

and Fig. 13). The varying occurrence of different cloud types with temperature – i.e. Type 1a clouds at higher temperatures (lower altitudes) and an increasing part of Type 2/WBF clouds with decreasing temperature (increasing altitude) – might correspond to different INP regimes. At temperatures below about $-20\,°\mathrm{C}$ (253 K), for example, efficient mineral dust INP might initiate the freezing process, while at warmer temperatures less frequently occurring biological particles most likely act as INP (Augustin-Bauditz et al., 2014; Kanji et al., 2017). In addition, the increasing fraction of Type 2/WBF clouds with decreasing temperature reflects the fact that the colder the environment, the higher the probability

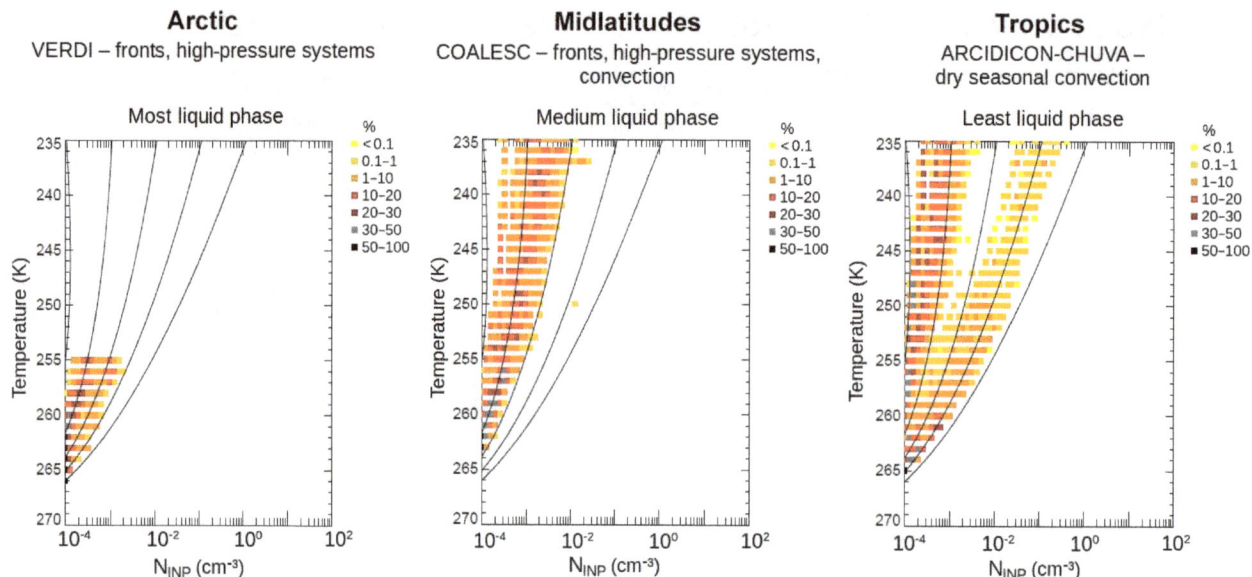

**Figure 13.** Frequencies of occurrence of INP concentrations ($N_{INP}$) vs. temperature during VERDI (Arctic), COALESC (midlatitudes) and ACRIDICON–CHUVA (tropics). INP number concentrations are estimated via aerosol concentrations for particles $> 0.6 \, \mu$m following DeMott et al. (2010). The black lines indicate INP concentrations for constant aerosol concentrations of 0.01 (leftmost line), 0.1, 1, 10 and 100 cm$^{-3}$ (rightmost line).

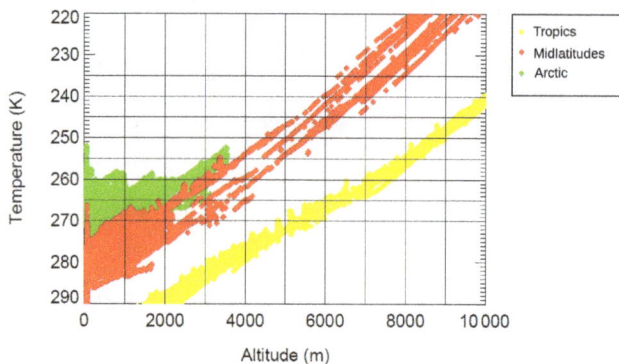

**Figure 14.** Temperature vs. altitude for the field campaigns VERDI and RACEPAC (Arctic), COALESC (midlatitudes) and ACRIDICON–CHUVA (tropics). The profile differs due to the varying latitudes.

that the RH$_w$ falls below 100 %: with decreasing temperature, more and more droplets freeze and exploit the gas phase water when they grow. As a consequence, less gas-phase water is available the colder the temperature is. In the transition range between predominantly Type 1a/mostly liquid and only Type 2/WBF clouds (temperatures between −20 and −10 °C or 253 and 263 K), Type 1b/coexistence clouds appear, which we interpret as clouds in which the freezing process has started, but in which the RH$_w$ is still above 100 % (blue curve slightly below RH$_w$ = 100 % in Fig. 2).

Type 1c/secondary ice clouds appear in midlatitude clouds more often than in the Arctic, which might reflect the in-

creased availability of initial ice. It is unlikely that these small particles are shattering artefacts, because they often occur in clouds with no or few large ice particles – these large particles, however, are those that usually shatter (Korolev et al., 2011). In addition, as discussed in Sect. 2.3, based on IAT analysis, shattering could be almost excluded in the measurements. In contrast, the majority of those clouds occur at temperatures between −5 and −13 °C (268 to 258 K), which is an indication of an efficient Hallett–Mossop process having altered the cloud at slightly warmer temperatures. Note that the classification is aimed at the result of cloud transforming processes, not the cloud transformation itself. Which process precisely took place before the cloud section was probed cannot be proven with this 1 Hz data set.

At midlatitudes, Type 1a/mostly liquid, Type 1b/coexistence and Type 2/WBF clouds show the same vertical velocity distributions (Fig. 12). The peak updraughts are slightly higher and the widths slightly narrower in comparison to the Arctic clouds. This is another hint that underscores the above-discussed dependence of the cloud categories on RH$_w$: within the same vertical velocity range, the relative humidity can vary strongly depending on the available amount of water and the cloud development stage (cloud particle nucleation, sedimentation, evaporation). The Type 1c/secondary ice clouds show a different updraught distribution with faster vertical velocities, which might indicate that these clouds occurred in more turbulent environments, which is consistent with the idea that the cloud particles need to collide during the rime-splintering process.

### 3.5.3 Tropical clouds

During the tropical field campaign ACRIDICON–CHUVA in convective towers, stronger updraughts and downdraughts were observed more frequently than during the other campaigns (Fig. 12, right panel). The records include extreme vertical velocities up to $-10$ and $+15\,\mathrm{m\,s^{-1}}$ (not shown here). However, these events were rarely observed, because due to flight safety, these cloud sections were mostly avoided. Velocities of 0.5 to $1.0\,\mathrm{m\,s^{-1}}$ were observed in more than 10 % of all data points. The wider distribution of vertical velocities shows that the cloud dynamics are much stronger in the tropical clouds than at midlatitudes and in the Arctic.

In comparison to the other regions, fewer Type 1a/mostly liquid clouds are found in the tropics, also for warmer temperatures. This might be a consequence of sedimenting ice, or it might indicate a higher concentration of INPs that are already ice active at comparably high temperatures, pointing to biological INPs. This seems to be plausible for tropical regions, but is only partially confirmed by the INP estimate (see Fig. 13, right panel). The probed clouds occurred in both very clean air with fewer INPs than in the midlatitudes and in heavily polluted areas over fire clearance regions. A more detailed study on how the aerosol concentration affects the cloud-type distribution during ACRIDICON–CHUVA was done by Cecchini et al. (2017), based also on NIXE–CAPS aspherical fractions. The study shows that clouds in polluted environments contained more and smaller liquid water droplets and less ice, while clouds in clean conditions held more ice crystals and few liquid water droplets.

As a consequence of the higher vertical velocities in the convective towers, more Type 1b/coexistence clouds are observed than at midlatitudes or in the Arctic. A small number of the liquid droplets < 50 µm survived down to the homogeneous drop freezing temperature ($\sim -38\,°C$, 235 K) in cases where the vertical velocity was high enough (see also Fig. 2, red).

However, the Type 2/WBF and large ice clouds (Fig. 15, right panel) are the most frequent at all temperatures. Those large cloud particles might stem from sedimentation out of the cloud anvils, which usually consist of mostly large aggregates, or might be transported downwards in the strong downdraughts within the convective clouds (compare Jäkel et al., 2017).

It is, nevertheless, important to note again that due to security restrictions, the in situ measurements were mostly restricted to cloud regions with small updraught velocities (see Fig. 12), i.e. to young developing clouds or edges of convective towers. Due to this flight pattern, we most probably have probed conditions that favour the WBF process (consistent with Fig. 2, dark blue), even if those conditions might not be representative for tropical convective clouds in general. This part of the analysis should therefore be seen as an incentive for further studies and not be used as a basis for cloud type statistics in tropical dry seasonal convection.

In the tropical data set, the cloud type 1c/secondary ice is scarce at the lower levels – as at midlatitudes – but prevalent at cold temperatures, i.e. at high levels. The high concentrations of small aspherical particles might indicate a population of frozen droplets that quickly develop complex shapes in supersaturation. Alternatively, other ice multiplication processes (e.g. ice splintering or plasma-induced particle shattering due to lightning; see Leisner et al., 2014) might take place. Again, as discussed in Sect. 3.5.2, shattering artefacts can be almost excluded as the reason for the high number of small aspherical particles: large ice crystals appear at all temperatures up to 0 °C (273 K); the secondary ice cloud type 1c is, however, only observed at temperatures between $-38$ and $-20\,°C$ (235 and 253 K). Additionally, an analysis of interarrival times of the secondary ice cloud sections did not show shorter interarrival times than in other parts of the data set.

## 4 Summary and conclusions

The study presented here gives an overview of typical cloud properties observed between 0 and $-38\,°C$ (273 to 235 K, mixed-phase temperature regime) and links the clouds at differing stages of glaciation to ice formation and evolution mechanisms. It gives hints to the relevance of cloud processes at different geographical locations and altitudes.

To this end, the cloud spectrometer NIXE–CAPS was deployed in four airborne field campaigns to conduct measurements of cloud particle sizes, number concentrations and, as an additional parameter, the cloud particles' asphericity. Based on the observations, which consist of 38.6 h within clouds, we developed algorithms based on the measurements of particle size distributions and aspherical fractions to identify four cloud types:

– Type 1a/mostly liquid refers to dense clouds consisting of mostly small droplets. All particles in the size range from 20 to 50 µm are spherical. The few large cloud particles > 50 µm might occasionally include ice crystals.

– Type 1b/coexistence is dense clouds consisting of mostly small particles with a low percentage (< 50 %) of small aspherical ice particles. Ice crystals > 50 µm are present. The coexistence of liquid droplets and ice crystals is most probably due to supersaturation over both water and ice caused by higher vertical velocities.

– Type 1c/secondary ice refers to dense clouds consisting of mostly small particles between 3 and 50 µm with a high percentage (> 50 %) of aspherical ice particles. The aspherical fractions found are comparable to those of cirrus clouds; we thus conclude that these clouds are completely glaciated. The large cloud particles > 50 µm are also frozen. The ice crystal numbers exceed the expected ice nuclei concentrations by several orders of magnitude, which suggests that the small crystals result

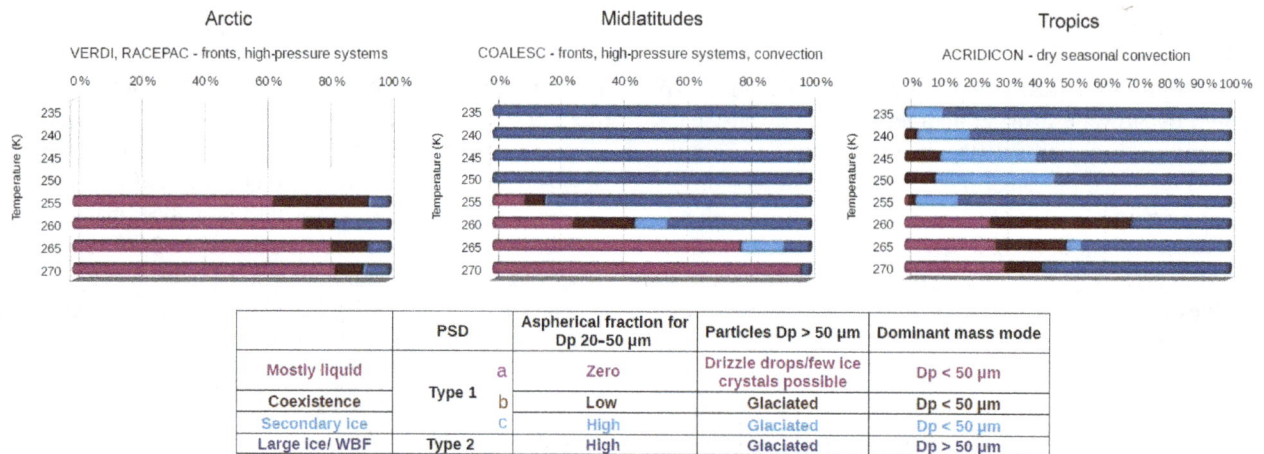

**Figure 15.** Occurrence of the cloud types defined in Sect. 3.3: Type 1a/mostly liquid clouds are dominated by small, exclusively spherical particles (for details on the particle size distributions (PSDs), see Sect. 3.1). They have high overall number concentrations. Type 1b/coexistence clouds are dense too, but do contain some small aspherical particles, indicating that a glaciation process has begun. The Type 1c/secondary ice cloud type is again very dense – the particle numbers exceed the INP concentration estimations by far (see Sect. 3.2). Here, most of the small particles in the size range between 20 and 50 μm are aspherical; the cloud must therefore consist of ice. In contrast, clouds in the category Type 2/WBF and large ice show low overall number concentrations. These clouds are dominated by large ice particles which may resume from the Wegener–Bergeron–Findeisen process or, especially in the tropics, be large, sedimenting ice aggregates from cumulonimbus anvils.

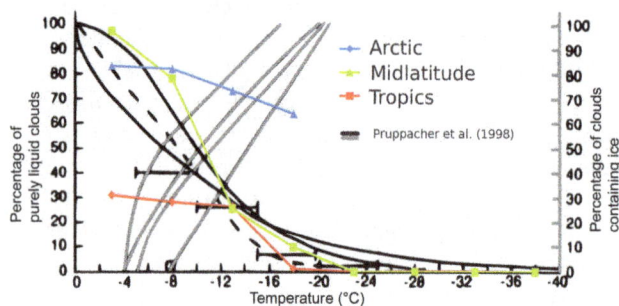

**Figure 16.** Percentage of clouds containing no ice (left $y$ axis) or ice (right axis); modified after Pruppacher et al. (1998). Black/grey lines: measurements reported in Pruppacher et al. (1998). The coloured lines refer to the left axis and represent measurements from this study. Blue: Arctic (VERDI/RACEPAC), green: midlatitude (COALESC), red: tropics (ACRIDICON–CHUVA).

from secondary ice production. Small ice crystal production by shattering can be almost excluded from IAT analysis of the specific situations.

– Type 2/WBF and large ice refers thin clouds with low number concentrations and a mass distribution dominated by large cloud particles > 50 μm. The aspherical fractions of the small particles are high and the large particles are frozen: these clouds are fully glaciated. The reduced number of small particles in comparison to the mostly liquid clouds can be explained by the WBF process. However, from the asphericity detection it is obvious that small ice crystals are present in WBF clouds

with higher concentrations than large ice crystals. Alternatively, these clouds might consist of sedimenting aggregates. The cloud particle number concentrations agree reasonably well with the estimated ice nuclei concentrations.

We quantified the occurrence of these cloud types for Arctic, midlatitude and tropical regions.

For the Arctic, we observed mpt clouds for temperatures higher than −20 °C (253 K). The most common were Type 1a/mostly liquid clouds, with a small percentage of Type 1b/coexistence and Type 2/WBF and large ice clouds. We hypothesise that this cloud type distribution is a result of low concentrations of ice active INPs, particularly biological INPs, during our field campaign in the Arctic. This hypothesis is in agreement with the low INP concentrations found for this region in a modelling study by Wilson et al. (2015), which is based on field measurements.

At midlatitudes, mpt clouds down to −40 °C (233 K) were probed, mostly in frontal systems with moderate updraughts between 0 and 0.5 m s$^{-1}$. Here, the glaciated Type 2/WBF and large ice clouds dominate most of the temperature range, pointing to a sufficient availability of INPs. Only at temperatures warmer than −20 °C (253 K) was an increasing fraction of Type 1b/coexistence clouds and Type 1c/secondary ice clouds found. The temperature range for the secondary ice clouds is consistent with the preconditions for the Hallett–Mossop process.

In the tropics, mostly moderate but also very strong vertical velocities were recorded. Correspondingly, the glaciated Type 2/WBF and large ice clouds dominate the measurements over all temperature ranges, but Type 1b/coexistence clouds are also observed down to $-40\,^{\circ}$C (233 K). The supercooled liquid droplets freeze homogeneously when transported to higher altitudes. Type 1c/secondary ice clouds are observed at colder temperatures (higher altitudes) than at midlatitudes, indicating that ice-splintering processes other than the Hallett–Mossop process might be active here.

Pruppacher et al. (1998) summarise several studies (see Fig. 16) that tracked the percentage of clouds containing no ice crystals (left $y$ axis) or the percentage of clouds containing ice crystals (right $y$ axis) as a function of temperature. Their findings agree well with our observations at midlatitudes (green line in Fig. 16). It is noteworthy, however, that our observations in the Arctic (blue line) show higher liquid fractions, while in the tropical observations (red line) more ice clouds are found.

In general, the analysis of small cloud particle aspherical fractions advises against the assumption that all cloud particles smaller than 50 µm are liquid. In contrast to previous assumptions, small particles were frequently found to be aspherical. The aspherical particle fractions are an important parameter for the identification of the four cloud types investigated here. Observations that contain this information (e.g. Mioche et al., 2017) can be used to extend the cloud statistics presented here. In case no small particle shapes are available, particle size distributions can be used to differ between the Type 1 cloud group (mostly liquid/coexistence/secondary ice clouds) and the Type 2 clouds (WBF and large ice clouds). A sufficiently large database would, for example, allow the quantification of the efficiency of the WBF process with regard to temperature and location. Along these lines, this study might serve as a starting point for a growing cloud type database in the mpt regime.

*Competing interests.* The authors declare that they have no conflict of interest.

*Special issue statement.* This article is part of the special issue "VERDI – Vertical Distribution of Ice in Arctic Clouds (ACP/AMT inter-journal SI)", resulting from the conference EGU2016, Vienna, Austria, 17 to 22 April 2016, as well as of the special issue "The ACRIDICON-CHUVA campaign to study deep convective clouds and precipitation over Amazonia using the new German HALO research aircraft (ACP/AMT inter-journal SI)".

*Acknowledgements.* This work was supported by the Max Planck society, the DFG (Deutsche Forschungsgemeinschaft, German Research Foundation) Priority Program SPP 1294, the German Aerospace Center (DLR), and the FAPESP (São Paolo Research Foundation). Heike Wex is currently funded by DFG within the Ice Nuclei research UnIT (INUIT, FOR 1525), WE 4722/1-2. We thank Martin Schnaiter and Emma Järvinen for the fruitful discussions during the RICE03 campaign.

We also thank two anonymous reviewers for their thorough comments which helped to improve the quality of this paper.

*Edited by: Corinna Hoose*

## References

Augustin-Bauditz, S., Wex, H., Kanter, S., Ebert, M., Niedermeier, D., Stolz, F., Prager, A., and Stratmann, F.: The immersion mode ice nucleation behavior of mineral dusts: A comparison of different pure and surface modified dusts, Geophys. Res. Lett., 41, 7375–7382, 2014.

Baumgardner, D., Jonsson, H., Dawson, W., O'Connor, D., and Newton, R.: The cloud, aerosol and precipitation spectrometer: a new instrument for cloud investigations, Atmos. Res., 59/60, 251–264, https://doi.org/10.1016/S0169-8095(01)00119-3, 2001.

Baumgardner, D., Newton, R., Krämer, M., Meyer, J., Beyer, A., Wendisch, M., and Vochezer, P.: The Cloud Particle Spectrometer with Polarization Detection (CPSPD): A next generation open-path cloud probe for distinguishing liquid cloud droplets from ice crystals, Atmos. Res., 142, 2–14, https://doi.org/10.1016/j.atmosres.2013.12.010, 2014.

Baumgardner, D., Abel, S. J., Axisa, D., Cotton, R., Crosier, J., Field, P., Gurganus, C., Heymsfield, A., Korolev, A., Krämer, M., Lawson, P., McFarquhar, G., Ulanowski, Z., and Um, J.: Cloud Ice Properties: In Situ Measurement Challenges; Chapter 9 of "Ice Formation and Evolution in Clouds and Precipitation: Measurement and Modeling Challenges", Meteorol. Monogr., 58, 9.1–9.23, https://doi.org/10.1175/AMSMONOGRAPHS-D-16-0011.1, 2017.

Boucher, O., Randall, D., Artaxo, P., Bretherton, C., Feingold, G., Forster, P., Kerminen, V.-M., Kondo, Y., Liao, H., Lohmann, U., Rasch, P., Satheesh, S. K., Sherwood, S., Stevens, B., and Zhang, X. Y.: Clouds and Aerosols, Climate Change 2013: The Physical Science Basis, Contribution of Working Group I to the Fifth Assessment Report of the Intergovernmental Panel on Climate Change, edited by: Stocker, T. F., Qin, D., Plattner, G.-K., Tignor, M., Allen, S. K., Boschung, J., Nauels, A., Xia, Y., Bex, V., and Midgley, P. M., cambridge University Press, Cambridge, UK and New York, NY, USA, https://www.ipcc.ch/report/ar5/ (last access: 5 October 2017), 2013.

Bühl, J., Ansmann, A., Seifert, P., Baars, H., and Engelmann, R.: Toward a quantitative characterization of heterogeneous ice formation with lidar/radar: Comparison of CALIPSO/CloudSat with ground-based observations, Geophys. Res. Lett., 40, 4404–4408, 2013.

Cecchini, M. A., Machado, L. A. T., Wendisch, M., Costa, A., Krämer, M., Andreae, M. O., Afchine, A., Albrecht, R. I., Artaxo, P., Borrmann, S., Fütterer, D., Klimach, T., Mahnke, C., Martin, S. T., Minikin, A., Molleker, S., Pardo, L. H., Pöhlker, C., Pöhlker, M. L., Pöschl, U., Rosenfeld, D., and Weinzierl, B.: Illustration of microphysical processes of Amazonian deep convective clouds in the Gamma phase space: Introduction and potential applications, Atmos. Chem. Phys. Discuss., https://doi.org/10.5194/acp-2017-185, in review, 2017.

Curry, J. A., Schramm, J. L., Rossow, W. B., and Randall, D.: Overview of Arctic Cloud and Radiation Characteristics, J. Clim., 9, 1731–1764, https://doi.org/10.1175/1520-0442(1996)009<1731:OOACAR>2.0.CO;2, 1996.

DeMott, P. J., Prenni, A. J., Liu, X., Kreidenweis, S. M., Petters,

M. D., Twohy, C. H., Richardson, M. S., Eidhammer, T., and Rogers, D. C.: Predicting global atmospheric ice nuclei distributions and their impacts on climate, Proc. Natl. Acad. Sci. USA, 107, 11217–11222, https://doi.org/10.1073/pnas.0910818107, 2010.

Ehrlich, A., Wendisch, M., Bierwirth, E., Gayet, J.-F., Mioche, G., Lampert, A., and Mayer, B.: Evidence of ice crystals at cloud top of Arctic boundary-layer mixed-phase clouds derived from airborne remote sensing, Atmos. Chem. Phys., 9, 9401–9416, https://doi.org/10.5194/acp-9-9401-2009, 2009.

Field, P., Lawson, R. P., Brown, P. R. A., Lloyd, G., Westbrook, C., Moisseev, D., Miltenberger, A., Nenes, A., Blyth, A., Choularton, T., Connolly, P., Buehl, J., Crosier, J., Cui, Z., Dearden, C., DeMott, P., Flossmann, A., Heymsfield, A., Huang, Y., Kalesse, H., Kanji, Z. A., Korolev, A., Kirchgaessner, A., Lasher-Trapp, S., Leisner, T., McFarquhar, G., Phillips, V., Stith, J., and Sullivan, S.: Ice Formationand Evolution in Clouds and Precipitation: Measurement and Modeling Challenges, Chapter 7: Secondary Ice Production – current state of the science and recommendations for the future, Meteorol. Monogr., 58, 58, 7.1–7.20, 2017.

Field, P. R., Heymsfield, A. J., and Bansemer, A.: Shattering and Particle Interarrival Times Measured by Optical Array Probes in Ice Clouds, J. Atmos. Ocean. Technol., 23, 1357–1371, https://doi.org/10.1175/JTECH1922.1, 2006.

Field, P. R., Lawson, R. P., Brown, P. R. A., Lloyd, G., Westbrook, C., Moisseev, D., Miltenberger, A., Nenes, A., Blyth, A., Choularton, T., Connolly, P., Buehl, J., Crosier, J., Cui, Z., Dearden, C., DeMott, P., Flossmann, A., Heymsfield, A., Huang, Y., Kalesse, H., Kanji, Z. A., Korolev, A., Kirchgaessner, A., Lasher-Trapp, S., Leisner, T., McFarquhar, G., Phillips, V., Stith, J., and Sullivan, S.: Chapter 7, Secondary Ice Production – current state of the science and recommendations for the future, Meteor. Mon., 0, https://doi.org/10.1175/AMSMONOGRAPHS-D-16-0014.1, 2015.

Findeisen, W., Volken, E., Giesche, A. M., and Brönnimann, S.: Colloidal meteorological processes in the formation of precipitation, Meteorol. Z., 24, 443–454, https://doi.org/10.1127/metz/2015/0675, 2015.

Frey, W.: Airborne in situ measurements of ice particles in the tropical tropopause layer, Ph.D. thesis, Johannes Gutenberg-Universität, 2011.

Hallett, J. and Mossop, C.: Production of secondary ice particles during the riming process, Nature, 249, 26–28, 1974.

Jäkel, E., Wendisch, M., Krisna, T. C., Ewald, F., Kölling, T., Jurkat, T., Voigt, C., Cecchini, M. A., Machado, L. A. T., Afchine, A., Costa, A., Krämer, M., Andreae, M. O., Pöschl, U., Rosenfeld, D., and Yuan, T.: Vertical distribution of the particle phase in tropical deep convective clouds as derived from cloud-side reflected solar radiation measurements, Atmos. Chem. Phys., 17, 9049–9066, https://doi.org/10.5194/acp-17-9049-2017, 2017.

Järvinen, E., Schnaiter, M., Mioche, G., Jourdan, O., Shcherbakov, V. N., Costa, A., Afchine, A., Krämer, M., Heidelberg, F., Jurkat, T., Voigt, C., Schlager, H., Nichman, L., Gallagher, M., Hirst, E., Schmitt, C., Bansemer, A., Heymsfield, A., Lawson, P., Tricoli, U., Pfeilsticker, K., Vochezer, P., Möhler, O., and Leisner, T.: Quasi-spherical Ice in Convective Clouds, J. Atmos. Sci., 73, 3885–3910, https://doi.org/10.1175/JAS-D-15-0365.1, 2016.

Kanji, Z., Ladino, L. A., Wex, H., Boose, Y., Burkert-Kohn, M., Cziczo, D. J., and Krämer, M.: Ice Formation and Evolution

in Clouds and Precipitation: Measurement and Modeling Challenges, Chapter 1: Overview of Ice Nucleating Particles, Meteorol. Mon., 58, 1.1–1.33, 2017.

Klingebiel, M., de Lozar, A., Molleker, S., Weigel, R., Roth, A., Schmidt, L., Meyer, J., Ehrlich, A., Neuber, R., Wendisch, M., and Borrmann, S.: Arctic low-level boundary layer clouds: in situ measurements and simulations of mono- and bimodal supercooled droplet size distributions at the top layer of liquid phase clouds, Atmos. Chem. Phys., 15, 617–631, https://doi.org/10.5194/acp-15-617-2015, 2015.

Knollenberg, R. G.: The Optical Array: An Alternative to Scattering or Extinction for Airborne Particle Size Determination, J. Appl. Meteorol., 9, 86–103, https://doi.org/10.1175/1520-0450(1970)009<0086:TOAAAT>2.0.CO;2, 1970.

Koop, T., Luo, B., Tsias, A., and Peter, T.: Water activity as the determinant for homogeneous ice nucleation in aqueous solutions, Nature, 406, 611–614, 2000.

Korolev, A.: Limitations of the Wegener-Bergeron-Findeisen Mechanism in the Evolution of Mixed-Phase Clouds, J. Atmos. Sci., 64, 3372–3375, https://doi.org/10.1175/JAS4035.1, 2007.

Korolev, A. and Field, P. R.: Assessment of the performance of the inter-arrival time algorithm to identify ice shattering artifacts in cloud particle probe measurements, Atmos. Meas. Tech., 8, 761–777, https://doi.org/10.5194/amt-8-761-2015, 2015.

Korolev, A. and Sussman, B.: A Technique for Habit Classification of Cloud Particles, J. Atmos. Ocean. Tech., 17, 1048–1057, https://doi.org/10.1175/1520-0426(2000)017<1048:ATFHCO>2.0.CO;2, 2000.

Korolev, A., Emery, E., Strapp, J., Cober, S., Isaac, G., Wasey, M., and Marcotte, D.: Small ice particles in tropospheric clouds: Fact or artifact? Airborne Icing Instrumentation Evaluation Experiment, B. Am. Meteorol. Soc., 92, 967–973, 2011.

Korolev, A., Emery, E., and Creelman, K.: Modification and Tests of Particle Probe Tips to Mitigate Effects of Ice Shattering, J. Atmos. Oceanic Technol., 30, 690–708, https://doi.org/10.1175/JTECH-D-12-00142.1, 2013.

Lachlan-Cope, T., Listowski, C., and O'Shea, S.: The microphysics of clouds over the Antarctic Peninsula – Part 1: Observations, Atmos. Chem. Phys., 16, 15605–15617, https://doi.org/10.5194/acp-16-15605-2016, 2016.

Lawson, R. P., Woods, S., and Morrison, H.: The Microphysics of Ice and Precipitation Development in Tropical Cumulus Clouds, J. Atmos. Sci., 72, 2429–2445, https://doi.org/10.1175/JAS-D-14-0274.1, 2015.

Leisner, T., Duft, D., Möhler, O., Saathoff, H., Schnaiter, M., Henin, S., Stelmaszczyk, K., Petrarca, M., Delagrange, R., Hao, Z., Lüder, J., Petit, Y., Rohwetter, P., Kasparian, J., Wolf, JP., and Wöste, L.: Laser-induced plasma cloud interaction and ice multiplication under cirrus cloud conditions, Proc. Natl. Acad. Sci. USA, 110, 10106–10110, 2014.

Lloyd, G., Choularton, T. W., Bower, K. N., Crosier, J., Jones, H., Dorsey, J. R., Gallagher, M. W., Connolly, P., Kirchgaessner, A. C. R., and Lachlan-Cope, T.: Observations and comparisons of cloud microphysical properties in spring and summertime Arctic stratocumulus clouds during the ACCACIA campaign, Atmos. Chem. Phys., 15, 3719–3737, https://doi.org/10.5194/acp-15-3719-2015, 2015.

Luebke, A. E., Afchine, A., Costa, A., Grooß, J.-U., Meyer, J., Rolf,

C., Spelten, N., Avallone, L. M., Baumgardner, D., and Krämer, M.: The origin of midlatitude ice clouds and the resulting influence on their microphysical properties, Atmos. Chem. Phys., 16, 5793–5809, https://doi.org/10.5194/acp-16-5793-2016, 2016.

Meyer, J.: Ice Crystal Measurements with the New Particle Spectrometer NIXE-CAPS, PhD, Bergische Universitaet Wuppertal, Jülich, http://juser.fz-juelich.de/record/22871 (last access: Please provide last acces date.), record converted from VDB: 12.11.2012; Wuppertal, University, 2012.

Mioche, G., Jourdan, O., Delanoë, J., Gourbeyre, C., Febvre, G., Dupuy, R., Szczap, F., Schwarzenboeck, A., and Gayet, J.-F.: Characterization of Arctic mixed-phase cloud properties at small scale and coupling with satellite remote sensing, Atmos. Chem. Phys. Discuss., https://doi.org/10.5194/acp-2017-93, in review, 2017.

Nichman, L., Fuchs, C., Järvinen, E., Ignatius, K., Höppel, N. F., Dias, A., Heinritzi, M., Simon, M., Tröstl, J., Wagner, A. C., Wagner, R., Williamson, C., Yan, C., Connolly, P. J., Dorsey, J. R., Duplissy, J., Ehrhart, S., Frege, C., Gordon, H., Hoyle, C. R., Kristensen, T. B., Steiner, G., McPherson Donahue, N., Flagan, R., Gallagher, M. W., Kirkby, J., Möhler, O., Saathoff, H., Schnaiter, M., Stratmann, F., and Tomé, A.: Phase transition observations and discrimination of small cloud particles by light polarization in expansion chamber experiments, Atmos. Chem. Phys., 16, 3651–3664, https://doi.org/10.5194/acp-16-3651-2016, 2016.

Osborne, S. R., Abel, S. J., Boutle, I. A., and Marenco, F.: Evolution of Stratocumulus Over Land: Comparison of Ground and Aircraft Observations with Numerical Weather Prediction Simulations, Bound.-Lay. Meteorol., 153, 165–193, https://doi.org/10.1007/s10546-014-9944-0, 2014.

Pruppacher, H. R., Klett, J. D., and Wang, P. K.: Microphysics of clouds and precipitation, Taylor & Francis, 1998.

Rossow, W. B. and Schiffer, R. A.: ISCCP cloud data products, B. Am. Meteorol. Soc., 72, 2–20, 1991.

Shupe, M. D. and Intrieri, J. M.: Cloud Radiative Forcing of the Arctic Surface: The Influence of Cloud Properties, Surface Albedo, and Solar Zenith Angle, J. Clim., 17, 616–628, https://doi.org/10.1175/1520-0442(2004)017<0616:CRFOTA>2.0.CO;2, 2004.

Shupe, M. D., Daniel, J. S., de Boer, G., Eloranta, E. W., Kollias, P., Luke, E. P., Long, C. N., Turner, D. D., and Verlinde, J.: A Focus On Mixed-Phase Clouds, B. Am. Meteorol. Soc., 89, 1549–1562, https://doi.org/10.1175/2008BAMS2378.1, 2008.

Storelvmo, T. and Tan, I.: The Wegener-Bergeron-Findeisen process – Its discovery and vital importance for weather and climate, Meteorol. Z., 24, 455–461, https://doi.org/10.1127/metz/2015/0626, 2015.

Taylor, J. W., Choularton, T. W., Blyth, A. M., Liu, Z., Bower, K. N., Crosier, J., Gallagher, M. W., Williams, P. I., Dorsey, J. R.,

Flynn, M. J., Bennett, L. J., Huang, Y., French, J., Korolev, A., and Brown, P. R. A.: Observations of cloud microphysics and ice formation during COPE, Atmos. Chem. Phys., 16, 799–826, https://doi.org/10.5194/acp-16-799-2016, 2016.

Wendisch, M. and Brenguier, J.-L.: Airborne measurements for environmental research: methods and instruments, John Wiley & Sons, 2013.

Wendisch, M., Yang, P., and Ehrlich, A.: Amplified climate changes in the Arctic: Role of clouds and atmospheric radiation, Sitzungsberichte der Saechsischen Akademie der Wissenschaften zu Leipzig, Mathematisch-Naturwissenschaftliche Klasse, 132, 1–34, 2013.

Wendisch, M., Pöschl, U., Andreae, M. O., Machado, L. A. T., Albrecht, R., Schlager, H., Rosenfeld, D., Martin, S. T., Abdelmonem, A., Afchine, A., Araùjo, A., Artaxo, P., Aufmhoff, H., Barbosa, H. M. J., Borrmann, S., Braga, R., Buchholz, B., Cecchini, M. A., Costa, A., Curtius, J., Dollner, M., Dorf, M., Dreiling, V., Ebert, V., Ehrlich, A., Ewald, F., Fisch, G., Fix, A., Frank, F., Fütterer, D., Heckl, C., Heidelberg, F., Hüneke, T., Jäkel, E., Järvinen, E., Jurkat, T., Kanter, S., Kästner, U., Kenntner, M., Kesselmeier, J., Klimach, T., Knecht, M., Kohl, R., Kölling, T., Krämer, M., Krüger, M., Krisna, T. C., Lavric, J. V., Longo, K., Mahnke, C., Manzi, A. O., Mayer, B., Mertes, S., Minikin, A., Molleker, S., Münch, S., Nillius, B., Pfeilsticker, K., Pöhlker, C., Roiger, A., Rose, D., Rosenow, D., Sauer, D., Schnaiter, M., Schneider, J., Schulz, C., de Souza, R. A. F., Spanu, A., Stock, P., Vila, D., Voigt, C., Walser, A., Walter, D., Weigel, R., Weinzierl, B., Werner, F., Yamasoe, M. A., Ziereis, H., Zinner, T., and Zöger, M.: The ACRIDICON-CHUVA campaign: Studying tropical deep convective clouds and precipitation over Amazonia using the new German research aircraft HALO, B. Am. Meteorol. Soc., 97, 1885–1908, https://doi.org/10.1175/BAMS-D-14-00255.1, 2016.

Wendisch, M., Brückner, M., Burrows, J. P., Crewell, S., Dethloff, K., Ebell, K., Lüpkes, C., Macke, A., Notholt, J., Quaas, J., Rinke, A., and Tegen, I.: Understanding Causes and Effects of Rapid Warming in the Arctic, EOS, 98 pp., https://doi.org/10.1029/2017EO064803, 2017.

Wilson, T. W., Ladino, L. A., Alpert, P. A., Breckels, M. N., Brooks, I. M., Browse, J., Burrows, S. M., Carslaw, K. S., Huffman, J. A., Judd, C., Kilthau, W. P., Mason, R. H. McFiggans, G., Miller, L. A., Najera, J. J., Polishchuk, E., Rae, S., Schiller, C. L., Si, M. Temprado, J. V., Whale, T. F., Wong, J. P. S., Wurl, O., Yakobi-Hancock, J. D., Abbatt, J. P. D., Aller, J. Y., Bertram, A. K., Knopf, D. A., and Murray, B. J.: A marine biogenic source of atmospheric ice-nucleating particles, Nature, 525, 234–238, 2015.

Yano, J.-I. and Phillips, V. T. J.: Ice-Ice Collisions: An Ice Multiplication Process in Atmospheric Clouds, J. Atmos. Sci., 68, 322–333, https://doi.org/10.1175/2010JAS3607.1, 2011.

# Permissions

# List of Contributors

**Bengt G. Martinsson, Johan Friberg and Oscar S. Sandvik**
Division of Nuclear Physics, Lund University, Lund, Sweden

**Markus Hermann**
Leibniz Institute for Tropospheric Research, Leipzig, Germany

**Peter F. J. van Velthoven**
Royal Netherlands Meteorological Institute (KNMI), De Bilt, the Netherlands

**Andreas Zahn**
Institute of Meteorology and Climate Research, Institute of Technology, Karlsruhe, Germany

**ZhiliWang**
State Key Laboratory of Severe Weather and Key Laboratory of Atmospheric Chemistry of CMA, Chinese Academy of Meteorological Sciences, Beijing, 100081, China

**Lei Lin**
School of Atmospheric Sciences and Guangdong Province Key Laboratory for Climate Change and Natural Disaster Studies, Sun Yat-sen University, Zhuhai, 519000, China

**Meilin Yang**
Institute of Urban Meteorology, China Meteorological Administration, Beijing, 100089, China

**Yangyang Xu**
Department of Atmospheric Sciences, Texas A&M University, College Station, Texas 77843, USA

**Jiangnan Li**
Canadian Centre for Climate Modelling and Analysis, Science and Technology Branch, Environment Canada, Victoria, V8P5C2, Canada

**Caihong Xu, Chao Zhu, Jiarong Li, Ganglin Lv, Xianmang Xu, Weijun Li, Bing Chen, Wenxing Wang and Qingzhu Zhang**
Environment Research Institute, School of Environmental Science and Engineering, Shandong University, Jinan 250100, China

**Lulu Zheng and Guodong Sui**
Shanghai Key Laboratory of Atmospheric Particle Pollution and Prevention (LAP3), Fudan Tyndall Centre, Department of Environmental Science & Engineering, Fudan University, Shanghai 200433, China

**Aijun Ding**
Institute for Climate and Global Change Research, School of Atmospheric Sciences, Nanjing University, Nanjing 210023, Jiangsu, China

**Jianmin Chen**
Environment Research Institute, School of Environmental Science and Engineering, Shandong University, Jinan 250100, China
Shanghai Key Laboratory of Atmospheric Particle Pollution and Prevention (LAP3), Fudan Tyndall Centre, Department of Environmental Science & Engineering, Fudan University, Shanghai 200433, China
Institute for Climate and Global Change Research, School of Atmospheric Sciences, Nanjing University, Nanjing 210023, Jiangsu, China

**Min Wei**
Environment Research Institute, School of Environmental Science and Engineering, Shandong University, Jinan 250100, China
College of Geography and Environment, Shandong Normal University, Jinan 250100, China

**Abdelwahid Mellouki**
Environment Research Institute, School of Environmental Science and Engineering, Shandong University, Jinan 250100, China
Institut de Combustion, Aérothermique, Réactivité et Environnement, CNRS, 45071 Orléans CEDEX 02, France

**Dipesh Rupakheti**
Key Laboratory of Tibetan Environment Changes and Land Surface Processes, Institute of Tibetan Plateau Research, Chinese Academy of Sciences, Beijing 100101, China
University of Chinese Academy of Sciences, Beijing 100049, China

**Bhupesh Adhikary, Puppala Siva Praveen and Arnico Kumar Panday**
International Centre for Integrated Mountain Development (ICIMOD), Kathmandu, Nepal

**Khadak Singh Mahata and Mark G. Lawrence**
Institute for Advanced Sustainability Studies (IASS), Potsdam 14467, Germany

**Manish Naja**
Aryabhatta Research Institute of Observational Sciences (ARIES), Nainital, India

**Qianggong Zhang**
Key Laboratory of Tibetan Environment Changes and Land Surface Processes, Institute of Tibetan Plateau Research, Chinese Academy of Sciences, Beijing 100101, China
Center for Excellence in Tibetan Plateau Earth Sciences, Chinese Academy of Sciences, Beijing 100085, China

**Maheswar Rupakheti**
Institute for Advanced Sustainability Studies (IASS), Potsdam 14467, Germany
Himalayan Sustainability Institute (HIMSI), Kathmandu, Nepal

**Shichang Kang**
University of Chinese Academy of Sciences, Beijing 100049, China
State Key Laboratory of Cryospheric Science, Cold and Arid Regions Environmental and Engineering Research Institute (CAREERI), Lanzhou 730000, China
Center for Excellence in Tibetan Plateau Earth Sciences, Chinese Academy of Sciences, Beijing 100085, China

**Guoshuai Zhang and Shaopeng Gao**
Key Laboratory of Tibetan Environment Changes and Land Surface Processes, Institute of Tibetan Plateau Research, Chinese Academy of Sciences, Beijing 100101, China

**Pengfei Chen**
State Key Laboratory of Cryospheric Sciences, Northwest Institute of Eco-Environment and Resources, Chinese Academy of Sciences, Lanzhou 730000, China

**Fangping Yan and Bin Qu**
Laboratory of Green Chemistry, Lappeenranta University of Technology, Sammonkatu 12, 50130 Mikkeli, Finland

**Ye Hong**
Institute of Atmospheric Environment, China Meteorological Administration, Shenyang 110166, China

**Zhejing Zhu, Jiwei Li and Bing Chen**
Environmental Research Institute, Shandong University, Jinan 250100, China

**Mika Sillanpää**
Laboratory of Green Chemistry, Lappeenranta University of Technology, Sammonkatu 12, 50130 Mikkeli, Finland
Department of Civil and Environmental Engineering, Florida International University, Miami, FL 33174, USA

**Shichang Kang**
State Key Laboratory of Cryospheric Sciences, Northwest Institute of Eco-Environment and Resources, Chinese Academy of Sciences, Lanzhou 730000, China
CAS Center for Excellence in Tibetan Plateau Earth Sciences, Beijing 100101, China

**Xiaowen Han**
Key Laboratory of Tibetan Environment Changes and Land Surface Processes, Institute of Tibetan Plateau Research, Chinese Academy of Sciences, Beijing 100101, China
University of Chinese Academy of Sciences, Beijing 100049, China

**Zhaofu Hu**
State Key Laboratory of Cryospheric Sciences, Northwest Institute of Eco-Environment and Resources, Chinese Academy of Sciences, Lanzhou 730000, China
University of Chinese Academy of Sciences, Beijing 100049, China

**Chaoliu Li**
Key Laboratory of Tibetan Environment Changes and Land Surface Processes, Institute of Tibetan Plateau Research, Chinese Academy of Sciences, Beijing 100101, China
Laboratory of Green Chemistry, Lappeenranta University of Technology, Sammonkatu 12, 50130 Mikkeli, Finland
CAS Center for Excellence in Tibetan Plateau Earth Sciences, Beijing 100101, China

**Masakazu Taguchi**
Department of Earth Science, Aichi University of Education, Kariya, 448-8542, Japan

**Jean-Christophe Raut, Jennie L. Thomas, Katharine S. Law, Tatsuo Onishi and Julien Delanoë**
LATMOS/IPSL, UPMC Univ. Paris 06 Sorbonne Universités, UVSQ, CNRS, Paris, France

**Jerome D. Fast, Larry K. Berg and Richard C. Easter**
Pacific Northwest National Laboratory, Richland, WA, USA

**Anke Roiger and Hans Schlager**
Deutsches Zentrum für Luft- und Raumfahrt (DLR), Institut für Physik der Atmosphäre, Oberpfaffenhofen, Germany

**Bernadett Weinzierl and Katharina Heimerl**
Deutsches Zentrum für Luft- und Raumfahrt (DLR), Institut für Physik der Atmosphäre, Oberpfaffenhofen, Germany
Ludwig-Maximilians-Universität, Meteorologisches Institut, Munich, Germany
University of Vienna, Aerosol Physics and Environmental Physics, Vienna, Austria

**Louis Marelle**
LATMOS/IPSL, UPMC Univ. Paris 06 Sorbonne Universités, UVSQ, CNRS, Paris, France

Center for International Climate and Environmental Research, Oslo, Norway

**Luca Pozzoli, Srdan Dobricic and Elisabetta Vignati**
European Commission, Joint Research Centre (JRC), Directorate for Energy, Transport and Climate, Air and Climate Unit, Ispra (VA), 21027, Italy

**Simone Russo**
European Commission, Joint Research Centre (JRC), Directorate for Competences, Modelling, Indicators and Impact Evaluation Unit, Ispra (VA), 21027, Italy

**Alberto Cazorla, Juan Andrés Casquero-Vera, Roberto Román, Juan Luis Guerrero-Rascado and Lucas Alados-Arboledas**
Andalusian Institute for Earth System Research, IISTA-CEAMA, University of Granada, Junta de Andalucía, Granada, Spain
Department of Applied Physics, University of Granada, Granada, Spain

**Carlos Toledano and Victoria E. Cachorro**
Grupo de Óptica Atmosférica (GOA), Universidad de Valladolid, Valladolid, Spain

**José Antonio G. Orza**
SCOLAb, Física Aplicada, Universidad Miguel Hernández, Elche, Spain

**Gloria Titos, Marco Pandolfi and Andres Alastuey**
Institute of Environmental Assessment and Water Research (IDAEA-CSIC), Barcelona, Spain

**Natalie Hanrieder**
German Aerospace Center (DLR), Institute of Solar Research, Plataforma Solar de Almería, Almería, Spain

**María Luisa Cancillo and Antonio Serrano**
Department of Physics, University of Extremadura, Badajoz, Spain
Institute of Water Research, Climate Change and Sustainability, IACYS, University of Extremadura, Badajoz, Spain

**Lili Xia and Alan Robock**
Department of Environmental Sciences, Rutgers University, New Brunswick, New Jersey, USA

**Peer J. Nowack**
Department of Chemistry, Centre for Atmospheric Science, University of Cambridge, Cambridge, UK

Grantham Institute and Department of Physics, Faculty of Natural Sciences, Imperial College London, London, UK

**Simone Tilmes**
Atmospheric Chemistry Observations and Modeling Laboratory, National Center for Atmospheric Research, Boulder, Colorado, USA

**Pierre Sicard**
ACRI-HE, Sophia Antipolis, France

**Alessandro Anav and Elena Paoletti**
Institute of Sustainable Plant Protection, National Research Council, Sesto Fiorentino, Italy

**Alessandra De Marco**
Italian National Agency for New Technologies, Energy and the Environment, C.R. Casaccia, Italy

**Anja Costa, Armin Afchine, Gebhard Günther and Martina Krämer**
Forschungszentrum Jülich GmbH, Jülich, Germany

**James R. Dorsey and MartinW. Gallagher**
Centre for Atmospheric Science, University of Manchester, Manchester, UK

**Andre Ehrlich and Manfred Wendisch**
Leipziger Institut für Meteorologie, Universität Leipzig, Germany

**Darrel Baumgardner**
Droplet Measurement Technologies, Longmont, CO 80503, USA

**Heike Wex**
Leibniz Institute for Tropospheric Research, Leipzig, Germany

**Jessica Meyer**
Forschungszentrum Jülich GmbH, Jülich, Germany
Bundesanstalt für Arbeitsschutz und Arbeitsmedizin, Dortmund, Germany

**Anna Luebke**
Forschungszentrum Jülich GmbH, Jülich, Germany
Atmosphere in the Earth System Department, Max Planck Institute for Meteorology, Hamburg, Germany

# Index

**A**

Accelerator-based Methods, 1, 3

Access Airborne Campaign, 97

Aerosol Forcings, 18-22, 28-29, 31

Aerosol Optical Depth (AOD), 2, 98, 104, 140-141

Aerosol Spectral Light Absorption, 45

Aerosol-induced Change, 18

Agro-residue Burning, 45, 52-53

Air Pollution Episodes, 45

Airborne Field Campaigns, 190, 193, 205

Ambient Concentrations of Key

Air Pollutants, 45-46

Ambient Fungi, 32-33, 37-38, 41

Ascomycota, 32-33

Asian Tropopause Aerosol Layer (ATAL), 1, 13

Atmospheric Circulation, 19, 26, 29, 124-126, 128-129, 131-132, 134-138

Atmospheric Distribution, 69

Atmospheric Ozone Concentrations, 156

Atmospheric Patterns, 124-129, 131-132, 135-136

**B**

Basidiomycota, 32-33

Bc-induced Fast Adjustment, 18

Bc-induced Slow Climate Adjustment, 18

Biomass Burning, 19, 30, 44, 48, 53-55, 60-67, 80, 97-98, 101-102, 104, 106, 109-112, 116, 118-123, 127, 129, 136, 141, 153-154, 168, 186

Birthplace of Buddha, 45

Black Carbon (BC), 46, 68-69, 77, 79, 97-98, 124-125

Boreal Forest Fires, 97-98, 117, 177

Buddhist Heritage, 45

**C**

Calipso, 1, 6, 11, 13, 16, 105, 141, 155, 207

Climate Adjustment, 18

Community Earth System Model, 18-19, 30, 156-157, 168, 170, 186

Community Earth System Model Version 1 (CESM1), 18-19

Concentration Pathways Emission Scenarios, 171

Concentration Profile, 1

**D**

Deposition Processes, 97-98, 111-112, 114, 118

Deposition Rates, 68, 70, 72, 74, 76-77, 177

Diurnal Characteristics, 45

Dynamical Tropopause, 1, 3, 9, 12-13

**E**

East Asian Summer Monsoon (EASM), 18

Extratropical Transition Layer (EXTL), 1, 10, 13

**F**

Fossil Fuel Combustion, 45, 60, 62

**G**

Geopotential Height Differences, 82

Global Atmospheric Chemistry Transport Models, 171, 183

**H**

High O3 Concentrations., 171

High Surface O3 Levels, 171-172

High-pressure Anomaly, 124, 131

Himalayas and The Tibetan Plateau (HTP), 68-69

**I**

Iagos-caribic Platform, 1-3, 12

Iberian Ceilometer Network (ICENET), 140-141

Impact From Volcanism, 1

Indo-gangetic Plain, 45-47, 65-67, 80

Internal Transcribed Spacer Region Sequencing, 32

International Air Pollution Measurement Campaign, 45, 47, 63, 66, 77

**J**

Junge Layer, 1, 10, 12-13

**K**

Kruskal-wallis Rank Sum Test, 32, 36, 39-40

**L**

Land-sea Surface Thermal Contrast, 18-19, 22, 24-26, 28-29

Large-scale Weather Patterns, 124, 127

Linear Regression Methodology, 1, 6

Low-pressure Anomaly, 124

Lowermost Stratosphere (LMS), 1, 12

**M**

Major Stratospheric Sudden Warmings (MSSWS), 82

Meridional Gradient, 97, 116, 118

Metastats Analysis, 32, 40

Meteorological Conditions, 45, 202

Mineral Dust (MD), 68-69

Model Assimilation And Evaluation, 140, 153

**N**

Near-surface Atmosphere, 32, 36, 42

Numerical Weather Prediction Model, 82-83, 93

**O**

Ocean Response, 18, 23, 29

Optical Aerosol Characterization, 140

Organic Carbon (OC) Levels, 68

**P**

Particle Elastic Scattering Analysis (PESA), 1

Photosynthetic Assimilation, 171-172

Particle-Induced X-Ray Emission (PIXE), 1

Polar Atmospheric Warming, 124

Polar Vortex, 82, 84, 86, 88, 90-94, 96

Primary Biological Aerosol

Particles, 32, 43

Public Health, 32, 45

**Q**

Quantitative Real-time Polymerase Chain Reaction (QPCR), 32

**R**

Regional Forest Fires, 45, 63

River Sediment Transport, 68

**S**

Satellite-borne Lidar, 1

Sea Surface Temperature (SST), 18-19

Seasonal Variation, 1, 8, 11, 13, 32-33, 36, 39, 41, 63, 66, 121

Secondary Ice Formation, 190

Signal-to-noise Ratio, 20, 140, 144, 154

Solar Insolation Reduction, 156

Solar Radiation Management (SRM), 156-157

Sulfur-rich Air, 1

Surface Ozone Concentrations, 156-157, 161-162, 167, 178, 186, 188

Suskat-abc (sustainable Atmosphere For The Kathmandu Valley - Atmospheric Brown Clouds)., 45

Svalbard Archipelago, 97, 108-109, 117

**T**

Total Equilibrium Response, 18-19, 27

Tropospheric Chemistry, 125, 127, 138, 156, 167

Tropospheric Thermodynamic, 18, 28

**U**

Unece, 16, 171-174, 185, 188

Upper Troposphere (UT), 1-2, 12

Ut Particulate Sulfur, 1, 3-4, 6, 8, 10-13

**W**

Wegener-bergeron-findeisen Process, 128, 206, 209

Wrf-chem Simulations, 97, 100, 102, 118

www.ingramcontent.com/pod-product-compliance
Lightning Source LLC
Chambersburg PA
CBHW080636200326

41458CB00013B/4646